International Series in Operations Research & Management Science

Volume 226

More information about this series at http://www.springer.com/series/6161

Harilaos N. Psaraftis
Editor

Green Transportation Logistics

The Quest for Win-Win Solutions

Springer

Editor
Harilaos N. Psaraftis
Department of Transport
Technical University of Denmark
Kgs. Lyngby, Denmark

ISSN 0884-8289 ISSN 2214-7934 (electronic)
International Series in Operations Research & Management Science
ISBN 978-3-319-17174-6 ISBN 978-3-319-17175-3 (eBook)
DOI 10.1007/978-3-319-17175-3

Library of Congress Control Number: 2015942145

Springer Cham Heidelberg New York Dordrecht London

Printed on acid-free paper

Springer International Publishing AG Switzerland is part of Springer Science+Business Media (www.springer.com)

Foreword and Acknowledgments

This book is a compilation of material on green transportation logistics. The material comes from various sources, mostly from projects that my associates and I have been involved in, but also from work of several external invited colleagues. An outline of the scope and contents of the book is presented in the Preface.

The trigger for writing the book came from EU project SuperGreen on green corridors (2010–2013), and, in fact, several chapters of this book are based on the results of this project. Sometime before the end of this project, and when I was still at the National Technical University of Athens (NTUA), it occurred to me that material developed in the course of the project might form the basis for a book. However, the decision to go on with the book, as well as the specifics of the book, did not come until after the SuperGreen project was completed and I had moved to the Technical University of Denmark (DTU).

The story of the SuperGreen project is an exciting one. The project spanned the period January 2010–January 2013, but its origins can be traced back to October 2007, when Atle Minsaas of Marintek (Trondheim, Norway), a colleague I had collaborated with since the mid-1990s in the context of several EU projects, visited me at NTUA. Atle asked me if I was interested in a specific call for proposals under EU's 7th Framework Programme for R&D, which addressed the issue of green corridors. Green corridors were one of the concepts introduced within the European Commission's "Freight Logistics Action Plan" of 2007. In that plan, a number of short- to medium-term actions were presented that would help Europe address its current and future challenges and ensure a competitive and sustainable freight transportation system. One among these actions was the so-called Green transport corridors for freight. These were characterized by a concentration of freight traffic between major hubs and by relatively long distances of transport. Green corridors would in all ways be environmentally friendly, safe, and efficient.

I responded positively to Atle's question, and in December of 2007 we decided that Marintek would take the lead in coordinating the proposal and that Atle would serve as Project Manager. A group of partners across Europe were assembled, and in May 2008 we submitted a proposal to the European Commission on the so-called

SuperGreen project, tasked to investigate this topic. This was a "coordinated action" comprising all stakeholders involved in the green corridors, and proposing to analyze and suggest, among other things, how the green corridors can be made even greener by green technologies and smarter utilization of Information and Communication Technologies (ICT). I was involved in the proposal as the main contact person for partner No. 2, NTUA, which was represented by the Laboratory for Maritime Transport, of which I was the Director.

Even though we received the proposal's evaluation report from the Commission as early as July 2008, we were not invited for contract negotiations until April 2009, as the Commission had to rank-order all proposals and then choose which among them could proceed. These negotiations were concluded in the fall of 2009 and the project set out to start in mid-January 2010. The SuperGreen consortium consisted of 22 partners from 13 countries in Europe and its official title was "Supporting EU's Freight Transport Logistics Action Plan on Green Corridors Issues" (Grant Agreement No. TREN/FP7TR/233573/"SUPERGREEN").

It was during these negotiations that a switch between partners No. 1 and 2 took place. The switch assigned to NTUA the role of Project Coordinator, and to me the role of Project Manager. This was due to Atle's assignment to manage the development of the new research infrastructure, the Ocean Space Centre, a 10+ year project in Norway dealing with maritime and ocean science and technology. Atle felt he could not undertake both managerial jobs and asked me if I could take over his role in SuperGreen. I accepted such role with simultaneous excitement and apprehension, as I knew that managing multi-partner EU projects could be nontrivial (at that point in time I had managed two such projects, but these were much smaller, both in scope and number of partners). Atle remained connected with the project in its entirety and I am glad that he accepted to co-author with me the final chapter of this book.

Complementary to SuperGreen, I also had several other projects at NTUA that were related to the interface between transportation logistics and emissions. All of these projects were maritime related. They were the following:

- "Ship Air Emissions Study," study funded by Hellenic Chamber of Shipping (January–June 2008).
- "Effective Bulk Transport," gift agreement funded by Det Norske Veritas (January 2008–December 2010).
- "Assessment of Environmental Impact in Marine Transportation and Related Activities," project funded by the American Bureau of Shipping (June 2008–May 2011).
- "Centre of Excellence in Ship Total Energy-Emissions-Economy," project funded by The Lloyds Register Foundation (February 2010–December 2015).
- "Envi-Shipping: Green Assessment of a Ship's Lifecycle," multi-partner project funded by the General Secretariat of Research and Technology, Greece (May 2011–November 2014).

The financial support of all of the above funding sources, including the European Commission, is gratefully acknowledged. As with SuperGreen, some of the work conducted under these projects eventually found its way into this book, as much as it was also published in journals and presented in conferences. The same was the case with my engagement with the International Maritime Organization (IMO), as an advisor of the Hellenic Chamber of Shipping. In the period between 2007 and 2013, I participated in several meetings of IMO's Marine Environment Protection Committee (MEPC) and in several expert groups on matters pertaining to Green House Gases (GHGs). I thought that some of this work was also relevant for the book.

Work in the 3 years of SuperGreen was as exciting as it was challenging. I still remember walking in a snowstorm in Kuopio, Finland, home of our partner Sito Ltd., to go to a meeting where we would preselect 15 corridors for the analysis. Or a stakeholder's workshop in Helsinki where we presented a proposal for the nine corridors we would use for our analysis. We even gave nicknames to the corridors: Brenner, Mare Nostrum, Two Seas, Nureyev, Cloverleaf, Strauss, Finis Terrae, Edelweiss, and Silk Way, the latter linking Europe to the Far East. Not to forget were also the stakeholder workshop in Napoli (Nola terminal) to come up with a list of sensible key performance indicators (KPIs), and the near-disaster stakeholder workshop in Antwerp on the same subject, as there was much confusion on what sensible KPIs might be. The turning point of the project came in the workshop in Malmö (March 2011), when the final list of KPIs was decided.

Beyond that important milestone, progress in the project went smoothly. When in the fall of 2011, the European Commission issued its proposal for the new Trans-European Transport Network (TEN-T) guidelines, something that evolved into a Regulation in 2013, we were intrigued that the proposed TEN-T Core Network had substantial overlap with the set of nine corridors we had chosen about a year earlier. Contact was initiated with the Commission's TEN-T Policy unit in 2012 and we maintained a fruitful dialog with them even after the end of the project.

Dissemination activities in the project were broad and diverse. Just within its official duration we had 8 papers in journals, book chapters, and scientific conferences, 3 plenary workshops, 4 regional workshops, and close to 50 presentations in other external events. Dissemination continued after January 2013, when the project officially closed. All project results and deliverables can be found in www.supergreenproject.eu.

A special challenge in SuperGreen was the evolving scene of related EU legislation on green corridors and related topics. When the proposal was being written in 2008, the 2001 White Paper on Transport and the 2007 Freight Logistics Action Plan (fresh out of the oven at the time) were the two main EU policy documents on the subject. Between that time and at the end of the SuperGreen project there were several additional and important policy developments, including the 2010 Regulation on an European rail network for competitive freight, the 2011 White Paper on Transport, and the 2011 proposal for new TEN-T guidelines. Even though the SuperGreen grant agreement, signed in late 2009, could not foresee these later developments, we felt that it was our obligation to provide input on them during the course the project. Even more, for the purposes of this book, we considered essential to include policy developments that materialized after the end of SuperGreen. These include the proposed Fourth Railway Package (2013), the proposed Regulation on Monitoring, Reporting, and Verification (MRV) of CO_2 emissions from maritime transport (2013), the Regulation on TEN-T guidelines (2013), the Strategy for reducing Heavy-Duty Vehicles' fuel consumption and CO_2 emissions (2014), the Directive on the Deployment of Alternative Fuels Infrastructure (2014), and several others.

When I moved to DTU in mid-2013, I developed a new graduate course, given at both the Master's and Doctoral levels, which was entitled "*Green Transport Logistics*." Much of the work of SuperGreen, plus some additional related material formed the basis for that course. It was around that time that the idea for a book on this subject resurfaced.

Advice on where to publish the book was given to me by colleague James Bookbinder of the University of Waterloo (Canada). Some SuperGreen colleagues and I had already written a chapter on SuperGreen for Jim's "Handbook of Global Logistics" (Springer, 2013), referring to work in the project circa 2011. So upon Jim's advice, I contacted Frederick Hillier of Stanford University, then Editor for Springer's International Series in Operations Research and Management Science and put the book idea to him. In the proposal for the book, six chapters were devoted to SuperGreen. But also eight other chapters were proposed on other subjects related to green transportation logistics, which was chosen as the book's title. I considered Fred Hillier an iconic figure, as his book with Gerald Lieberman "Introduction to Operations Research" that I had purchased back in the late 1970s when I was a graduate student at MIT was critical in my subsequent interest in Operations Research and Management Science. I was pleased that Fred liked the book proposal, and eventually an agreement was reached for Springer to be the publisher of this work. Fred subsequently stepped down as Series Editor and I worked with Camille Price, Neil Levine, Christine Crigler, and Matthew Amboy at Springer. I would like to thank all of them for the production of this book.

In January 2014, we were pleased to see the European Commission Communication COM (2013) 940 final "Building the Transport Core Network: Core Network Corridors and Connecting Europe Facility" citing the SuperGreen project in the context of measuring the sustainability of the TEN-T Core Network corridors. And in May 2014, we were informed that the SuperGreen project had been selected by the European Commission as a success story for a Policy Brochure on logistics for the Transport Research and Innovation Portal (www.transport-research.info). The main task of a Policy Brochure is to demonstrate the synergy between EU-funded research and EU policy in a certain area. For each Policy Brochure, three "success stories" are presented, i.e., research projects whose results are recognized as highly successful in supporting EU policy. Recognition is always gratifying, and in this particular case I think it reflects on every individual who contributed to the project.

So it is my duty to acknowledge several people for contributing to the success of the SuperGreen project. I apologize in advance for any omissions. First and foremost, my thanks go to Rein Jüriado and Fleur Breuillin, Project Officers at the European Commission (Directorate General for Mobility and Transport,

DG-MOVE), as well as to Pawel Stelmaszczyk, then Head of Unit B-3 (Logistics, Co-modality, Motorways of the Sea & Marco Polo) at DG-MOVE for their technical and administrative support and for their advice in general throughout the project. I also want to thank Gudrun Schulze, Policy Officer from the same Directorate General, for her constructive feedback on issues pertaining to TEN-Ts.

Then I would like to thank

- George Panagakos, Christos Kontovas, Nikos Ventikos, Stefanos Chatzinikolaou, Dimitris Lyridis, Panagiotis Zacharioudakis, Stylianos Iordanis, and Panos Tsilingiris of NTUA (Lab for Maritime Transport), Greece
- Atle Minsaas, Even Ambros Holte, Aud Marit Wahl, and Beate Kvamstad of Marintek, Norway
- Ilkka Salanne and Sanni Rönkkö of Sito, Finland
- Chara Georgopoulou and Nikos Kakalis of Det Norske Veritas, Norway (Piraeus, Greece office)
- Sergio Barbarino, Indrek Ilves, and Maximo Martinez of Procter and Gamble Eurocor, Belgium
- Anders Sjöbris, Christopher Pålsson, and Niklas Bengtsson of IHS Fairplay, Sweden
- Juha Schweighofer and Andreas Bäck of via Donau, Austria
- Valerio Recagno, Sara Fozza, and Francesca Vio of D'Appolonia, Italy
- Humberto Moyano, Beatriz Galan, and Ainhoa Puebla of the Port of Gijon, Spain
- Jorge d' Almeida and Rui Pinto of PSA Sines, Portugal
- Konrad Pütz of the Norwegian Public Roads Administration, Norway
- Alkis Corres and Maro Varvate of the Hellenic Short-Sea Association, Greece
- Christiane Geiger, Agnes Eiband, and Uwe Clausen of the Technical University of Dortmund, Germany
- Andrea Schön and Monica Holm of DB Schenker, Germany
- Paulus Aditjandra, Dewan Islam, Eero Vanaale, and Tom Zunder of the University of Newcastle (Newrail), UK
- Elina Multanen and Rolf Sandberg of Straightway, Finland
- Ruth Sanchez and Rachid Berd of Consultrans, Spain
- Taneli Antikainen of the Finnish Transport Agency, Finland
- Vesa Stenvall and Otto Lehtipuu of VR Group, Finland
- Kenan Ülkü and Recai Kilik of the Turkish Railways, Turkey
- Carole Yche, Monica Grosso, and Chiara Rebuffi of Captrain Italia, Italy
- Andriy Vartsaba of TES Consult, Ukraine

Indrek Ilves must be credited for the depiction of the nine SuperGreen corridors in metro format and Christos Kontovas for the SuperGreen logo and for the maintenance of the SuperGreen web site.

I also want to thank the members of the project's external Advisory Committee, which included (alphabetically) Herman de Meester, Karin de Schepper, Jacques Dirand, Philippe Domergue, Fuensanta Martinez, Pawel Mickiewicz, Manfred

Reuter, Algirdas Sakalys, Jerker Sjögren, Nicolette van der Jagt, and Peter Wolters. Their advice and support was invaluable.

A great number of other people, too many to mention by name, participated in the stakeholder and other workshops of the project and contributed to the debate on green corridors issues. I acknowledge, and this was a lesson to be learned, that their constructive consultation has contributed to the outcome of the project.

Needless to say, I would like to thank all authors who accepted my invitation and contributed to the book. Especially, I am grateful to George Panagakos and Christos Kontovas, formerly of NTUA and now of DTU, for their overall contribution and assistance. I am also grateful to those (external) authors who were not connected to SuperGreen or to any other of my projects but who accepted my invitation to write chapters in the book. They were chosen to cover areas such as vehicle routing (Tolga Bektaş, Emrah Demir, and Gilbert Laporte), air transportation (Antony Evans), and inland navigation (Gernot Pauli), all from a green perspective. The book is definitely richer because of their contributions.

My own work on the production of this book was kindly supported in part (a) from an internal grant provided by the President of DTU, (b) from the DKMK (Danish Maritime Cluster) project, and (c) from the GreCOR project of the Interreg IVB North Sea Region Programme (EU). The DKMK project spanned the period 2011–2014 and was financed by the European Social Fund and the Capital Region of Denmark (Growth Fund). GreCOR spanned the period 2012–2015 and its full name is "Green Corridor in the North Sea Region: Oslo-Randstad." GreCOR (www.grecor.eu) is a green corridor project that, among other things, has extended and applied some of the methodologies of SuperGreen. In that context, GreCOR has also partially supported the Ph.D. of George Panagakos at DTU, in continuation of the work of SuperGreen. This work will be reported in future publications.

Last but not least, I am grateful to my wife Alexandra Manousaki and to my children Anastasia and Nikos for their support and especially for putting up with the myriad of travels I took to attend all the meetings, workshops, and conferences related to work reported in this book.

Kgs. Lyngby, Denmark
January 2015

Harilaos N. Psaraftis

Preface

Scope

The title of this book is "Green Transportation Logistics," and the subtitle is "The Quest for Win–Win Solutions." It is therefore fair that in this Preface we start by trying to explain what we mean by both.

The traditional analysis of transportation logistics problems has been in terms of cost-benefit, economic, or other optimization criteria from the point of view of the logistics provider, carrier, shipper, or other end user. Such traditional analysis by and large either ignores environmental issues or considers them of secondary importance. Green transportation logistics tries to bring the environmental dimension into the problem, by analyzing various trade-offs and exploring "win–win" solutions. In doing so, criteria for the benefit of the private end user may give their place to criteria that are more relevant from a societal point of view.

Interest in green transportation logistics has grown in recent years due to increased emphasis by both private industry and policy-makers to make transportation more environment-friendly. The objective to attain a green performance of the overall transportation supply chain is and is likely to be a central goal for both industry and policy-makers in the years ahead. To that effect, various analyses of many aspects of the problem have been and are being carried out and a spectrum of environment-friendly measures is being contemplated. These measures may be technological, logistics-based or market-based, and may have important side-effects as regards the economics and logistics of the supply chain.

There can be many definitions of the word "green," and a definition can be critical as regards the subsequent approach and measures to achieve whatever goal is set. For instance, if by green we mean minimizing emissions from transportation, and we subsequently strive to apply a series of technological measures that would achieve that goal, a conceivable outcome might be that transportation may become unprofitable and various undesirable side-effects may occur, including reduction of trade, relocation, or even shutdown of production, and possibly others. It is clear that one can always minimize emissions from A to B if trade from A to B is minimized. In the extreme case that trade from A to B ceases to exist because no operator would make a profit engaging in that trade, emissions would drive down to zero. But that's not a desirable outcome.

So things may be more complex than they appear at first glance, and in fact the goal of greening the logistical supply chain may involve several trade-offs that are at stake, and which have to be analyzed and evaluated if a desirable solution is to be achieved. The purpose of this book is to take a critical look at these trade-offs, take stock at models that can be used to assess them, and discuss possible relevant measures and policies. In the long road towards a sustainable global transportation system, a sound knowledge of the balances between economic and environmental objectives, and of the factors that may affect these balances, is a necessary condition. We believe that the material in this book will help improve such knowledge.

In a strict sense, what we mean by *green transportation logistics* will hopefully become clearer to the reader after this book has been read. However, to set the stage and give the reader an idea of what will follow, below is a working definition of the phrase "green transportation logistics" in this book:

- *Green transportation logistics is an attempt to attain an acceptable environmental performance in the transportation supply chain, while at the same time respecting traditional economic performance criteria.*

Social criteria are often embedded in the above definition, either on their own right, or as part of the set of economic criteria. It is clear that the weights among the various criteria vary among stakeholders, a private operator assigning more weight to economic criteria, an environmental organization more weight to environmental criteria, and others perhaps preferring social criteria. Whatever it is, achieving the above is what we call a "win–win" scenario, and the pursuit of win–win solutions is the underpinning concept in the book. As we will see throughout the book, a win–win outcome may not always be achievable. The word "sustainable" is often used to denote a similar outcome, and *sustainable transportation logistics* is often meant to imply a transportation system that combines acceptable economic, environmental, and social performances.

The above definition also implies that there exists a well-defined set of criteria to assess the various facets of performance of the logistical system under consideration. These criteria are often called key performance indicators (KPIs). Selecting appropriate and meaningful KPIs is a very important step and one that may be more difficult than it seems at first glance. Difficulties may be due to a variety of reasons, as will be seen later.

Part of this book will draw from the recently completed EU project "SuperGreen" on green corridors,[1] whose purpose has been to assist EU policy-makers to analyze policy alternatives as regards attaining a good performance both from an economic and from an environmental perspective. This combined goal is

[1] http://www.supergreenproject.eu/

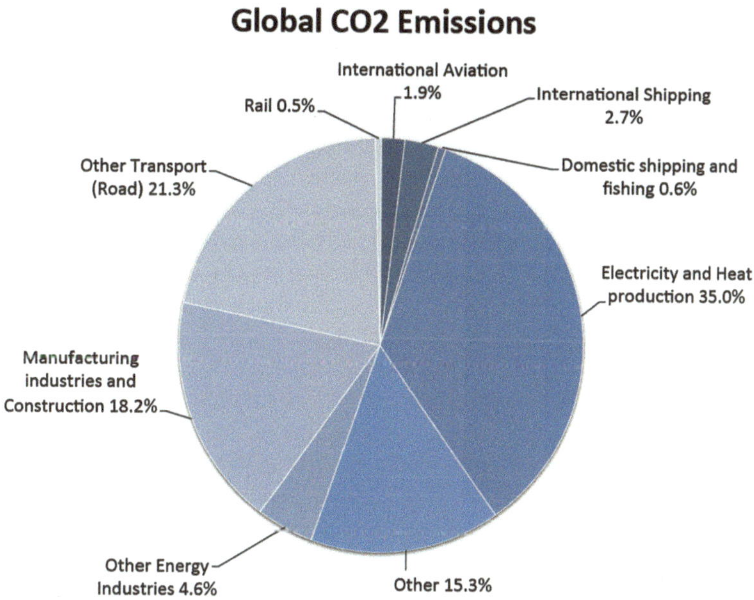

Fig. 1 Global CO_2 emissions, 2007 baseline year. Adapted from Buhaug et al. (2009)

central in a variety of recent EU policy documents and will form the basis for future development of the Trans-European Network (TEN-T). In addition, we supplement the book by several chapters drawn from work outside the above, covering both the methodological base and the application context.

Several clarifications need to be made at the outset, on the scope, and therefore contents of the book:

First, and as regards the primary focus, what we mean in this book by "accept able environmental performance" is mainly *acceptable level of emissions*. This is so due to the increased attention anthropogenic emissions have been getting in recent years, both at a global and a regional level. Among them, certainly carbon dioxide (CO_2) and other greenhouse gas (GHG) emissions have attracted much of the focus from a climate change perspective and the world community has set ambitious goals to mitigate them. To put things into perspective, Fig. 1 shows the global breakdown of emissions among major energy producing industries (2007 baseline year). It can be seen that the top CO_2 producer is electricity and heat production (35 %). Transportation activities account for 27 % of the total, and among them the top CO_2 producer is road (21.3 % of total and 78.8 % of transportation emissions).

Other types of emissions, such as sulphur oxides (SOx), nitrogen oxides (NOx), and others are also important.

It should be recognized that, other than emissions, there are certainly additional environmental attributes of transportation that may create undesirable effects. These include noise, hazardous substances, oil pollution, ballast water, dust, residues, garbage, and others. Apart from some individual considerations (see for instance noise in aviation, Chap. 13), the book will not focus on such other attributes.

What may be a second bias in the book is that it has a distinct, yet nonexclusive, *European* tint. Again this stems in part from the SuperGreen project, which gave us a chance to look at a broad spectrum of EU R&D and regulatory policies on green transportation issues. We believe that this causes no loss of generality and even that some of these activities may serve as models for other parts of the world. On selected cases, such as for instance maritime transportation (Chaps. 8–11) and air transportation (Chap. 13), a more global perspective is taken.

A third orientation of the book is its main focus on *freight* as opposed to passenger transportation. This stems from the fact that much of the material of the book (Chaps. 1, 3, 4, 5, 6, and 12, as well as Annex I) is based on the results of SuperGreen, which had a freight perspective. In addition, material on maritime transportation (Chaps. 8–11) and inland navigation (Chap. 14) is mostly on freight. However, the chapters on green vehicle routing (Chap. 7), which is on a road setting, on being green on sulphur (Chap. 10), and on green air transportation (Chap. 13) may also concern passenger transportation. In addition, the chapters in which the TEN-T is discussed (Chaps. 1 and 4) refer to both freight and passengers, as the TEN-T is designed for both. In many instances, freight and passengers share the same infrastructures, and this has to be kept in mind. Green *urban* transportation logistics, a significant topic in and of itself, and which involves, among other things, city logistics, last mile logistics, public transportation logistics, grocery logistics, electric vehicle logistics, and bicycle transportation, is by and large outside the main scope of the book. The interested reader may refer to, among other sources, Gonzalez-Feliu, Semet, and Routhier (2014) and Meyer and Meyer (2013), and also to a number of R&D projects that approach the above subjects from a sustainability perspective. These include Citylog,[2] Enclose,[3] LaMiLo,[4] and Sugar.[5]

The rest of this Preface is structured as follows. The section that follows discusses some of the challenges in green tranpsortation logistics. Then we outline how the rest of this book is organized. We finally comment on the intended audience of the book, including what the reader is expected to get out of it.

Challenges

Achieving a green transportation supply chain may involve several nontrivial challenges. Below we present some of them, with the understanding that the rest of this book will provide a more detailed picture.

---

[2] http://www.city-log.eu/

[3] http://www.enclose.eu/

[4] http://www.lamiloproject.eu/

[5] http://www.sugarlogistics.eu/

Ambitious Environmental Goals

As a first example, we outline some of the main challenges in Europe.

The EU 2011 White Paper on Transport[6] aims, at a high-level target, at reducing by year 2050 transportation-related GHG emissions by at least 60 % with respect to 1990 levels. Lower-level targets that are related to the sustainability of transportation include the following (see Chap. 1 for more details):

- By 2030, halve the use of "conventionally fuelled" cars in urban transport; by 2050, phase them out in cities.
- By 2030, achieve essentially CO_2-free city logistics in major urban centers.
- By 2030, 30 % of road freight over 300 km should shift to other modes such as rail or waterborne transport, and more than 50 % by 2050, facilitated by efficient and green freight corridors. To meet this goal will also require appropriate infrastructure to be developed.
- By 2050, reduce EU CO_2 emissions from maritime bunker fuels by 40 % (if feasible by 50 %).
- By 2050, increase the use of low-carbon sustainable fuels in aviation to 40 %.
- Move towards full application of "user pays" and "polluter pays" principles and private sector engagement to eliminate distortions, including harmful subsidies, generate revenues, and ensure financing for future transportation investments.

Challenges in other parts of the developed world (including North America, Japan, and Australia) are quite similar. They may be even more pronounced in developing economies in Asia, South America, and Africa. Many of the latter countries question the basic premise that they should be subject to the same kinds of environmental guidelines as in developed economies, on the ground that this may impede their own economic development. International bodies such as the United Nations Framework Conference on Climate Change (UNFCCC) and others are routinely presented with arguments centering on what is known as the Common But Differentiated Responsibilities (CBDR) principle, which gives developing countries ground for such a position (more on CBDR on Chap. 8).

Whatever the viewpoint, the main challenge in all cases seems to be the following: how can international transportation grow and be profitable in the face of such ambitious environmental goals?

Number of Stakeholders

The number of stakeholders in problems in this area is significant and may typically include (list is not exhaustive):

[6] [COM (211) 144] 'Roadmap to a Single European Transport Area—Towards a competitive and resource efficient transport system.'

- Transportation operators
- Terminal and warehouse operators
- Infrastructure operators
- Cargo owners (shippers)
- Non governmental organizations (NGOs)
- Environmental organizations
- Authorities responsible for social and spatial planning
- Public officials and politicians
- Other industries (e.g., manufacturing, retailing, recycling)
- R&D organizations and universities

Each of the above stakeholders may have their own agenda and objectives that are many times conflicting with the objectives of other stakeholders. It thus may be difficult to reach consensus solutions, and political considerations may sometimes prevail. Determining the final set of corridor KPIs in the SuperGreen project involved several stakeholder workshops and extensive consultation with these stakeholders. Adopting the Energy Efficiency Design Index (EEDI) for maritime CO_2 emissions at the International Maritime Organization (IMO) revealed widely different views between industrialized and developing countries and the solution obtained was not a consensus solution.

Elusive Data

In the quest to reduce emissions, one may pose the naïve question; can we at least measure them? It is clear that to reduce anything, one should be able to measure it first. However, it turns out that emissions from various sources are not being measured directly and the only data that exist are *estimates* of these emissions. Even for past emissions, these estimates are produced by specific methods, most of which involve modeling and various assumptions on model inputs such as fuel consumptions and speeds of vehicles, activity profile of fleet, fuel sales, and others. These estimates can vary significantly, depending on the method. This problem has been recognized and significant regulatory activity is in place to monitor, report, and verify emissions, as a basis for further action to reduce them.

Perhaps more fundamentally, lack of data, or data of questionable quality, is also a problem. This is mainly as regards freight flow data, especially in multimodal scenarios. This may come as a surprise to Operations Research/Management Science (OR/MS) analysts if they assume that the data that is necessary to feed their OR/MS model is readily available. Nothing can be further from the truth in many cases. Origin-Destination (OD) data, transshipment data, or simply flow data in the links of a network or through the nodes of that network may be either elusive or of questionable quality. Some other data may be considered proprietary by carriers or other stakeholders. Then the question is, if you do not have the data, what do you do? As with emissions, one solution is modeling, that is, try to estimate, that is, generate data that emulates the missing data by running and calibrating a model. Many times the methodology that is used depends on the kinds of data that are available.

The flow data availability problem has been recognized as such in at least the EU, and various efforts to address it have been made through the years. Progress has been questionable, and has been exacerbated by some institutional developments. For instance, the abolition of customs in intra-EU road border crossings has removed a monitoring checkpoint along the supply chain, and no adequate replacements have been found as of yet. It is hoped that the use of Information and Communication Technologies (ICT) will help alleviate this problem, but thus far this has not materialized in any substantial way.

Win–Win and the "Push-Down, Pop-Up" Principle

"Win–win" is a nice set of words. The only problem is that finding win–win solutions may not always be easy. More often than not, the "push-down, pop-up" principle applies: if you push a certain button down, at least another one will pop up somewhere else. Speed reduction in maritime transportation is a prime example: if ones make the world fleet go slower, one will reduce emissions, will reduce fuel costs, and will take care of vessel overcapacity, which is important when the market is depressed. That seems like killing three birds in one stone, so it looks pretty good, or in fact a win–win–win proposition. But is that really the case?

The answer is, it depends. Reducing ship speed may have other ramifications, which may not be beneficial. For instance, more ships will be needed to produce the same transportation throughput. But this will entail some costs. Also, cargo in-transit inventory costs will generally increase. This is due to the delay in the arrival of the cargo. The inventory costs are proportional to the value of the cargo, so if you really have high-value goods, hauling them at a lower speed may entail significant costs.

Another push-down, pop-up effect is that in the short run, freight rates will go up once the overall transportation supply is reduced because of slower speeds. At a minimum, the rates will not go down as much, and this may help the market, but shippers will foot the bill. This fact is seldom mentioned in any of the discussions on green maritime policies. The extent of the rate increase would depend on the particular scenario.

Yet another push-down, pop-up effect concerns effects that reduction in ship speeds may have on other modes of transportation, to the extent these are alternatives to sea transportation. This is the situation as regards many intra-European destinations, but may also be true in North America, if coastal shipping is contemplated to relieve highways from congestion. If ships are made to go slower, shippers may be induced to prefer land-based alternatives, mostly road, and that may increase overall GHG production. Road is certainly worse than maritime in terms of GHG emissions per tonne-km.

A similar "push-down, pop-up effect" may very well occur because of another green policy initiative is followed. The use of lower sulphur fuel within designated sulphur emissions control areas (also known as SOx ECAs, or SECAs) as of 1 January 2015 may have a reverse impact on the stated European Transport policy goal to shift cargo from land to sea, by making short-sea shipping less favorable to road transportation, something that might ultimately lead to more CO_2 pollution. Currently in Europe the Baltic Sea, the North Sea, and the English Channel are designated SOx ECAs, and so is the entire North American and US Caribbean coast, which also includes restrictions on nitrogen oxides (NOx).

In the search for environmentally friendly policies, it is clear that a holistic approach is necessary, one that looks into and optimizes the overall supply chain instead of its individual components. Otherwise, the solutions are likely to be suboptimal, both cost-wise and environment-wise.

Another example that comes to mind concerns the push for the widespread use of electric power for surface transportation vehicles, whether these are passenger cars, buses, railway locomotives, or even trucks and bicycles. The EU goal to achieve essentially CO_2-free city logistics in major urban centers by 2030 depends critically on the successful use of electric technologies in urban vehicles. Yet, a basic premise that does not usually appear in public discussion is that the extra energy necessary to power these electric vehicles should produce less emissions than the emissions of the conventionally fueled vehicles that are replaced. This is true if this extra energy is produced by nuclear, hydro, or solar power, but not necessarily true if it is produced by a coal plant or a plant using fossil fuels.

The same is true for "cold ironing," that is, the provision of electricity to a ship by plugging into a port's electricity supply system so as to switch off the ship's auxiliary engines at port. This is an idea that originated in the ports of Los Angeles and Long Beach (California, USA) and is likely to be the norm for many world ports in the future. The rationale is minimizing in-port emissions. But again, the question is what emissions will be produced by the generation of the extra shore electricity, and if that is less than the emissions saved by switching off the ship's auxiliary power at port.

Book Organization

The rest of this book is organized as follows:

Chapter 1 by Panagakos presents the policy context behind green transportation logistics, mainly from a European perspective.

Chapter 2 by Kontovas and Psaraftis present some basics on transportation emissions, including estimating emissions, environmental policy measures, how one can define cost-effectiveness and possible barriers that may exist.

Chapter 3 by Panagakos presents the green corridors basics, including definitions, benchmarking, KPIs, and other methodological aspects, as developed in the SuperGreen project.

Chapter 4 by Panagakos discusses the concept of the TEN-T and explores the relation of the TEN-T Core Network to green corridors.

Chapter 5 by Georgopoulou et al. engages in a benchmarking exercise for SuperGreen corridors by means of conventional technologies: the question is how can these corridors become greener by such technologies.

Chapter 6 by Geiger explores the potential role of ICT in making green corridors greener. ICT cannot directly reduce emissions, but judicious use of it can do so.

Chapter 7 by Bektaş et al. introduces the green vehicle routing problem as a vehicle routing problem with environmental attributes and develops models and algorithms to solve variants of this problem.

Chapter 8 by Psaraftis presents various market-based measures that are under discussion for reducing maritime CO_2 emissions.

Chapter 9 by Psaraftis and Kontovas discusses speed and route optimization as a potential win–win proposition in maritime transportation. Fixed and flexible route scenarios are examined.

Chapter 10 by Kontovas et al. discusses the sulphur problem in SOx ECAs and what the possible side-effects on modal split and other attributes including emissions might be.

Chapter 11 by Chatzinikolaou and Ventikos presents the concept of life cycle (cradle to grave) emissions in a maritime setting and present models and approaches on that subject.

Chapter 12 by Aditjandra et al. discusses green rail transportation and goes over relevant issues towards making railways more sustainable.

Chapter 13 by Evans discusses green air transportation in terms of technologies, policies, and other measures that can make aviation greener.

Chapter 14 by Pauli discusses green inland navigation and the spectrum of issues that are at play to make it greener as regards emissions.

Finally, Chap. 15 by Minsaas and Psaraftis presents possible areas for further research in this area.

All chapters are to a great extent self-contained, with cross-referencing among them wherever appropriate. For the SuperGreen-related chapters, it is recommended that Chap. 3, and, to a lesser extent, Chaps. 1 and 4, be read before Chaps. 5, 6, and 12. In these chapters, references are often made to SuperGreen public deliverables that provide full details of work carried out. All of these deliverables can be found by visiting this link: http://www.supergreenproject.eu/info.html.

Intended Audience

The intended audience of this book consists of several groups:

- Faculty, students, and researchers active in transportation logistics and interested in the environmental dimension

- Carriers, shippers, infrastructure managers, and other logistics providers who aim at improving their environmental performance while staying in business
- Technology designers and providers
- Policy-makers at the national and international level
- Other stakeholders, environmental, or other

We believe that those who will read this book will be able to (among other things):

- Understand the main criteria and trade-offs in green transportation logistics
- Analyze concepts such as internal vs. external costs, marginal abatement costs, the polluter pays principle, and others, and how these can influence system performance
- Learn how green transportation corridors can be benchmarked in terms of specific Key Performance Indicators (KPIs)
- Go over some optimization models and recent literature for this class of problems
- See how one can formulate logistics optimization problems with environmental criteria taken on board
- Examine the effects of technical, operational, and market-based measures
- Review and discuss relevant policy initiatives
- Get a flavor of directions for further research in this area.

References

Buhaug, Ø., Corbett, J. J., Endresen, Ø., Eyring, V., Faber, J., Hanayama, S., et al. (2009). *Second IMO GHG study 2009*. IMO document MEPC59/INF. 10.

Gonzalez-Feliu, J., Semet, F., & Routhier, J.-L. (Eds.). (2014). *Sustainable urban logistics: Concepts, methods and information systems*. Heidelberg: Springer.

Meyer, A., & Meyer, D. (Eds.). (2013). *City Logistics Research. A Transatlantic Perspective. Summary of the First EU-U.S. Transportation Research Symposium. Transportation Research Board, The National Academies, Washington, DC.*

Kgs. Lyngby, Denmark Harilaos N. Psaraftis

Contents

Contributors

Paulus T. Aditjandra NewRail—Newcastle Centre for Railway Research, School of Mechanical and Systems Engineering, Newcastle University, Newcastle upon Tyne, UK

Tolga Bektaş University of Southampton, Southampton, Highfield, UK

Stefanos D. Chatzinikolaou Laboratory for Maritime Transport, School of Naval Architecture and Marine Engineering, National Technical University of Athens, Athens, Greece

Emrah Demir Eindhoven University of Technology, Eindhoven, The Netherlands

Antony Evans UCL Energy Institute, London, UK

Sara Fozza D'Appolonia S.p.A., Genova, Italy

Christiane Geiger Faculty of Mechanical Engineering/Institute of Transport Logistics, TU Dortmund University, Dortmund, Germany

Chara Georgopoulou DNV GL Maritime Research & Development, Aitolikou 5, Piraeus, Greece

Dewan Md Zahurul Islam NewRail—Newcastle Centre for Railway Research, School of Mechanical and Systems Engineering, Newcastle University, Newcastle upon Tyne, UK

Nikolaos M.P. Kakalis DNV GL Maritime Research & Development, Aitolikou 5, Piraeus, Greece

Christos A. Kontovas Technical University of Denmark, Lyngby, Denmark

Gilbert Laporte HEC Montreal, Montreal, QC, Canada

Atle Minsaas Norwegian Marine Technology Research Institute (Marintek), Valentinlyst, Trondheim, Norway

Roberto Palacin NewRail—Newcastle Centre for Railway Research, School of Mechanical and Systems Engineering, Newcastle University, Newcastle upon Tyne, UK

George Panagakos Technical University of Denmark, Lyngby, Denmark

Gernot Pauli Central Commission for the Navigation of the Rhine (CCNR), Strasbourg, France

Harilaos N. Psaraftis Technical University of Denmark, Lyngby, Denmark

Valerio Recagno D'Appolonia S.p.A., Genova, Italy

Erini Stamatopoulou National Technical University of Athens, Athens, Greece

Nikolaos P. Ventikos Laboratory for Maritime Transport, School of Naval Architecture and Marine Engineering, National Technical University of Athens, Athens, Greece

Thomas H. Zunder NewRail—Newcastle Centre for Railway Research, School of Mechanical and Systems Engineering, Newcastle University, Newcastle upon Tyne, UK

About the Editor

Harilaos N. Psaraftis is Professor of Transport Optimization at the Department of Transport of the Technical University of Denmark (DTU). He has a diploma from the National Technical University of Athens (NTUA, 1974), and two M.Sc. degrees (1977) and a Ph.D. (1979) from the Massachusetts Institute of Technology (MIT). He was a faculty member at MIT from 1979 to 1989 (Department of Ocean Engineering), receiving tenure there in 1985. In the period from 1989 to 2013, he was Professor of Maritime Transport at the School of Naval Architecture and Marine Engineering and Director of the Laboratory for Maritime Transport at NTUA.

Psaraftis's main interests are in transportation logistics, with a focus on sustainability issues in recent years. At MIT he headed some 11 research projects on various subjects including ship routing and scheduling, dial-a-ride systems, and oil spill response. At NTUA he led some 30 projects in practically all areas of maritime transportation. Among the 20 or so EU projects he was involved in, he has coordinated 3 multi-partner EU consortia, including project SuperGreen on European Green Corridors (2010–2013). At DTU he leads a new project on mitigating and reversing the impact of environmental legislation on the Ro-Ro sector in Northern Europe, funded by the Danish Maritime Fund (2015–2017).

Psaraftis's published work includes 3 books, 18 book chapters, over 135 articles in refereed journals and conferences, over 150 talks in various other conferences, and over 100 other publications. He has chaired several international conferences and organized several conference sessions and clusters on topics such as logistics, vehicle routing and scheduling, maritime and intermodal transportation, ports, and maritime safety, security, and environmental protection, and is on the editorial board of several journals, including *Transportation Science* since 1987. He is a member of INFORMS since 1977 and a Fellow of SNAME since 2009. He has received various academic and industry awards.

In addition to his academic duties, Psaraftis served as CEO of the Piraeus Port Authority (OLP) from Aug 1996 to March 2002. His tenure has been linked with such developments that include (a) Pireaus putting itself on the list of the top 50 world container ports (No. 41 in 1998) and becoming the top hub of the Eastern Mediterranean, with traffic more than doubling from 575,000 TEU (1996) to 1,160,000 TEU (2001), and (b) OLP being transformed from a public law undertaking into a corporation (1999). He has been the third longest serving CEO since OLP's inception in 1930. In 2006, he became a member of the Greek delegation to the International Maritime Organization (IMO), where he has served as chairman of various correspondence and working groups and as a member of various expert groups.

Chapter 1
The Policy Context

George Panagakos

Abstract The purpose of this chapter is to present the policy framework of the 'green' freight logistics, thus, setting the scene for the individual subjects of the following chapters. Its coverage is distinctly (albeit not exclusively) European, as it is mostly based on work undertaken under the SuperGreen project. The term 'green' is used in the sustainability context, meaning that it features economic and social dimensions in addition to the usual environmental one. The most important EU transport policy documents are reviewed and briefly presented by transportation mode. Horizontal documents covering all modes are reviewed first. The material spans a 15-year period, from the Sustainable Development Strategy of May 2001 to the Directive 2014/94/EU of October 2014 on the deployment of alternative fuels infrastructure.

Abbreviations

ATM	Air Traffic Management
CARE	Community Road Accident Database
CNG	Compressed natural gas
CO	Carbon monoxide
CO_2	Carbon dioxide
DG-MOVE	Directorate-General for Mobility and Transport, European Commission
DWT	Deadweight (of a ship)
EC	European Commission
ECA	Emission control area
ECSA	European Community Shipowners' Association
EEDI	Energy efficiency design index
EEOI	Energy efficiency operational indicator
EP&C	European Parliament & Council

G. Panagakos (✉)
Technical University of Denmark, Lyngby, Denmark
e-mail: geopan@transport.dtu.dk

H.N. Psaraftis (ed.), *Green Transportation Logistics*, International Series in Operations Research & Management Science 226,
DOI 10.1007/978-3-319-17175-3_1

ERA	European Railway Agency
ERTMS	European Rail Traffic Management System
ETCR	Regulation in Energy, Transport and Communications (OECD)
ETS	Emissions Trading System
EU	European Union
FTLAP	Freight Transport Logistics Action Plan
FUTRE	Future prospects on transport evolution and innovation challenges for the competitiveness of Europe (7th Framework Programme)
GDP	Gross domestic product
GHG	Greenhouse gas
GRT	Gross registered tonnage
HC	Non-methane hydrocarbon
HDV	Heavy duty vehicle
ICS	International Chamber of Shipping
ICT	Information and Communication Technology
IMO	International Maritime Organization
ITS	Intelligent Transport Systems
IWT	Inland Waterway Transport
LNG	Liquefied natural gas
LPG	Liquefied petroleum gas
LPI	Logistics performance index, the World Bank
LRIT	Long range identification and tracking
MARPOL	International convention for the prevention of pollution from ships
MBM	Market based measure
MEPC	Marine Environment Protection Committee, IMO
MoS	Motorway of the sea
MRV	Monitoring, reporting and verification of CO_2 emissions
NOx	Nitrogen oxides (NO and NO_2)
OECD	Organization for Economic Co-operation and Development
PM	Particulate matter
PSC	Public service contract
RFC	Rail freight corridor
RIS	River Information Services
RMMS	Rail Market Monitoring Scheme
Ro-Ro	Roll on—roll off (for ships)
SDS	Sustainable Development Strategy
SECA	Sulphur Emission Control Area
SEEMP	Ship Energy Efficiency Management Plan
SESAR	Single European Sky ATM Research
SO_2	Sulphur dioxide
SSN	SafeSeaNet (a VTMIS)
TEN-T	Trans-European Transport Network
UIC	International Union of Railways
UN	United Nations

UNECE	United Nations Economic Commission for Europe
VECTO	Vehicle energy consumption calculation tool
VTMIS	Vessel Traffic Monitoring and Information System
WEF	World Economic Forum
WTO	World Trade Organization

1.1 Introduction

The purpose of this chapter is to delineate the term 'green' when used in the context of freight transportation logistics. This will be done by reviewing a number of relevant policy documents. At the same time, this review will set the scene for presenting the individual subjects of the following chapters.

The material presented here is mainly based on work performed under the "SuperGreen" project financed by the EC's 7th Framework Programme of Research and Technological Development. This is the reason for the distinctly European coverage of this chapter. It should be noted that material from the SuperGreen project has been updated to reflect developments after the end of the project (January 2013) and until October 2014. Also, the chapter provides some references to related policy documents issued by institutions like the International Maritime Organization (IMO). By the same token, coverage is limited to the regional scope of the EU, which usually reflects a negotiated compromise between the national views of the Member States.

Very often the term 'green' is used to refer to merely environmental protection features. In this book, by 'green' we mean 'sustainable,' thus, adding economic and social attributes to the usual environmental ones (Fig. 1.1).

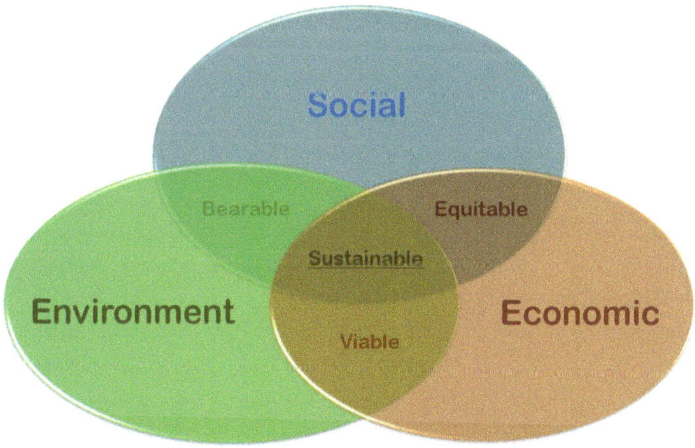

Fig. 1.1 The three dimensions of sustainable development. *Source*: The Sustainable Leader (2014)

Presenting a subject as wide and complex as the EU transport policy in the limited space of a book chapter is not an easy task. In addition, the broader perspective shown in Fig. 1.1 creates the need for reviewing a much wider range of EU policies. We are forced, then, to be very selective in our presentation and focus only on those documents that have a direct relevance to the search for win-win solutions of the following chapters. Although there is no guarantee that the author's personal biases have been left out entirely, every effort has been made to cover as many aspects of policy making as possible always in the context of green freight logistics.

The present chapter basically follows a modal structure. Road, rail and waterborne transportation are each covered in a separate section. Aviation is absent as the relevant material is presented in Chap. 13. Pipelines are outside the scope of the book. Within each mode (section), the documents reviewed are presented in chronological order.

Section 1.2 covers basic material published before 2010. After a brief discussion on the sustainability concept, the section presents the EU action plan on freight logistics and the so-called 'Greening transport package.' Section 1.3 is devoted to more recent documents but still horizontal in nature. It outlines the transportation strategy of the EU for the next decade, its new transportation infrastructure policy, its initiatives on alternative fuels and the newly introduced transport scoreboard. Section 1.4 relates to road transportation and presents the EC policy on ITS deployment and its proposals on revised dimensions and strategy formulation for reducing fuel consumption and CO_2 emissions from trucks. Section 1.5 deals with EC efforts to liberalize rail transportation and increase the priority of international freight trains. The last section of the chapter concerns waterborne transportation and more specifically IMO and EU initiatives addressing greenhouse gas (GHG) and SOx emissions of ships.

1.2 The Background

The first appearance of the term 'green' in the context of EU policy on transportation logistics took place in 2007, when the Freight Transport Logistics Action Plan introduced the 'green corridors.' Therefore, this document can serve as our point of departure. But before departing, it is necessary to look briefly into the way the European policy makers view the concept of sustainability with emphasis placed on sustainable transportation.

In relation to the external costs of transportation, the European Parliament asked the Commission, in 2006, to present "*a generally applicable, transparent and comprehensible model for the assessment of all external costs… and a strategy for a stepwise implementation of the model for all modes of transport*". In response to this request, the Commission prepared the 'Greening transport package', which was adopted in July 2008 (EC, 2008a). It basically consists of:

- the *Greening Transport Inventory* that describes the actions already taken by the EU to make transportation greener, and
- the *Strategy to Internalize the External Costs of Transport* accompanied by a proposal for introducing road tolls for trucks and track access charges for rail differentiated according to the environmental impact of train operation.

Both these documents will be briefly reviewed in this section, too.

1.2.1 The European Sustainable Development Strategy

Building on the traditional "Brundtland Commission" definition of sustainable development, i.e. "*development that meets the needs of the present without compromizing the ability of future generations to meet their own needs*," the EU developed its own Sustainable Development Strategy (SDS) in 2001. The SDS called for a society where **economic growth**, **social cohesion** and **environmental protection** go hand in hand, and laid down long-term objectives and priority actions in six policy areas (EC, 2001):

- climate change and clean energy,
- public health,
- social exclusion, demography and migration,
- management of natural resources,
- sustainable transport, and
- global poverty and development challenges.

In terms of sustainable transportation, SDS set the headline objectives of:

- decoupling transportation growth from GDP growth in order to reduce congestion and other negative side-effects of transportation,[1] and
- bringing about a shift in transportation use from road to rail, water and public passenger transportation,

and identified a number of priority actions, two of which found their way to the conclusions of the subsequent Gothenburg Council of June 2001 (Council, 2001):

- adopt revised guidelines for trans-European transport networks with a view to giving priority, where appropriate, to infrastructure investment for public and for railways, inland waterways, short sea shipping, intermodal operations and effective interconnection, and

[1] The indicator adopted for monitoring SDS implementation in terms of sustainable transport is: Energy consumption of transport relative to GDP.

- propose a framework to ensure that by 2004 the price of using different modes of transportation better reflects costs to society.

A revised SDS was adopted at the Brussels European Council of June 2006 (Council, 2006). The renewed SDS now rests on four separate pillars—**economic prosperity**; **social equity and cohesion**; **environmental protection**; and **global governance**—and is based on a long list of guiding principles: promotion and protection of fundamental rights, solidarity within and between generations, the guarantee of an open and democratic society, involvement of citizens, involvement of businesses and social partners, policy coherence and governance, policy integration, use of best available knowledge, the precautionary principle and the polluter-pays principle. Furthermore, a seventh policy area—sustainable consumption and production—is added to the previous six.

The overall objective of sustainable transportation is now:

- ensuring that our transportation systems meet society's needs whilst minimizing their undesirable impacts on the economy, society and the environment,

while the corresponding operational targets for freight transportation are:

- decoupling economic growth and the demand for transportation,
- achieving sustainable levels of transportation energy use and reducing GHGs,
- reducing pollutant emissions,
- achieving a balanced shift towards environment friendly transportation modes, and
- reducing transportation noise both at source and through mitigation measures.

Talking about sustainability, it should be mentioned that **sustainable growth**, i.e. promoting a more resource efficient, greener and more competitive economy, comprises one of the three mutually reinforcing priorities of *Europe 2020*, the strategy aimed at dragging Europe out of the 2008–2009 economic crisis (EC, 2010); the other two being **smart growth** (developing an economy based on knowledge and innovation) and **inclusive growth** (fostering a high-employment economy delivering social and territorial cohesion). The relevant targets set for 2020 by this document are:

- reduce GHG emissions by at least 20 % compared to 1990 levels or by 30 %, if the conditions are right,
- increase the share of renewable energy sources in EU's final energy consumption to 20 %, and
- increase energy efficiency by 20 %.

Moreover and in order to catalyse progress, *Europe 2020* has put forward seven flagship initiatives among which, the most relevant to the subject of this book is "Resource efficient Europe" helping to: decouple economic growth from the use of resources, support the shift towards a low carbon economy, increase the use of renewable energy sources, modernize our transport sector and promote energy efficiency.

1.2.2 Freight Transport Logistics Action Plan

In 2007, the Freight Transport Agenda (EC, 2007a) was launched by the EC to broaden the focus on freight transportation policy through a set of policy initiatives. The first one among them was the Freight Transport Logistics Action Plan (FTLAP), which introduced a number of short- to medium-term actions aimed at integrating transportation modes (EC, 2007b). The most important among these actions were:

- **Measuring performance of integrated systems:** The plan suggested the identification and monitoring of operational, infrastructural and administrative bottlenecks, the establishment of a core set of generic indicators that would measure and record performance (e.g. sustainability, efficiency etc.) in freight transportation logistics chains, and the elaboration of a set of generic (dynamic and static) benchmarks for multimodal terminals.
- **Exchange of information through interoperable ICT systems:** The plan introduced the concept of e-Freight denoting the vision of a seamless electronic flow of information associating the physical flow of goods with a paperless trail built by Information and Communication Technology (ICT) regardless of transportation mode, and called for the development of an Action Plan for deploying Intelligent Transport Systems (ITS) in road transportation.
- **Easing regulatory requirements for the exchange of information between modes:** The plan called for the assessment of establishing a single transportation document for all carriage of goods, irrespective of mode, of introducing within the EU of a standard (fall-back) liability clause and of establishing a single window (single access point) and one stop-administrative shopping for administrative procedures in all modes.
- **Introducing 'green corridors':** The plan introduced the concept of 'green corridors,' denoting corridors of highly dense freight traffic and of relatively long transportation distances equipped with adequate transshipment facilities at strategic locations. Industry should be encouraged along these corridors to rely on co-modality and on advanced technology in order to accommodate rising traffic volumes, while promoting environmental sustainability and energy efficiency.
- **Enhancing the urban dimension of integrated transportation solutions:** The plan introduced a holistic vision paying attention to aspects of land use planning, environmental considerations and traffic management.

It is interesting to note that an action plan on transportation logistics concentrated solely on mode integration issues. In fact, two of the five actions listed above (green corridors and urban distribution) relate to cargo flows, two (exchange of information and administrative procedures) concern information flows, while the fifth one (performance indicators) applies to both.

The green corridors introduced by the FTLAP play a key role in this book not only because "... *[they] are ideal environments for the development and introduction of solutions that help promote environmental sustainability and energy efficiency, so that they may become showcases of 'green' freight transport*," as stated by the Impact Assessment document accompanying the FTLAP (EC, 2007c), but because they comprise a vehicle that can address wider policy objectives of the EU, including modal integration, simplification of administrative formalities, internalization of external costs, and harmonization of safety, security and social legislation (Panagakos et al., 2013). Green corridors are studied in detail in Chaps. 3–6 of this book.

The e-Freight concept introduced by FTLAP also deserves special attention, as the exchange of information is a basic pillar of supply chain integration. It has a positive effect on, among others, demand planning, capacity and production planning, performance management, and inventory control. It is also vital in applications related to international safety and security. Information integration is considered as one of the most prominent future trends in supply chain management. Along this line, the e-Freight initiative of the EU aims at:

- enhancing interoperability between freight transportation information systems,
- enabling operators to enter information only once in the whole multimodal supply chain for planning, execution, monitoring and reporting purposes,
- developing interoperable information and booking tools (such as multimodal journey planners for freight) for an optimised use of multimodal transportation possibilities, and
- developing a structure for the use of information coming from tracking and tracing technologies as well as from intelligent cargo applications.

More specifically, the reference framework for ICT in transportation logistics that e-Freight seeks to establish would enable the provision of services like:

- a single transportation document, as an electronic waybill across modes,
- a single window for information sharing across modes, for business-to-administration, administration-to-administration and administration-to-business purposes at national, between national and at EU level,
- a standard description of transportation services and the issuing of transportation instructions.

1.2.3 Greening Transport Inventory

This document (EC, 2008b) compiles a list of measures that were in place in 2008 to reduce the negative impacts of transportation and more specifically those related to climate change, regional and local pollution, noise pollution, congestion and accidents. The most important of them are mentioned below:

Multiple-impact measures

- Motor fuel taxation: Minimum rates are set depending on the type of fuel. Alternative fuels, such as LPG and natural gas, are treated favourably, as are biofuels.
- State aids (subsidies, tax breaks etc.): Can be allowed for environmental purposes in specific cases. Transportation projects with clear environmental benefits can be funded through the TEN-T, the Cohesion and European Regional Development Funds. Special financing (Marco Polo programme) is available for projects that stimulate traffic avoidance or modal shift from road to rail, short-sea shipping and inland waterway transportation.[2]
- Environmental assessment: Required for projects over a certain size in all transportation modes, as well as for policy plans and programmes setting the framework for future infrastructure projects.
- Research and technology: Actions on transportation, the environment and energy, as well as ICTs which have an impact on all of these areas are funded through the 7th Framework Research Programme.[3]

Climate change

- Commercial aviation to be included in the EU Emissions Trading System (ETS).
- Limitation or reduction of GHG emissions from ships is to be achieved through the IMO. In the absence of sufficient progress at the IMO, the Commission will propose EU measures.
- A strategy to reduce the CO_2 emissions from light duty road vehicles (i.e. cars and vans) was adopted in 2007.[4]
- Minimum energy performance measures to be put in place when buildings with a useful floor area of more than 1,000 m^2 (including airports, port terminals, stations and car parks) are renovated or constructed.

Local pollution

- EU rules set maximum levels for sulphur in both diesel fuel and gas oil and for polycyclic aromatic hydrocarbons in diesel fuel.
- International rules establish a maximum worldwide level of sulphur content in fuel oil burned by ships. They also set up Sulphur Emission Control Areas (SECAs) where more stringent limits apply (refer to Chap. 10).
- EU rules set limits on the sulphur content of gas oil and marine gas oil, which are commonly used for inland navigation to 0.1 %.

---

[2] The last Marco Polo II call of the 2007–2013 financial period was launched in March 2013.

[3] The 7th Framework Programme was the EU's research and innovation funding programme for the period 2007–2013. The current programme is Horizon 2020, but there are many projects funded under the 7th Framework Programme that are still running.

[4] This strategy led to Regulation (EU) No 510/2011, which sets emission performance standards for new light commercial vehicles.

- International rules limit the NOx emissions from new marine diesel engines over a certain size.
- EU rules set limit values for emissions of CO, HC, NOx and PM from new engines for locomotives and inland waterway vessels sold in the EU.
- EU measures limit emissions of various pollutants including CO, HC, NOx, PM, smoke and ammonia (the "EURO" standards) from road vehicles.
- EU rules exist to limit the emissions of volatile organic compounds during the storage, loading, distribution and unloading of petrol.
- Specific EU rules exist for the collection and disposal of waste oils, used and shredded tyres, batteries and accumulators from automotive sources etc.
- International rules on the discharge of ballast water from ships have been adopted, aiming to prevent the transfer of harmful aquatic organisms and pathogens.

Noise pollution

- Member States are required to monitor and map noise, as well as draw up action plans to prevent and reduce noise.
- EU rules require all non-passenger vessels with a deadweight of more than 350 ton which travel on inland waterways to not exceed 75 dB(A) when moving and 65 dB(A) when stationary.
- EU rules limit noise emissions from both conventional and high-speed rail. New rolling stock for conventional rail should have low-noise brake blocks which reduce noise emissions by 50 %.
- EU rules set the maximum permissible noise emission levels for all new motor vehicles except tractors. There are separate EU requirements for noise from passenger car tyres and from van, bus and truck tyres.
- Limits also exist for aircraft, and more stringent restrictions can be put in place at certain EU airports.

Congestion

- EU measures have helped financing increased and alternative infrastructure capacity.
- Since March 2003, all new high speed lines must be equipped with ERTMS (the European Rail Traffic Monitoring System) and, since September 2006, all new sections of conventional priority projects. ERTMS will allow increased capacity on the railways through reducing congestion.
- All sectors will benefit from the possibilities that Galileo (Global Navigation Satellite System) will offer for congestion avoidance through optimizing transportation routes.

Accidents

- There are numerous international and EU safety requirements concerning the design, construction and maintenance of road and rail vehicles, inland waterway vessels, ships and aircraft.

- EU rules set out the maximum dimensions (height, width and length) and minimum turning circles for trucks in international and national traffic, as well as the maximum weights for trucks in international traffic.
- All trucks must have speed limiters fitted to be used on the road; they must be set at 90 km/h.
- EU rules exist aiming to improve safety of the transportation of dangerous goods by all transportation means.
- EU rules on tunnel safety require all tunnels longer than 500 m and belonging to the TEN-T road network to meet minimum safety requirements.

1.2.4 The Strategy to Internalize the External Costs of Transportation

The aim of this document was to propose a common methodology for the internalization of transportation-related external costs (EC, 2008c). Internalization intends to give the right price signal; so that users bear the costs they create and thus have an incentive to change their behaviour in order to reduce these costs.

In theory, the "social marginal cost charging," i.e. the additional short-term cost created by one extra person using the infrastructure, is the appropriate price setting mechanism that does not lead to overexploitation of resources (through under-pricing), and at the same time does not damage the transportation sector or ultimately the economy (through over-pricing). However in practice, marginal costs cannot be calculated easily, as they vary according to time and place. Furthermore, for some costs, such as those relating to noise, the method for estimating the marginal costs is very complex, and average cost pricing is used instead.

It should be mentioned that external costs, which are internalized according to the 'polluter pays' principle, should not be confused with infrastructure costs that are funded according to the 'user pays' principle.

After setting the principles, the document proposed a methodology adapting the overall strategy of external cost internalization to the characteristics of each mode of transportation.

For the **road sector**, Directive 1999/62/EC on the charging of Heavy Duty Vehicles (HDVs) precluded incorporating any of the external costs when calculating tolls. It was amended by Directive 2006/38/EC to allow different tariffs to be applied depending on vehicles' environmental characteristics. However, with the exception of mountainous regions, and then only in certain circumstances, toll revenues could not exceed infrastructure costs. This was the case even in more congested regions or regions with higher levels of pollution. The Commission, therefore, proposed to revise Directive 1999/62/EC in order to enable Member States to integrate in tolls levied on HDVs an amount which reflects the cost of air pollution and noise pollution caused by traffic. During peak periods, it would also allow tolls to be calculated on the basis of the cost of congestion imposed upon other vehicles. The amounts would vary with the travelled distance, location and time of use of roads to better reflect these external costs (EC, 2008d).

An interesting feature of the proposed revision was that the proceeds would have to be used by Member States for making transportation more sustainable through projects such as research and development on cleaner and more energy efficient vehicles, mitigating the effect of road transportation pollution or providing alternative infrastructure capacity for users. The charge would have to be collected through electronic systems which do not impede the free flow of traffic and which can be extended to other part of the network at a later stage without significant additional investments.

In addition, the proposal extended the scope of the current Directive beyond the TEN-T network to avoid inconsistent pricing schemes between major corridors and other interurban roads.[5] The same charging principles could also be extended to private cars.

For the **rail sector**, Community action was suggested to reduce the exposure of citizens to rail noise by retrofitting freight wagons with low-noise brakes. To overcome the financial burden of retrofitting, the Commission analysed different measures and concluded that a combination of noise-differentiated track access charges, noise emission ceilings and voluntary commitments is the most appropriate solution (EC, 2008e).

- In the framework of a revised Directive 2001/14/EC12, which harmonises charging principles including noise, a system of **noise-differentiated track access charges** could be introduced. Three basic models could be used as an incentive:
  - a cost-neutral bonus-malus system with reduced charges for silent wagons and higher charges for noisy ones,
  - a bonus system in the form of economic incentives for the wagon owners to retrofit their wagons in the start-up phase, and
  - a malus system consisting of increased charges for noisy wagons.

Infrastructure managers would be in charge of the installation of identification systems and the necessary ICT tools.[6]

[5] The proposal was adopted in 2011 as Directive 2011/76/EU amending Directive 1999/62/EC on the charging of heavy goods vehicles for the use of certain infrastructures.

[6] Directive 2012/34/EU establishing a single European railway area (recast of the first railway package) was adopted on 21 November 2012. A provision for non-mandatory noise-differentiated track access charges is included as Art. 31(5). In addition, the Commission shall adopt implementing measures by 2015 setting out the charging modalities for the cost of noise effects enabling the differentiation of infrastructure charges to take into account, among others, the sensitivity of the area affected.

- The **noise emission ceiling** limits the average emissions within a determined period at a certain location along the line. Such schemes leave it to the rail sector to find optimal solutions and can comprise the second step after the initial retrofitting programmes have been completed.
- **Voluntary commitments** by the rail sector can guarantee the effectiveness of differentiated track access charges and help to speed up their implementation even before legal requirements enter into force.

As for the **maritime transportation**, the Commission expressed its wish to include it in the post-2012 agreement on preventing climate change. If IMO would not make sufficient progress, the Commission suggested taking action at European level; with one of the possible options being to include the maritime sector in the EU ETS.

Before changing subject, it should be mentioned that in order for internalization to be effective, the transportation user must be price sensitive. Sometimes this is not possible for specific reasons, such as the lack of credible alternatives, insufficient competition with regard to a particular mode of transportation, insufficient incentive to innovate and switch to clean vehicles, etc. Internalization is a necessary step in itself, but it must be accompanied by other measures intended to create greater elasticity of demand, i.e. greater sensitivity to price variations, to make the supply of certain services more attractive or to speed up technological innovation. In order to reduce the external costs, we therefore need a strategy that includes various other elements in addition to internalization, elements such as providing infrastructure, encouraging technological innovation, competition policy and setting standards.

1.3 Horizontal Policies

This section presents more recent policy documents which, due to their horizontal nature, cannot be allocated to one of the modal sections that comprise the remainder of this chapter.

The highest-level strategic document presenting the EC's vision for the future of the EU transportation system and defining the policy agenda for the following decade is usually contained in a White Paper issued at the beginning of each decade, followed by its mid-term revision. The 2011 White Paper on transport, which is fully compatible with the Europe 2020 strategy and its "Resource Efficient Europe" flagship initiative, presented in Sect. 1.2.1, is the latest such document and will be briefly presented here.

The section will also present the EU policies in relation to transportation infrastructure and the deployment of alternative fuels in the transportation sector. The recently introduced EU transport scoreboard, comparing the performance of the Member States in a number of transportation-related issues completes the section.

1.3.1 The White Paper on Transport

The 2011 White Paper on Transport (EC, 2011a) is the single most important document in EU transport policy, as it describes the EC's vision of future transportation and the corresponding strategy for the next decade. More specifically, it takes a global look at developments in the transportation sector, at its future challenges and at the policy initiatives that need to be considered in the period until 2020 in order to meet the long-term requirement for limiting climate change to 2 °C. This general objective is translated into the following specific objectives:

(a) a reduction of transport-related GHG emissions by approximately 60 % by 2050 compared to 1990,
(b) a drastic decrease in the oil dependency of transport-related activities by 2050, and
(c) limiting the growth of congestion.

According to the document, the Commission's vision of future transport is:

> a system that underpins European economic progress, enhances competitiveness and offers high quality mobility services while using resources more efficiently. Curbing mobility is not an option. New transport patterns must emerge, according to which larger volumes of freight are carried jointly to their destination by the most efficient (combination of) modes. Individual transport is preferably used for the final miles of the journey and performed with clean vehicles. Information technology provides for simpler and more reliable transfers. Transport users pay for the full costs of transport in exchange for less congestion, more information, better service and more safety.

Alternatively, this vision is expressed through three strands, which are listed below together with ten related benchmarks for achieving the GHG emissions reduction target:

- Improving the energy efficiency performance of vehicles across all modes; developing and deploying sustainable fuels and propulsion systems.

  1. Halve the use of 'conventionally-fuelled' cars in urban transport by 2030; phase them out in cities by 2050; achieve essentially CO_2-free city logistics in major urban centers by 2030.
  2. Low-carbon sustainable fuels in aviation to reach 40 % by 2050; also by 2050 reduce EU CO_2 emissions from maritime bunker fuels by 40 % (if feasible 50 %).

- Optimizing the performance of multimodal logistic chains, including by making greater use of inherently more resource-efficient modes, where other technological innovations may be insufficient (e.g. long distance freight).

  3. 30 % of road freight over 300 km should shift to other modes such as rail or waterborne transport by 2030, and more than 50 % by 2050, facilitated by efficient and green freight corridors. To meet this goal will also require appropriate infrastructure to be developed.

4. By 2050, complete a European high-speed rail network. Triple the length of the existing high-speed rail network by 2030 and maintain a dense railway network in all Member States. By 2050 the majority of medium-distance passenger transport should go by rail.
5. A fully functional and EU-wide multimodal TEN-T 'core network' by 2030, with a high quality and capacity network by 2050 and a corresponding set of information services.
6. By 2050, connect all core network airports to the rail network, preferably high-speed; ensure that all core seaports are sufficiently connected to the rail freight and, where possible, inland waterway system.

- Using transport and infrastructure more efficiently through use of improved traffic management and information systems, and advanced logistics and market measures.

7. Deployment of the modernised air traffic management infrastructure (SESAR) in Europe by 2020 and completion of the European Common Aviation Area. Deployment of equivalent land and waterborne transport management systems (ERTMS, ITS, SSN and LRIT, RIS). Deployment of the European Global Navigation Satellite System (Galileo).
8. By 2020, establish the framework for a European multimodal transport information, management and payment system.
9. By 2050, move close to zero fatalities in road transport. In line with this goal, the EU aims at halving road casualties by 2020. Make sure that the EU is a world leader in safety and security of transport in all modes of transport.
10. Move towards full application of "user pays" and "polluter pays" principles and private sector engagement to eliminate distortions, including harmful subsidies, generate revenues and ensure financing for future transport investments.

The above mentioned targets shall be met through the following 4-tier strategy:

- **Internal market:** Create a genuine single European transport area by eliminating all residual barriers between modes and national systems, easing the process of integration and facilitating the emergence of multinational and multimodal operators.
- **Innovation:** EU research needs to address the full cycle of research, innovation and deployment in an integrated way through focusing on the most promising technologies and bringing together all actors involved.
- **Infrastructure:** The EU transport infrastructure policy needs a common vision and sufficient resources. The costs of transportation should be reflected in its price in an undistorted way.
- **International:** Opening up third country markets in transport services, products and investments continues to have high priority. Transportation is included in all trade negotiations with European participation (WTO, regional and bilateral).

Furthermore, a total of 131 actions, organised in 40 concrete initiatives, are proposed by the document for the materialization of this strategy.

1.3.2 The New TEN-T Policy

In line with the 2011 White Paper of the previous section and in view of persisting obstacles at EU level, like:

- missing links, in particular at cross-border sections,
- considerable and enduring infrastructural bottlenecks, in particular with respect to the east-west connections,
- fragmented transportation infrastructure between modes,
- significant investments in transportation infrastructure needed in order to achieve the GHG emission reduction target, and
- interoperability problems due to different operational rules and requirements by the Member States, adding to the transportation infrastructure barriers and bottlenecks,

the Commission has redefined its long-term transportation infrastructure policy up to 2030/2050 through revising the so-called 'TEN-T guidelines' (EP&C, 2013a), which set out priorities and provide implementation measures for the trans-European transport network (TEN-T).

The main objective, i.e. the establishment and development of a complete TEN-T, consisting of infrastructure for railways, inland waterways, roads, maritime and air transportation, is pursued through two fields of action.

The first one concerns the 'conceptual planning' of the network for which a dual-layer approach has been selected, consisting of a comprehensive and a core network. The comprehensive network constitutes the basic layer of the TEN-T and is, in large part, derived from the corresponding national networks. It should be in place by 2050 at the latest. The core network overlays the comprehensive network and consists of its strategically most important parts. It constitutes the backbone of the multimodal mobility network and concentrates on those components of TEN-T with the highest European added value: cross border missing links, key bottlenecks and multimodal nodes. The core network is to be in place by 2030 at the latest.

It is worth mentioning that the guidelines (Article 39) lay down specific requirements for the core network, in addition to the requirements for the comprehensive network. The most prominent among them is the necessity to provide 'alternative clean fuels' for all transportation modes. This term includes fuels such as electricity, hydrogen, biofuels (liquids), synthetic fuels, methane (CNG, LNG and biomethane) and LPG, which serve, at least partly, as a substitute for fossil oil sources in the supply of energy to transport and contribute to its decarbonization. For rail transportation, this requirement is further defined as full electrification of the line tracks and sidings. Furthermore, new railway lines should have a nominal track gauge of 1,435 mm (with certain exceptions), while ERTMS should be fully deployed on all new and existing lines. In addition, the freight lines of the core network should be able to accommodate at least 22.5 ton axle loads, 100 km/h line speeds and running trains with a length of 740 m. For motorways, emphasis is placed on the development of rest areas approximately every 100 km.

The second field of action concerns the implementation instruments. The Commission has developed the concept of 'core network corridors', taking due account of the rail freight corridors introduced with Regulation No 913/2010 (refer to Sect. 1.5.2), as an instrument for the coordinated implementation of the core network. Core network corridors (Article 43):

- cover the most important cross-border long-distance flows in the core network,
- are multimodal in nature and involve at least three transportation modes,
- cross at least two borders, and
- include Motorways of the Sea[7] (MoS), where appropriate.

Annex I to Regulation No 1316/2013 (EP&C, 2013b), establishing the Connecting Europe Facility, which finances EU priority infrastructure in transportation, energy and digital broadband, lists nine core network corridors. They are shown in Fig. 1.2.

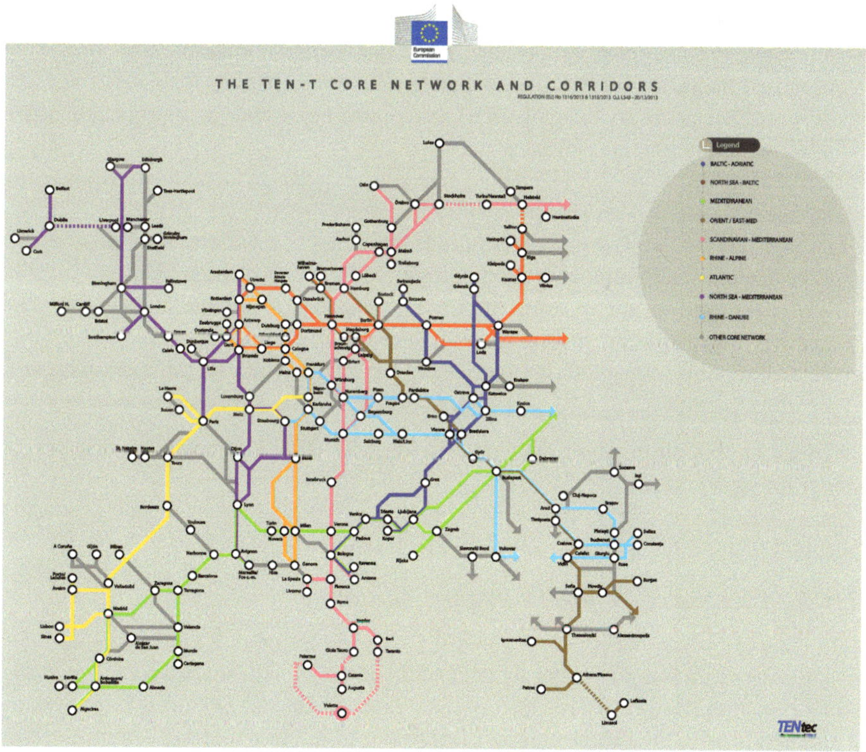

Fig. 1.2 The nine TEN-T core network corridors and other connections. *Source*: EC (2014a)

[7] MoS represent the maritime dimension of the TEN-T and consist of maritime links between maritime ports of the comprehensive network including the related facilities and infrastructure for direct land and sea access (Article 21).

In terms of governance, the new TEN-T guidelines provide for European Coordinators to be designated by the EC in agreement with the Member States concerned. A European Coordinator shall be assigned to each and every core network corridor, while two additional Coordinators shall be designated for implementing the horizontal ERTMS and MoS respectively. Acting in the name and on behalf of the EC, the European Coordinators shall facilitate the coordinated implementation of the core network corridors. They will be assisted in this task by a secretariat and by a consultative forum (the Corridor Forum), established for each corridor. The European Coordinators shall chair the Corridor Fora, the composition of which shall be agreed with the relevant Member States.

A central task of the European Coordinator is drawing up a corridor work plan and monitoring its implementation, in consultation with the Corridor Forum and the relevant Member States. The work plan shall include (Article 47):

- a description of the characteristics of the core network corridor including its cross-border sections,
- a list of objectives and priorities to be pursued,
- a plan for the removal of physical, technical, operational and administrative barriers between and within transportation modes,
- a deployment plan of interoperable traffic management systems,
- proposed measures to enhance resilience to climate change,
- proposed measures to mitigate GHG emissions, noise and, as appropriate, other negative environmental impacts,
- a list of projects for the extension, renewal or redeployment of transportation infrastructure,
- an analysis of the investment required, including the various funding sources envisaged, at international, national, regional, local and Union levels,
- where appropriate, measures to improve the capacity to design, plan, implement and monitor major transportation projects, and
- details of public consultations supporting the development of the work plan and its implementation.

Based on this information the Commission will adopt implementing acts (decisions) for each corridor.

1.3.3 Clean Power for Transport Initiative

Transportation in Europe is heavily dependent on oil, which counted for 94 % of the sector's energy needs in 2010 (EC, 2013a). The fact that 84 % of it is imported, in combination with the recent political instability of major exporting regions, raises genuine security of supply concerns. The cost of oil imports for transportation was close to €1 billion a day in 2011 and this figure should be viewed in conjunction with increasing volatility and uncertainty (crude oil prices have left their historic range of $10–$30 per barrel, and rose to nearly $150 per barrel before the global downturn in 2008). Furthermore, mitigating the environmental impact of transportation has already been documented in the previous sections as a primary objective of the EU transportation policy. Alternative fuels are, therefore, urgently needed to switch to a post-oil economy.

Research has led to the successful development of alternative fuel solutions for all transportation modes. However, their market uptake is slower than usual, mainly due to the fact that the use of alternative fuels requires the gradual build-up of charging and refuelling infrastructures and, thus, significant investments. The relationship between vehicles capable of running on alternative fuels and the appropriate refuelling infrastructure is often described as a 'chicken and egg' problem, requiring state intervention.

Faced with this challenge, the EC adopted in January 2013 the so-called 'Clean Power for Transport' package aiming to facilitate the development of a single market for alternative fuels for transportation in Europe. The package consisted of:

- a comprehensive European alternative fuels strategy for the long-term substitution of oil as energy source in all modes of transportation (EC, 2013a),
- an action plan for a broad market uptake of LNG in the shipping sector (EC, 2013b), and
- a proposal for a Directive on the deployment of alternative fuels recharging and refuelling infrastructure, accompanied by its impact assessment.

Following the inter-institutional negotiations, the above proposal led to Directive 2014/94/EU of 22 October 2014 (EP&C, 2014), which:

- requires Member States to adopt national policy frameworks for the market development of alternative fuels and the deployment of the relevant infrastructure,
- sets minimum coverage and timetable for each use of alternative fuels in accordance with Table 1.1 below,
- ensures the use of common technical specifications for recharging and refuelling stations, and
- paves the way for setting up appropriate labelling of alternative fuels, as well as for providing information that enables sound price comparisons by the end users.

Member States have to submit their national policy frameworks to the Commission within 2 years and report on their implementation on 3-year intervals thereafter. The Commission will assess and report on those national policy frameworks in order to ensure coherence at Union level.

1.3.4 The EU Transport Scoreboard

In April 2014, the European Commission published for the first time a scoreboard on transport in the EU. It compares the performance of the Member States in 21 transportation-related categories and highlights the five top and bottom

Table 1.1 Coverage and timetable of alternative fuel uses (Directive 2014/94/EU)

Alternative fuel	Coverage	Timing
Electricity for motor vehicles in urban/suburban and other densely populated areas	Appropriate number of publically accessible recharging points	By end 2020
CNG for motor vehicles in urban/suburban and other densely populated areas	Appropriate number of refuelling points	By end 2020
CNG for motor vehicles along the TEN-T core network	Appropriate number of refuelling points	By end 2025
Shore-side electricity supply for seagoing and IWT vessels	Ports of the TEN-T core network and other ports	By end 2025
Hydrogen for motor vehicles in the Member States which choose to develop it	Appropriate number of refuelling points	By end 2025
LNG at maritime ports	Ports of the TEN-T core network	By end 2025
LNG at inland ports	Ports of the TEN-T core network	By end 2030
LNG for heavy-duty vehicles	Appropriate number of refuelling points along the TEN-T core network	By end 2025

performers for most of these categories. It aims at helping Member States identify shortcomings and define priorities for investment and policies.

The scoreboard builds on the World Bank's Logistics Performance Index (LPI) which, since 2007, assists countries benchmark their performance on trade logistics. It draws data from a variety of sources (Eurostat, the European Environment Agency, the World Bank and the OECD) and can be consulted either by mode of transportation (road, rail, waterborne, air) or by one of the following categories:

- **Single market:** It assesses the level of market integration for each mode of transportation:
  - *Regulation of road freight transportation*, based on the OECD indicator of regulation in energy, transportation and communications (ETCR), which considers entry barriers and price control by authorities.
  - *Market share of all but the principal railway undertakings*, separately for freight and passenger transportation, on the basis of RMMS (Rail Market Monitoring Scheme) data.
  - *Maritime cabotage transportation of goods*, based on Eurostat data (no ranking is provided for this indicator, which simply exhibits the volume of national transportation of goods by sea).
  - *Regulation of air passenger transportation*, based on OECD's ETCR which, for air passenger transportation, considers entry barriers and public ownership.

- **Infrastructure:** It assesses the quality of infrastructure for each mode of transportation:
  - *Motorway density*, expressed by the ratio between the total length of motorways and the population (in millions), on the basis of data from Eurostat, UNECE and national sources.
  - *Quality of rail infrastructure*, rating based on a survey by the World Economic Forum (WEF).
  - *Quality of port infrastructure*, rating based on a WEF survey of business executives' perception of their country's port facilities.
  - *Quality of air transportation infrastructure*, rating based on a WEF survey.
- **Environmental impact:** Indicators are provided only for road and rail transportation:
  - *Average CO_2 emissions from new passenger cars*, on the basis of European Environment Agency data (in gCO_2/km).
  - *Electrified railway lines*, expressed as a percentage of electrified railway lines over total lines in use, on the basis of data from the International Union of Railways (UIC) and national sources.
- **Safety:** Once again only road and rail transportation indicators are provided:
  - *Road fatalities*, defined as persons deceased within 30 days of a road accident per million inhabitants, on the basis of information from the CARE database of DG MOVE.
  - *Railway victims*, defined as persons (including workers, passengers, crossing users and unauthorised persons) deceased or seriously injured in railway accidents in relation to the overall rail transportation activity (in million train-km), calculated using Eurostat and ERA data.
- **Transposition of EU law:** Percentage of EU transportation directives for which Member States have notified transposition measures to the Commission by 31 December 2013, even with delays (total number of directives to be transposed: 115).
- **Infringements of EU law:** According to DG MOVE, on 31 December 2013, the Commission was dealing with a total of 202 infringement proceedings in the area of transportation (cases of a Member State not applying an EU law properly). The scoreboard presents the number of cases separately for each mode of transportation, while an additional category deals with cases that are not mode-specific, in particular concerning passenger rights.
- **Research and innovation:** This horizontal category covers two aspects:
  - *Private investment in transportation research and development*, defined as investment by transportation companies in research and development, as percentage of GDP. It includes manufacturing of motor vehicles, other transportation equipment, air/spacecraft, railway locomotives and rolling stock, transportation and storage. It is based on information from FUTRE project.

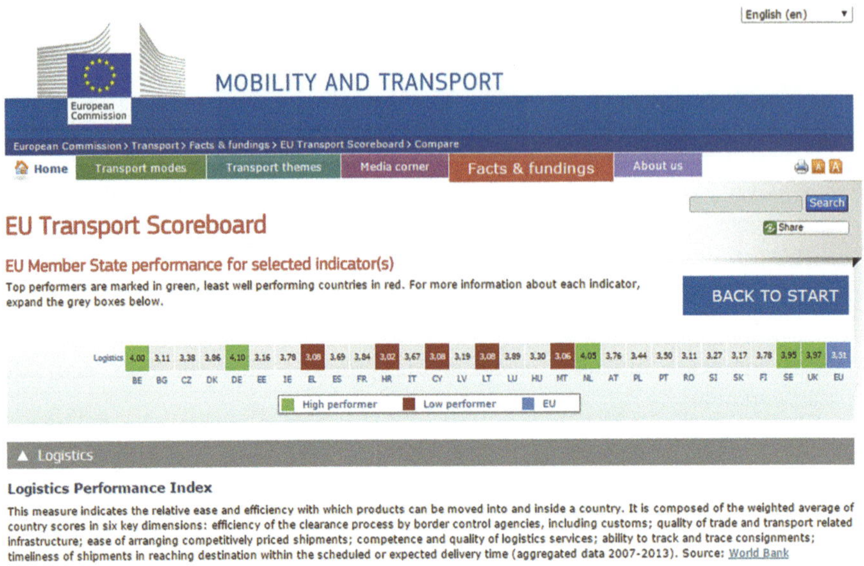

Fig. 1.3 Indicative scoreboard screen exhibiting the performance of EU Member States in relation to Logistics. *Source*: EU Scoreboard

- *Innovative transportation companies*, defined as the percentage of companies that replied positively to the question 'do you innovate?' of the 2010 Community Innovation Survey of Eurostat.

- **Logistics:** The World Bank's Logistics Performance Index, rating the relative ease and efficiency with which products can be moved into and inside a country (refer to Fig. 1.3).

The Commission intends to further refine the above indicators, in dialogue with Member States, industry and other stakeholders.

1.4 Road Transportation

Among the challenges facing the transportation sector today, the following are the most important ones that pertain to road transportation:

- High congestion levels already seriously affect road transportation in several Member States, while by 2030 it is expected to obstruct the inter-urban network as well.
- The share of CO_2 emissions from EU transportation as a proportion of all EU emissions continues increasing and road transportation accounts for 82 % of the energy consumption of the transportation sector.
- Whilst road fatalities are in regression their number is still unacceptably high.

A number of initiatives have been taken by the EC to address them. The deployment of ITS, the revision of the maximum authorised dimensions and weights for HDVs and the recently adopted strategy for reducing fuel consumption and CO_2 emissions are the most interesting among them.

1.4.1 Deployment of Intelligent Transport Systems

ICT systems play a key role in the development and evolution of transportation operations, as they identify and alleviate bottlenecks and release latent demand and supply for transportation services exploiting in full the capacity of infrastructure, vehicles and equipment. In this respect, they improve the efficiency of using the transportation infrastructure and equipment, reduce transportation costs, improve the quality of transportation services, and enhance the environmental sustainability of the sector through improved traffic management, reduced congestion and emissions, optimised operations, lower externalities etc.

In 2008, the EC adopted its *Action plan for the deployment of Intelligent Transport Systems in Europe* to create the momentum necessary to speed up market penetration of rather mature ITS applications and services in Europe. It was prepared on the basis of input provided by a wide consultation of stakeholders. Traffic management, congestion relief on freight corridors and in cities, promotion of co-modality, in-vehicle safety systems, real time traffic and travel information and an open in-vehicle platform to integrate applications were among the priority issues identified.

The Action Plan outlined the following six priority areas for action:

Action Area 1. Optimal use of road, traffic and travel data
Action Area 2. Continuity of traffic and freight management ITS services on European transport corridors and in conurbations
Action Area 3. Road safety and security
Action Area 4. Integration of the vehicle into the transportation infrastructure
Action Area 5. Data security and protection, and liability issues
Action Area 6. European ITS cooperation and coordination.

As a result of this Action Plan, Directive 2010/40/EU establishing a framework for the deployment of ITS in the field of road transportation (the 'ITS Directive') was adopted on 7 July 2010 to accelerate the deployment of these innovative applications across Europe (EP&C, 2010b). Aiming to establish interoperable and seamless ITS services while leaving Member States the freedom to decide which systems to invest in, it is an important instrument for the coordinated implementation of ITS in Europe.

Under the ITS Directive, the EC has to adopt within 7 years specifications (i.e. functional, technical, organizational or services provisions) to address the compatibility, interoperability and continuity of ITS solutions across the EU. The first priorities are traffic and travel information, the e-Call emergency system and intelligent truck parking.

1.4.2 New Dimensions and Weights

In April 2013, the EC communicated its proposal to amend the maximum authorised dimensions and weights for heavy duty road vehicles, which have been in force since 1996, in order to allow more energy efficient vehicles to be put on the market (EC, 2013d). The proposal intends to:

- Grant derogations from the maximum dimensions of vehicles for the addition of aerodynamic devices to the rear of vehicles or to redefine the geometry of the cabs for tractors. While ensuring compliance with road safety rules and the constraints imposed by infrastructure and traffic flow, these derogations aim to open up new prospects for manufacturers of tractors, trucks and trailers, provided that the load capacity of the vehicles is not increased. In addition to reducing fuel consumption and GHG emissions, the new designs of tractor cabs are expected to improve the drivers' field of vision, increasing the drivers' comfort and safety and reducing road accidents.
- Authorize a weight increase of 1 ton for vehicles with an electric or hybrid propulsion, to take account of the weight of batteries or the dual motorization, without prejudice to the load capacity of the vehicle.
- Facilitate the development of intermodal transportation by allowing a derogation of 15 cm in the length of trucks carrying 45-ft containers, which are increasingly used in intercontinental and European transportation. This minor adjustment is sufficient to permit an extra EUR-pallet to fit in a 45-ft palletwide container (8½-ft wide), adding about 3 % load efficiency to the usual 32 EUR-pallet arrangement, while improving safety by reducing empty spaces.
- Confirm that cross-border use of longer vehicles is lawful for journeys that only cross one border, if the two Member States concerned already allow it and if the existing infrastructure and the road safety situation allow it.
- Enable the inspection authorities to better detect infringements through the use of either weighing systems built into the road or by means of onboard sensors in vehicles which communicate remotely with roadside inspectors. These technologies will allow a better filtering of the vehicles, so that only vehicles strongly suspected of infringement are stopped for manual inspection. Furthermore and in order to encourage the spread of such devices, the Commission plans to define the technical standards for onboard weighing devices, particularly the standards for the electromagnetic communication interface. Such systems offer the additional advantage of enabling drivers to better control the weight of their vehicles.

1.4.3 Strategy for Reducing Fuel Consumption and CO_2 Emissions

Despite the existence of several technical solutions that improve the fuel efficiency of a Heavy Duty Vehicle (HDV), their market uptake is very slow. Even solutions that can be implemented at a net profit are often not adopted. Aarnink, Faber, and den Boer (2012) have identified a number of market barriers that hamper the implementation of such measures, including split incentives (i.e. the owner of a vehicle does not benefit from fuel savings when this is operated by a separate entity), limited access to finance and the practice of manufacturers offering fuel saving technologies as optional rather than standard features of a new vehicle. However, the most important barrier found was the lack of information on the fuel savings associated with individual technical measures. It appears that the freight transportation industry is more focused on operational improvements for fuel savings than on new technologies, which are perceived as more costly. This knowledge gap results from the fact that HDV CO_2 emissions are not measured, certified and recorded when new vehicles are registered.

In May 2014, the EC issued its strategy to improve HDV performance and cut CO_2 emissions through measures that address the knowledge gap and unlock a large part of the existing potential (EC, 2014b).

With the exception of transportation demand, which is linked to economic activity and lies outside the scope of the document, the proposed strategy is built around the other main drivers of HDV fuel consumption and CO_2 emissions: modal split, fuel GHG intensity, vehicle energy efficiency and operation of HDV fleets.

In terms of modal split, the new TEN-T policy (refer to Sect. 1.3.2) aims to reverse the trend of increasing share of road transportation. The development of multimodal freight corridors enhanced by the e-freight initiative (refer to Sect. 1.2.2) is also expected to influence modal split.

The revised TEN-T guidelines are expected to have a positive impact in reducing the fuel GHG intensity, too, through the requirement for alternative fuel availability along the core network corridors for all modes, including road transportation. The Clean Power for Transport Initiative of Sect. 1.3.3 will further support this development. The proposed inclusion of a CO_2 element in fuel taxation can further enhance the share (~6 % in 2010) of alternative fuels in the energy use of road transportation.

In the area of HDV fleet operation, the on-going review of road user charging legislation aims to take measures improving load factors, accelerating the renewal of fleets and creating conditions for greater co-modality (refer to Sect. 1.2.4 on internalization of external costs). The ITS Directory of Sect. 1.4.1 will further improve the efficiency of using the road infrastructure and vehicles, as well as the interfaces with other modes of transportation. The review of the remaining restrictions on road cabotage and the inclusion of eco-driving requirements in the truck drivers' examinations can also help make road transportation more efficient.

As for supporting the deployment of more energy efficient vehicles, the proposed revision of the maximum authorised dimensions of HDVs to improve their aerodynamics (refer to Sect. 1.4.2) is one of the measures foreseen. Others include the funding of research under the 'Green Car Initiative' and the 'Horizon 2020' programmes, as well as the EU legislation on the procurement of more environment friendly vehicles by public entities.

However, no standards have been set at EU level in relation to the fuel consumption and CO_2 emissions of HDVs. A prerequisite to address these issues is to measure and monitor them. This is exactly the focus of the short-term actions of the proposed strategy. Unlike the approach selected for the waterborne transportation (refer to Sect. 1.6.3), the actions foreseen for road transportation are:

- **Completion of the VECTO simulation tool.** The Vehicle Energy Consumption Calculation Tool (VECTO) is a simulation tool that is being developed by the EC in cooperation with industry stakeholders since 2009. It is used for measuring total vehicle emissions including emissions due to the vehicle's motor and transmission, aerodynamics, rolling resistance, and auxiliaries. The simulation approach has been selected for addressing the identified knowledge gap because CO_2 testing on the basis of a testing cycle (as is the case with cars and vans) is not appropriate for HDVs due to the diversity of existing models and tasks.
- **Legislative action for certifying and reporting CO_2 emissions.** The methodology for determining fuel consumption and CO_2 emissions (VECTO calculations) needs to be included in the relevant type approval legislation.

On the basis of the findings of these short-term actions, medium-term policy options, including the setting of mandatory CO_2 emission standards for newly registered HDVs would be considered in order to assist meeting the environmental targets of the EU transportation policy.

1.5 Rail Transportation

In its effort to strengthen the position of railways vis-à-vis other transportation modes, the EC has been very active during the last 25 years in restructuring the rail transportation market, basically through interventions in three areas:

- **Opening of the rail transportation market to competition**, addressing the structure of state monopolies that characterised European railways until not very long ago.
- **Improving the interoperability and safety of national networks**, addressing the patchwork of different rail systems that exist (differences range across a wide spectrum, including at least four different rail gauges, at least four different electricity systems, at least a dozen different signalling systems, various clearance profiles, various technical specifications of locomotives and other rolling stock, and many other differences, not the least of which is that trains in some countries run on the left and in some other countries on the right side).
- **Developing rail transportation infrastructure**, addressing bottlenecks due to insufficient capacity and/or poor quality of existing rail networks.

The latter point is dealt with the new TEN-T policy that offers preferential treatment to railway infrastructure, which features persistently in all TEN-T core network corridors (refer to Sect. 1.3.2). Furthermore, ERTMS, the European approach to handling interoperability problems in the rail transportation, is prescribed as a requirement for all TEN-T core network corridors, which to a large extent supersede the so-called 'ERTMS corridors,' introduced by the relevant deployment plan.[8] That leaves the liberalization of rail transportation as the only topic that needs to be discussed further in this section. Special attention will also be given to the Rail Freight Corridor concept, which was introduced together with the green corridors and paved the way for the TEN-T core network corridors that were adopted later on.

1.5.1 *Liberalization of Railway Markets*

Community involvement in the sector came in 1991 with a Directive requiring separate accounts to be kept for railway infrastructure management and the provision of railway transportation services. Ten years later, in February 2001, the 'first railway package' was adopted aiming to enable rail operators to have access to the trans-European network on a non-discriminatory basis. The Commission underlined the need to improve the distribution of train paths, establish a tariff structure that reflects relevant costs, reduce delays at borders and introduce quality criteria.

The 'second railway package' of 2004 accelerated the liberalization of rail freight services by fully opening the rail freight market to competition as from 1 January 2007. In addition, the package created the European Railway Agency situated in Valenciennes (France), introduced common procedures for accident investigation and established Safety Authorities in each Member State.

In October 2007, the 'third railway package' was adopted opening up the international passenger transportation market including cabotage by 2010. Since then, operators may pick up and set down passengers at any station on an international route, including at stations located in the same Member State. Furthermore, the third railway package regulated the rail passenger rights and the certification of train crews.

In 2012, Directive 2012/34/EU (a recast of the first railway package) establishing a single European railway area, reinforced existing provisions on competition, regulatory oversight and the financial architecture of the rail sector (EP&C, 2012a). However, a number of remaining regulatory and market failures have been identified basically related to the full implementation and enforcement of EU legislation by Member States. In many cases infrastructure managers and

[8] Commission Decision of 22.7.2009 amending Decision 2006/679/EC as regards the implementation of the technical specification for interoperability relating to the control-command and signalling subsystem of the trans-European conventional rail system, C(2009) 5607, Brussels, 22.7.2009.

operators are not fully independent and the effectiveness of the regulatory oversight of market functioning remains problematic.

In view of these problems, the EC adopted in January 2013 the 'fourth railway package' comprising of legislative proposals in the following four areas:

Market access

- Open by 2019 the domestic rail passengers market to competition either by offering competing commercial services (open access) or through bidding for public service contracts (PSCs), which account for some 90 % of EU rail journeys and will now be subject to mandatory tendering.
- Introduce an obligation for competent authorities to take the financial risk of the residual value of rolling stock at the end of a PSC by appropriate means (i.e. assume ownership of the rolling stock, provide a bank guarantee for the purchase of new, set up a leasing company).
- Establish national integrated ticketing systems on a voluntary basis, subject to non-discrimination requirements.

Market structure

- Separate infrastructure managers from any transportation operator running the trains (albeit vertically integrated 'holding structures,' formed prior to the current legislation's entry into force, may be accepted provided that all safeguards ensuring the legal, financial and operational independence are in place).
- Strengthen infrastructure managers so that they perform all functions related to the development, operation and maintenance of the infrastructure, including traffic management (albeit subcontracting of specific renewal or maintenance works to railway undertakings is still possible).
- Establish a Coordination Committee which will allow all infrastructure users to express their needs and ensure that the difficulties they encounter are properly addressed.
- Create a Network of Infrastructure Managers to ensure that issues of cross-border and pan-European nature are properly addressed in a coordinated manner.

Harmonised standards and approvals

- Reinforce the role of the European Railway Agency (ERA) to become a 'one stop shop,' issuing EU wide vehicle authorizations in the form of "vehicle passports" as well as EU wide safety certificates for operators.

Rail workforce

- Allow Member States to protect rail workers beyond the general EU requirements by requiring new contractors to take them on when PSCs are transferred.
- Oblige pan-European railway undertakings to create European Works Councils and to take part in the Railway Social Sectoral Dialogue Committee.

1.5.2 *Rail Freight Corridors*

As part of the 2007 Freight Transport Agenda (EC, 2007a), which also included the Freight Transport Logistics Action Plan of Sect. 1.2.2, the Commission issued a Communication on a freight-oriented rail network (EC, 2007d), which aimed at making rail freight more competitive, in particular by ensuring lower transit times and increasing rail's reliability and responsiveness to customer requirements. The following actions were proposed:

- Creation of freight-oriented corridors
- Measures on improving service quality along a corridor
- Increasing the infrastructure capacity of a corridor
- More coordination and more priority to international freight trains
- Priority rules applying in the case of traffic disturbance
- Improving ancillary rail services (especially terminals and marshalling yards)
- Monitoring of the measures proposed.

This initiative eventually led to the adoption of Regulation No 913/2010 (EP&C, 2010a), which lays down rules for the establishment, organization and management of international rail corridors with a view to developing a European rail network for competitive freight.

The nine initially designated Rail Freight Corridors (RFCs) appear in Fig. 1.4. A process of capacity allocation to freight trains with better coordination of priority

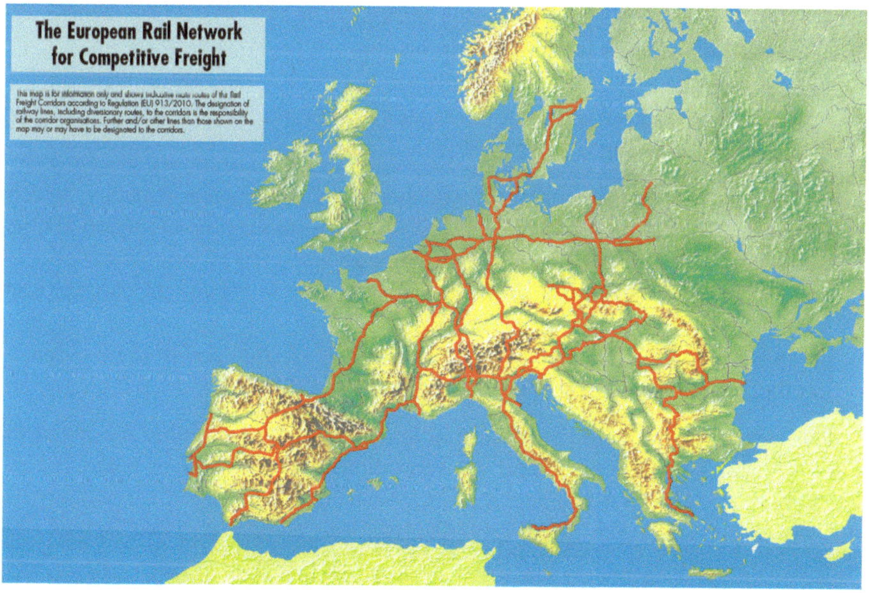

Fig. 1.4 The European Rail Network for Competitive Freight. *Source*: EC (2011b)

rules and prioritizing, among freight trains, those that cross at least one border is described in the Regulation for the RFCs.

It further sets up detailed rules for the governance of each RFC through:

- an executive board composed of representatives of the authorities of the Member States concerned,
- a management board composed of the infrastructure managers concerned and, where relevant, the allocation bodies,
- an advisory group made up of managers and owners of the terminals of the RFC including, where necessary, sea and inland waterway ports, and
- a further advisory group made up of railway undertakings interested in the use of the freight corridor.

The measures for implementing the RFC, described by the Regulation, include:

- drafting and periodically updating a transportation market study relating to the existing and expected traffic conditions on the RFC,
- drawing up an implementation plan describing:
  - the characteristics of the freight corridor (including bottlenecks),
  - the programme of measures necessary for creating the freight corridor,
  - the objectives for the RFC, in particular in terms of the quality of service and the capacity of thc corridor,
- drawing up and periodically reviewing an investment plan providing details of:
  - indicative medium- and long-term investment for infrastructure and its equipment along the corridor,
  - the relevant financial requirements and sources of finance,
  - a deployment plan relating to the interoperable systems along the freight corridor, and
  - a plan for the management of the capacity of freight trains which may run on the freight corridor,
- setting up an one-stop-shop for application for infrastructure capacity, which would also display infrastructure capacity available at the time of request and its characteristics in accordance with pre-defined parameters,
- monitoring the performance of rail freight services on the freight corridor and publishing the results of this monitoring once a year, and
- organizing a satisfaction survey of the users of the freight corridor and publishing the results of it once a year.

The governance structures of transportation corridors are further discussed in Chap. 4, and more on green rail transportation can be found in Chap. 12 of this book.

1.6 Waterborne Transportation

The international character of shipping makes the regulatory environment of this sector more efficient if agreed, adopted and implemented on a global basis. The International Maritime Organization (IMO), the standard-setting UN agency for the safety, security and environmental performance of international shipping, is the forum at which this process takes place. The promotion of sustainable shipping and sustainable maritime development is one of the major priorities of IMO in recent years.

IMO's drive to reduce GHG emissions from ships has followed thus far two quasi-parallel tracks. One track relates to setting energy efficiency standards for new ships and has led to the adoption of the Energy Efficiency Design Index (EEDI) in July 2011 at the 62nd session of IMO's Marine Environment Protection Committee (MEPC 62). The EEDI is discussed in Sect. 1.6.1 below.

The other track concerns Market Based Measures (MBMs), of which more in Chap. 8 of this book. However, the proposed in June 2013 EU Regulation on monitoring, reporting and verification of CO_2 emissions, constituting a first step towards an MBM, is presented in Sect. 1.6.3.

Meanwhile, in November 2012 the EU adopted Directive 2012/33/EU transposing into European law the IMO standards on maximum sulphur content of marine fuels adopted in 2008. This is the subject of Sect. 1.6.2.

1.6.1 *The Adoption of EEDI and SEEMP*

The IMO's Energy Efficiency Design Index (EEDI) is a benchmarking scheme aiming to provide an indication of a merchant ship's CO_2 output in relation to its transport work. Adoption of EEDI is the first step of IMO's drive to reduce CO_2 emissions from shipping. The EEDI compares design-level CO_2 emissions and transport work of a vessel and benchmarks this ratio against an IMO-set requirement.

For a given ship, the EEDI is provided by the following formula:

$$\frac{\left(\prod_{j=1}^{M} f_j\right)\left(\sum_{i=1}^{nME} P_{ME(i)} C_{FME(i)} \cdot SFC_{ME(i)}\right) + \left(P_{AE} \cdot C_{FAE} \cdot SFC_{AE}{}^{*}\right) + \left(\left(\prod_{j=1}^{M} f_j \cdot \sum_{i=1}^{nPTI} P_{PTI(i)} - \sum_{i=1}^{n_{eff}} f_{eff(i)} \cdot P_{AEeff(i)}\right) C_{FAE} \cdot SFC_{AE}\right) - \left(\sum_{i=1}^{n_{eff}} f_{eff(i)} \cdot P_{eff(i)} \cdot C_{FME} \cdot SFC_{ME}\right)}{f_i \cdot Capacity \cdot V_{ref} \cdot f_w}$$

There is no need to explain all these symbols here. The numerator in the formula is a function of all power generated by the ship (main engine and auxiliaries), and the denominator is a product of the ship's deadweight and the ship's 'reference speed', appropriately defined as the speed corresponding to 75 % of the Maximum Continuous Rating of the ship's main engine. The units of EEDI are grams of CO_2 per tonne mile.

The EEDI of a new ship is to be compared with the so-called '*EEDI (reference line)*,' which is defined as:

$$EEDI\ (reference\, line) = aDWT^{-c} \tag{1.1}$$

where DWT is the deadweight of the ship and a and c are positive coefficients determined by regression from the world fleet database, per major ship category.

For a given new ship, the attained EEDI value should be equal or less than the required EEDI value which is provided by the following formula:

$$Attained\ EEDI \leqq Required\ EEDI\ = (1 - X/100)\ aDWT^{-c} \tag{1.2}$$

where X is a 'reduction factor' specified for the required EEDI compared to the EEDI (reference line).

The values of X specified by the IMO are as follows:

- $X = 0\ \%$ for ships built from 2013 to 2015
- $X = 10\ \%$ for ships built from 2016 to 2020
- $X = 20\ \%$ for ships built from 2020 to 2025 and
- $X = 30\ \%$ for ships built from 2025 to 2030.

This means that it will be more stringent to be EEDI-compliant in the years ahead. If a ship's attained EEDI is above the required value, the ship is not allowed to operate until and unless measures to fix the problem are taken.

The reference line parameters a and c in Eqs. (1.1) and (1.2), which have been finalised by regression analyses after a long debate within the IMO are presented in Table 1.2 below, although they are subject to revision.

For Ro-Ro ferries the basic concept seems the same at first glance, but the EEDI (reference line) formula is more complex in that its various coefficients are not constant.

The basic philosophy of EEDI, which applies to all ships of 400 GRT and above, is to build ships that are more energy efficient, that is, reduce emissions (numerator) per unit of transport work (denominator). Measures to achieve this end are intended to be mostly technological.

Table 1.2 EEDI reference line parameters a and c for various ship types

Ship type	a	c
Bulk carrier	961.79	0.477
Gas carrier	1,120.00	0.456
Tanker	1,218.80	0.488
Container ship	174.22	0.201
General cargo ship	107.48	0.216
Reefer	227.01	0.244
Combination carrier	1,219.00	0.488

Source: IMO

In contrast to EEDI, which relates to the design of new ships, IMO adopted in July 2011 the Ship Energy Efficiency Management Plan (SEEMP), which addresses energy saving at the operational stage and applies to all (existing and new) ships above 400 GRT. SEEMP takes the form of a mandatory management plan and aims to establish a mechanism for a shipping company and/or a ship to improve the energy efficiency of ship operations through four steps: planning, implementation, monitoring, and self-evaluation and improvement.

The Energy Efficiency Operational Indicator (EEOI) has been proposed by the IMO as a monitoring tool in the SEEMP. The EEOI is calculated by the following formula, in which a smaller EEOI value means a more energy efficient ship:

$$EEOI = \frac{actual\ CO_2\ emission}{performed\ transport\ work}$$

The intention was to develop a formula enabling the continuous monitoring of individual ships in operation and thereby quantifying the impact of any change made to the ship or its operation. However, it should be clarified that ships operate under a broad variety of different conditions, some of which are beyond the control of their operators. As such, although EEOI has been adopted as an indicator to be used for assessing the performance of individual ships in the framework of SEEMP, industry circles consider its use for comparisons between ships to be flawed (ICS, 2013).

1.6.2 The Sulphur Directive

In addition to GHG, IMO regulates the emission of air pollutants from ship exhausts, including NOx and SOx emissions. These regulations are contained in the MARPOL Annex VI protocol which, in addition, designates specific geographic areas as Emission Control Areas (ECAs), where more stringent requirements apply. An ECA can be designated for NOx and PM, or SOx, or all three types of emissions from ships (the term SECA is used for a SOx ECA). The existing ECAs appear in Fig. 1.5, while their entry into force date is shown in Table 1.3.

The latest revision of MARPOL Annex VI was adopted in October 2008. Its basic provisions that relate to SOx emissions include:

- a reduction in the global limit of sulphur content in fuel to 3.5 % by mass (from 4.5 %) effective from 1 January 2012; then to 0.5 %, effective from 1 January 2020 subject to a feasibility review to be completed no later than 2018 (it can be postponed to 1 January 2025 if the review reveals that not enough fuel with a sulphur content of 0.5 % is available for global shipping in 2020),

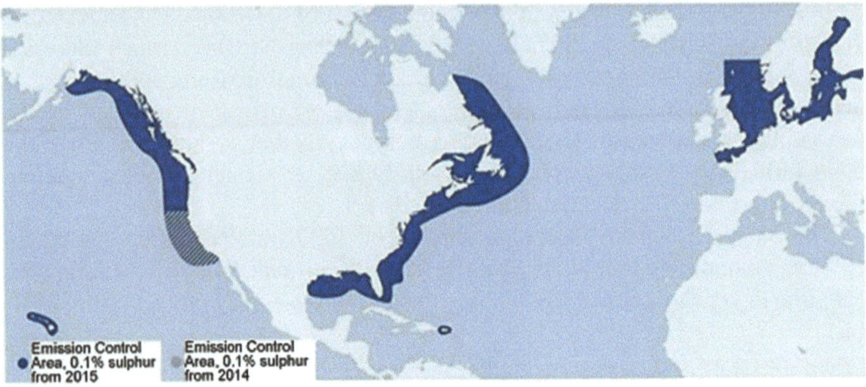

Fig. 1.5 The emission control areas. *Source*: CIW (2014)

Table 1.3 The adoption, entry into force and effective dates of ECAs

Adoption, entry into force and date of taking effect of special areas			
Special areas	Adopted on	Date of entry into force	In effect from
Baltic Sea (SOx)	26 Sept 1997	19 May 2005	19 May 2006
North Sea (SOx)	22 Jul 2005	22 Nov 2006	22 Nov 2007
North American (SOx, NOx and PM)	26 Mar 2010	1 Aug 2011	1 Aug 2012
United States Caribbean Sea ECA (SOx, NOx and PM)	26 Jul 2011	1 Jan 2013	1 Jan 2014

Source: IMO (2014)

- a reduction in sulphur limits for fuels in SECAs to 1 %, beginning on 1 July 2010 (from 1.5 %); being further reduced to 0.1 %, effective from 1 January 2015,
- the possibility of using suitable abatement equipment as an alternative to fuel switching requirements on the basis that equivalent SOx emissions are achieved on a continuous basis.

The timing of the above sulphur content limits are represented graphically in Fig. 1.6.

At European level, these provisions were not transposed into European law until November 2012, when Directive 2012/33/EU was adopted (EP&C, 2012b).[9] The Directive aligns to the IMO regulations and brings the 0.5 % limit into force on 1 January 2020 for all EU sea territory, even if on global scale this limit gets postponed to 2025. Furthermore, the Commission's proposal for passenger ships to

[9] The previous IMO limits were applied by Directive 2005/33/EC which, in addition, imposed a 1.5 % sulphur limit for fuels used by passenger vessels on regular services between EU ports from 11 August 2006, and a 0.1 % sulphur limit on fuel used by inland waterway vessels and by seagoing ships at berth in EU ports, from 1 January 2010.

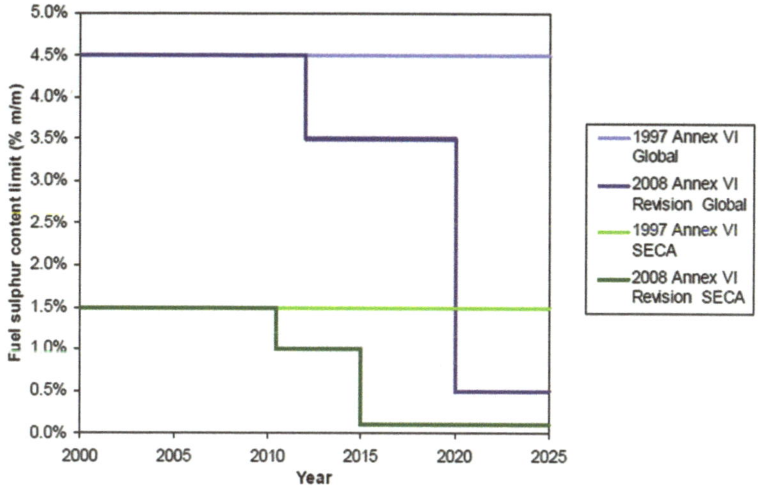

Fig. 1.6 Revised MARPOL Annex VI—fuel sulphur limits. *Source*: Entec (2010)

follow the SECA limits of 0.1 % also outside the SECA area from 2020 onwards was not approved, and the current 1.5 % limit will be lowered to 0.5 % in 2020 as for all shipping within the EU.

The 0.1 % limit, effective as of 1 January 2015 within SECAs, can only be achieved by fitting expensive exhaust scrubbers, consuming LNG, or burning Marine Gas Oil, which is currently around $300 per tonne more expensive that Heavy Fuel Oil 1.0 %S. This is expected to have adverse effects on shipping and ports in SECAs, as well as the industries that depend on their services (refer to Chap. 10 for a more detailed discussion on this issue). However, the focus of the shipping industry has now moved to concerns about the effective enforcement of these rules, which is far from trivial.

1.6.3 Monitoring, Reporting and Verification of CO_2 Emissions

The IMO work on Market Based Measures (MBMs, see also Chap. 8 of this book) was suspended in May 2013 in the wake of a clash between developed and developing Member States at MEPC 65. One month later the European Commission issued its proposal for a Regulation on monitoring, reporting and verification of CO_2 emissions, the so-called MRV proposal, as a first step towards setting GHG reduction targets and taking further measures, including an MBM (EC, 2013c).

The immediate objective of the MRV proposal is to produce accurate information on the CO_2 emissions of large ships using EU ports and incentivize energy efficiency improvements by making this information publicly available. In this way, the Commission sets the ground for possible future MBMs or efficiency standards, while at the same time attempts to address one of the market barriers found to prevent the implementation of cost-effective abatement measures by the industry, namely the lack of reliable information on fuel efficiency of ships.[10] Yet, another stated objective of introducing an MRV system is the securing of more time to discuss emission reduction targets and relevant measures, particularly at global level in IMO.

The proposed MRV system applies to ships above 5,000 GRT, regardless of flag, and covers intra-EU, incoming (from the last non-EU port to the first EU port of call) and outgoing (from an EU port to the next non-EU port of call) voyages. It concerns the CO_2 emissions only.

Following the preparation of an emission monitoring plan by the ship-owning company and its approval by an accredited verifier, information on fuel consumption, distance travelled, time at sea and cargo carried is collected by the company for each ship and each journey falling under the Regulation. Actual fuel consumption for each voyage can be calculated using one of the following methods, provided that the method selected is pre-defined in the monitoring plan and, once chosen, is applied consistently:

- Bunker Fuel Delivery Notes and periodic stocktakes of fuel tanks,
- Bunker fuel tank monitoring on board,
- Flow meters and applicable combustion processes, and
- Direct emissions measurements.

Based on these parameters, a number of energy efficiency/emissions indicators are calculated and reported on an annual basis. The annual reports are submitted to the Commission and the flag state after their approval by the verifiers, who issue conformity documents that need to be kept on board the ships covered by the system. Conformity is to be checked by the flag state and through the port state control system. Sanctions are foreseen for the failure to comply, including in certain cases the expulsion of a ship, i.e. banning its entry to EU ports until the compliance problem has been resolved. The energy efficiency performance of the ships falling within the scope of the Regulation is made publicly available by the Commission every year.

As is usually the case, the proposal has attracted criticism from both directions. The environmental groups consider the proposal exceptionally mild, while the shipping interests argue that it imposes unnecessary obligations to an industry that suffers already from excessive administrative burdens (refer also to Sect. 8.6.2).

In November 2014, an informal agreement by the EU legislators on the MRV proposal was announced according to which, no major modifications on the final

[10] The other market barriers relate to: (a) the split incentives between ship owners who invest into efficiency improvements and ship operators who reap the benefits of such investments through lower fuel bills, and (b) the lack of access to finance for these investments.

text should be expected. Both the European Community Shipowners Association (ECSA) and the International Chamber of Shipping (ICS) expressed their concern that with the MRV proposal, which is expected to be fully operational by 2018, the EU may pre-empt negotiations taking place at IMO. Furthermore, ICS drew attention to the need to handle data on cargo carried by ships with particular sensitivity because of the suspicion that this could lead to the development of a mandatory operational efficiency index, like the EEOI of Sect. 1.6.1, whose mandatory application for benchmarking different vessels was considered inappropriate by IMO on technical grounds (GreenPort, 2014).

Acknowledgement Work reported herein was supported in part by EU project SuperGreen and an internal grant at the Technical University of Denmark.

References

Aarnink, S., Faber, J., & den Boer, E. (2012, October). *Market barriers to increased efficiency in the European on-road freight sector*. CE Delft report, Delft.

CIW. (2014, November 30). *What impact will new low-sulphur limits have?* Container Insight Weekly, Drewry Maritime Research. Retrieved December 4, 2014, from http://ciw.drewry.co.uk/release-week/2014-22/

Council. (2001, June 15–16). *Presidency conclusions*. Council of the European Union, Gothenburg.

Council. (2006, June 9). *Review of the EU sustainable development strategy (EU SDS)—Renewed strategy*. Council of the European Union, Brussels.

Entec. (2010, July). *Study to review assessments undertaken of the revised MARPOL annex VI regulations*. Prepared for the shipowner associations of Belgium, Finland, Germany, Holland, Sweden and UK and endorsed by ECSA and ICS.

EU Scoreboard. (2014). Retrieved November 21, 2014, from http://ec.europa.eu/transport/index_en.htm

European Commission. (2001, May 15). *A sustainable Europe for a better world: A European Union Strategy for sustainable development*. COM(2001) 264 final, Brussels.

European Commission. (2007a, October 18). *The EU's freight transport agenda: Boosting the efficiency, integration and sustainability of freight transport in Europe*. COM(2007) 606, Brussels.

European Commission. (2007b, October 18). *Freight Transport Logistics Action Plan*. COM (2007) 607, Brussels.

European Commission. (2007c, October 18). *Impact assessment of the Freight Transport Logistics Action Plan*. SEC(2007) 1320, Brussels.

European Commission. (2007d, October 18). *Towards a rail network giving priority to freight*. COM(2007) 608, Brussels.

European Commission. (2008a, July 8). *Greening transport*. COM(2008) 433, Brussels.

European Commission. (2008b, July 8). *Greening Transport Inventory*. SEC(2008) 2206, Brussels.

European Commission. (2008c, July 8). *Strategy for the internalization of external costs*. COM (2008) 435, Brussels.

European Commission. (2008d, July 8). *Proposal for a Directive of the European Parliament and of the Council amending Directive 1999/62/EC on the charging of heavy goods vehicles for the use of certain infrastructures*. COM(2008) 436, Brussels.

European Commission. (2008e, July 8). *Rail noise abatement measures addressing the existing fleet*. COM(2008) 432, Brussels.
European Commission. (2010, March 3). *EUROPE 2020—A strategy for smart, sustainable and inclusive growth*. COM(2010) 2020, Brussels.
European Commission. (2011a, March 28). *WHITE PAPER. Roadmap to a single European transport area—Towards a competitive and resource efficient transport system*. COM(2011) 144, Brussels.
European Commission. (2011b, June 30). *Handbook on the regulation concerning a European rail network for competitive freight (regulation EC 913/2010)*. A DG MOVE staff working document, Brussels.
European Commission. (2013a, January 24). *Clean power for transport: A European alternative fuels strategy*. COM(2013) 17, Brussels.
European Commission. (2013b, January 24). *Actions towards a comprehensive EU framework on LNG for shipping*. SWD(2013) 4, Brussels.
European Commission. (2013c, June 28). *Proposal for a regulation of the European Parliament and of the Council on the monitoring, reporting and verification of carbon dioxide emissions from maritime transport and amending regulation (EU) no 525/2013*. COM(2013) 480 final, Brussels.
European Commission. (2013d, April 15). *Proposal for a directive of the European Parliament and of the Council amending Directive 96/53/EC of 25 July 1996 laying down for certain road vehicles circulating within the community the maximum authorised dimensions in national and international traffic and the maximum authorised weights in international traffic*. COM(2013) 195 final, Brussels.
European Commission. (2014a). Retrieved November 27, 2014, from http://ec.europa.eu/transport/infrastructure/tentec/tentec-portal/site/maps_upload/metro_map2013.pdf
European Commission. (2014b, May 21). *Strategy for reducing heavy-duty vehicles' fuel consumption and CO_2 emissions*. COM(2014) 285 final, Brussels.
European Parliament & Council. (2012a, November 21). *Directive 2012/34/EU of 21 November 2012 establishing a single European railway area (recast)*. Strasbourg.
European Parliament & Council. (2012b, November 21). *Directive 2012/33/EU of 21 November 2012 amending Council Directive 1999/32/EC as regards the sulphur content of marine fuels*. Strasbourg.
European Parliament & Council. (2010a, September 22). *Regulation (EU) no 913/2010 of 22 September 2010 concerning a European rail network for competitive freight*. Strasbourg.
European Parliament & Council. (2010b, July 7). *Directive 2010/40/EU of 7 July 2010 on the framework for the deployment of Intelligent Transport Systems in the field of road transport and for interfaces with other modes of transport*. Strasbourg.
European Parliament & Council. (2013a, December 11). *Regulation (EU) no 1315/2013 of 11 December 2013 on union guidelines for the development of the trans-European transport network and repealing decision no 661/2010/EU*. Strasbourg.
European Parliament & Council. (2013b, December 11). *Regulation (EU) no 1316/2013 of 11 December 2013 establishing the connecting Europe facility, amending regulation (EU) no 913/2010 and repealing regulations (EC) no 680/2007 and (EC) no 67/2010*. Strasbourg.
European Parliament & Council. (2014, October 22). *Directive 2014/94/EU of 22 October 2014 on the deployment of alternative fuels infrastructure*. Strasbourg.
GreenPort. (2014). Industry speaks out over CO2 monitoring. Retrieved December 4, 2014, from http://www.greenportcom/news101/europe/industry-speaks-out-about-co2-reporting-plans?mkt_tok=3Rk MMJWWfF9wsRoivqrLZKXonjHpfsX87eokWbHr08Yy0EZ5VunJEUWy2oAFStQ%2FcOedCQkZHblFnVQASa2xS7kNoqwE
ICS. (2013, October 10). *Preliminary ICS comments on the draft EU regulation on MRV*. An International Chamber of Shipping brief for EU Member States and Members of the European Parliament.

IMO. (2014). *Special areas under MARPOL*. Retrieved December 4, 2014, from http://www.imo.org/OurWork/Environment/PollutionPrevention/SpecialAreasUnderMARPOL/Pages/Default.aspx

Panagakos, G., Moyano, H., Psaraftis, H., Lyridis, D., Mpaliami, T., Webb, S., et al. (2013, February 25). *Policy recommendations—Version 2*. A SuperGreen project document.

The Sustainable Leader. (2014). Sustainable development. Retrieved October 30, 2014, from http://thesustainableleader.org/wp-content/uploads/2013/02/sustainable_development.jpg

Chapter 2
Transportation Emissions: Some Basics

Christos A. Kontovas and Harilaos N. Psaraftis

Abstract Transportation is the backbone of international trade and a key engine driving globalization. However, there is growing concern that the Earth's atmospheric composition is being altered by human activities, including transportation, which can lead to climate change. Air pollution from transportation and especially carbon dioxide emissions are at the center stage of discussion by the world community through various international treaties, such as the Kyoto Protocol. The transportation sector also emits non-CO2 pollutants that have important effects on air quality, climate, and public health. The main purpose of this chapter is to introduce some basic concepts that are relevant in the quest of green transportation logistics. First, we present the basics of estimating emissions from transportation activities, the current statistics and future trends, as well as the total impact of air emissions and its contribution to climate change. In addition, this chapter presents the basics of environmental policy measures. In that context, we describe a way to measure the cost-effectiveness of various measures through the so-called Marginal Abatement Cost (MAC). Finally, the chapter deals with the topic of the energy efficiency gap and examines why governments and companies may forego cost-effective investments in energy efficiency, even though they could significantly reduce energy consumption at a lower cost.

Abbreviations

ACEA	European Automobile Manufacturers Association
AR5	Fifth Assessment Report
CAC	Command and control approach
CBDR	Common but differentiated responsibilities
CH_4	Methane
CO	Carbon monoxide
CO_2	Carbon dioxide
CO_2eq	Carbon dioxide equivalent

C.A. Kontovas (✉) • H.N. Psaraftis
Technical University of Denmark, Lynby, Denmark
e-mail: kontova@transport.dtu.dk; hnpsar@transport.dtu.dk

H.N. Psaraftis (ed.), *Green Transportation Logistics*, International Series in Operations Research & Management Science 226,
DOI 10.1007/978-3-319-17175-3_2

Defra	UK Department for Environment, Food and Rural Affairs
EC	European Commission
EEA	European Environment Agency
EEDI	Energy efficiency design index
EIA	Energy Information Administration
EPA	US Environmental Protection Agency
EU	European Union
GHG	Greenhouse gas
gr	Grams
Gtons	Gigatonnes
HFO	Heavy fuel oil
ICAO	International Civil Aviation Organization
IEA	International Energy Agency
Ifeu	Institute for Energy and Environmental Research
IMO	International Maritime Organization
IPCC	Intergovernmental Panel on Climate Change
kg	Kilograms
km	Kilometers
LCA	Life cycle assessment
MAC	Marginal abatement cost
MBM	Market based measures
MDO	Marine diesel oil
MGO	Marine gas oil
MOVES	MOtor vehicle emission simulator
MT	Metric tonnes
NOx	Nitrogen oxides
NPV	Net present value
O_3	Ozone
PM	Particulate matter
PV	Present value
RCP	Representative concentration pathways
RENFE	Red Nacional de los Ferrocarriles Españoles
RF	Radiative forcing
SBB	Schweizerische Bundesbahnen
SEAP	Sustainable Energy Action Plan
SNCB	Société Nationale des Chemins de fer Belges
SNCF	Société Nationale des Chemins de Fer Français
SO_2	Sulphur dioxide
SRES	Special Report on Emissions Scenario
tn	Tonnes
UK	United Kingdom
UN	United Nations
UNFCCC	United Nations Framework Conference on Climate Change
US	United States (of America)
VOC	Volatile organic compounds

2.1 Introduction

Throughout the course of history, the economic growth of our world has been closely tied to efficient methods of transportation. The transportation sector involves the movement of people and goods by cars, trucks, ships, airplanes, trains, and other vehicles. Many of the vehicles used are equipped with internal combustion engines that combust petroleum-based products. Even electric vehicles require energy that may be produced by fossil fuels or even coal. It is well known that fossil fuels contain a high percentage of hydrocarbons and the burning of these fuels produces carbon dioxide (CO_2), which is a greenhouse gas (GHG). Thus, as freight and passenger traffic constantly continues to grow, a major challenge is how to ensure the long-term sustainability of such growth. More important, and as Chap. 1 of this book has shown, policy documents such as the EU 2011 White Paper on Transport stipulate ambitious 'decarbonization' goals in reducing transport-related emissions by 60 % by 2050 as compared to 1990 levels. Challenges such as these are playing an increasingly important part in the policy debate on trade and development, environmental sustainability and energy security. If left unchecked, unsustainable patterns are likely to prevail, undermining the progress that already has been made on sustainable development and growth.

As discussed in the Preface of this book, the traditional analysis of transportation logistics problems has been in terms of cost-benefit, economic or other optimization criteria, which by and large either ignore environmental issues, or consider them of secondary importance (See Psaraftis and Kontovas, 2010). Green transportation logistics tries to bring the environmental dimension into the problem, by analyzing various trade-offs and exploring 'win-win' policies and solutions. The main purpose of this chapter is to introduce some basic concepts on transportation emissions that are relevant for the scope of this book. To that effect, we start by presenting the basics of estimating emissions from transportation activities, the current statistics and future trends, as well as the total impact of air emissions and its contribution to climate change. We next present the basics of environmental policy measures. In that context, we describe a way to measure the cost-effectiveness of various measures through the so-called Marginal Abatement Cost (MAC). Finally we deal with the topic of the energy efficiency gap and examines why governments and companies may forego cost-effective investments in energy efficiency even though they could significantly reduce energy consumption at a lower cost.

At a macro level, and according to the United Nations Framework Conference on Climate Change (UNFCCC), CO_2 contributes to *global warming* which is defined as an increase in the average temperature of the Earth's near-surface air and oceans (UNFCC, 1997). There is a growing concern that the Earth's atmospheric composition is being altered by human activities which can lead to climate change. This view has led the UNFCCC to adopt the objective *to achieve stabilization of greenhouse gas concentrations in the atmosphere at a level that would prevent dangerous anthropogenic interference with the climate system* (see Article 2 of the UNFCCC (UNFCCC, 1997)). The stabilization of concentrations of atmospheric CO_2 will require significant reductions in global emissions of CO_2 in the future, but the resultant temperature from stabilizing these concentrations at various levels (e.g., 450, 550 ppm, etc.) depends on many factors. Models estimate that the global mean surface temperature arising from a doubling of CO_2 concentrations is between 2 and 4.5 °C (IPCC, 2007). Due to the many uncertainties involved with climate change, the Intergovernmental Panel on Climate Change (IPCC), a scientific intergovernmental body, was tasked to evaluate the risk of climate change caused by human activity. The reports produced by IPCC made it clear that differing viewpoints within the scientific community do exist.

The so-called Kyoto Protocol is an international treaty which extends the 1992 UNFCCC that commits State Parties to reduce greenhouse gases emissions, based on the premise that (a) global warming exists and (b) man-made CO_2 emissions have caused it. The Kyoto Protocol was adopted in Kyoto, Japan, on 11 December 1997 and entered into force on 16 February 2005. Currently there are 191 States-Parties to the Protocol. The Protocol's first commitment period started in 2008 and ended in 2012. Negotiations are currently under way to agree on a post-Kyoto legal framework. In accordance with Articles 4 and 12 of the Climate Change Convention, and the relevant decisions of the Conference of the Parties, countries that are Parties to the Convention submit national GHG inventories to the Climate Change secretariat. Emissions from international aviation and maritime transportation (also known as international bunker fuel emissions) should be calculated as part of the national GHG inventories of the Parties, but should be excluded from national totals and reported separately.

Thus, emissions from bunker fuels are not subject to the limitation and reduction commitments under the Convention and the Kyoto Protocol. Emissions from aviation and marine 'bunker' fuels form a significant part of the global climate problem—almost 10 %. The Kyoto Protocol assigned responsibility for reducing bunker greenhouse gas emissions to developed (the so-called 'Annex I') countries working through the International Civil Aviation Organization (ICAO) and International Maritime Organization (IMO), both United Nations agencies.

The IPCC is the most authoritative international body on climate science and IPCC's Assessment Reports provide a comprehensive summary of climate change. The IPCC's latest report—the Fifth Assessment Report (AR5)—is the most comprehensive assessment of climate change undertaken. The third installment of the AR5 is Climate Change 2014: Mitigation of Climate Change by Working Group III (WG III) and Chap. 8 presents a comprehensive assessment of transportation. According to this report, *reducing global transport greenhouse gas (GHG) emissions will be challenging since the continuing growth in passenger and freight activity could outweigh all mitigation measures unless transport emissions can be strongly decoupled from GDP growth* (IPCC, 2014).

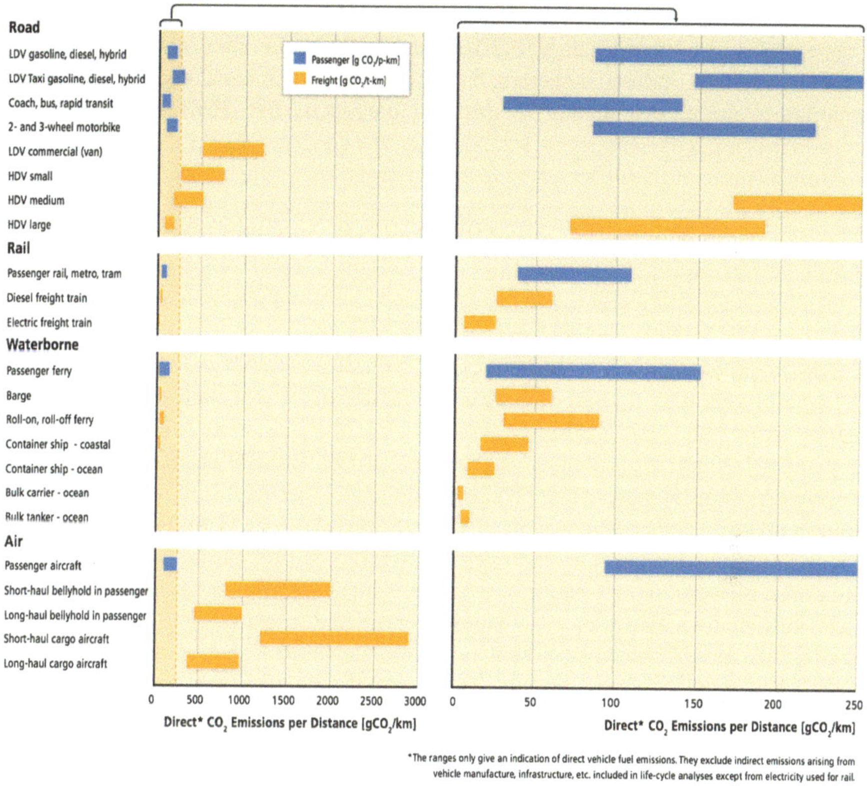

Fig. 2.1 Typical ranges of direct CO_2 emissions per passenger-kilometer and per tonne-kilometer for freight, for the main transportation modes when fuelled by fossil fuels. *Source*: IPCC (2014)

According to the statistics presented, the transportation sector produced 7.0 gigatonnes (Gt) of carbon dioxide equivalent[1] (CO_2eq) of direct GHG emissions (including non-CO_2 gases) in 2010 and hence was responsible for approximately 23 % of total energy-related CO_2 emissions. The report notes that there has been a growth in emissions although there was consensus among the experts that more efficient vehicles are being used and despite relevant policies being adopted. However, direct vehicle CO_2 emissions per kilometer vary widely for each mode, see Fig. 2.1.

Readers may recall the truism *you can't control what you can't measure*. Thus, estimating CO_2 emissions (but also other types of emissions) has been very

[1] Carbon dioxide equivalents (CO_2 eq) provide a standard of measurement against which the impacts of releasing (or avoiding the release of) different greenhouse gases can be evaluated. According to IPCC (2007) every greenhouse gas (GHG) has a Global Warming Potential (GWP), a measurement of the impact that particular gas has on 'radiative forcing'; that is, the additional heat/energy which is retained in the Earth's ecosystem through the addition of this gas to the atmosphere.

important recently partly because countries that are Parties to the Climate Change Convention must submit national GHG inventories to the Climate Change secretariat. In addition, industrial sectors, businesses and also individuals are now more sensitive to the environmental footprint of their activities. However, estimating emissions is not a trivial task, as Sect. 2.2.1 of this chapter will elaborate. A fortiori, it is much more difficult to estimate future emissions, where there is significant uncertainty for the future use of energy. Many scenarios for future GHG emissions are based on assumptions on global development in the IPCC Special Report on Emissions Scenario (SRES) storylines.

The transportation sector also emits non-CO_2 pollutants that have important effects on air quality, climate, and public health. These include methane (CH_4), which is another GHG, volatile organic compounds (VOCs), nitrogen oxides (NOx), sulphur dioxide (SO_2), carbon monoxide (CO), particulate matter (PM), black carbon, and non-absorbing aerosols (Ubbels, Rietveld, & Peeters, 2002). Beyond any doubt, air pollutants such as SOx, NOx, and PM negatively affect the environment and human health. There are concerns that exposure to high concentrations of SO_2 is associated with effects on breathing, respiratory illness, alterations in pulmonary defenses, and aggravation of existing cardiovascular disease. Similarly, nitrogen oxides can irritate the lungs and lower resistance to respiratory infections. In the air, it is a potentially significant contributor to a number of environmental effects such as acid rain and eutrophication in waters. Actually, both SO_2 and NO_X, which are not GHGs, are the major precursors to acid rain, which is associated with the acidification of lakes and streams, accelerated corrosion of buildings and monuments, and reduced visibility. In addition, these emissions are also responsible for Climate Change, however, their contribution is much more uncertain and complicated (see Sect. 2.2.4 for more details).

One could go into deeper details on every one of the uncertainties described above or even into more generic ones. For example, Schelling (2007) poses some questions that expose the relevant uncertainties. How much carbon dioxide may join the atmosphere in a 'business as usual' scenario? How much average warming is to be expected from a specific increase in the concentration of GHGs? How will this average warming translate into climate change and what the effects will be in 50 or 150 years from now? These are not very easy questions to answer.

From the environmental economics' point of view, climate change is *the greatest and widest-ranging market failure ever seen*, presenting a unique challenge for economics (Stern, 2006). To use international shipping as an example, and based on the principles of equal treatment and level playing field, developed countries are urging all Member-States to quickly adopt emission reduction regulation, noting however that most of the world tonnage is registered in non-Annex I countries. On the other hand, many developing countries including China, Brazil, India and others are totally against the application of such regulations to these countries. These countries argue that the 'Common But Differentiated Responsibilities' (CBDR) principle also discussed in the Kyoto Protocol should be applied. According to CBDR, developing countries should not be subject to the same emissions reduction goals as developed countries, on the ground of their economic development. This is in line with Stiglitz (2006) which states that the biggest problem with Kyoto is *to bring the developing countries within the fold.*

A number of measures, operational, technical and others have been stipulated to curb emissions. The IPCC report (IPCC, 2014) concludes that direct GHG emissions from passenger and freight transportation can be reduced by, inter alia:

(a) avoiding journeys where possible by, for example, sourcing localized products, restructuring freight logistics systems, and utilizing advanced information and communication technologies (ICT);
(b) encouraging modal shift to lower-carbon transportation systems, also by increasing investment in public transportation
(c) infrastructure, and modifying roads, airports, ports, and railways to become more attractive for users and minimize travel time and distance;
(d) lowering energy intensity (that is energy used per cargo kilometer)—by enhancing vehicle and engine performance, using lightweight materials, increasing freight load factors and passenger occupancy rates and deploying new technologies;
(e) reducing carbon intensity of fuels (carbon equivalent emissions per mass of energy used) by substituting oil- based products with sustainable energy sources (e.g. natural gas, bio-methane, biofuels etc.)

The rest of this chapter is structured as follows. Section 2.2 discusses possible ways to estimate emissions, presents emission statistics from transportation activities and its future trends as well as their impact to climate change. Section 2.3 presents the basics of environmental policy measures. Section 2.4 presents an index to measure the cost effectiveness of emission reduction measures and the concept of Marginal Abatement Cost (MAC). In addition, this section explains the reasons why some reduction measures that seem to be cost-effective are not adopted in practice. Finally, Sect. 2.5 concludes this chapter.

2.2 Calculating Emissions

There are generally two main methods that can be used to produce fuel consumption and emission estimates for transportation activities. The first method is called the 'top-down' method, or 'fuel-based' method, and uses fuel sales to estimate emissions. This would be the most reliable method of estimating total fuel consumption and emissions if the figures of fuels sales that are reported are absolutely reliable. Fuel sales figures are mainly collected from energy databases published by the Energy Information Administration (EIA), the International Energy Agency (IEA) and the UNFCCC. However, due to difficulties mainly because of unreliable or inaccurate fuel sale statistics, an alternative method has emerged. This is the so-called 'bottom up' method, or 'activity-based' method. This is an approach based on 'fleet activity,' that is, information on movements and vehicle characteristics (vehicle type and size, engine type and age, fuel type, etc.), as well as the corresponding fuel consumption figures and emission factors.

Below we describe each method and also comment on additional methods to calculate emissions.

2.2.1 The Top-Down or Fuel-Based Approach

According to this method, to estimate emissions from transportation one should multiply the energy or fuel used by an appropriate 'emissions factor.' The following equation may be used:

$$\text{Emissions} = (\text{Energy or Fuel Consumption}) \times \text{Emissions Factor} \quad (2.1)$$

where

Energy or Fuel Consumption is the energy or fuel source consumption related to the operational profile of the engine used and is typically expressed as hp-h, kW-h, or MW-h (energy), or gallons, kg or tonnes (fuel consumption).

Emissions Factor is the ratio of emissions produced per unit energy or unit fuel consumed. In general this factor is a function of type of emissions, type of fuel and type of engine and is typically expressed in g/hp-h, g/kW-h, or g/MW-h (energy); or, lb/gal, g/kg or tonne/tonne (fuel consumption).

If a number of different fuels are used by various engines or energy sources on the vehicle and fuel consumption is used as an input, Eq. (2.1) translates into

$$E_F = \sum_{i=1}^{N_F} F_i \times f_{F_i} \quad (2.2)$$

where E_F are the total emissions produced
F_i is the emissions factor of fuel type i ($i = 1,\ldots, N_F$)
f_{F_i} is the fuel consumption used by each engine and
N_F is the number of different fuels used.

Emission estimates based on *real* fuel consumption data are the most accurate. In this case, one may use empirically-based emissions factors to estimate the emissions. As said earlier, these factors depend on the fuel type (in the case of CO_2), the sulphur content (for SO_2) or the engine itself (for NOx). There are a wide variety of sources for emission factors that can be used to estimate emissions from transportation activities. Much work has been done related to GHG emissions basically due to mandatory reporting as per Kyoto Protocol. For non-GHGs, the literature is very fragmented as many national authorities (such as the US Environmental Protection Agency and the UK Department for Environment, Food and Rural Affairs) provide guidance documents on how to estimate non-GHG emissions. A reliable source of emission factors is the 'Air pollutant emission inventory guidebook' published by the European Environment Agency. See Table 2.1 for some typical factors.

Emissions are also indirectly produced if the vehicle is driven by electrical motors, even though obviously there are no emissions generated by the vehicle itself. This is due to the fact that the source that generates the electrical energy that

Table 2.1 Emission factors (in g/kg fuel)

Sector/fuel type	CO_2	NO_X	PM_{10}	SO_2
Aviation (jet kerosene)	2,600	8.3	0	0.8
Railways (gas oil/diesel)	3,140	52.4	1.44	
Shipping (bunker fuel oil)	3,170	79.3	6.2	20
Shipping (marine diesel/marine gas oil)	3,140	78.5	1.5	20
Road (passenger car—gasoline)	3,180	8.73	0.03	0.08
Road (heavy duty vehicle—diesel)	3,140	33.37	0.94	0.016
Road (bus—compressed natural gas)	2,750	13	0.02	–

Compiled from EMEP/EEA (2013)

Table 2.2 Electricity emissions factors (g/KWh)

	CO_2	NO_X	PM10	SO_2
Electricity mix (UK)	479	0.7	0.01	0.5
Natural-gas	402	0.35	0	0.016
Coal	902	0.41	0.003	0.37

is used by the vehicle will generally produce emissions. Then emissions are calculated as follows:

$$E_E = \sum_{j=1}^{N_E} F_E \times f_{E_j} \tag{2.3}$$

where again E_F are the total emissions produced, and
F_E is the emissions factor (in g/kWh),
f_{E_j} is electrical energy used by each motor j (in kW-h) and
N_E is the number of different electricity sources used.

Note that the electrical energy is the energy **input to** the equipment and is usually taken from utility bills or electricity meters. To estimate the emissions produced when using electricity one must know the mix that is used to produce this energy. Table 2.2 presents the emission factors to estimate emissions for producing electricity from various sources based on the so-called RAINS model.

Note that in many studies emissions from producing electricity by using hydro, wind, nuclear or solar energy are supposed to be zero. However, in reality, it would be more appropriate to use LCA (Life Cycle Assessment) emission factors, which take into consideration the overall life cycle of the energy carrier and includes not only the emissions of the final combustion, but also all emissions of the whole supply chain (exploitation, transportation, processing, etc.). LCA Emission factors for CO_2 are given in the Covenant of Mayors' Sustainable Energy Action Plan (SEAP), see European Commission (2010). Thus, emission factors are heavily based on the energy mix used to produce electricity. As a random example, the CO_2 emission factor for the energy mix in Greece is 1.149 g/kWh (due to the widespread use of lignite) whereas the EU-27 average is 460 g/kWh. LCA emission factors are 1.167 and 578 g/kWh respectively. More on the LCA approach for emissions can be found in Chap. 11 of this book.

2.2.2 The Activity Based or Bottom Up Approach

Sometimes, fuel consumption data is either not available or is unreliable in order to estimate emissions using the "top-down" approach. In this case we use a different approach to estimating emissions, the so-called "activity based", or "bottom-up" method. In activity-based approaches, one tries to estimate emissions based on modeling of the transportation activity or by using conversion factors that convert the available data into emissions.

There are various ways to estimate emissions based on the approach that uses the conversion factors and the approach used depends on the available data and inventory purpose. In order to report the emissions associated with a transportation activity, the 'activity data' such as distance travelled must be converted into emissions by using an appropriate conversion factor. For instance in the case that distance travelled is available, conversion factors in terms of kg of CO_2 per vehicle-km can be used. In this case, an average payload is assumed since the cargo transported is unknown. If the amount of cargo transported (e.g. in tonnes) is available, conversion factors in terms of kg of emissions per tonne-km can be used to estimate emissions.

As an example, for CO_2 emissions of a certain transportation mode one can use the following equation:

$$CO_2\ emissions\ (in\ tonnes) = V \times D \times F \tag{2.4}$$

where
V is the transportation volume (in tonnes)
D is the average transportation distance (in kilometers) and
F is the average CO_2-emission factor per tonne-km

The aforementioned conversion factors allow organizations, companies and individuals to calculate emissions from a range of activities, including transportation activities, in a simple and acceptable way. More information on conversion factors for transportation activities can be found in the *2014 Government GHG Conversion Factors for Company Reporting: Methodology Paper for Emission Factors* of the UK Department for Environment, Food and Rural Affairs (Defra), see Defra (2013).

Estimating emissions using the activity-based approach contains many uncertainties and is therefore not a trivial task. For instance, emissions from road freight or passenger transportation depend on the vehicle (e.g. engine type and age, tires, fuel used, the condition of the vehicle) but also on the way the vehicle is used (e.g. the payload, which may vary along the route, the traffic conditions like congestion and weather, the driver's behavior, etc). Chapter 7 of this book on green vehicle routing provides, among other things, some modeling details on possible fuel consumption functions in a road setting. Emissions from ships depend on the characteristics of the vessel and its engine, but also on the operating profile, mainly speed and payload. Weather is also an important factor. For aviation,

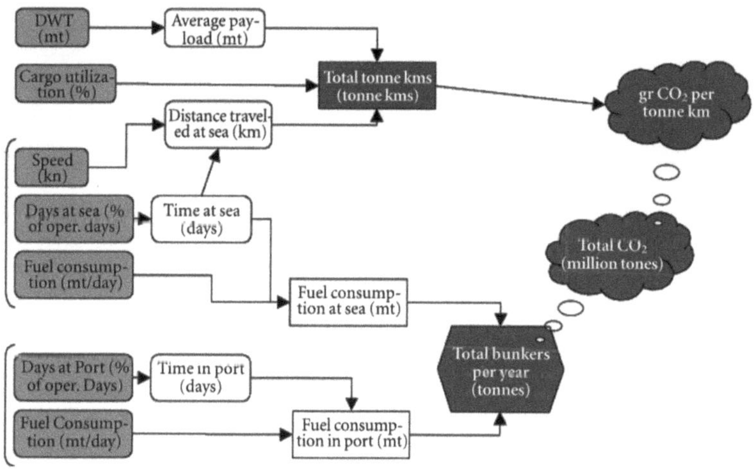

Fig. 2.2 Emissions calculation flowchart. *Source*: Psaraftis and Kontovas (2009)

payload, weather conditions, cruising speed and altitude are important parameters in estimating emissions.

As an example, Fig. 2.2 is a flowchart representation of the logic of an activity-based model to estimate emissions from international shipping, and how each of the emissions statistics is computed (see Psaraftis and Kontovas (2009) for more details). Total fuel consumption is the sum of the consumptions at sea and in port. Bunker consumption at sea is estimated based on an average consumption per day and the total time spend at sea. Total tonne-km's are computed by multiplying the average payload carried by the ship when at sea by the total sea kilometers traveled by the ship in a year.

Detailed modeling methodologies of emissions are available for all transportation modes. The interested reader is referred to the US Environmental Protection Agency (EPA) methods for calculating on-road (e.g. cars, trucks and motorcycles), non road (e.g. cargo handling equipment) and off-road (e.g. commercial marine, locomotives, and aircraft). The EPA provides both software solutions and detailed methodologies on how to model emissions and the relevant emission factors. MOVES (MOtor Vehicle Emission Simulator) is EPA's current official model for estimating air pollution emissions from cars, trucks and motorcycles (EPA, 2014a). The so-called NONROAD emission inventory model (EPA, 2014b) can be used to predict emissions of hydrocarbons, carbon monoxide, oxides of nitrogen, particulate matter, and sulphur dioxides from small and large nonroad vehicles, equipment, and engines. The NONROAD model does not include aircraft and aircraft engines, locomotives, or commercial marine vessels but detailed information on how to model emissions from these activities can be found at EPA's website at http://www.epa.gov/otaq/invntory.htm.

The European Environment Agency (EEA) publishes a very detailed air pollutant emission inventory guidebook (formerly called the EMEP/CORINAIR emission inventory guidebook) that provides guidance on estimating emissions from both anthropogenic sources, including transportation activities. Although this guidebook is mainly used to provide technical guidance to prepare national emission inventories, it provides information on how to model emissions and presents emission factors also for individual transportation activities and is one of the most recognized sets of emission estimation methods used in air pollution studies in Europe. Part B of the '2013 EMEP EEA air pollutant emission inventory guidebook', see EMEP/EEA (2013) deals with emissions from combustion of energy for aviation (Sect. 1.A.3.a), road transportation (Sect. 1.A.3.b), railways (Sect. 1.A.3.c), navigation (shipping) (Sect. 1.A.3.d), pipeline transportation (Sect. 1.A.3.e) and non-road mobile sources and machinery (Sect. 1.A.4).

An interesting question is, are there differences between emissions estimates based on top-down versus bottom up? The answer is yes, and Fig. 2.3 is illustrative for international shipping. It can be seen that bottom up estimates are higher than equivalent top-down figures, and that differences can be important. These differences add to the overall uncertainty of estimating emissions. They can be attributed to a number of reasons, including inaccurate or unreliable fuel sales statistics for the top-down approach and the great number of modeling assumptions and uncertain data for the bottom-up approach.

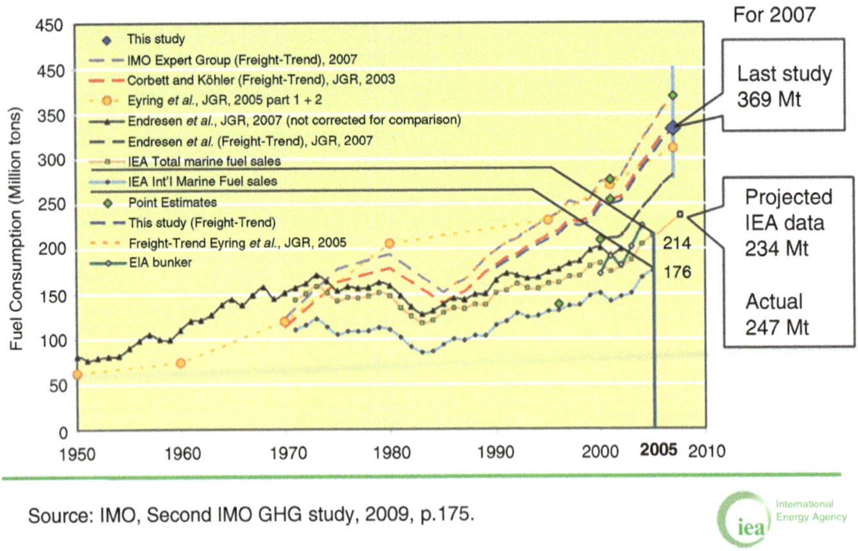

Fig. 2.3 Differences between top-down and bottom-up emissions estimates for international shipping. IEA statistics are much lower than bottom-up estimates. *Source*: IEA, adapted from Buhaug et al. (2009)

2.2.3 Calculation of GHG Emissions Based on the EN 16258 Standard

As stated above, one of the main problems with measuring or estimating the footprint of transportation is that it is a complicated process and there has been no standard way of doing it. Different sectors use different ways to estimate emissions and even within the same sector differences may exist between different companies. Much work has been done related to GHG emissions basically due to the mandatory reporting (see Kyoto Protocol) but the literature related to non-GHG emissions is more fragmented. The publication of the European standard EN 16258 (CEN, 2012), covering a *Methodology for calculation and declaration of energy consumption and GHG emissions of transport services (freight and passengers)* in 2012 was welcomed by many logistics companies. This standard sets out the methodology and requirements for calculating and reporting energy consumption and greenhouse gas (GHG) emissions in transportation services. This first edition of the standard was primarily focused on energy consumption and emissions associated with vehicles during the operational phase of the lifecycle, the so-called "tank-to-wheels" assessment. However, when calculating the energy consumption and emissions associated with vehicles, account is also taken of the energy consumption and emissions associated with energy processes for fuels and/or electricity used by vehicles (including for example production and distribution of transportation fuels). This ensures the EN 16258 standard takes a "well-to-wheel" approach when undertaking calculations, and when making declarations to transportation service users. Thus, in the so-called "well-to-wheels assessment" estimations are related to both vehicle and energy processes.

Table 2.3 is taken from the EN 16258 standard and presents the GHG emission factors for the main transportation fuels.

2.2.4 The EcoTransIT World Emissions Calculator

A variety of transportation emissions calculators exist. Among them we mention the so-called 'EcoTransIT World' calculator.

This is a web-based tool developed by the Institute for Energy and Environmental Research (ifeu), the Öko-Institut, the Rail Management Consultants GmbH (RMCon/IVE mbH), all from Germany, in order to quantify the emissions from freight transportation. This project was initiated by five European railway companies in 2000—DB Schenker Rail, the Schweizerische Bundesbahnen (SBB), Green Cargo AB, Trenitalia S.p.A, the Société Nationale des Chemins de Fer Français (SNCF), the Red Nacional de los Ferrocarriles Españoles (RENFE) and the Société Nationale des Chemins de fer Belges (SNCB). Although this is basically a rail consortium, the tool covers all surface modes.

Table 2.3 Transportation fuels: density, energy factor and GHG emission factor

Fuel type description	Density (d)	GHG emission factor					
		Tank-to-wheels (g_t)			Well-to-wheels (g_w)		
	kg/l	$kgCO_2e/MJ$	$kgCO_2e/kg$	$kgCO_2e/l$	gCO_2e/MJ	$kgCO_2e/kg$	$kgCO_2e/l$
Gasoline	0.745	75.2	3.25	2.42	89.4	3.86	2.88
Ethanol	0.794	0	0	0	58.1	1.56	1.24
Gasoline/ethanol blend 95/5	0.747	72.6	3.08	2.30	88.4	3.74	2.80
Diesel	0.832	74.5	3.21	2.67	90.4	3.90	3.24
Bio-diesel	0.890	0	0	0	58.8	2.16	1.92
Diesel/bio-diesel blend 95/5	0.835	71.0	3.04	2.54	88.8	3.80	3.17
Liquefied petroleum gas (LPG)	0.550	67.3	3.10	1.70	75.3	3.46	1.90
Compressed natural gas (CNG)		59.4	2.68		68.1	3.07	
Aviation gasoline (AvGas)	0.800	70.6	3.13	2.50	84.8	3.76	3.01
Jet gasoline (Jet B)	0.800	70.6	3.13	2.50	84.8	3.76	3.01
Jet kerosene (Jet A1 and Jet A)	0.800	72.1	3.18	2.54	88.0	3.88	3.10
Heavy fuel oil (HFO)	0.970	77.7	3.15	3.05	84.3	3.41	3.31
Marine diesel oil (MDO)	0.900	75.3	3.24	2.92	91.2	3.92	3.53
Marine gas oil (MGO)	0.890	75.3	3.24	2.88	91.2	3.92	3.49

Source: Table A.1—EN 16258:2012 standard, CEN (2012)

EcoTransIT identifies the environmental impacts of freight transportation in terms of direct energy consumption and emissions during the operation of vehicles during the transport of products. Moreover, the calculation covers the indirect energy consumption and emissions related to production, transportation and the distribution of energy required for operating the vehicles. Although one basic mode of operation of the tool has some average built-in parameters, the user can also alter the inputs and other parameters of the EcoTransIT application, according to a company's individual conditions.

More on this tool can be found at the following link: http://www.ecotransit.org/about.en.html.

2.2.5 *Allocation of Transportation Emissions*

Among the many human activities that produce GHGs, the use of energy, 80 % of which still comes from fossil fuels, represents by far the largest source of emissions. Within the energy sector, CO_2 dominates total GHG emissions. In addition, the energy sector includes emissions from "fuel combustion" and "fugitive emissions", which are intentional or unintentional releases of gases resulting from production, processes, transmission, storage and use of fuels (e.g. CH_4 emissions from coal mining). Figure 2.4 shows direct GHG emissions of the transportation sector broken down by transportation mode for the period 1970–2010.

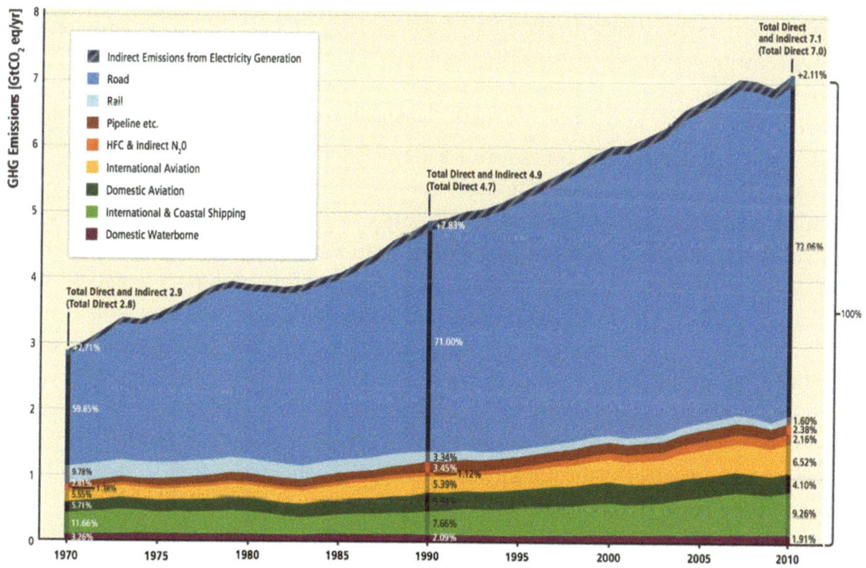

Fig. 2.4 Direct GHG emissions of the transportation sector (shown here by transportation mode) rose 250 % from 2.8 Gt CO_2eq worldwide in 1970 to 7.0 Gt CO_2eq in 2010. *Source*: IPCC AR5 report—IPCC (2014)

According to the 2014 highlights on CO_2 emissions from fuel combustion, transportation is responsible for 23 % of global CO_2 emissions. As for transportation, the fast emissions growth was driven by emissions from the road sector, which increased by 64 % since 1990 and accounted for about three quarters of transportation emissions in 2012. It is interesting to note that despite efforts to limit emissions from international transportation, emissions from marine and aviation bunkers, 65.8 % and 86.4 % higher in 2012 than in 1990 respectively, grew even faster than those from road. According to the IEA, international marine bunkers are responsible for 602.2 $MtCO_2$ in 2012 and aviation bunkers for 477.86 $MtCO_2$ of a total of 31,734.3 $MtCO_2$ of emissions due to fuel combustion.

As discussed previously, the Kyoto Protocol addresses emissions from fuel used for international aviation and maritime transportation in its Article 2, paragraph 2. Article 2.2 of the Kyoto Protocol states that the Parties included in Annex I shall pursue limitation or reduction of emissions of greenhouse gas emissions not controlled by the Montreal Protocol from aviation and marine bunker fuels, working through the ICAO and the IMO, respectively. It is generally accepted that the emission estimates of transportation activities presented by the UNFCCC and IEA are reliable for most modes except aviation and shipping. Bunker emissions from these two sectors are reported by the respective UN agencies.

According to the Third IMO GHG study (Smith et al. (2014)), with year 2012 as base year, total international shipping emissions were approximately 949 million tonnes CO_2 and 972 million tonnes CO_2eq for all GHG gases. International shipping accounts for approximately 2.2 % (and 2.1 %) of global CO_2 (and GHG) emissions on a CO_2eq basis, respectively. In absolute numbers, international shipping emissions for 2012 are estimated to be 796 million tonnes CO_2 and 816 million tonnes CO_2eq for GHGs.

According to the same study, for the period 2007–2012, on average, international shipping accounted for approximately 3.1 % of annual global CO_2 and approximately 2.8 % of annual GHGs on a CO_2eq basis using 100-year global warming potential conversions from the AR5. This study estimates 2007–2012 average annual totals of 20.9 million and 11.3 million tonnes for NOx (as NO_2) and SOx (as SO_2) from all shipping, respectively, representing about 15 % and 13 % of global NOx and SOx from anthropogenic sources reported in the latest IPCC Assessment Report (AR5), respectively. This study uses also IEA data to estimate emissions based on fuel consumption (top-down method). International shipping fuel consumption ranged between approximately 200 million and 270 million tonnes per year, depending on whether consumption was defined as fuel allocated to international voyages (top-down) or fuel used by ships engaged in international shipping (bottom-up), respectively.

In general we should note that different studies produce different results. There is a large difference between the IEA estimates, the UNFCCC submissions and the

2014 IMO GHG study estimations. The need for a consensus because on this subject is due to the fact that if we do not know what is the level of emissions we cannot set any meaningful emissions reduction targets.

2.2.6 *Future Emissions and the Total Contribution to Climate*

As much as it is difficult to estimate current or even past emissions, a fortiori it must be much more difficult to estimate future emissions. Many scenarios for future GHG emissions from transportation, for example those presented for maritime transportation in Buhaug et al. (2009), are based on assumptions on global development in the IPCC Special Report on Emissions Scenario (SRES) storylines. Since projections of climate change depend heavily upon future human activity, climate models are run against scenarios. There are 40 different scenarios, each making different assumptions for future greenhouse gas pollution, land-use and other driving forces and divided into six categories: A1FI, A1B, A1T, A2, B1, and B2. Assumptions about future technological development as well as the future economic development are thus made for each scenario and most include an increase in the consumption of fossil fuels. These scenarios have been superseded by the so-called 'Representative Concentration Pathways' (RCPs), which are four greenhouse gas concentration (not emissions) trajectories adopted by the IPCC for its fifth Assessment Report (AR5) in 2014. They describe four possible climate futures, all of which are considered possible depending on how much greenhouse gases are emitted in the years to come. The four RCPs, RCP2.6, RCP4.5, RCP6, and RCP8.5, are named after a possible range of radiative forcing values in the year 2100 relative to pre-industrial values (+2.6, +4.5, +6.0, and +8.5 W/m^2, respectively). 'Radiative forcing' is the additional heat/energy which is retained in the Earth's ecosystem through the addition of a gas to the atmosphere.

The transportation sector emits non-CO_2 pollutants that are also climate forcers. These include CH_4, NOx, SO_2, CO, F-gases, black carbon, and non-absorbing aerosols (Ubbels et al., 2002). Although most current climate policies focus on GHGs, which have a relatively well-known behavior and radiative forcing of climate, there is strong evidence that the other emissions and mechanisms also play an important role for the transportation sector.

Quantifying these effects is a complex scientific undertaking because of the broad mix of substances and physical/chemical processes involved. Without getting into details, there are emissions that contribute to Global Warming and others that have a cooling effect. Several studies have investigated this interaction by specifically isolating the climate forcing from transportation (Fuglestvedt,

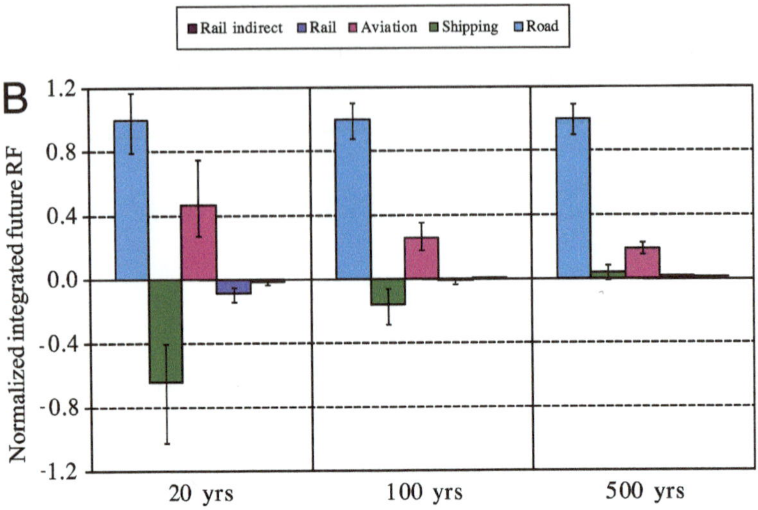

Fig. 2.5 Integrated global mean net RF per sector due to 2,000 transportation emissions, normalized to the values for road transportation for various time horizons (20, 100, and 500 years). Uncertainty ranges are given as one SD. *Source*: Fuglestvedt et al. (2008)

Berntsen, Myhre, Rypdal, & Skeie, 2008). For example, Fuglestvedt et al. (2008) showed that the transportation sector contributes significantly to man-made radiative forcing and that current emissions from transportation are responsible for 16 % of the integrated net forcing from all current anthropogenic emissions over the next 100 years. Sulfate emissions from shipping and rail result in a negative impact (see also Fig. 2.5).

International shipping has been a fast growing sector of the global economy and its share on total anthropogenic emissions has increased lately but the nature of the contribution to climate change is complex. In contrast to global warming induced by CO_2 emissions, ship emissions of SO_2 cause cooling through effects on atmospheric particles and clouds, while NO_x increase the levels of the greenhouse gas ozone (O_3) and reduce CH_4, causing warming and cooling, respectively and the result is a net global mean radiative forcing from the shipping sector that is currently strongly negative (Eyring et al., 2009). However, due to new regulations on SO_2 and NO_x, their emissions will decrease and after 50 years the net global mean effect of current emissions will be close to zero (Buhaug et al., 2009; Eyring et al., 2009; Fuglestvedt et al., 2008). Eyring et al. (2010), a paper co-authored by some of the same authors involved in estimation of GHG inventories and also in the second IMO study state that in 2005 the total radiative force from shipping effect measured in W/m^2 was −0.408, which means that currently shipping causes indeed cooling.

2.3 Environmental Policy

The number of uncertainties described earlier should not be a reason for inaction. This is in line with the so-called 'Precautionary Principle' as expressed in Article 3 of the Convention on Climate Change which reads as follows: *The parties should take precautionary measures to anticipate, prevent or minimize the causes of climate change and mitigate its adverse effects. Where there are threats of serious or irreversible damage, lack of full scientific research should not be used as a reason for postponing such measures, taking into account the policies and measures to deal with climate change should be cost-effective so as to ensure global benefits at the lowest possible cost*" (UNFCCC, 1997).

Indeed policy makers facing a panel of experts with divergent beliefs have to deal the situation as a case of decision-making under pure uncertainty. According to Woodward and Bishop (1997), their restatement of the Arrow-Hurwicz theory demonstrates that rational choice criteria under this kind of uncertainty have to focus on the extremes of the state space and not on the average as typically assumed under the expected utility hypothesis. That means that we should not look for averages or compromises rather than consider the best and the worst of the credible scenarios. Given that in future more information will be available, one of the two sides of the climate change debate will turn out to be right. If the more pessimistic scenario is right then acting on the low case will be a failure of today's decision making. On the other hand, if we act on the extreme scenario and in future we realize that the low case is right then it means that the society has overinvested (Ackerman & Heinzerling, 2004). The cost may be substantial and may also effect future generations.

Under this pressure, various industry sectors have been considering many possible measures including operational, technical and market-based options. From the environmental economics' point of view, climate change is the greatest and widest-ranging market failure ever seen, presenting a unique challenge for economics. Without entering into detail, markets may fail to achieve the optimal outcome when an externality exists, that is when the actions of a firm impact on those not directly involved. That is exactly what happens in the case of air pollution. In environmental policy-making, policies are often classified in market-based, command-and-control and voluntary instruments. This section will focus on the possible ways to internalize the cost of externalities according to standard economic theory. The role of incentives such as taxes, trading schemes and command-and control will be analyzed. Operational and technical measures are not discussed here as they are covered in other chapters of this book.

2.3.1 Emission Standards

Historically, emission standards have been the most popular approach to control environmental pollution. Emission standards do create incentives for research and development in emissions control and can have a positive effect in the problem that we are discussing but are not the most effective policy measures. However, we believe that they have to be the first approach to deal with the problem of carbon dioxide emissions as it will be easier to be applied by most IMO-member states.

In public policy, a Command and Control (CAC) approach is one where the regulator mandates the behavior in law to what is thought to be socially desirable (Field & Field, 2009). As the name implies, this approach consists of a 'command', which sets a standard, and a 'control', which monitors and enforces the standard (Asafu-Adjaye, 2005). There are three main types of standards: ambient, technology and emission. In brief, ambient standards are environmental quality levels in the ambient environment, such as a city or a port, and are usually expressed as average concentration level over some period of time. On the other hand, technology standards specify the technologies or techniques that should be adopted and do not specify some end result, such as a threshold level (Field & Field, 2009). For example, the requirement that all ships should be equipped with scrubbers in order to lower SO_x emissions is a technology standard. Furthermore, the regulator may specify operational measures, such as a mandatory speed reduction measure. The third type of standards is the so-called emission standards (or performance standards) and regulate the level of emissions allowed. Standards may impose a ceiling on total emissions in a period or a maximum allowable emissions rate, something that IMO has set in the case of NOx emissions.

An advantage of standards is that they are the most widely understood form of environmental policy and the most pragmatic approach in the case of environmental protection under uncertainty. Furthermore, this is the most favorite form of environmental policy for politicians since it has the lowest political cost, way lower compared to market based instruments (Hanley, Shogren, & White, 2006). On the other hand, disadvantages of standards are that the threshold is difficult to be determined and that under a CAC approach firms have no incentive to reduce emissions beyond the standards. Note also that this approach is effective only when the penalties are high and the enforcement methods are strong enough (Asafu-Adjaye, 2005; Field & Field, 2009).

In most cases, emissions standards are firstly tested as voluntary agreements and when it seems that they work then become mandatory. For example, the so called ACEA agreement was an agreement between the European Commission (EC) and the European Automobile Manufacturers Association (ACEA) that was signed in 1998 and sought to achieve an average of 140 g/km of CO_2 by 2008 for new passenger vehicles sold in the EU. The ultimate target is to reach an average of 130 g/km by 2015. Being a voluntary agreement this system was a failure although some reduction was achieved. In April 2009, the European Commission published Regulation No 443/2009 which sets the average CO_2 emissions for new passenger cars at 130 g CO_2/km, by means of improvement in vehicle motor technology (European Parliament, 2009). This is in line with the NO_x standards already being used in shipping.

Similarly, on September 15, 2009, the EPA and the US Department of Transportation's National Highway Safety Administration (NHTSA) proposed a National Program (EPA, 2010) that could reduce emissions and improve fuel economy for new cars and trucks sold in the United States according to which they propose a limit of an average CO_2 emissions at approximately 155 g CO_2/km (250 g CO_2/mile).

The ICAO is also developing a global carbon dioxide standard for new aircraft types, among other recommendations from the organization's committee on aviation environmental protection (CAEP). CAEP met at ICAO headquarters from 1 to 12 February 2010, in Montreal, Canada, and announced the ICAO's commitment to the development of a CO_2 standard for commercial aircraft by 2013. More on aviation can be found in Chap. 13 of this book.

For the maritime mode, in 2011 the IMO has adopted the Energy Efficiency Design Index (EEDI) for new ships as an amendment of MARPOL's Annex VI. The concept behind the EEDI has been described in Sect. 1.6.1 of this book and basically EEDI tries to minimize the ratio of a ship's CO_2 emissions divided by that ship's transportation work. In addition to GHGs, IMO regulates the emission of other pollutants from ship exhausts, including NOx and SOx emissions. These regulations are contained in the MARPOL Annex VI protocol, which, amongst others, designates specific geographic areas as Emission Control Areas (ECAs), where more stringent requirements apply. Section 6.2 of Chaps. 1 and 10 of this book provide more details on the issue of sulphur regulations.

2.3.2 *Incentive-Based Strategies*

Incentive-based environmental policies require that public authorities set the targets of emission reductions and the rules and leave the polluters adopt cost-effective emission control measures. There are basically two categories of incentive policies: (1) charges and subsides and (2) transferable emission permits. Polluters, in general, do not take into account the damage to the environment that their emissions cause. In an emission charge system, polluters are allowed to discharge any amount of emissions they want, but are required to pay a certain amount of money for each unit they emit and in a subsides system they receive money for each unit that they do not emit. Thus, the basic idea behind incentives is to design a system with private flexibility to achieve the desired public objectives set by the regulators.

In the quest to reduce emissions, economists mainly favor emissions taxes following the idea of Pigou (1920) that by charging for every unit of emissions released, polluters will tend to reduce their emissions. Note that obtaining all necessary information to impose the ideal tax is quite costly and, in practice, regulators determine the charge by using a trial-and-error process. The most important problem of such a system is effective monitoring.

On the other hand, tradable emission permits allow the voluntary transfer of the right to emit from one firm to another. In this system firms are allocated a number of emission permits and are entitled to emit one unit per permit but these permits are transferable. A market for these permits will eventually develop and firms that can reduce emissions at a low cost may prefer to sell its permit to a firm that can reduce pollution only at a high cost (Field & Field, 2009). 'Cap-and-trade' programs work a little bit different since the first step is to make a centralized decision on the aggregate quantity of total emissions. The permits are then distributed among the emitters and, in general, some emitters, if not all, receive less permits than their actual emissions. The permits are traded in an overall market and will flow from firms with relatively low marginal abatements cost to those with higher marginal costs. Thus, there will be a constant trading among emitters and an incentive for them to look for ways to reduce emissions.

Those emitters who can reduce their emissions more cheaply are able to sell extra allowances to others who would otherwise have to pay more to comply and because of this, a cap-and-trade system helps assure that we can achieve an overall cap at the lowest possible cost.

One may notice that these two systems lead to equivalent results in the long term but with different uncertainty for the outcome see Figs. 2.6 and 2.7. A comparison between a carbon tax and the cap-and-trade approach comes down to the issue of certainty. A tax provides for cost certainty; the cost is fixed because of the Pigovian tax. Trading permits, on the other hand, provides for environmental certainty. What's fixed is the cap itself—and it is based on an assessment of the level of

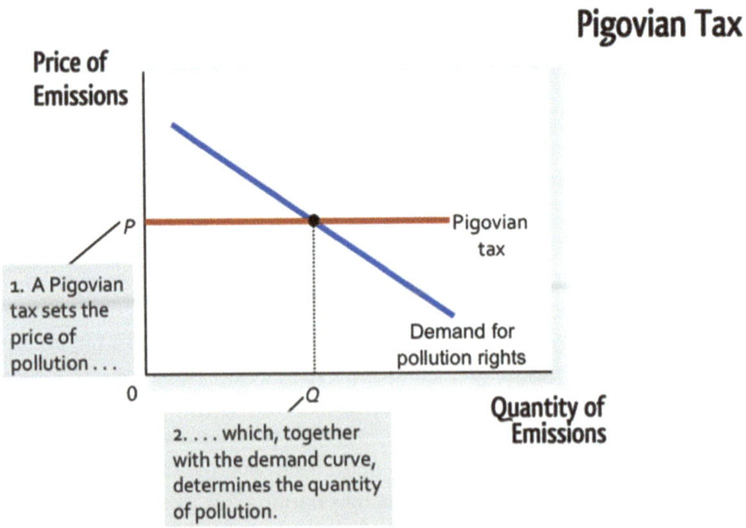

Fig. 2.6 Illustration of Pigovian Taxes – Adapted from Mankiw (2006)

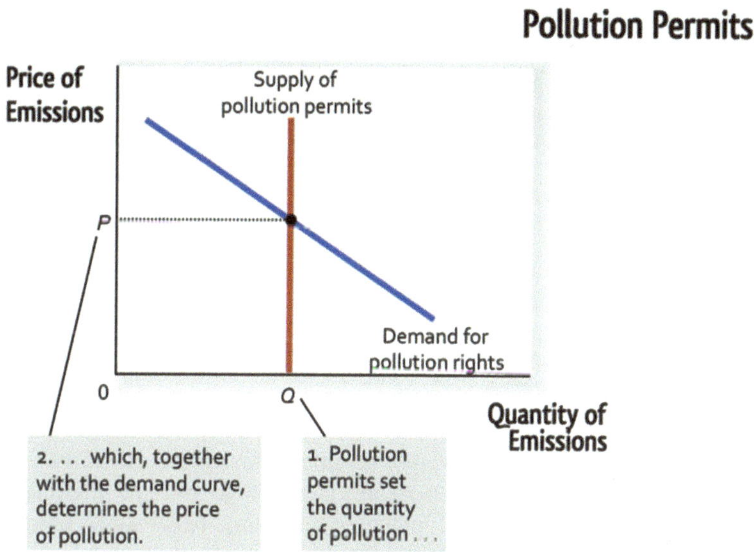

Fig. 2.7 Illustration of Pollution Permits – Adapted from Mankiw (2006)

emissions you need to get to in order to protect the climate. In that sense, if the cap is set too high, permit prices will be low and the incentive effect will be weak. If the cap is set too low, permit prices will be very high and that can lead to the disruption of economy and trade.

For a practical application of the policies described above the reader is referred to Chap. 8 of this book that introduces the concept of Market Based Measures (MBMs) to reduce GHG emissions from ships, and reviews several distinct MBM proposals that have been under consideration by the IMO.

2.4 Measuring Cost Effectiveness

There are several technical measures and practices that have a great potential to reduce CO_2 emissions, see for instance Chap. 5. It is obviously very important to evaluate the various reduction measures based on their environmental effectiveness, that is, the proposed measure should be able to contribute to achieving a particular stabilization level or, consequently, a specific reduction target to avoid dangerous climate change in accordance with Article 2 of the UNFCCC. However, emission reductions come at a cost, which in some cases is very high. Thus, cost-effectiveness plays a very important role in decision-making. In line with the above, Article 3(3) of the Convention states that the

> Parties should take precautionary measures to anticipate, prevent or minimize the causes of climate change and tak[e] into account that policies and measures to deal with climate change should be cost-effective so as to ensure global benefits at the lowest possible cost.

2.4.1 *Cost Effectiveness Index*

Without entering into details, cost effectiveness is measured by a simple ratio, usually referred to as the Marginal Abatement Cost (MAC), which is defined as the ratio of the net present value (NPV) of a specific abatement measure, divided by the corresponding emissions reduction for the entire lifetime of the project, see Eq. (2.8). The net present value of the emissions reduction measure takes into account the costs and the benefits.

In general, the **cost** component (present value of costs) consists of the one-time (initial) and running costs of the measure, cumulating over the lifetime of the system. The **benefit** part (present value of monetary benefits) is much more intricate. In addition, benefits and costs occurring in different time periods within the lifetime of the project have to be aggregated to obtain the net present value (NPV). All actions have an associated flow of costs and benefits during their life time (T years) that have to be added to obtain the NPV.

The net present value (NPV) of implementing an abatement measure is calculated using the following equation:

$$NPV = \sum_{t=0}^{T} \frac{C_t - B_t}{(1+r)^t} \tag{2.5}$$

where
B_t are benefits in period t;
C_t the costs in period t;
r is rate used for discounting (per period); and
T the number of periods (usually years) the project will last.

To discount a flow of n equal amounts A (can be costs or benefits) that incur at regular intervals (for instance at the end of each period for a total of T periods) assuming that the discounting rate is constant, it can be easily shown (see for example Brealey and Meyers (2003)) that the present value (PV) of this money flow (cost or benefit) is:

$$PV = A \cdot \left[\frac{1}{r} - \frac{1}{r(1+r)^T}\right] = A \cdot \left[\frac{(1+r)^T - 1}{r(1+r)^T}\right] = \sum_{t=1}^{T} \frac{A}{(1+r)^T} \tag{2.6}$$

The denominator is the mass of emissions averted. For the case of GHGs the notion of CO_2eq can be used.

2.4.1.1 MAC for a Specific Emissions Reduction Measure: An Illustrative Example

Based on the previous definition we now illustrate a way to estimate the cost effectiveness of a reduction measure by using the following example. Assume a GHG reduction measure is repeated every L years (L is the life of the measure) during the N years, which is the lifetime of the project. The times that the measure is repeated is λ and is equal to the nearest integer rounded down of the fraction N/L. The annual costs (AC) of the project are supposed to be known and occur every year except the years when the measure is applied and there exist only capital costs (C).

Furthermore, the benefits are supposed to be equal to the bunker savings (although this may be extended to include other benefits too). The benefits are estimated as follows:

$$B = p \cdot \alpha \cdot FC$$

where p is the bunker price ($ per tonne), α is the abatement potential (%) and FC is the annual fuel consumption (without the measure).

The Net Present Value can be estimated as follows

$$NPV = C + \sum_{i=1}^{N} \frac{AC - B}{(1+r)^i} + \sum_{j=1}^{\lambda} \frac{C - AC}{(1+r)^{jL}} \tag{2.7}$$

The potential CO_2 abatement for this project is equal to

$$\Delta R = N \cdot a \cdot f \cdot FC$$

where f is the CO_2 emissions factor.

Therefore, the marginal abatement cost of this project is equal to

$$MAC = \frac{NPV}{\Delta R} \tag{2.8}$$

2.4.2 Marginal Abatement Cost (MAC) Curves

A MAC curve is a graph that shows the cost-effectiveness index defined above of all the available measures (each measure represented as a bar) that are ordered in a function of increasing abatement costs, see Fig. 2.8. The height of each bar represents the average marginal cost of avoiding a ton of CO_2-eq given that all measures on its left are already applied and the width represents the potential of that

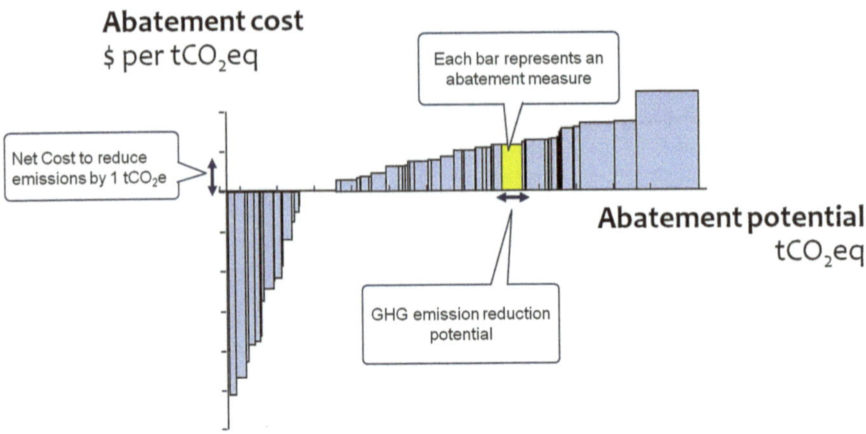

Fig. 2.8 Schematic example of a marginal abatement cost curve

measure to reduce emissions. Note that the word 'marginal' denotes that all available reduction measures are mutually exclusive.

The MAC curves have been popularized by the consulting firm McKinsey & Company, which used them on a global and country-wide scale to target areas for carbon abatement.

Figure 2.9 presents an example of a marginal abatement cost (MAC) curve published in a global study by McKinsey in 2007 (Enkvist, Nauclér, & Rosander, 2007). It shows the annual abatement needed to achieve stable atmospheric greenhouse gas concentrations of 500 ppm (parts per million), 450 and 400 ppm of CO_2-equivalents. For example, a global emissions reduction of 26 Gtons of CO_2-eq per year would stabilize greenhouse gas concentrations at 450 ppm of CO_2-eq, and that reduction would need all the abatement measures up to a cost of €40 per ton of CO_2eq.

It is interesting to note that according to Fig. 2.9 the most cost-effective abatement measure is building insulation. There are indeed many measures that have a negative abatement cost. This means they carry no net life cycle cost, and they come free of charge. However, although the low cost measures are efficient and cost-effective, in general they may not be sufficient to deliver the required emissions reductions by themselves. If so, we will have to move up the MAC curve progressively, adopting more and more expensive measures, until the desired emissions reduction is achieved. Thus, the role of policy makers is to enforce or provide incentives to adopt these measures.

2.4.3 Caveats of MAC Curves

For all their usefulness, MAC estimates have some weak points. First of all, MAC prices depend heavily on emission reduction targets and stabilization targets

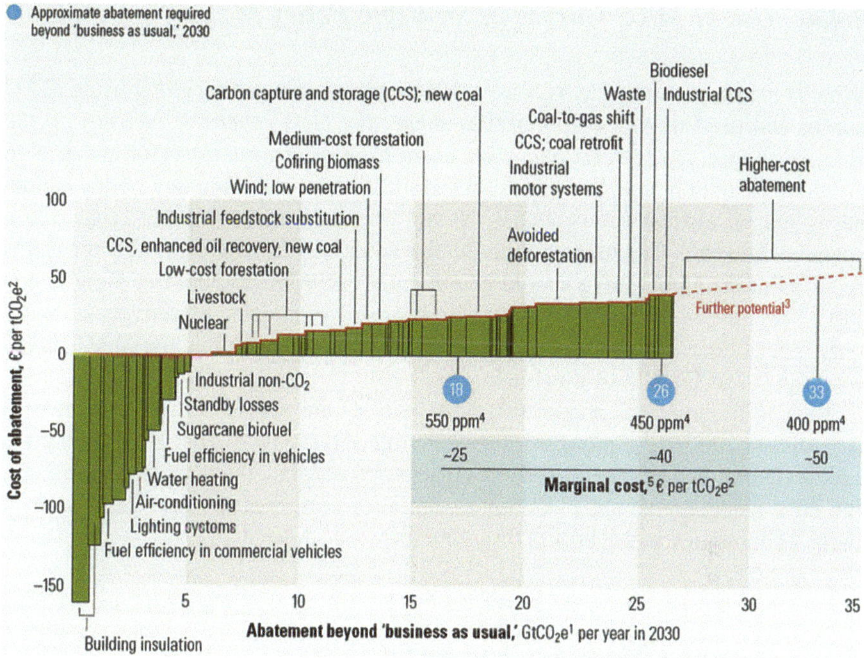

Fig. 2.9 Cost Curve for GHG abatement measures. *Source*: Enkvist et al. (2007)

and, therefore, differ among sectors and countries. Furthermore, they do change massively over time as innovation kicks in with more cost-effective measures (FOEI, 2008).

Ekins, Kesicki, and Smith (2011) discuss the various caveats and methodological problems associated with the MAC curves. Obviously, the easily digestible form of the graphic MACC led to these problems being overlooked, placing a strong confidence in the results. The general shortcomings presented in their work include the focus on emission reduction without considering ancillary benefits such as health improvement which can not be easily monetized, and a static representation of costs which fails to consider path dependency. Another problem has to do with hidden costs, including transaction and monitoring costs. In addition, these authors note the lack of full disclosure of the assumptions and the non-consideration of interdependencies and "intersectoral, intertemporal, behavioral, macroeconomic, and international interactions" which can lead to problems defining the emissions baseline especially in a time horizon of 20 years or even more.

Finally, another problem is related to the fact that MAC curves cannot capture well the related uncertainties. Thus, we should look beyond the estimated marginal abatement cost and obviously pay attention to the assumptions behind.

2.4.4 Use of MAC Curves in Strategic Policy Making

What is the proper use of MAC curves? If the carbon price derived by using MAC curves and used in policy appraisal *is simply the most expensive measure in the strategy to meet the budgets and other overall government targets* (for instance an emission reduction target, or emission stabilization levels), then any policy can be easily judged against others (FOEI, 2008). Therefore, measures that cost up to a specific threshold should be proposed for adoption. In this case, the measures reported in the MAC curve should be mutually exclusive. Otherwise, the graph is just a list of the MAC of different measures sorted by their cost-effectiveness. MAC curves have been prepared for various industries, see for example the MAC curves for shipping in Chap. 8 of this book (Fig. 2.10).

Recent literature has shown that MAC curves are very sensitive to numerous assumptions, including, among others, the projected (up to 2020 or 2030) fuel price and fleet size, the abatement measures that are considered, the discount rate used in the NPV calculations, the reduction potential of each measure and the uptake of these technologies in the future.

2.4.5 Negative Marginal Abatement Costs

It is interesting to note that in many MAC curves there are measures that have a negative MAC. These measures are at the left-most part of the MAC curve, as the

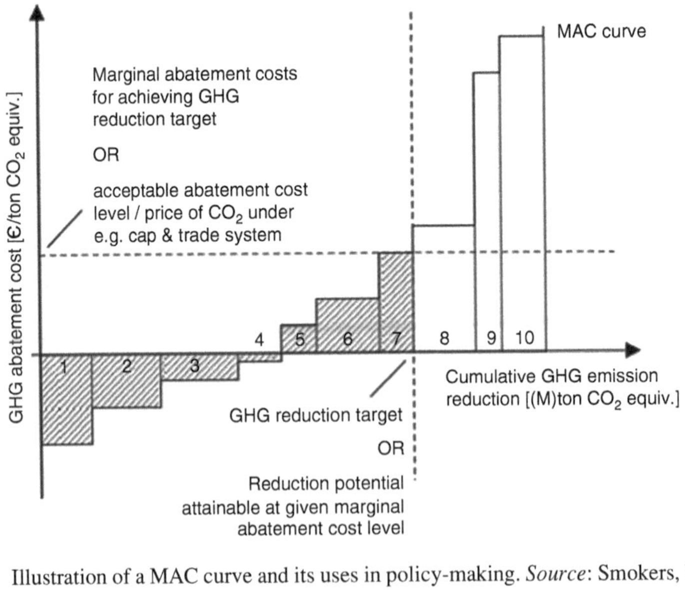

Fig. 2.10 Illustration of a MAC curve and its uses in policy-making. *Source*: Smokers, Buck, and van Valkengoed (2009)

curve is constructed by non-decreasing order of MAC. A negative MAC means that the benefit of adopting an emissions reduction measure in monetary terms is more that its cost. This means that whoever contemplates the measure would have an economic incentive to apply it, and, in that sense, a win-win solution would be achieved. Conversely, a measure with a positive MAC would provide no economic incentive for its adoption and would have to be mandated by regulation if deemed appropriate.

A negative MAC is likely to be caused by a variety of factors. Perhaps the easiest to understand is a high fuel price. If a measure (such as for instance a more fuel-efficient engine) results in reduction of fuel consumption (and hence emissions), and fuel price is high, most of the measure's economic benefits will come in the form of savings in fuel cost. If these savings are high enough, the associated MAC may be negative.

A caveat here is a MAC that is calculated as negative but in reality is positive because of incorrect calculation of its components. For instance one may have an incorrect or incomplete estimate of the abatement cost, and/or an incorrect estimate of the private (or social) discount rate. Errors of such nature may be common for technologies that are at the design stage and cost data may be inaccurate or elusive. Errors may also inevitably occur in the benefits side. An example is in the estimation of future fuel prices, which are probably the single most influential factor in determining the benefits of a certain measure and hence its MAC. All such errors may seriously misrepresent a measure's cost-effectiveness.

Another caveat concerns the existence of barriers, as is described in the section that follows.

2.5 Barriers to Energy Efficiency

2.5.1 Definitions and Background

Sorrell et al. (2004) define a barrier to energy efficiency as *a postulated mechanism that inhibits a decision or behavior that appears to be both energy efficient and economically efficient*.

To determine the existence of barriers empirically we need to ask (Sorrell et al. 2011; Weber, 1997):

1. *What is the barrier?* This can be persons or organizations, regulations, patterns of behaviors and attitudes, hidden costs, risk, lack of capital, lack of information, inadequate financial incentives etc.
2. *Who or what is it an obstacle to?* This can be consumers including firms, public organizations, departments within organizations, individuals, etc.
3. *What does it prevent?* This can be purchasing of more efficient equipment, retrofitting, decreeing an energy tax or establishing a monitoring and targeting scheme.

The barriers to be discussed later on are based on the answers to the following questions (Sorrell et al. (2004, 2011):

- Why do organizations impose very stringent investment criteria for projects to improve energy efficiency?
- Why do organizations neglect projects that appear to meet these criteria?
- Why do organizations neglect energy-efficient and apparently cost-effective alternatives when making broader investment, operational, maintenance, and purchasing decisions?

There is a large body of literature on energy efficiency barriers, which according to Schleich (2009) draws on concepts from neo-classical economics, institutional economics (principal-agency theory and transaction cost economics), behavioral economics, sociology and psychology. Barriers to energy efficiency are very diverse and are classified in a variety of ways, see for example Jaffe and Stavins (1999), IEA (2009) and Sorrell et al. (2004, 2011).

The above literature tries to explain why organizations fail to invest in energy efficiency even though it is or, more specifically, it is perceived to be profitable under current economic conditions to do so—a "phenomenon" that has also been referred to as the "energy efficiency gap" (Jaffe & Stavins, 1994).

Jaffe and Stavins (1994) make a distinction between market barriers and failures. *Market barriers* refer to any factor, which explains why technologies, which appear cost effective at current prices, are not taken up. *Market failures* to those barriers correspond to the instances indicated above and which therefore might justify a public policy intervention to improve energy efficiency. In addition, there may be market failures that do not explain the energy efficiency gap, but which may nevertheless justify public intervention. The obvious example is environmental externalities, which are currently not reflected in energy prices (Sorrel, 2004).

'Energy efficiency gap' is the lag between the time an energy efficient technology is available and the time that it gets implemented. On this basis, debate has raged about the extent to which there are low-cost or no-cost options for reducing fossil energy use through improved energy efficiency. According to Jaffe and Stavins (1994) this is a debate between the so called 'economists' and 'technologists'. They characterize 'technologists' as believing that there are many opportunities for low-cost, or even 'negative-cost' improvements in energy efficiency, and that realizing them will require a market intervention to help overcome barriers in order to use more efficient technologies. On the other hand, most 'economists', acknowledge that there are 'market barriers' to the adoption of various technologies but that only some of these barriers represent real 'market failures' that reduce economic efficiency. This view emphasizes the tradeoffs between economic efficiency and energy efficiency. This perspective suggests that reducing emissions is more costly than the technologists argue, and it puts relatively more emphasis on market-based emission control policies like taxes and tradable permit systems.

Thus, their analysis suggests two distinct notions of economic potential and two distinct notions of social optimum. First of all, by using 'economic potential' they describe the degree of energy efficiency that would be achieved if various economic barriers were removed. Further, they describe the scenario in which we can eliminate market failures in the energy technology market as the economists' economic potential. If we were also to remove non-market failure market barriers, we would achieve the technologists' economic potential.

Concluding, to understand the basic elements of the debate, one should distinguish first between energy efficiency and economic efficiency as illustrated in Fig. 2.11, which is taken from Jaffe, Newell, and Stavins (2004). The horizontal axis measures the increasing economic efficiency (decreasing overall economic cost per unit of economic activity) and the vertical one measures the increasing energy efficiency (decreasing energy use per unit of economic activity). Possible energy-using technologies can be shown as points in this diagram and indicated by their energy and economic efficiency.

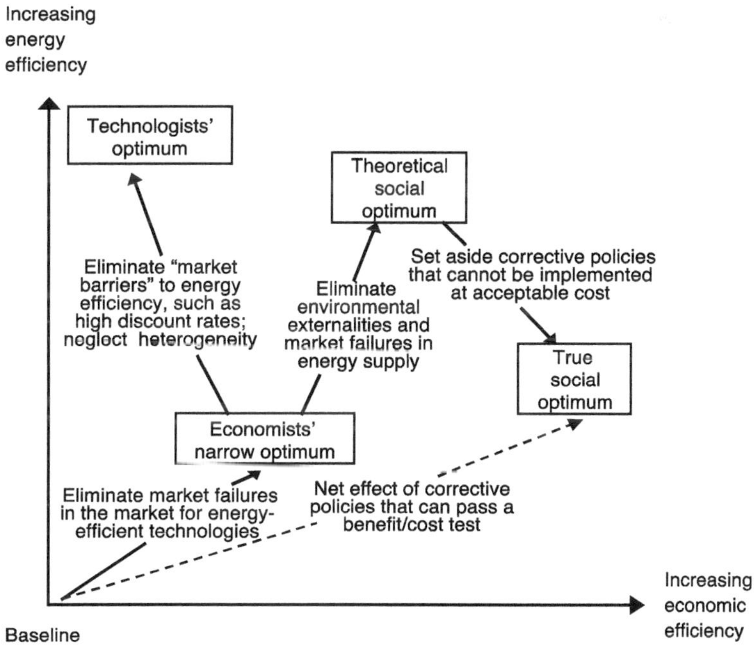

Fig. 2.11 Alternative notions of the energy efficiency gap. *Source*: Jaffe et al. (2004)

2.5.2 The Nature of Barriers: A Taxonomy Based on Sorrell et al. (2004, 2011)

As discussed above, barriers to energy efficiency are very diverse and are classified in a variety of ways, see Table 2.4. Relying on the taxonomy developed in Sorrell et al. (2004, 2011), this section provides a brief summary of the main concepts of barriers to energy efficiency, which is heavily based on Sorrell et al. (2004, 2011) and Schleich (2009). It is important to stress out that some of the barriers are interrelated and also there may be an interaction between them.

We next comment on each of these barriers.

Table 2.4 A taxonomy of barriers to energy efficiency

Barrier	Claim
Risk	The short paybacks required for energy efficiency investments may represent a rational response to risk. This could be because such investments represent a higher technical or financial risk than other types of investment, or that business and market uncertainty encourages short time horizons
Imperfect information	Lack of information on energy efficiency opportunities may lead to cost-effective opportunities being missed. In some cases, imperfect information may lead to inefficient products driving efficient products out of the market
Hidden costs	Engineering-economic analyses may fail to account for either the reduction in utility associated with energy efficient technologies, or the additional costs associated with them. As a consequence, the studies may overestimate energy efficiency potential. Examples of hidden costs include overhead costs for management, disruptions to production, staff replacement and training, and the costs associated with gathering, analyzing and applying information
Access to capital	If an organization has insufficient capital through internal funds, and has difficulty raising additional funds through borrowing or share issues, energy efficient investments may be prevented from going ahead. Investment could also be inhibited by internal capital budgeting procedures, investment appraisal rules and the short-term incentives of energy management staff
Split incentives	Energy efficiency opportunities are likely to be foregone if actors cannot appropriate the benefits of the investment. For example, if individual departments within an organization are not accountable for their energy use they will have no incentive to improve energy efficiency
Bounded rationality	Owing to constraints on time, attention, and the ability to process information, individuals do not make decisions in the manner assumed in economic models. As a consequence, they may neglect energy efficiency opportunities, even when given good information and appropriate incentives

Source: Sorrell et al. (2004)

2.5.3 Risk

Stringent investment criteria (high discount rates for efficient investments) and the rejection of particular energy-efficient technologies may represent a rational response to perceived risk (Schleich, 2009; Sorrell et al., 2004). Within the context of implementation barriers, risk may fall into three categories (Sorrell et al., 2011). The first one is the *external risk*, which is related to the overall economic trends (e.g. recessions), expected fluctuations in energy use and energy prices and political changes and government policies. Uncertainty in energy prices can be related to fluctuation of exchange rates and stochastic prices for emission allowances under an emissions trading scheme (Schleich, 2009). Another category is the *business risk* which has to do with the economic trends of the individual firm or the sector and the financing risk. The last one is the *technical risk,* which is associated with the technical performance of the investment, measured for instance by its effectiveness and reliability. This is mainly the case with new and unfamiliar technologies which are perceived to be risky and, thus, the operation risk may overweight the potential benefits from reduced cost. Sutherland (1991) also suggests that these investments require higher hurdle rates because they are 'illiquid' and irreversible. In these cases the regulator or the industrial associations should aim to *increase confidence and disseminate information and awareness among potential adopters* (Schleich, 2009).

2.5.4 Imperfect Information

Huntington, Schipper, and Sanstad (1994) state that information problems *taking different forms are the principal source of market failures that account for the 'gap' in energy efficiency investments.* Sorrell et al. (2004) present the several dimensions to imperfect information based on Golove and Eto (1996) who distinguish imperfect information to 'lack of information' as ignorance on the energy performance of different technologies; 'cost of information', which includes the costs associated with *searching and acquiring information on the energy performance of technologies* and 'accuracy of information' as accurate information *may be difficult to obtain, since sellers of technologies may have incentives to exaggerate or manipulate performance data* and given that unbiased information may *be available from other sources but this may be more costly.* According to Sorrell et al. (2011) the information we are interested in, within the scope of the energy efficiency gap, is related to information on current energy consumption, on energy-specific investment opportunities and on energy consumption of new and refurbished buildings, process plant and purchased equipment. The lack of this information leads to making suboptimal decisions and in particular under-investing in energy efficiency.

Note that information has the attribute of a 'public good' and once created it can be used by many people at little or no additional cost. Jaffe et al. (1999) state that it is difficult for a firm that invests in information creation to prevent others who do not pay for it from using this information especially given that the act of adopting a new technology is, by itself, a source of important information for others.

Finally, a special form of the imperfect information problem arises when parties to a transaction have access to different levels of information (asymmetric information). This is mainly the case in the principal-agent problems, as when the owner chooses the level of investment in energy efficiency in a system, but the energy bills are paid by the tenant. This issue will be discussed in the 'split incentive' barrier.

2.5.5 Hidden Costs

Hidden costs are the explanatory variable for the 'gap'; the argument against the 'efficient gap' hypothesis (DeCanio & Watkins, 1998). In fact, cost effective measures are not cost-effective when such hidden costs are included. According to Nichols (1994) technical studies fail to account for either the reductions in benefits associated with the investments or the additional costs associated with them and as a result they overestimate the efficiency potential. Sorrell et al. (2004) present the following three broad categories of hidden costs: general overhead costs of energy management, costs specific to a technology investment and loss of benefits associated with an efficient technology.

2.5.6 Access to Capital

Usually technologies that are energy-efficient are more expensive to purchase than alternative ones (Hirst & Brown, 1990). This means that these technologies require a substantial upfront cost. According to Jaffe et al. (2004) there is a behavioral bias that causes purchasers to focus more on this cost than they do on the lifetime operating costs of investments. There are, then, two possibilities to obtain this access to capital, either the firm has sufficient capital though internal funds or it has to borrow them.

The financing barrier, also called the liquidity constraint, refers to the restrictions on capital availability for potential borrowers and in practice small firms are frequently unable to borrow as a results of their economic status or 'credit-worthiness' (Golove & Eto, 1996) and also due to the lender's difficulty in estimating the performance of the investment, both of which also involve the hidden cost to acquire the necessary information. Energy efficiency projects tend to be evaluated based on payback periods rather than discounted cashflow analyses (Schleich, 2009). The required rate of return implied by short payback periods exceeds those for business development projects, which is obviously the case for most energy-efficient investments. As a result, firms usually have a priority list to allocate the available finance and energy efficient projects in many cases come at the bottom of the list mainly due to Information gaps, institutional barriers, short time horizons, and non-separability of energy equipment may also be apparent (Brown, 2001).

2.5.7 *Split Incentives*

It is quite usual that costs and benefits of energy efficient investments accrue to different agents, splitting the incentive to invest. The most well-known example for split incentives is the landlord/tenant or user/investor dilemma, see IEA (2009) for more. For example, in construction of new system, builders may have difficulty conveying the benefits of energy conserving technologies to prospective buyers, because these technologies (and their future energy use consequences) are not observable, which is usually the case of asymmetric information. In addition, the adopter of the energy efficient technology may not be able to recover all of the value of such investments (in the form of higher rents) in the case where the renter pays the energy bills but the owner makes the investment; and tenants who make these investments in cases where the landlord pays the energy bill may not be able to get reduced rents (Jaffe & Stavins, 1994). The latter is also the case where the tenant is likely to move out before fully benefited from the reduced energy bills (Schleich, 1990). In theory (Jaffe and Stavings, 1994), this so-called principal-agent problem could be avoided if the informer can transmit the information of the future benefits (which then transforms this to an 'imperfect information' barrier) and all transaction cost (the case of 'hidden costs' barrier) are taken into account.

2.6 Discussion and Conclusions

As discussed above, there is a growing concern that the Earth's atmospheric composition is being altered by human activities, which can lead to climate change. Although there exist many uncertainties and differing viewpoints within the scientific community, the scientific evidence for warming of the climate system due to anthropogenic activities is unequivocal. Transportation is responsible for roughly one quarter of total anthropogenic CO_2 emissions. In addition, fuel consumption leads to other types of air emissions. Quantifying the total effects is a complex scientific undertaking because of the broad mix of substances and physical/chemical processes involved. This chapter presented some of the basic statistics including current and future trends, see IPCC (2014) for more. A number of measures, operational, technological and others has been stipulated to curb emissions.

Traditional analysis by and large either ignores environmental issues, or considers them of secondary importance. Thus, quantifying the environmental impact of transportation activities is an important task. Other chapters of this book present a variety of measures that may be used to reduce emissions. Some of these solutions are difficult to be implemented. For instance, due to the fact that consumption and production of goods happens in different places, avoiding long journeys is not an easy option. There should be a drastic change e.g. sourcing localized products or restructuring the whole logistic chain. Utilizing advanced Information and Communication technologies (as per Chap. 6), can contribute to this direction. An easier option is to encourage modal shifts to lower-carbon transportation systems and the investment in more fuel-efficient transportation vehicles. Shifting to 'greener' fuels (e.g. natural gas, bio-methane, biofuels etc.) can also be an option although in most cases this would lead to high investment and operational costs. Exploring 'win-win' solutions is also not trivial, as one has to weight between the economic cost and the environmental benefit. The present chapter presents a way to estimate the cost effectiveness of abatement measures.

There are indeed some measures that can be easily applied and they are, or seem to be, cost-effective. For instance increasing the freight load factor and passenger occupancy rates are obvious measures that lead to better energy efficiency. Investing in new technologies is more expensive that adopting an operational measure such as optimizing the operational profile of a vehicle. For example, significant fuel savings can be achieved by encouraging drivers to maintain a consistent speed and restrict their speed (eco-driving). In maritime transportation, reducing speed is by far the most promising emission reduction measure. The reason for many companies having their vessels running in slower speeds is mainly twofold: reduce fuel costs and emissions. Given that fuel costs and emissions are directly proportional to one another (both being directly proportional to fuel used), it would appear that reducing both would be a straightforward way towards a "win-win" solution. In shipping parlance this is known as "slow steaming". However, some of these operational measures, although effective in meeting environmental objectives, may have non-trivial side effects on the economics of the logistical supply chain; for instance, reductions of speed and changes in the number of ships in the fleet and possibly on things such as in-transit inventory and other costs (for more details see Psaraftis and Kontovas (2013) and Chap. 9 of this book).

One of the major questions posed in this chapter is if there are indeed measures that come at a negative MAC and how can we identify possible barriers to adopt energy efficient measures. To achieve these win-win results, we should determine the existence of barriers and identify ways to remove them. The first step to energy efficiency is to be able to identify the barriers that exhibit transportation activities. To that extend, we have presented a taxonomy of barriers based on the work of Sorrell et al. (2011) and a review of the literature on barriers as they appear in the industry. A number of measures may not be implemented due to lack of information among owners/operators about the existence of the measure. Sometimes the information received is not well trusted (lack of independent data). In other cases, there are high transaction costs involved in gathering reliable information on fuel saving technologies as real information can mainly be received only when technologies are applied in practice (full-scale experiments). In addition, there may also be hidden costs, for example more energy efficient vehicles may require modifications to the infrastructure. Thus, by removing the relevant barriers we are able to adopt cost effective measures and thus reduce the environmental impact of transportation at a lower cost.

Acknowledgments Work on this chapter has been supported in part by various sources, including the Lloyd's Register Foundation (LRF) in the context of the Centre of Excellence in Ship Total Energy-Emissions-Economy at the National Technical University of Athens (NTUA), the authors' former affiliation. LRF helps to protect life and property by supporting engineering-related education, public engagement and the application of research. This work has also been supported in part by an internal grant at the Technical University of Denmark (DTU).

References

Ackerman, F., & Heinzerling, L. (2004). *Priceless: On knowing the price of everything and the value of nothing*. New York: The New Press.

Asafu-Adjaye, J. (2005). *Environmental economics for non-economists: Techniques and policies for sustainable development* (2nd ed.). Hackensack, NJ: World Scientific.

Brealey, & Meyers. (2003). *Principles of corporate finance*. Boston: McGraw-Hill.

Brown, M. (2001). Market failures and barriers as a basis for clean energy policies. *Energy Policy, 29*, 1197–1207.

Buhaug, Ø., Corbett, J. J., Endresen, Ø., Eyring, V., Faber, J., Hanayama, S., et al. (2009). *Second IMO GHG study 2009, International Maritime Organization (IMO) London, UK, April 2009.*

CEN. (2012), Standard EN 16258 "Methodology for calculation and declaration of energy consumption and greenhouse gas emissions of transport services", European Committee for Standardization.

DeCanio, S. J., & Watkins, W. E. (1998). Investment in energy efficiency: Do the characteristics of firms matter? *Review of Economics and Statistics, 80*, 95–107.

Defra. (2013). Greenhouse gas conversion factors for company reporting 2013: Methodology paper for emission factors. UK Department for Environment, Food and Rural Affairs (Defra).

Ekins, P., Kesicki, F., & Smith, A. Z. (2011). *Marginal abatement cost curves: A call for caution*. London: UCL Energy Institute.

EMEP/CORINAIR. (2002, October). The Core Inventory of Air Emissions in Europe (CORINAIR), Atmospheric Emission Inventory Guidebook, EMEP Co-operative Programme for Monitoring and Evaluation of the Long Range Transmission of Air Pollutants in Europe (3rd ed.).

EMEP/EEA. (2013). EMEP/EEA air pollutant emission inventory guidebook 2013. *European Environment Agency*. Retrieved from http://www.eea.europa.eu/publications/emep-eea-guidebook-2013

Enkvist, P., Nauclér, T., & Rosander, J. (2007). A cost curve for greenhouse gas reduction. *Then McKinsey Quarterly, 1*, 35–45.

EPA. (2010). *EPA and NHTSA to propose greenhouse gas and fuel efficiency standards for heavy-duty trucks; begin process for further light-duty standards: Regulatory announcement*. US Environmental Protection Agency. Retrieved from http://www.epa.gov/otaq/climate/regulations/420f10038.htm

EPA. (2014a). *MOVES (Motor Vehicle Emission Simulator)*. Office of Transportation and Air Quality (OTAQ), US Environmental Protection Agency. Retrieved from http://www.epa.gov/otaq/models/moves/index.htm

EPA. (2014b). *NONROAD Model (nonroad engines, equipment, and vehicles)*. US Environmental Protection Agency. Retrieved from http://www.epa.gov/otaq/nonrdmdl.htm

European Parliament. (2009). *Reduction of CO_2 emissions from light-duty vehicles: emission performance standards for new passenger cars (1753/2000/EC)*. Final legislative act, European Parliament. Retrieved from http://www.europarl.europa.eu/oeil/file.jsp?id=5582632

European Commission. (2010). *Mayors' sustainable energy action plan (SEAP)*. Retrieved from http://www.covenantofmayors.eu/IMG/pdf/seap_guidelines_en-2.pdf

Eyring, V., Isaksen, I. S. A., Terje, B., Collins, W. J., Corbett, J., Endresen, O., et al. (2010). Transport impacts on atmosphere and climate: Shipping. *Atmospheric Environment, 44*(34), 4735–4771.

Field, B. C., & Field, M. K. (2009). *Environmental Economics* (5th ed.). New York: McGraw-Hill.

FOEI. (2008). *The price of carbon: What should it be and why? Friends of the Earth seminar, Mercer Suite, Royal Society, 8th July 2008*.

Fuglestvedt, J., Berntsen, T., Myhre, G., Rypdal, K., & Skeie, R. B. (2008). Climate forcing from the transport sectors. *Proceedings of the National Academy of Sciences of the United States of America, 105*, 454–458.

Golove, W. H., & Eto, J. H. (1996). *Market barriers to energy efficiency: A critical reappraisal of the rationale for public policies to promote energy efficiency*. Berkeley, CA: Lawrence Berkeley Laboratory. LBL-38059.

Hanley, N., Shogren, J., & White, B. (2006). *Environmental economics: In theory and practice* (2nd ed.). London: Palgrave Macmillan.

Hirst, E., & Brown, M. (1990). Closing the efficiency gap: Barriers to the efficient use of energy. *Resources, Conservation and Recycling, 3*, 267–281.

Huntington, H. G., Schipper, L., & Sanstad, A. H. (1994). Editor's introduction. *Energy Policy, 22* (10), 795–797.

IEA. (2009). *Energy technology transitions for industry: Strategies for the next industrial revolution*. Paris: International Energy Agency.

IPCC. (2007). *Climate change 2007: The physical science basis—Contribution of Working Group I to the fourth assessment, report of the intergovernmental panel on climate change*. Cambridge, UK: Cambridge University Press.

IPCC. (2014). Climate change 2014: Mitigation of climate change. In O. Edenhofer, R. Pichs-Madruga, Y. Sokona, E. Farahani, S. Kadner, K. Seyboth, A. Adler, I. Baum, S. Brunner, P. Eickemeier, B. Kriemann, J. Savolainen, S. Schlömer, C. von Stechow, T. Zwickel, & J. C. Minx (Eds.), *Contribution of Working Group III to the Fifth Assessment Report of the Intergovernmental Panel on Climate Change*. Cambridge, England: Cambridge University Press.

Jaffe, A., Newell, R., & Stavins, R. (2004). The economics of energy efficiency. In C. Cleveland (Ed.), *Encyclopedia of energy* (pp. 79–90). Amsterdam: Elsevier.

Jaffe, A. B., & Stavins, R. N. (1994). The energy-efficiency gap: What does it mean. *Energy Policy, 22*(10), 804–810.

Mankiw, G. N. (2006). *Brief principles of microeconomics*. Mason, OH: Thomson South-Western.

Nichols, A. L. (1994). Demand-side management: Overcoming market barriers or obscuring real costs? *Energy Policy, 22*(10), 840–847.

Pigou, A. C. (1920). *The economics of welfare* (4th ed.). London: Macmillan.

Psaraftis, H. N., & Kontovas, C. A. (2009). CO2 emissions statistics for the world commercial fleet. *WMU Journal of Maritime Affairs, 8*(1), 1–25.

Psaraftis, H. N., & Kontovas, C. A. (2010). Balancing the economic and environmental performance of maritime transportation. *Transportation Research Part D: Transport and Environment, 15*(8), 458–462.

Psaraftis, H. N., & Kontovas, C. A. (2013). Speed models for energy-efficient maritime transportation: A taxomomy and survey. *Transportation Research Part C: Emerging Technologies, 26*, 331–351.

Schelling, T. (2007). Climate change: the uncertainties, the certainties and what they imply about action. *The Economists' Voice, 4(3)*, Article 3.

Schleich, J. (2009). Barriers to energy efficiency: A comparison across the German commercial and services sector. *Ecological Economics, 68*, 7.

Smith, T. W. P., Jalkanen, J. P., Anderson, B. A., Corbett, J. J., Faber, J., Hanayama, S., et al. (2014). *Third IMO GHG Study 2014*. London: International Maritime Organization (IMO).

Smokers, R., Buck, A., & van Valkengoed, M. (2009). *GHG reduction in transport: An expensive option? Marginal abatement costs for greenhouse gas emission reduction in transport compared with other sectors*. Delft, The Netherlands: CE Delft.

Sorrell, S., Mallett, A., & Nye, S. (2011, October). *Barriers to industrial energy efficiency: A literature review* (Working Paper). United Nations Industrial Development Organization (UNIDO): Development policy, statistics, and research branch.

Sorrell, S., O'Malley, E., Schleich, J., & Scott, S. (2004). *The economics of energy efficiency—Barriers to cost-effective investment*. Cheltenham, UK: Edward Elgar.

Stern, N. (2006). *Stern review on the economics of climate change. Executive summary*. London: HM Treasury.

Stiglitz, J. (2006). A new agenda for global warming. *The Economists' Voice, 3(7)*, Article 3.

Sutherland, R. J. (1991). Market barriers to energy efficiency investments. *The Energy Journal, 12*(3), 15–34.

Ubbels, B., Rietveld, P., & Peeters, P. (2002). Environmental effects of a kilometre charge in road transport: An investigation for the Netherlands. *Transportation Research Part D: Transport and Environment, 7*, 255–264.

UNFCCC. (1997). Kyoto protocol to the United Nations Framework Convention on Climate Change. Retrieved from http://unfccc.int/resource/docs/convkp/kpeng.htm

Weber, L. (1997). Some reflections on barriers to the efficient use of energy. *Energy Policy, 25*(10), 833–835.

Woodward, R., & Bishop, R. (1997). How to decide when experts disagree: Uncertainty-based choice rules in environmental policy. *Land Economics, 73*, 492–507.

Chapter 3
Green Corridors Basics

George Panagakos

Abstract The purpose of this chapter is to introduce the concept of 'green corridors' as a means to develop integrated, efficient and environmentally friendly transportation of freight between major hubs and by relative long distances. The basis of this material is work conducted in the context of the EU SuperGreen project, which aimed at advancing the green corridor concept through a benchmarking exercise involving Key Performance Indicators (KPIs). The chapter discusses the available definitions of green corridors and identifies the characteristics that distinguish a green corridor from any other efficient surface transportation corridor. After providing examples of green corridor projects in Europe, it focuses on the KPIs that have been proposed by various projects for monitoring the performance of a freight corridor. Emphasis is given to the SuperGreen KPIs, covering the economic, technical, environmental, social and spatial planning aspects of freight logistics, as they have been scrutinized extensively by stakeholders in order to keep their number within practical and operable limits. In addition, the chapter presents the performance monitoring methodology that was developed by SuperGreen in an effort to close the gap of earlier works. The lessons learned from SuperGreen lead to a revised methodology suitable for monitoring the performance of a corridor.

Abbreviations

API	Aggregate performance indicator
BGLC	Bothnian Green Logistics Corridor
CO_2	Carbon dioxide
CO_2-eq	Carbon dioxide equivalent unit
CPI	Consumer price index
DEFRA	Department for Environment, Food and Rural Affairs (UK)
EC	European Commission

G. Panagakos (✉)
Technical University of Denmark, Lyngby, Denmark
e-mail: geopan@transport.dtu.dk

H.N. Psaraftis (ed.), *Green Transportation Logistics*, International Series in Operations Research & Management Science 226,
DOI 10.1007/978-3-319-17175-3_3

EDU	Equivalent Delivery Unit
EP&C	European Parliament & Council
ERTMS	European Rail Traffic Management System
EU	European Union
EWTC	East-West Transport Corridor
GHG	Greenhouse gas
HC	Non-methane hydrocarbon
HES	Household Expenditure Survey
ICT	Information and Communication Technology
ITS	Intelligent Transport Systems
JRC	Joint Research Centre, European Commission
KPI	Key performance indicator
LCA	Life cycle assessment
LTM	Lands Trafik Modellen (Danish National Transport Model)
MoU	Memorandum of Understanding
NECL	North East Cargo Link
NOx	Nitrogen oxides (NO and NO_2)
NTM	Network for Transport Measures
PM	Particulate matter
RNE	RailNetEurope
RTD	Research and Technological Development
SCANDRIA	Scandinavian-Adriatic corridor for innovation and growth
SGCI	Swedish Green Corridors Initiative
SO_2	Sulphur dioxide
SoNorA	South-North-Axis corridor in central Europe
STEEP	Socio-cultural, Technological, Economic, Ecological, and Political
TEN-T	Trans-European Transport Network
TOE	Tonne of oil equivalent

3.1 Introduction

The purpose of this chapter is to introduce the concept of green corridors and present a method for monitoring the performance of a freight transportation corridor in order to: (i) obtain a better understanding of its present status, (ii) identify areas for improvements, (iii) observe changes over time, and (iv) compare with benchmarks.

The basis of this material is work conducted in the context of the EU SuperGreen project and therefore the geographical setting of the chapter is Europe. Much of the material of the chapter is an expanded version of the so-called *Green Corridors Handbook—Vol. I* (Moyano et al., 2012) and *Vol. II* (Panagakos et al., 2012) published by SuperGreen. Shorter versions of this material have appeared in Panagakos and Psaraftis (2014) and Panagakos, Psaraftis, and Holte (2015).

The general objective of the SuperGreen project has been to support the development of sustainable transportation networks by fulfilling requirements covering environmental, technical, economic, social and spatial planning aspects. More specifically the project aimed at:

- giving overall support and recommendations on green corridors to the EU's *Freight Transport Logistics Action Plan*,
- encouraging co-modality for sustainable solutions,
- benchmarking green corridors based on selected Key Performance Indicators (KPIs) covering all aspects of transportation operations and infrastructure (emissions, internal and external costs),
- conducting a programme of networking activities between stakeholders (public and private),
- delivering policy recommendations at a European level for advancing the development of green corridors, and
- providing recommendations concerning new calls for R&D proposals to support the development of green corridors.

This chapter attempts to clarify the concept of a green corridor as much as possible and present a methodology for monitoring its performance through a set of selected KPIs. Other work in the project is presented in other book chapters, and the basic identity of the project appears in Annex I to this book.

In terms of scope, it has to be clarified that the chapter deals only with surface freight transportation, noting however that the quality of transportation and logistics services is also affected by passenger transportation competing for route capacity. Aviation is outside the scope of this analysis (for this see Chap. 13), as is the use of pipelines for liquid cargoes.

The rest of this chapter is structured as follows. Section 3.2 discusses the physical and functional elements of a transportation corridor. Section 3.3 presents the available definitions of a green transportation corridor and explains the benefits associated with this concept in the search for win-win solutions. Section 3.4 provides a brief presentation of the most important green corridor projects in Europe. Sections 3.5 and 3.6 are devoted to monitoring a corridor's performance. The former presents the KPIs that SuperGreen and other projects have suggested, while the latter focuses on the relevant benchmarking methodology. The chapter ends with a set of guidelines for corridor benchmarking.

3.2 Transportation Corridors

Despite being used for years as a concept, there is no precise definition for a 'transportation corridor'. The World Bank publication *Best Practices in Management of International Trade Corridors* (Arnold, 2006) provides a descriptive definition that suits the way this term is used here. According to this definition, transportation corridors have both physical and functional elements. In terms of their physical dimension:

- Transportation corridors include one or more routes that connect centers of economic activity.
- The routes have different alignments but common transfer points and common end points, which are gateways that allow traffic to enter or exit the corridor.
- The routes are composed of the links over which the transportation services travel and the nodes that interconnect the transportation services.
- Some corridors are uni-modal, but most involve multiple modes.
- Some corridors are relatively short and defined by a principal gateway like a port; others are defined by the region they serve; still others are defined as part of a network serving a larger region.

As for their functional dimension:

- Transportation corridors provide transportation and other logistics services that promote trade among the cities and countries along the corridor. In fact, most transportation corridors are developed to support regional economic growth. It is for this reason that many transportation corridors are associated with corresponding trade and economic corridors.
- Transportation corridors can be domestic or international.
- A domestic corridor is a designated set of routes within the national transportation network that is used to distribute goods within the country. It includes links and nodes for the various modes as well as nodes that connect different modes and different service areas.
- An international transportation corridor may serve the foreign trade of a single country or several neighbouring countries. It may also connect countries that are separated by one or more transit countries or provide a landlocked country with access to the sea.

In relation to this last distinction, it should be mentioned that the international transportation corridors consist of a number of national ones. As such, they are often characterized by competing functions, conflicting objectives, multiple jurisdictions and different funding schemes for their development and maintenance. On the other hand, they are usually associated with larger volumes of cargo and greater impact on the economies involved.

Corridor A, the corridor from Rotterdam to Genoa is a good example of an international transportation corridor in the European context (refer to Fig. 3.1). It stretches from the sea ports of Rotterdam, Zeebrugge and Antwerp to the port of Genoa, right through the heart of the EU along the so-called "Blue Banana". This is the most heavily industrialized North-South route in Central Europe and connects Europe's prime economic regions.

The "Blue Banana" includes economically strong urban centers such as Rotterdam, Amsterdam, Duisburg, Cologne, Frankfurt, Mannheim, Basle, Zurich, Milan and Genoa. All these centers are served and connected by the corridor, also indirectly including London and Brussels. The countries directly involved are The Netherlands, Belgium, Germany, Switzerland and Italy.

Fig. 3.1 Rail Corridor A serving the "Blue Banana" region. *Source*: Corridor Rhine-Alpine

This outstanding position together with the resulting fact that this corridor carries by far the greatest transport volume in Europe, makes the Rotterdam-Genoa route with its branch to Zeebrugge and Antwerp the pioneer for international rail freight transportation in Europe.

3.3 The 'Green corridor' Concept

'Green corridors' are not defined any more precisely than transportation corridors are, in fact one of the most important contributions of ongoing research on the topic would be to develop an explicit and workable definition of the term.

The concept was introduced in 2007 by the *Freight Transport Logistics Action Plan* of the European Commission (EC, 2007). According to this document:

> ... [green] transport corridors are marked by a concentration of freight traffic between major hubs and by relatively long distances ...
>
> ... Industry will be encouraged along these corridors to rely on co-modality[1] and on advanced technology in order to accommodate rising traffic volumes, while promoting environmental sustainability and energy efficiency ...
>
> ... Green transport corridors will ... be equipped with adequate transhipment facilities at strategic locations ... and with supply points initially for bio-fuels and, later, for other forms of green propulsion ...
>
> ... Green corridors could be used to experiment with environmentally-friendly, innovative transport units, and with advanced Intelligent Transport Systems (ITS) applications ...
>
> ... Fair and non-discriminatory access to corridors and transhipment facilities should be ensured in accordance with the rules of the Treaty.

Some years later, the Swedish Logistics Forum worked out a more structured definition (Fastén & Clemedtson, 2012). According to them:

> Green Corridors aim at reducing environmental and climate impact while increasing safety and efficiency. Characteristics of a green corridor include:
>
> - sustainable logistics solutions with documented reductions of environmental and climate impact, high safety, high quality and strong efficiency,
> - integrated logistics concepts with optimal utilization of all transport modes, so called co-modality,
> - harmonized regulations with openness for all actors,
> - a concentration of national and international freight traffic on relatively long transport routes,
> - efficient and strategically placed transhipment points, as well as an adapted, supportive infrastructure, and
> - a platform for development and demonstration of innovative logistics solutions, including information systems, collaborative models and technology.

A direct comparison between the two definitions reveals the following differences:

- The Swedish definition includes 'high safety' in the list of characteristics, referring to social acceptance, the third pillar of sustainability as it appears in the strategic document *Europe 2020* (EC, 2010). On the contrary, the EU definition confines itself to the other two dimensions of sustainability; those of economic and environmental efficiency.
- The Swedish definition makes reference also to harmonized regulations as a necessary feature of a green corridor.
- Although both definitions mention technology as a green corridor element, only the EU one makes direct reference to alternative fuels and green propulsion.

[1] In the EU transport policy documents, the term co-modality is used to refer to the "use of different transport modes on their own and in combination" in the aim of obtaining "an optimal and sustainable utilization of resources".

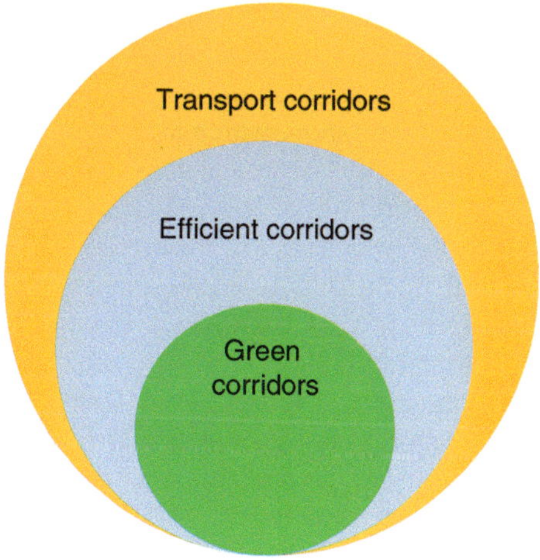

Fig. 3.2 Green corridors as a subset of efficient corridors. *Source*: Panagakos and Psaraftis (2014)

Despite their differences, the two definitions share an important aspect of green corridors: these corridors are more than just economically efficient and they are more than just environmentally sustainable; they are both economically efficient and environmentally sustainable. It is for this reason that green corridors enjoy a central position in green freight logistics and also a central role in this book's search for win-win solutions.

If, for simplicity purposes, we consider safety as a pre-condition constraining economic efficiency, then green corridors comprise a subset of the efficient ones. Figure 3.2 depicts this notion schematically.

What are, then, the specific characteristics that distinguish a green corridor from an otherwise efficient one? To answer this question, one has to merge the two lists of characteristics presented above into a single one and exclude the features that pertain to any efficient corridor. The following green characteristics result from this exercise:

(a) Reliance on co-modality, i.e. the efficient use of different modes on their own and in combination, which in turn requires:
   - adequate transhipment facilities at strategic locations; and
   - integrated logistics concepts.
(b) Reliance on advanced technology allowing use of alternative clean fuels (in addition to energy efficiency that can be viewed as a characteristic of an efficient corridor anyway).
(c) Development and demonstration capabilities of environmentally-friendly and innovative transportation solutions, including advanced telematics applications.
(d) Collaborative business models.

The last question to address in this section relates to the expected benefits of this new concept. What is it that makes the green corridors so special?

The basic principle relates to the consolidation of large volumes of freight for transportation over long distances, in between the so-called first and last miles. This is a prerequisite for improving the competitiveness of modes like rail and waterborne transportation, which are environmentally friendlier than trucks, on the one hand, and exhibit spare capacity, on the other. Increased competitiveness leads to higher possibilities of engaging trains and ships in freight logistics. In turn, the shift of cargoes away from European roads is expected to alleviate the serious congestion problem that this transportation mode faces, producing positive externalities to the other users of the road network through improvements in reliability and reduction of transportation time.

Furthermore, the scale and length of such freight corridors enable further optimization in terms of energy use and emissions for these long-hauls, resulting in additional environmental and financial (due to lower operating costs) gains. The feasibility of investments associated with establishing a network of refuelling stations for alternative fuels (biofuels, electricity, LNG, etc.) along such corridors would be improved, while the use of more energy efficient vehicles/vessels (trucks with better aerodynamic performance and new engines, longer trains, LNG-fuelled vessels, etc.) would be boosted.

Advanced ICT applications like automatic guidance systems would further improve the utilization and performance of existing infrastructure through minimizing congestion and accidents. ICT would also help integrating regular rail, sea and inland waterway services with road transportation which will maintain the predominant role over short and medium distances. Applications would include cargo tracking and tracing, schedule optimization and simplification of formalities related to multimodal freight transportation.

In addition, the international character of the corridors (involve at least three Member States) addresses the fragmented nature of transportation networks, especially rail, dealing with the haunting interoperability issues in geographical terms. At the same time, focusing on a subset of the network improves the chances of identifying workable solutions by limiting the overwhelming scale of the problem.

The realization of international multimodal corridors cannot be implemented without appropriate corridor structures. These structures will bring together the Commission, Member States, the regions, the local authorities, but also the infrastructure owners and managers, transportation operators, shippers, financiers and, when appropriate, neighbouring countries. The involvement of such structures is absolutely necessary in promoting multimodal logistics, where lack of coordination comprises probably the most persisting problem.

The systematic exchange of information between national authorities would further enable the uniform enforcement of common safety, security, environmental and social legislation which, in turn, would benefit the users of transportation services and their providers through full market opening and the provision of a level playing field.

Last but not least, the establishment of corridors that enhance the efficiency of transportation modes (alone and in combination) through better utilization of resources is expected to limit the considerable investments needed for expanding the capacity of the transportation networks in an environment of budgetary consolidation and increasing public opposition to major transportation infrastructure projects especially in the vicinity of urban areas.

3.4 Green Corridor Projects in Europe

Those who follow the evolution of the EU transport policy cannot escape noticing that the corridor approach gains more and more importance as a response to the new and old challenges that the common transport policy faces in Europe (refer to the Preface for a discussion on these challenges).

- In March 2005, the European Commission and the railway sector agreed on a MoU referring to the implementation of the European Rail Traffic Management System (ERTMS—a signalling system that will replace all those currently in use throughout Europe) on six corridors to define a European migration strategy for the deployment of ERTMS (refer to Fig. 3.3).
- In October 2007, The European Commission published its *Freight Transport Logistics Action Plan*, which introduced the concept of 'green corridors' as a means to improve the efficiency and sustainability of freight transportation in Europe.

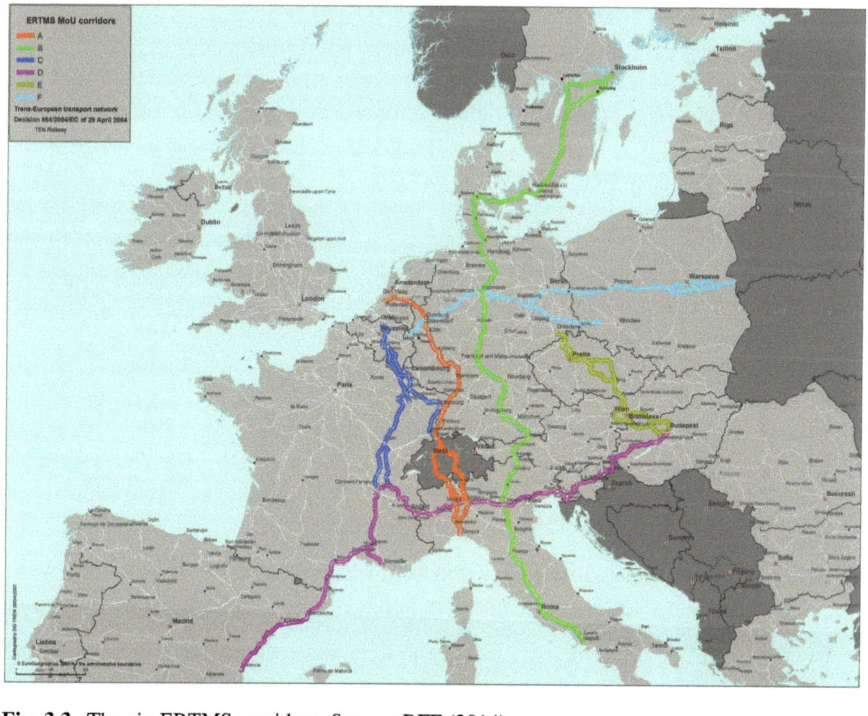

Fig. 3.3 The six ERTMS corridors. *Source*: RFF (2014)

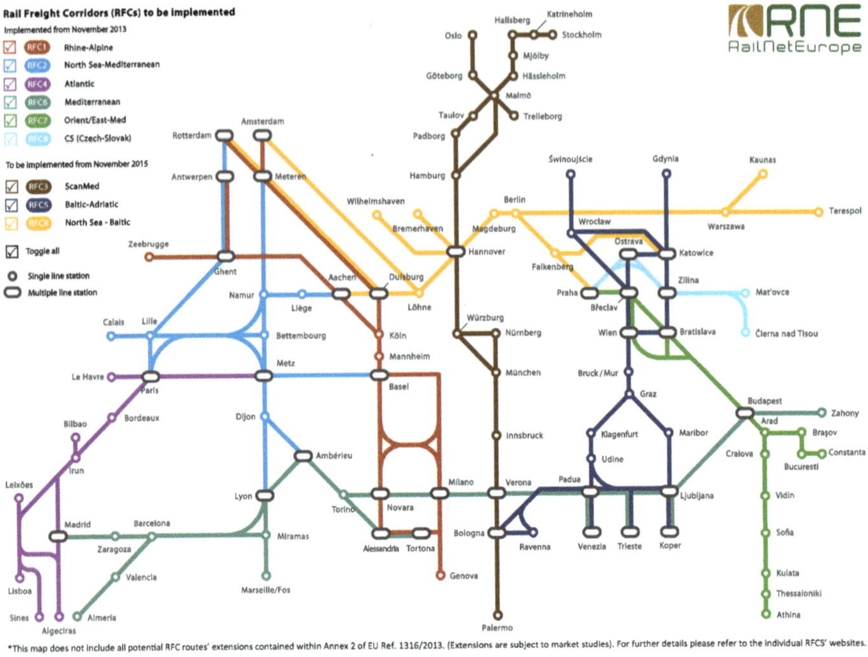

Fig. 3.4 The nine Rail Freight Corridors. *Source*: RNE (2014)

- In November 2010, the European Parliament and the Council adopted the EU Regulation No 913/2010 concerning a European rail network for competitive freight (EP&C, 2010). This Regulation defines nine initial Rail Freight Corridors (RFCs) along which, sufficient priority among freight trains, is given to those crossing at least one border (refer to Fig. 3.4).
- In March 2011, the European Commission in describing its vision of future transport and the corresponding strategy for the next decade, included in the latest White Paper on transport 'multimodal freight corridors' as a means to improve governance and to support pilot projects for innovative and clean transportation services (EC, 2011a).
- In December 2013, the European Parliament and the Council adopted the EU Regulation No 1315/2013 on Union guidelines for the development of the Trans-European Transport Network (TEN-T), which introduced the concept of 'core network corridors' as an instrument to facilitate the coordinated implementation of the parts of the TEN-T with the highest strategic importance (EP&C, 2013). The nine TEN-T core network corridors are shown in Fig. 3.5.

At a lower level, the initiatives listed below comprise a selection of the most important among a wide range of corridor applications in Europe:

- In December 2002, Germany, Austria and Italy adopted the Brenner Action Plan aiming at a significant and sustainable increase in intermodal volume along the Brenner corridor, one of the most trafficked international transit corridors,

Fig. 3.5 The nine TEN-T core network corridors. *Source*: EC (2014)

where—on a length of only 448 km between Munich and Verona—3 countries and thus railway infrastructures and the Alps are being bridged (Mertel & Sondermann, 2007).

- In January 2003, the Ministries of Transport of The Netherlands, Germany, Switzerland and Italy agreed on a MoU establishing an international working group to develop a comprehensive action plan aiming at bringing about numerous quantitative and qualitative improvements on the rail corridor from Rotterdam to Genoa (Corridor A/IQ-C, 2011). The so-called Corridor A was born (refer also to Sect. 3.2).

Fig. 3.6 The East-West Transport Corridor. *Source*: Fastén and Clemedtson (2012)

- In 2006, 42 partners (local, regional and national authorities, universities, harbours and private stakeholders) from Denmark, Lithuania, Russia and Sweden joined forces to strengthen transportation development along the so-called "East-West Transport Corridor—EWTC" through infrastructure improvements, new solutions for business, logistics and cooperation between researchers (refer to Fig. 3.6). The success of EWTC led to the follow up project EWTC II, which aims at transforming the EWTC into a green corridor in line with the EU policy.
- In 2008, the Swedish Green Corridors Initiative (SGCI) was introduced focusing on transportation routes and collaboration among shippers, forwarders, industry and haulers in order to optimize the use of available transportation capacity (Wålhberg et al., 2012). Two green corridors were established by this initiative:
  - The Oslo–Randstad corridor that follows one of Northern Europe's most important freight routes (Fig. 3.7). GreCOR, an Interreg IVB project running in the period 2012–2015, aims to: (i) improve knowledge about the logistic needs and conditions along this corridor, and (ii) implement the first green corridor in the North Sea Region (Hansson & Hansson, 2014). The project uses a collaborative approach to enhance co-modality and influence infrastructural development in the region, including the hinterland of the corridor's hubs. Among other results, GreCOR will develop a methodology for assessing the environmental performance of a corridor and a web-based market place for route planning.

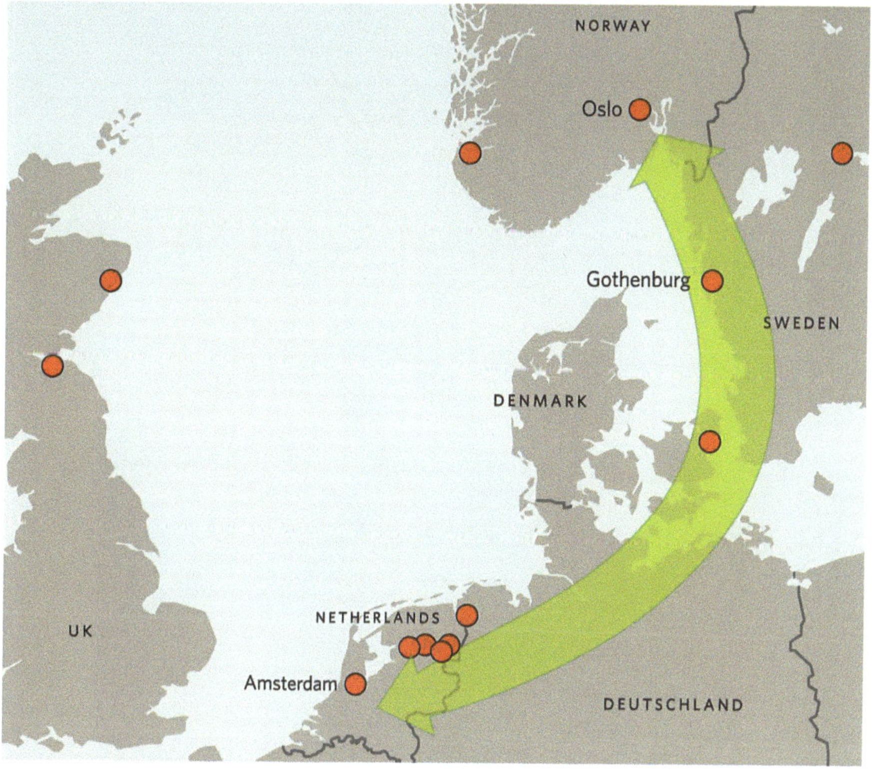

Fig. 3.7 The GreCOR corridor. *Source*: Hansson and Hansson (2014)

- The Bothnian Green Logistic Corridor (BGLC). Twenty-nine partners across five countries—Sweden, Finland, Norway, Germany and Poland—were working during 2012–2014 within different fields to develop BGLC (Fig. 3.8) into an efficient, reliable and green transportation corridor, connecting northern Scandinavia's raw materials with the markets in Europe (Södergren, Sorkina, Kangevall, Hansson, & Malmquist, 2012). Project activities involved: mapping cargo flows and future needs, elimination of bottlenecks, introduction of new intermodal solutions for increased flexibility, examination of the regional and economic effects of corridor development, identification of strategically important nodes, and the design of innovative business models and pilot projects enhancing collaboration between private and public stakeholders.

• In 2009, the Scandria project was introduced, covering the corridor from Region of Halland (Sweden), via Zealand (Denmark) to Mecklenburg-Vorpommern and Berlin (Germany). During 3 years, 19 partners and 16 associated partners from business, national, regional and local administration, and research institutions fostered green and innovative development between Scandinavia and Eastern Germany.

Fig. 3.8 The BGLC corridor. *Source*: Södergren et al. (2012)

The project also cooperated with SoNorA (South-North-Axis corridor in central Europe), extending coverage from Berlin to the Adriatic Sea (Friedrich, 2012).

- In 2009, the TransBaltic project was also introduced covering corridors across the Baltic Sea. The overall objective of this 3-year project was to provide regional level incentives for the creation of a sustainable multimodal transportation system in the Baltic Sea Region, through joint transportation development measures and jointly implemented business concepts (TransBaltic, 2012).
- In 2010, the Midnordic Green Transport Corridor project of NECL (North East Cargo Link) was initiated with the aim to address obstacles along the transportation corridor that stretches through the middle parts of Norway, Sweden and Finland (Fig. 3.9). Other objectives included carrying out pre-investment

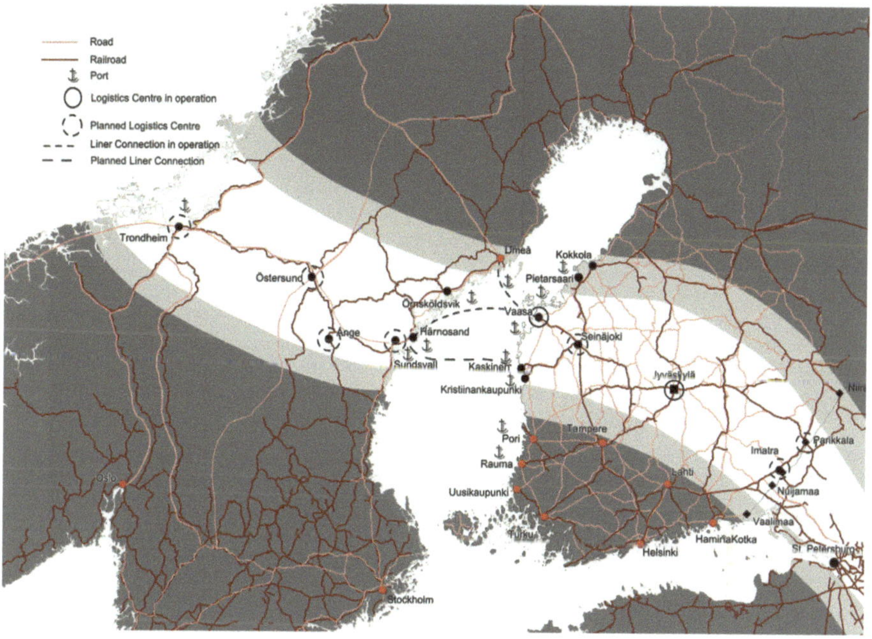

Fig. 3.9 The Midnordic transport network. *Source*: Kokki (2013)

studies, developing transportation solutions, marketing of the corridor on a macro region level and developing an ICT application (portal) in close cooperation with the national transportation authorities and industry over the national borders (Kokki, 2013).

- In 2010, SuperGreen, a Coordination and Support Action co-financed by the EC's 7th Framework Programme of Research and Technological Development (RTD), was launched. The main objective of this 3-year project was to assist in further defining and developing the green corridor concept. Basic information on this project can be found in Annex I to this book. Its central activity was the development of a corridor benchmarking methodology using a set of Key Performance Indicators (KPIs) that are suitable for monitoring the sustainable development goals of the European Union. The results of this activity will be presented in the following sections of this chapter; the discussion here will be confined to the SuperGreen corridors selected for applying the proposed methodology.

The project compiled an initial list of 60 potential corridors on the basis of the TEN-T priority projects, the pan-European transport network and proposals made by the project's industrial partners. After 2 consolidation rounds, the number of candidate corridors was reduced to 30. A survey was carried out to gather information on these 30 corridors. Based on the information gathered and criteria like corridor length, population affected, freight volume, types of goods transported, number and seriousness of bottlenecks, transport and information technology used,

Table 3.1 The nine SuperGreen corridors

Nicknames	Acronym	Corridor description
Brenner	BerPal	Malmö-Trelleborg-Rostock/Sassnitz-Berlin-Munich-Salzburg-Verona-Bologna-Naples-Messina-Palermo
		Branch A: Salzburg-Villach-Trieste (Tauern axis)
		Branch B: Bologna-Ancona/Bari/Brindisi-Igoumenitsa/Patras-Athens
Finis Terrae	MadPar	Madrid-Gijon-Saint Nazaire-Paris
		Branch A: Madrid-Lisboa
Cloverleaf	CorMun	Cork-Dublin-Belfast-Stranraer
		Branch A: Munchen-Friedewald-Nuneaton
		Branch B: West Coast Main line
Edelweiss	HelGen	Helsinki-Turku-Stockholm-Oslo-Göteborg-Malmö-Copenhagen (Nordic triangle including the Oresund fixed link)—Fehmarnbelt—Milan—Genoa
Nureyev	RotMos	Motorway of Baltic sea
		Branch: St. Petersburg-Moscow-Minsk-Klapeida
Strauss	RhiDan	Rhine/Meuse-Main-Danube inland waterway axis
		Branch A: Betuwe line
		Branch B: Frankfurt-Paris
Two Seas	AthDre	Igoumenitsa/Patras-Athens-Sofia-Budapest-Vienna-Prague-Nurnberg/Dresden-Hamburg
Mare Nostrum	SinOde	Odessa-Constanta-Bourgas-Istanbul-Piraeus-Gioia Tauro-Cagliari-La Spezia-Marseille-(Barcelona/Valencia)-Sines
		Branch A: Algeciras-Valencia-Barcelona-Marseille-Lyon
Silk Way	CNHam	Shanghai-Le Havre/Rotterdam-Hamburg/Gothenburg-Gdansk-Baltic ports-Russia Branch: Xangtang-Beijing-Mongolia-Russia-Belarus-Poland-Hamburg

Source: Salanne et al. (2010)

and assessment of the supply chain management, a pre-selection of 15 corridors was made. A geographic and modal balance was ensured among these pre-selected corridors. The aim at this stage was to select the ones with the highest "greening potential" rate.

Further information was collected on these 15 pre-selected corridors and a deeper analysis was performed taking into consideration land use aspects like the percentage of corridor surface comprising urban and environmentally sensitive areas. The analysis resulted in a recommendation of nine corridors for final selection, which was presented to a stakeholder workshop especially arranged for this purpose. In line with comments received during the workshop, the selected corridors were modified by adding segments that exhibit advanced "greening" characteristics.

These nine corridors were given nicknames and can be depicted in Table 3.1. Figure 3.10 presents this set of corridors in metro format.

In addition to being geography- and mode-wise balanced, the resulting set of corridors comprised a mix of environmentally advanced ones on one hand, and those exhibiting a high "greening potential" on the other, thus constituting a suitable field

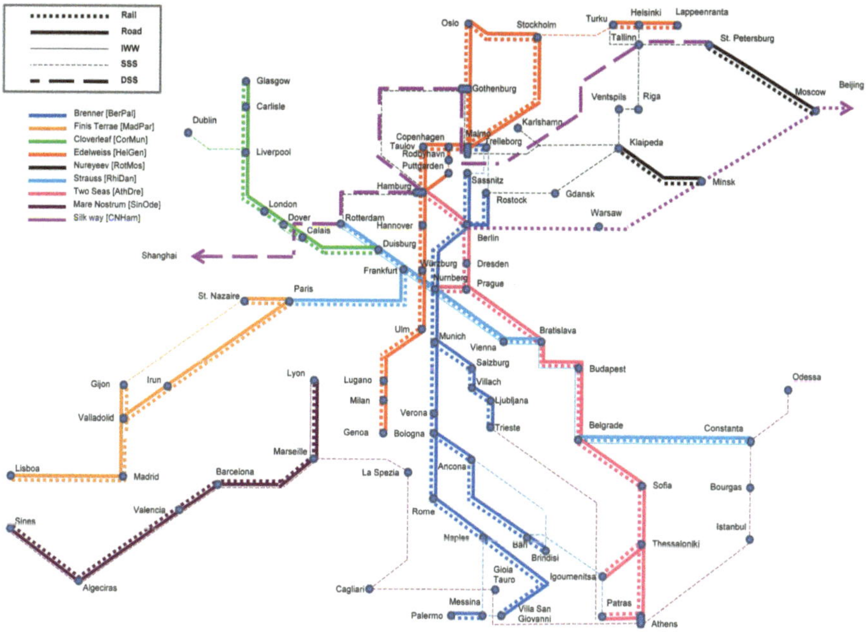

Fig. 3.10 The SuperGreen corridors in metro format. *Source*: Ilves et al. (2011)

for testing the benchmarking methodology and KPIs.[2] More details on SuperGreen corridor selection can be found at Salanne, Rönkkö, and Byring (2010).

- In 2011, the Green STRING Corridor project was launched, scheduled to run for 3 years. Its aim was to promote the potential of innovative transportation and logistics solutions for developing a green transportation corridor between the Öresund Region and Hamburg, capitalizing on the benefits of the forthcoming fixed Fehmarn Belt link. The project identified the conditions and challenges that a green transportation corridor sets for the distribution and logistics strategies of private companies, as well for cross-border planning among public authorities at a local, regional and national level (Stenbæk, Kinhult, & Hæstorp Andersen, 2014) (Fig. 3.11).

[2] It should be clarified that the selection of these corridors was made only for the purposes of the SuperGreen project and by no means has this implied any direct or indirect endorsement, either by the SuperGreen consortium or by the European Commission, of these corridors vis-à-vis any other corridor, with respect to any criteria, environmental, economic, or other.

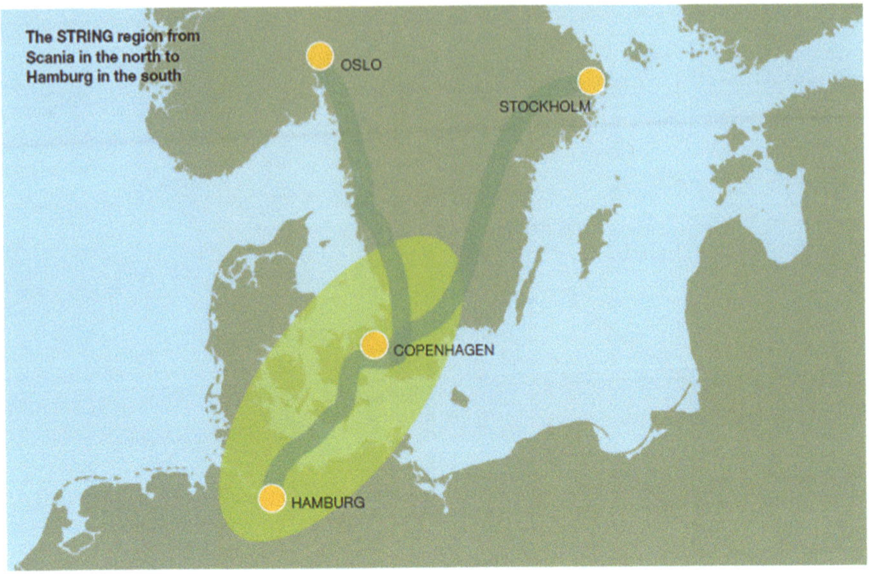

Fig. 3.11 The STRING green corridor. *Source*: Stenbæk et al. (2014)

3.5 Key Performance Indicators

Monitoring the performance of the relevant transportation corridors is a common need of all the projects of Sect. 3.4. Their performance needs to be assessed in terms of pre-specified qualities that correspond to the objectives pursued by the corridor management. Monitoring is achieved through a set of indicators which is defined either explicitly (Brenner corridor, Corridor A, EWTC, SGCI, GreCOR, BGLC, SuperGreen) or implicitly (Scandria, TransBaltic, STRING).

For example, the quality objectives of the BRAVO project (Brenner corridor) were punctuality, reliability, flexibility, customer information, employment rate of rolling stock, and reliability of transportation documents. The management of Corridor A (Rotterdam-Genoa) has selected indicators concerning traffic volume, modal split, punctuality and commercial speed. On a more theoretical basis, the World Bank proposes the use of cost, time, reliability and flexibility as corridor performance indicators (Arnold, 2006).

When discussing indicators, it should be kept in mind that KPIs ought to be:

- relevant (there should be a clear link between indicators and objectives),
- quantifiable[3] (assessed by certain units that have a numerical value attached),

[3] In cases of policy- and process-benchmarking, quantifiable indicators might need to be complemented by qualitative ones.

- clear (defined in a way that precludes misinterpretations and enables meaningful comparisons),
- simple (easy to use and compute in terms of data availability and cost),
- robust (resistant to manipulation by those responsible),
- sensitive to classified information,
- mutually exclusive and, to the extent possible, collectively exhaustive.

The three sets of indicators presented below are indicative of the different perspectives and level of detail employed.

3.5.1 *The SGCI Criteria*

The Swedish Green Corridors Initiative compiled a list of criteria for selecting, comparing and evaluating green corridor projects (SGCI, 2012). The document identifies two distinct aspects affecting a corridor's performance: (i) the framework that enables the provision of transportation services (policies and regulations, infrastructure, ICT applications, organizational issues, etc.) and (ii) the operational characteristics of the transportation services. It selects, however, to focus on the second one on the assumption that a sufficiently good framework is provided. Furthermore, the term 'green' is seen from a purely ecological perspective and the selected criteria cover the environmental dimension only.

It is interesting to note that all environmental criteria are described in both absolute and relative terms (refer to Table 3.2). The absolute measurement indicates actual emissions caused by a transportation activity and reflect the volume of transport work, while relative data describes the environmental efficiency of the transport activity.

3.5.2 *The EWTC KPIs*

In relation to SGCI, the East-West Transport Corridor (EWTC) project has advanced the KPI selection in two important ways: Firstly, the term 'green' now combines all three sustainability dimensions (economic, environmental and social efficiency). Secondly, the grouping of indicators into operational and enabling ones, which was only touched upon by SGCI, has now been strengthened. Operational indicators aim at optimizing cargo flows in the short run with regard to their overall sustainability and address the perspectives of transportation service providers, shippers and the corridor managers. On the other hand, enabling indicators aim to optimize long term development of the corridor framework and are relevant to infrastructure managers, policy makers and the corridor managers.

The EWTC scheme of KPIs appears in Table 3.3. It needs to be mentioned that, unlike the operational ones, the enabling indicators are monitored through a corridor dashboard. The dashboard highlights the need to eliminate bottlenecks that may occur either within or outside the immediate corridor region.

Table 3.2 The SGCI indicators

Transport service/technique			
		Max performance levels	
Performance area	Performance indicator	Year 2010	Year 2014
Environment			
GHG emissions$_{tot}$	CO2e [ton/year]	−x tonne CO2e/ year	−y tonne CO2e/ year
GHG productivity	CO2e [g/tkm]	x kg CO2e/tkm	y kg CO2e/tkm
NOx emissions$_{tot}$ (e.g. regulated)	NOx emissions [kg/year]	−x kg NOx/year	−y kg NOx/year
NOx emission productivity	NOx emissions [g/tkm]	x kg NOx/tkm	y kg NOx/tkm
SO_2 emissions$_{tot}$ (e.g. regulated)	SO_2 emissions [kg/year]	−x kg SO_2/year	−y kg SO_2/year
SO_2 emission productivity	SO_2 emissions [g/tkm]	x kg SO_2/tkm	y kg SO_2/tkm
HC emissions$_{tot}$ (e.g. regulated)	HC emissions [kg/year]	−x kg HC/year	−y kg HC/year
HC emission productivity	HC emissions [g/tkm]	x kg HC/tkm	y kg HC/tkm
PM emissions$_{tot}$ (e.g. regulated)	PM emissions [kg/year]	−x kg PM/year	−y kg PM/year
PM emission productivity	PM emissions [g/tkm]	x kg PM/tkm	y kg PM/tkm
Resources			
Energy use$_{tot}$	Energy use [kWh/year]	x kWh/year	y kWh/year
Energy use productivity	Energy use [kWh/tkm]	x kWh/tkm	y kWh/tkm
Requisite criteria			
Follow-up systems	Systematic plan	According to guidelines	According to guidelines
Vulnerability/redundancy plans	Systematic plan	According to guidelines	According to guidelines
Maintenance plans	Systematic plan	According to guidelines	According to guidelines
Corridor service (sum of total transport services included in the corridor)			
		Max performance levels	
Performance area	Performance indicator	Year 2010	Year 2014
Environment			
ΣGHG emissions$_{tot}$	CO2e [ton/year]	−x tonne CO2e/ year	−y tonne CO2e/ year
ΣGHG productivity	CO2e [g/tkm]	x kg CO2e/tkm	y kg CO2e/tkm
ΣNOx emissions$_{tot}$ (e.g. regulated)	Emissions [kg/year]	−x kg NOx/year	−y kg NOx/year

(continued)

Table 3.2 (continued)

Corridor service (sum of total transport services included in the corridor)			
ΣNOx emissions productivity	Emissions [g/tkm]	x kg NOx/tkm	y kg NOx/tkm
ΣSOx $emissions_{tot}$ (e.g. regulated)	SOx emissions [kg/year]	−x kg SO_2/year	−y kg SO_2/year
ΣSOx emission productivity	SOx emissions [g/tkm]	x kg SO_2/tkm	y kg SO_2/tkm
ΣHC $emissions_{tot}$ (e.g. regulated)	HC emissions [kg/year]	−x kg HC/year	−x kg HC/year
ΣHC emission productivity	HC emissions [g/tkm]	x kg HC/tkm	x kg HC/tkm
ΣPM $emissions_{tot}$ (e.g. regulated)	PM emissions [kg/year]	−x kg PM/year	−y kg PM/year
ΣPM emission productivity	PM emissions [g/tkm]	x kg PM/tkm	y kg PM/tkm
Resources			
ΣEnergy use_{tot}	Energy-use [kWh/year]	x kWh/year	y kWh/year
ΣEnergy use productivity	Energy use [kWh/tkm]	x kWh/tkm	y kWh/tkm
Requisite criteria			
Follow-up systems	Systematic plan	According to guidelines	According to guidelines
Vulnerability/ redundancy plans	Systematic plan	According to guidelines	According to guidelines
Maintenance plans	Systematic plan	According to guidelines	According to guidelines

Source: SGCI (2012)

Table 3.3 The EWTC KPIs

Performance areas	Operational indicators	Enabling indicators
Economic efficiency	Total cargo volumes	Corridor capacity
	On time delivery	
Environmental efficiency	Total energy use	Alternative fuels filling stations
	Greenhouse gases, CO2e	
	Engine standards	
	ISO 9001 dangerous goods	
Social efficiency	ISO 31 000	Safe truck parking
	ISO 39 000	Common safety rating
		Fenced terminals

Source: Fastén and Clemedtson (2012)

3.5.3 The SuperGreen KPIs

The KPI selection part of the SuperGreen project was a cumbersome procedure that drew heavily on stakeholder input. It was performed in two distinct phases. During the first phase, a process involving the compilation of a gross list of performance indicators, their categorization into five different groups and their filtering during detailed discussions among project partners resulted in an initial set of KPIs. The five KPI groups (efficiency, service quality, environmental sustainability, infrastructural sufficiency, and social issues) were formed so as to combine all three sustainability dimensions with the adequacy of the capacity, condition and administrative framework of the corridor infrastructure (the so-called 'enabling indicators' of SGCI and EWTC). These initial KPIs are presented in Table 3.4 along with their respective definition.

Table 3.4 Initial set of SuperGreen KPIs

KPIs	Units
Efficiency	
Absolute cost	€/tonne
Relative cost	€/tonne-km
Service quality	
Transport time	Hours
Reliability (time precision)	% of shipments delivered on time (within acceptable window)
Frequency of service	Number of services per year
ICT applications – Cargo tracking, availability – Cargo tracking, integration and functionality – Other ICT serv., availability – Other ICT serv., integration and functionality	Graded scale (1–5) Graded scale (1–5) Graded scale (1–5) Graded scale (1–5) Graded scale (1–5)
Cargo security	Number of incidents per total number of shipments
Cargo safety	Number of incidents per total number of shipments
Environmental sustainability[a]	
CO_2-eq	g/tonne-km
SO_2	g/1,000 tonne-km
NOx	g/1,000 tonne-km
PM_{10}	g/1,000 tonne-km
Infrastructural sufficiency	
Congestion	Average delay (hours) per tonne-km
Bottlenecks – Geography – Infrastructure capacity – Infrastructure condition – Administration	Graded scale (1–5) based on list of bottlenecks per category, accompanied by list of projects aiming at their removal/ mitigation

(continued)

Table 3.4 (continued)

KPIs	Units
Social issues	
Corridor land use	
– Urban areas	% of buffer zone[b] covered by urban areas
– Sensitive areas	% of buffer zone[b] covered by environmentally sensitive areas
Traffic safety	Sum of fatalities and serious injuries per year per million ton-km
Noise	% of corridor length above 50/55 dB

Source: Moyano et al. (2012)
[a]Well-to-wheel approach
[b]Shaped by a radius of 20 km around the median line of the corridor

Table 3.5 Revised set of SuperGreen KPIs

Indicator	Unit
Out-of-pocket costs (excluding VAT)	€/tonne-km
Transport time (or average speed)	Hours (or km/h)
Reliability of service (in terms of time precision)	% of shipments delivered within acceptable window
Frequency of service	Number of services per year
CO_2-eq emissions	g/tonne-km
SOx emissions	g/tonne-km

Source: Ilves et al. (2011)

With the aim of soliciting feedback, this initial set (together with the proposed benchmarking methodology that will be presented in the following section) was presented in three events: two regional stakeholder workshops (in Napoli, Italy and in Antwerp, Belgium) and a meeting of the project's Advisory Committee. The general consensus was that in broad terms the proposed KPIs cover all basic facets of the problem. However, there was also a general sense that the indicators were too ambitious and there was a need to simplify them so that the set be practical. In that sense, reducing the set of KPIs to a more manageable one was considered as a desirable outcome.

Following an internal round of KPI screening, a revised set was presented to a third regional SuperGreen workshop, organized in Malmö, Sweden and hosted by the Swedish Transport Administration. The aim was to set a basis for collaboration with the numerous green corridor initiatives in the Baltic region and take advantage of an audience directly or indirectly exposed to the green corridor concept. The KPI set that resulted from this process is the one of Table 3.5. This set was reaffirmed at a fourth regional stakeholder workshop of the project in Sines, Portugal.

It is worth noting that four of the six indicators concern economic efficiency (transportation costs accompanied by three KPIs related to quality of service—time, reliability and frequency), while the remaining two reflect environmental concerns (GHG and sulphur emissions). The social aspects are absent, probably signifying a secondary role that stakeholders attach to them when it comes to freight logistics.

3.6 Corridor Benchmarking

3.6.1 Early Works

Unlike KPIs, corridor benchmarking is not a very popular topic in the literature. Most benchmarking work stops at the transport chain level. The few exceptions found in the bibliography are presented below.

The World Bank's *Best Practices in Management of International Trade Corridors* contains a first attempt in assessing the performance of a corridor (Arnold, 2006). On the basis that a corridor is generally composed of several alternative routes, the method focuses on measuring the performance of each route. Refer to Fig. 3.12 for a schematic depiction of the methodology.

In the event that no information on market segments, commodity groups, shipment types and modal split is available (which is normally the case), the analysis starts with the construction of a sample. The paper does not specify the sample's configuration. However, the need to compute cost, time and reliability indicators for the sample, which comprises the next step of the methodology, makes

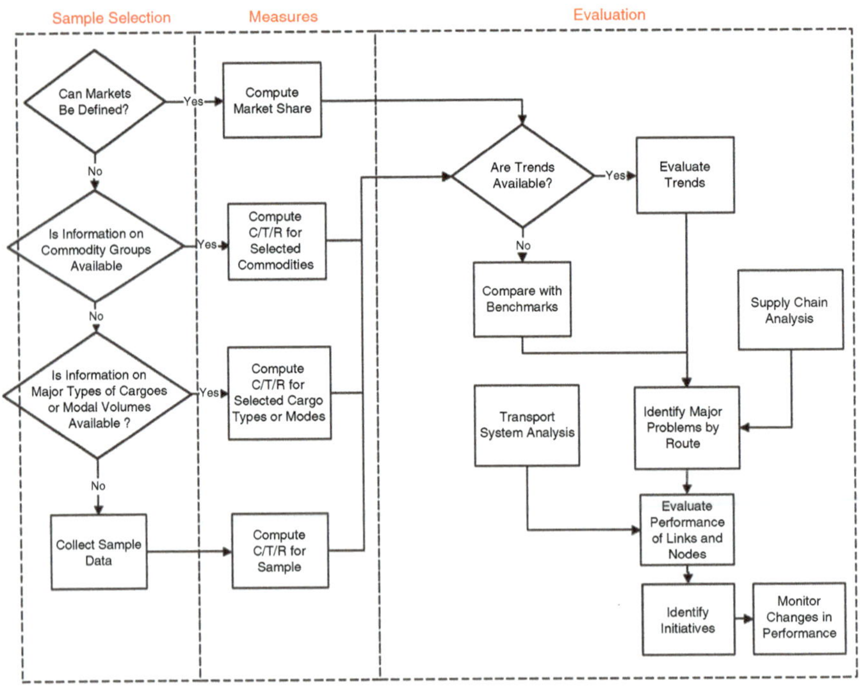

Fig. 3.12 Evaluation of corridor performance. *Source*: Arnold (2006)

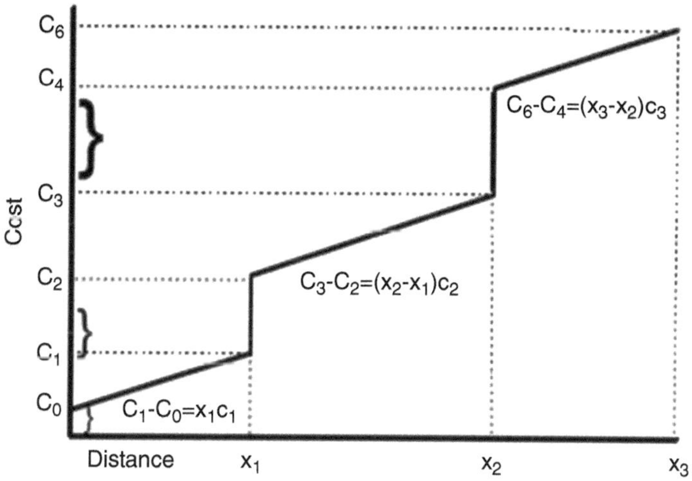

Fig. 3.13 Transport cost for a 3-leg chain. *Source*: Arnold (2006)

one infer that the sample is composed of transport chains.[4] After considering trends, the comparison with benchmarks leads to the identification of problems on a route basis. No details are given on how the chain-level indicators are transformed into route-level ones; a reference to supply chain analysis might be relevant. As a next step, route problems are translated into performance deficiencies at the links and nodes. No attempt is made to compute indicators at the corridor level. The absence of environmental considerations from the analysis is also noticeable.

An interesting contribution of this World Bank publication relates to the way cost and time figures of links and nodes are combined to form chain- and corridor-level indicators.

The cost of a transport chain consists of all out of-pocket costs plus either the insurance costs or any loss or damage to cargo while en route. The costs incurred in a transport link can be described as a combination of a fixed cost and a variable cost that depends on the distance travelled. The average transit cost for a transport chain consisting of three links can then be depicted as in Fig. 3.13. The vertical lines represent the costs incurred at the nodes and any fixed costs associated with using the subsequent link. The sloping lines represent the costs incurred while transiting a link with the slope proportional to the average variable cost.

Similarly, time can be shown in the form of the graph of Fig. 3.14, as a function of distance along the chain. The average transport time of a chain is defined as the time needed to complete all activities essential for moving from the origin to the destination of the chain. The sloping lines represent the time spent moving along a link; the slope is inversely proportional to the average link speed. The vertical lines

[4] It is worth noting that the flexibility indicator that has been proposed as a KPI earlier in the chapter does not enter the methodology, presumably due to its rather qualitative nature.

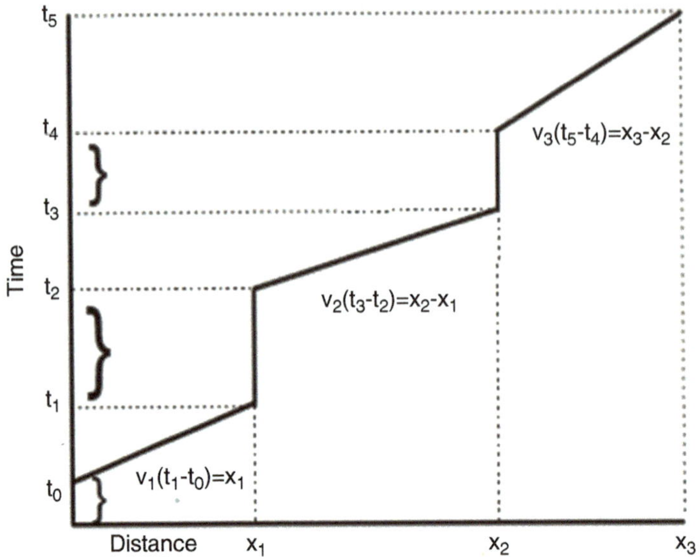

Fig. 3.14 Transport time for a 3-leg chain. *Source*: Arnold (2006)

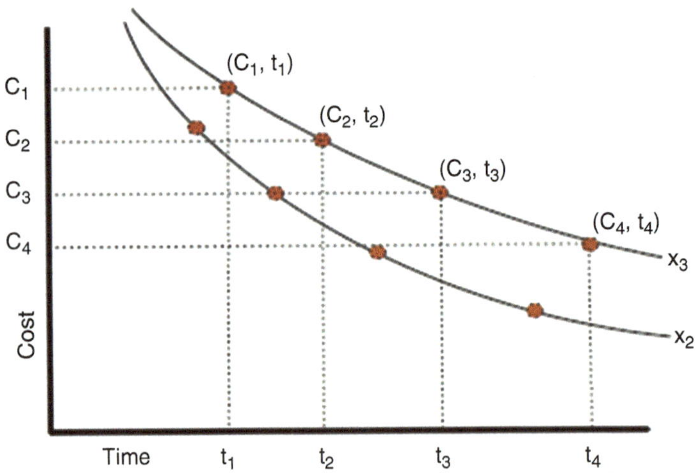

Fig. 3.15 Corridor time-cost options. *Source*: Arnold (2006)

represent the time spent at the nodes and include the delays associated with the frequency of services, with congestion at the nodes and with other required activities like cargo handling, transhipment, vehicle/cargo inspection, etc.

Each transport chain, then, can be represented by its average cost and time for transit. A corridor, consisting of different combinations of routes, modes, and chains can be represented by either the average time and cost for transiting the corridor or by a curve like the graph of Fig. 3.15, which combines the time-cost pairs of all transport chains that are available in the corridor. It is conceivable that an intervention in the corridor that improves both time and cost shifts the corridor frontier down and to the left.

A different approach was followed by the BE LOGIC project a few years later. In addition to developing a methodology for benchmarking transport chains through KPIs (Kramer et al., 2009), BE LOGIC went one step further by attempting to assess the performance of the freight transportation system at a strategic level through a set of Aggregate Performance Indicators (APIs). They are higher-level characteristics than the KPIs and are expressed at a modal level, as opposed to the company/terminal/transport chain level of the KPIs. A STEEP (Socio-cultural, Technological, Economic, Ecological, and Political) analysis was used for their assessment, which was purely qualitative (BE LOGIC, 2009). The APIs proposed by BE LOGIC for the transportation services were:

- operating cost per unit of transportation activity (e.g. €/tonne-km),
- energy consumed per unit of transportation activity (e.g. toe/tonne-km),
- emissions produced per unit of transportation activity (e.g. kg of CO_2/tonne-km),
- reliability (ability of mode to offer services punctual and according to the published schedule or promised delivery date and time),
- flexibility (ability of mode to adapt to changes in demand/volume/size/timetable and to cope with serious disruptions like cancellations, strikes, etc.),
- frequency (ability of mode to offer frequent services in line with the respective demand).

Although the BE LOGIC's APIs can be modified to address all desired criteria in monitoring the performance of a corridor, they would be suitable for benchmarking purposes only if estimated on a quantitative basis which, however, was not the case.

A quantitative but equally infeasible suggestion comes from the Swedish Green Corridors Initiative presented in Sect. 3.5.1. In the lower part of Table 3.2, the chain-level SGCI indicators are summed over all transport chains using the corridor to form the corridor-level KPIs. However, as can be seen from the corridors of Sect. 3.4, they are usually defined along broad lines making it difficult to identify the flows and services that need to be examined. Even if the corridors were more precisely defined, it is certain that the necessary data does not exist or if it did, the cost of extraction and manipulation would soon exceed the expected benefits of such undertaking.

This problem was spotted by the East-West Transport Corridor project, which suggests that the KPI analysis should be limited to a number of services along the corridor that need to be wisely selected[5] (Fastén & Clemedtson, 2012). In fact, EWTC went on to offer the following advice concerning this selection:

- Always keep in mind the purpose of the analysis.
- Select corridor sections with few parallel operations enabling effective monitoring.

[5] The East-West Transport Corridor II project run in parallel with SuperGreen and a certain degree of cross-fertilization took place between them.

- Identify large and stable flows, usually connected to large industries.
- Select operations run by organizations that are willing to share information.
- Take advantage of existing systems for data collection including relevant ICT applications like fleet monitoring systems, electronic toll systems, etc.
- Focus on known difficulties in meeting sustainability criteria, e.g. trade imbalances, old vintage engines etc.

The methodology proposed by EWCT includes the following steps:

Step 1. Produce a clear goal statement defining the purpose of the analysis. It should also describe the intended use of the results in meeting the stated goal.
Step 2. Define the scope of the analysis in terms of the objects to be monitored. These objects need to be described in detail in order to ensure consistency.
Step 3. Select a set of KPIs that reflect the purpose of the study and serve the monitoring needs of the selected objects.
Step 4. Set system boundaries in relation to: (i) the geographical coverage and physical boundaries of the system under examination, (ii) the activities of the transport services that comprise the sample, (iii) the activities accounted for when calculating energy consumption (e.g. life cycle), and (iv) the time period covered.
Step 5. Collect data including through secondary data sources and expert judgments in case of missing information.
Step 6. Calculate KPIs.

The approach suggested by EWTC is sensible and practical. Its only weakness relates to the fact that, as explicitly stated by Fastén and Clemedtson (2012), the proposed methodology aims to assess selected corridor components (services) rather than the corridor as such.

3.6.2 Benchmarking of the SuperGreen Corridors

A methodology aiming at benchmarking a corridor in its entirety was suggested by SuperGreen (Ilves et al., 2010, 2011). It was built around the concept of:

- decomposing the corridor into transport chains,
- benchmarking these chains using a set of KPIs, and then
- aggregating the chain-level KPIs to corridor-level ones using proper weights for the averaging.

Initially, the methodology included the following steps:

Step 1. Select one of the nine SuperGreen corridors to be used as pilot case for testing the methodology. The corridor with the best coverage in terms of data availability should be selected.

Step 2. Identify the 'critical' segment of the corridor involving a major link that cannot be bypassed due to geographical constraints. Examples are the Brenner passage of the Brenner corridor (link between Munich and Verona), the channel crossing of the Cloverleaf corridor (link between Calais and Dover) or the Pyrenees crossing of the Finis Terrae corridor (link between Valladolid and Irun). The rationale was that these segments are usually better studied than others improving the probability of securing the necessary data.

Step 3. Analyse cargo flows along the critical segment in terms of:

- origin/destination,
- types of cargoes moved,
- modes used,
- routes taken,
- trade imbalances (empties), etc.

Step 4. Select 4–5 typical cargoes being transported along the critical segment of the corridor. Unitized (containerized) cargoes should be given emphasis due to the importance of co-modality in green corridor projects. Part load break bulk is also suggested due to the special logistics requirements imposed by this type of cargo. Dry bulk and liquid bulk commodities can be selected due to their high volume and different supply chain organization. In general, the selection should be based on the relevant importance of each type of cargo and the special requirements that it imposes on the transportation means and the supply chains.

Step 5. Select 1–2 typical transport chains for each selected type of cargo. The origin/destination of the cargo could be any pair of nodes within or outside the corridor, provided that the routes/modes used are among those defined for the corridor. At this point the analysis moves away from the critical segment to cover the entire corridor. All branches of the corridor and all modes involved should be covered. Transport chains involving more than one mode are highly desirable. For sea-based corridors, transport chains should be selected based on:

- typical cargoes using each port in the corridor (use of port statistics)
- existing connections between ports in the corridor
- relative importance of connections in terms of volumes of cargo
- connections to land-based corridor segments
- types of vessels used
- 'best practice' cases identified in literature.

The output of Step 5 is a set of 10–15 transport chains that need to be analysed in terms of the selected KPIs.

Step 6. Locate the proper data sources for estimating the KPI values. Take into consideration that KPI estimation requires detailed information on the types of vehicles used, the technologies applied and other operational characteristics of the chains under examination.

Step 7. Estimate one set of KPIs for each chain selected under Step 5. Due to the length of the SuperGreen corridors, it is probable to have segments with different "green" qualities along a single corridor. It is thus preferable to do the analysis in segments to the extent possible.

Step 8. Identify obstacles in KPI estimation. A KPI re-engineering process might be needed for obstacles that can be addressed. KPIs running into unsolvable obstacles should be dropped. It is conceivable at this stage that segments of the corridor for which sufficient data is not available need to be dropped from further examination.

Step 9. Transform the KPI values estimated at the chain level to a single set of KPI values at the corridor level. Most probably weighted averages would have to be employed, using appropriate weights like cargo volumes, transport work, number of shipments, etc. It is, thus, important to come up with reliable information enabling calculation of the respective weights.

Step 10. Transform the set of KPI values derived under Step 9 to a single corridor rating. Relative weights should be assigned to each KPI. It is expected that different stakeholders would propose different weights for this calculation. A flexible approach of user specified weights should be considered as an alternative.

Step 11. Once the methodology suggested above has passed the applicability test successfully, it can be applied for the remaining SuperGreen corridors.

In applying this methodology, the Brenner corridor, extending from Malmö (SE) to Palermo (IT) with branches from Salzburg (AT) to Trieste (IT) through the Tauern axis, and from Bologna (IT) to Athens/Thessaloniki (GR) through the Italian and Greek Adriatic ports, was selected to be examined as a pilot case. The following steps were followed:

- the Brenner pass (Munich—Verona) was selected as the corridor's critical segment;
- the cargo flows along this critical segment were identified in literature;
- a small number (15) of typical transport chains concerning typical cargoes were selected;
- detailed information concerning these transport chains (type of vehicles used, load factors, etc.) was collected from studies and interviews with transport service providers; and
- the selected KPIs were evaluated for each one of these transport chains (emissions were estimated through the EcoTransIT World web based tool).

The chains examined for the Brenner corridor and the corresponding KPI values are presented in Table 3.6. It is noted that the KPIs on ICT tools, cargo security, cargo safety, NO_x and PM_{10} emissions were later on dropped from the analysis.

It soon became evident that the aggregation of Step 9, i.e. from chain-level KPIs to corridor-level ones for each and every segment of the corridor, would be problematic due to limited reliability on the grounds that:

Table 3.6 The Brenner corridor chains

Transport chain identity				Key performance indicators (KPIs)										
No	Origin-destination	Mode	Annual vol. (t)	Cost (€/tkm)	Time (h)	Reliability (%)	Frequency (no/year)	ICT (%)	Security (%)	Safety (%)	Emissions (g/tkm) CO$_2$	NOx	SOx	PM$_{10}$
1	Verona-Naples	Train	61,000	–	12	92	260	100	0	0	17.61	0.02	0.09	0.006
2	Verona-Numberg	Train	500,000	0.80	9	50	260	100	0	0	14.87	0.01	0.05	0.004
3	Verona-Numberg	Train	2,700,000	0.05	9	100	572	100	0	0	14.87	0.01	0.05	0.004
4	Verona-Berlin	Road	1,100	0.07	25	50	2,600	0	0	0	71.86	0.51	0.08	0.013
5	Rome-Numberg	Road	32,000	0.05	48	80	104	100	0	0	62.08	0.47	0.07	0.013
6	Rome-Palermo	SSS	1,500	0.04	24	100	52	100	0	0	16.99	0.25	0.12	0.018
7	Verona-Trelleborg	Interm.	13,000	0.04	50	98.8	624	100	0.5	2	10.62	0.01	0.02	0.002
8	Bari-Athens	Interm.	10,000	0.04	72–96	95	52	100	<0.5	0	27.28	0.18	0.08	0.008
9	Bari-Thessaloniki	Interm.	3,000	0.03	72–96	95	26	100	<0.5	0	42.11	0.29	0.10	0.011
10	Trieste-Munich	Train	81,000	–	12	85	416	100	1	1	12.53	0.01	0.04	0.003
11	Trieste-Salzburg	Train	652,500	–	8	90	208	100	1	1	9.49	0.01	0.05	0.003
12	Trieste-Villach	Train	135,500	–	4	95	364	100	1	1	16.36	0.02	0.09	0.006
13	Berlin-Thessaloniki	Interm.	437	0.09	76	99	104	0	<1	1	27.11	0.19	0.06	0.006
14	Bari-Berlin	Road	24,000	0.05	72	99	1,040	100	0	0	46.51	0.11	0.05	0.004
15	Bari-Athens	Interm.	8,500	0.05	24	99	520	100	0	0	25.41	0.25	0.14	0.024

Source: Ilves et al. (2011)

Table 3.7 KPI values for the Brenner corridor

KPIs	Intermodal	Road	Rail	SSS
Cost (€/tkm)	0.03–0.09	0.05–0.07	0.05–0.80	0.04
Av. speed (km/h)	9–41	19–40	44–98	23
Reliability (%)	95–99	50–99	50–100	100
Frequency (no/year)	26–624	104–2,600	208–572	52
CO_2 (g/tkm)	10.62–42.11	46.51–71.86	9.49–17.61	16.99
SOx (g/tkm)	0.02–0.14	0.05–0.08	0.04–0.09	0.12

Source: Ilves et al. (2011)

- the sample was very thin (for some segments there was only one observation) and the resulting figure would have limited statistical value, if any;
- not all of the chains reflected the entire door-to-door transport as needed to ensure comparability; some of them covered only terminal-to-terminal operations; and
- most data was collected through interviews and reflected personal assessments without strict validation.

It was, thus, decided to express corridor benchmarks as ranges of values that resulted from the transport chain data, i.e. minimum and maximum values of all chain-level KPIs. Table 3.7 summarizes the KPI values of the Brenner corridor presented by transportation mode.

The most important conclusion of this exercise is the width of the fluctuation range of some KPIs. Even after taking into consideration the drawbacks mentioned above, one would expect more concise estimates.

Furthermore, the aggregation of Step 10 of the initial methodology involving the transformation of all KPIs into a single corridor rating proved overoptimistic. The rationale for such a rating was to cope with interactions between different KPI groups, as is for example the case where measures introduced to improve performance in relation to one area might have adverse effects on another. However, this approach was later considered as an unnecessary complication given that:

- the weights needed for such calculation very much depend on the user (different users will propose different weights),
- it is a political issue best left for policy makers to decide,
- weights, if assigned, might lead to wrong interpretations,
- weights change over time (e.g. social issues might become more significant in the future), and
- weights would not reflect country specific characteristics of transportation operations.

The issue was discussed extensively in a SuperGreen workshop organized in Napoli, Italy and a decision was reached to exclude such attempt from the methodology. The decision was later confirmed by the project's Advisory Committee.

Table 3.8 Benchmarking results (all corridors)

Corridor	Mode	Cost (€/tkm)	Av. speed (km/h)	Reliability (%)	Frequency (no/year)	CO_2 (g/tkm)	SOx (g/tkm)
Brenner	Intermodal	0.03–0.09	9–41	95–99	26–624	10.62–42.11	0.02–0.14
	Road	0.05–0.07	19–40	50–99	104–2,600	46.51–71.86	0.05–0.08
	Rail	0.05–0.80	44–98	50–100	208–572	9.49–17.61	0.04–0.09
	SSS	0.04	23	100	52	16.99	0.12
Cloverleaf	Road	0.06	40–60	80–90	4,680	68.81	0.09
	Rail	0.05–0.09	45–65	90–98	156–364	13.14–18.46	0.01–0.02
Nureyev	Intermodal	0.10–0.18	13–42	80–90	156–360	13.43–33.36	0.03–0.15
	SSS	0.05–0.06	15–28	90–99	52–360	5.65–15.60	0.07–0.14
Strauss	IWT	0.02–0.44	–	–	–	9.86–22.80	0.01–0.03
Mare Nostrum	SSS	0.003–0.20	17	90–95	52–116	6.44–27.26	0.09–0.40
	DSS		–	–	–	15.22	0.22
Silk Way	Rail	0.05	26	–	–	41.00	–
	DSS	0.004	20–23	–	–	12.50	–

Source: Ilves et al. (2011)

The methodology, as it resulted from the pilot exercise, was applied for benchmarking five other corridors (Cloverleaf, Nureyev, Strauss, Mare Nostrum and Silk Way). Lack of data combined with time and resource restrictions did not permit the examination of the remaining three corridors (Finis Terrae, Two Seas and Edelweiss). The results are summarized in Table 3.8.

It is important to note that the emission KPIs of Table 3.7 were produced by the EcoTransIt World web emission calculator (EcoTransIt, 2014), while the remaining indicators are based on self-reported figures from interviewees and literature review. As such, they are only indicative. Using other tools and methods might have led to different results. The accuracy problem identified in the Brenner corridor was confirmed.

Table 3.8 leads to the following observations:

- The comparison of rail transportation attributes across corridors shows very high variance of cost and reliability for the Brenner corridor, which requires further investigation.
- The very low speed and high emissions of the trans-Siberian service is also noticeable, albeit expected due to the diesel traction and the gauge incompatibility problem along this route.
- The wide fluctuation of intermodal transportation attributes is also impressive and can be explained by the different nature of schemes examined in each case.

The more general conclusions stemming from the SuperGreen benchmarking work are summarized below:

- Corridor benchmarking is possible but we need to standardize both the process and the KPIs, if we want to make it operational.
- Even then, comparisons across corridors are problematic due to the fact that no consideration is given to corridor specific characteristics. It is certain that the

attributes of the logistical solutions employed in crossing the Baltic Sea are much different than those used for crossing the Alps. This type of risk is eliminated when comparing a time series of KPI values for the same corridor.

- The construction of sample chains on the basis of the 'critical segment' flows proved difficult in some cases, and in any event the characteristics of the critical segment might be totally irrelevant for other remote segments of the same corridor. Another solution should be conceived.
- Data collection proves to be a serious problem. Relevant obligations imposed by the corridor management might be a solution. The formation of corridor specific stakeholder groups can be helpful in this regard. Automated ICT applications, able to provide cargo flow data without causing physical disruptions of the vehicle flows or other administrative bottlenecks, can also be of particular importance.
- Aggregating chain-level KPIs to a single set of corridor- or segment-level ones is possible provided that an adequate sample of transport chains is examined under the same conditions. Otherwise, the use of value ranges is suggested.
- Aggregating corridor-level KPIs to an overall corridor rating should be omitted because there are problems associated with the weights needed for such calculation and the issue is a political one best left for policy makers to decide.

3.6.3 Guidelines for Corridor Benchmarking

In place of the usual concluding remarks, this section provides a set of guidelines for effective corridor benchmarking that takes into consideration the experiences of SuperGreen and other projects in this field.

Benchmarking goal: Monitoring the performance of a transportation corridor can serve several purposes. Obtaining a better understanding of the present conditions, identifying problems to be addressed, observing developments over time and comparing with benchmarks are some of them. Also important is the perspective of the analysis. A multiplicity of actors is involved in a corridor and their priorities do not always coincide. A corridor consists of various types of services offered by competing operators through organized supply chains over a multimodal infrastructural network within an international regulatory and administrative framework. In a complex system like this, setting the exact purpose of the analysis and its intended use is essential. A clear goal statement will assist decision making throughout the analysis and will affect all subsequent tasks. In general, it should be kept in mind that due to resource limitations, there is a trade-off between the width and the depth of analyses of this sort.

Corridor description: The next task cannot be different than defining the corridor under investigation. As can be inferred from Sect. 3.4, corridors tend to be described by locations that represent rather broad geographical areas/places where the corridors start, end or pass through. This has to be translated into a more detailed definition that includes the modes to be examined and the routes comprising the corridor. Each route should be described as a set of designated links, terminals and supporting facilities. Only existing major links should be designated to a route. Parallel secondary links or by-passes should be mentioned only as enhancing the resilience of a corridor. As for terminals, all uni- and/or multi-modal terminals should be designated to a route, except if irrelevant to the corridor traffic or unwilling to take part in it.

KPI selection: After extended consultation with stakeholders, the SuperGreen project proposes the following set of KPIs for corridor benchmarking applications:

- Out-of-pocket costs (excluding VAT), measured in €/tonne-km,
- Transport time, measured in hours (or average speed, measured in km/h, depending on the application),
- Reliability of service (in terms of timely deliveries), measured in percentage of consignments delivered within a pre-defined acceptable time window,
- Frequency of service, measured in number of services per year,
- CO_2-eq emissions, measured in g/tonne-km, and
- SO_x emissions, measured in g/tonne-km.

Among them, the cost indicator is the most difficult one to calculate due to scarcity of relevant data. In such cases, the volume of cargo moved along the corridor can serve as a proxy for describing its efficiency.

Other projects suggest different indicators. It needs to be emphasized that KPIs should be selected by the corridor management on the basis of the objectives being pursued.

Methodological principles: The methodology is built around the principle described by the following four steps:

Step 1. Disintegrate the corridor into transport chains.
Step 2. Select a representative set of typical transport chains.
Step 3. Estimate KPI values for each and every chain selected in Step 2.
Step 4. Aggregate these values into corridor level KPIs by using appropriate weights and methods.

Sample construction: In view of the problems encountered with the 'critical segment' notion applied in SuperGreen, it is suggested to construct a 'basket' of typical transport chains on the basis of traffic model results. Alternatively, the information of the 'Transport Market Study' foreseen by Reg. No 913/2010 for the Rail Freight Corridors and, through them, for the TEN-T core network corridors can be used for the sample construction (Panagakos, 2012). The proposed methodology resembles the functionalities of the Consumer Price Index (CPI) calculated by the statistical bureaus around the world. In the CPI context, the basket of goods and services used for CPI calculations is selected on the basis of the so-called Household Expenditure Survey (HES) that provides information on the spending habits of the population. In the context of green corridors, a traffic model can play the HES role.

The international character of a green corridor calls for a model covering effectively all of its routes. The European TRANS-TOOLS model (Ibánez-Rivas, 2010) is an ideal source of information, provided that its updating is successfully completed. Until then, national transport models can be used, but care should be taken to ensure compatibility.[6]

In selecting typical chains coverage of:

- all segments of the corridor,
- all modes of transport participating in the analysis,
- all possible types of transport chains examined by the model, and
- all types of vehicles examined by the model should be ensured.

Data collection: The task relates to the information needed for calculating KPI values for each and every transport chain of the basket. Readily available information from official statistics and other sources should be exploited to the extent possible. More detailed information should be solicited directly from stakeholders willing to take part in such an effort. To this end:

- a sample of transportation providers and major shippers should be formed for soliciting information,
- a questionnaire should be prepared for gathering the necessary information,
- follow-up actions should be foreseen for data collection including interviews if necessary, and
- a procedure addressing missing observations and quality adjustments should be designed.

As a general rule, the reported values should be:

- **Consistent:** The methodology employed should be consistent to allow for meaningful comparisons over time. Any changes to data, system boundaries, methods or any other relevant factor in the time series has to be clearly documented.
- **Transparent:** All relevant issues need to be addressed in a factual and coherent manner. The underline assumptions, calculation methodologies and data sources used have to be disclosed.
- **Accurate:** Ensure that uncertainties are reduced as far as practicable. Values reported should be of sufficient accuracy to enable users to make decisions with reasonable assurance as to the integrity of the reported information.

Emission estimation: When it comes to emissions, the definition of system boundaries is crucial in fulfilling all three criteria mentioned above (consistency, transparency and accuracy). Swahn (2010) defines four system boundaries (refer to Fig. 3.16):

- System boundary A includes traffic and transportation related activities regarding engine operation for the propulsion and equipment for climate control of goods, as well as losses in fuel tanks and batteries. This includes the traffic-related terminal handling, i.e. when goods do not leave their vehicle/vessel.

[6] The author of this chapter has used the Danish National Transport Model (LTM-Lands Trafik Modellen) for applying this methodology to the GreCOR corridor.

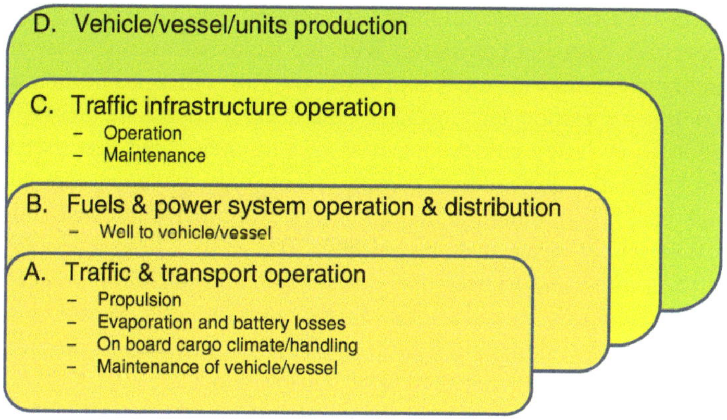

Fig. 3.16 Definition of system boundaries. *Source*: Swahn (2010)

- System boundary B includes in addition the supply of energy from energy source to the tank, battery and electric motor (trains). This is the minimum required system boundary for performance of comparisons between different modes of transportation.
- System boundary C includes in addition traffic infrastructure operation and maintenance.
- System boundary D includes in addition vehicle, vessel, load units production and scrapping (life cycle approach).

Although the introduction of the Life Cycle Assessment (LCA) methodology in decision making happens to be one of the policy recommendations that resulted from the SuperGreen project, it is essential to keep things as simple as possible in the early stages of a green corridor development. It is for this reason that the system boundary B is recommended to begin with. Later on, the boundary can be expanded to reach level D. Chapter 11 of this book deals with LCA considerations, with a focus on maritime transportation.

Another comment relates to the type of carbon emissions measured. In discussions of emissions, lots of terms are used—carbon emissions, carbon dioxide, greenhouse gases (GHGs). In fact, climate change is caused by a range of gases, known collectively as 'greenhouse gases'. Of these, the most common is carbon dioxide (CO_2). However, other GHGs are emitted from vehicle exhausts (i.e. nitrous oxide and methane), and their reporting is also valuable. This is done through CO_2-equivalent (CO_2-eq) units expressing GHGs as if they had the same climate change effects as CO_2. The choice between CO_2 and CO_2-eq depends on the availability of data and/or the capabilities of the emissions calculator used. CO_2-eq, if available, is preferable to CO_2.

In general, a specialized emission calculator is needed for estimating the emission KPIs. In SuperGreen, the web-based tool EcoTransIT World has been used but, as long as certified footprint calculators are not available, any other model could be used in its position, provided that a relevant qualification escorts the results. In the framework of the BGLC project, Öberg (2013) compared EcoTransIT World with NTM, a Swedish emission calculator, with inconclusive results. The announced cooperation between the two models towards creating synergies in their methodological expertise on carbon accounting is welcomed (EcoTransIt, 2014).

In relation to emission calculators, it should be mentioned that user specified inputs are preferred to any model's default values, only when they are adequately verified and there is consistency across all chains examined. Otherwise, it is safer to use the default values of the selected model.

Finally, it is important to note that in a multi-load multi-drop vehicle trip the allocation of emissions to specific loads becomes quickly almost unworkably complex, requiring far more data than is likely to be available. A simplification is suggested by DEFRA (UK) according to which, emissions are allocated on the basis of the number of EDUs (Equivalent Delivery Units) transported for each customer. Generally speaking, the choice of EDU should reflect the limiting factor on the loading of the vehicle. If the load is typically limited by volume, then a volume-based EDU such as pallets or cube should be used. If the load is more often limited by weight, then a weight-based EDU such as tonnes will be more appropriate and provide more accurate results.

KPI aggregation: The weights needed for aggregating chain-level KPIs into corridor-level ones depend on the relative significance of each chain in the route it belongs and in the entire corridor. As such, they have to be determined by using the model results that were considered in constructing the chain basket. These weights should be relatively fixed to permit historical comparisons.

It is noted that normally the weights for aggregating unit costs, CO_2 and SO_x emissions should be in tonne-km units. Transport time can only be aggregated if expressed as average speed, unless all chains examined concern a single origin-destination pair. The volume of cargo is probably the most suitable weight for aggregating transport time (or speed) and reliability. As for frequencies, one needs to be careful to avoid adding pears with apples. As a general rule of thumb, in serial services it is the least frequent one that determines the frequency of the chain.

Data verification: Before closing, it is necessary to alert the reader on the data verification issue. Verification is an independent assessment of the accuracy and completeness of data. Confidence in the quality and integrity of the data supports internal operations and decision making, by revealing existing problems or points for potential improvement. It can, thus, lead to improved performance, reliability and quality of operations. Another common reason for verifying data is to increase external stakeholder confidence. For example it may reassure a transport operator that they can include the green corridor data in what they report about their services, by demonstrating:

- credibility and reliability of the corridor data,
- consistency and accuracy of performance monitoring approach, and
- completeness of assessment.

Furthermore, verification can provide confidence that the data reported is fit for the purpose for which it is intended, for example, target setting or service benchmarking.

In general, it is not always necessary to get an external party to verify the reported data if reasonable and transparent processes are established. However, in the case of monitoring a complex system such as a transportation corridor, the engagement of an external verifier seems unavoidable. In such cases it is particularly important to be sure that the reported information is genuine and based on a consistent and accurate approach to measurement over time.

It is, thus, suggested the verification to be undertaken by a third party accredited by an internationally recognized body. Especially for GHG emission reporting, there are a number of internationally recognized standards and protocols that can be applied, like:

- ISO14064—Greenhouse gas accounting
- ISO14065—Requirements for greenhouse gas validation and verification bodies for use in accreditation or other forms of recognition.
- IEN 16258—The methodology and requirements for calculating and reporting energy consumption and greenhouse gas emissions in transport services.

Benchmarking frequency: The frequency of monitoring the performance of a corridor depends on the objectives set by the corridor management. As far as transportation services are concerned, an annual benchmarking is both feasible and practical, especially if customer satisfaction needs to be reported which happens to be the case with Rail Freight Corridors (Reg. 913/2010). Infrastructural developments can be reported on a less frequent basis.

A relevant issue relates to the periodical adjustments needed to account for changes in the composition of cargoes and transport chains using the corridor. As such changes would affect the model results (and the corresponding chain basket and weights), they can only be accounted for whenever the model is updated. In the CPI context, the HES is usually updated every 5–7 years.

General qualification: The method outlined above permits monitoring of the performance of a single corridor over time. It is not suitable for comparisons between corridors, as it does not consider differences in corridor characteristics that can be decisive in the overall performance of a corridor. This statement excludes the parameters determined by the Handbook on Reg. 913/2010 concerning railway transportation (EC, 2011b), as they have been aligned with the reports on train performance management of RNE in order to ensure a consistent quality of performance monitoring reports.

Acknowledgement Work reported herein was supported in part by EU project SuperGreen and an internal grant at the Technical University of Denmark.

References

Arnold, J. (2006, December). *Best practices in Management of International Trade Corridors.* Trade Logistics Group, Transport Papers TP-13, The World Bank Group, Washington, DC.

BE LOGIC, (2009, September 25). *Analysis of the impact of the external environment, supply chain and freight transport trends on the performance of the freight transport system.* A BE LOGIC project report.

Corridor A/IQ-C. (2011, August). *6th progress report.* Executive Board ERTMS Corridor A and International Group for Improving the Quality of Rail Transport in the North-South Corridor.

Corridor Rhine-Alpine. (2014). Retrieved October 16, 2014, from http://www.corridor-rhine-alpine.eu/about-us.html

EcoTransIt. (2014). Retrieved October 18, 2014, from http://www.ecotransit.org/

European Commission. (2007, October 18). *Freight Transport Logistics Action Plan.* Communication from the Commission COM(2007) 607, Brussels.

European Commission. (2010, March 3). *EUROPE 2020—A strategy for smart, sustainable and inclusive growth.* Communication from the Commission, COM(2010) 2020, Brussels.

European Commission. (2011a, March 28). *WHITE PAPER. Roadmap to a single European transport area—Towards a competitive and resource efficient transport system.* COM(2011) 144, Brussels.

European Commission. (2011b, June 30). *Handbook on the regulation concerning a European rail network for competitive freight (Regulation EC 913/2010).* A DG MOVE staff working document, Brussels.

European Commission. (2014). Retrieved October 18, 2014, from http://ec.europa.eu/transport/themes/infrastructure/news/ten-t-corridors_en.htm

European Parliament & Council. (2010, September 22). *Regulation (EU) no 913/2010 of the European Parliament and of the Council of 22 September 2010 concerning a European rail network for competitive freight.* Strasbourg.

European Parliament & Council. (2013, December 11). *Regulation (EU) no 1315/2013 of the European Parliament and of the Council of 11 December 2013 on Union guidelines for the development of the trans-European transport network and repealing Decision No 661/2010/EU.* Strasbourg.

Fastén, G., & Clemedtson, P. O. (2012, July 6). *Green Corridor Manual.* An East West Transport Corridor II report, NetPort.Karlshamn.

Friedrich, S. (2012). *Action programme on the development of the SCANDRIA corridor.* A SCANDRIA project report.

Hansson, A., & Hansson, N. (2014, August). *Green Corridor in the North Sea region—Inventory report.* A GreCOR project report.

Ibánez-Rivas, N. (2010). *Peer review of the TRANS-TOOLS reference transport model.* A JRC technical notes report.

Ilves, I., Holte, E. A., Georgopoulou, C., Rönkkö, S., Aditjandra, P., Vanaale, E., et al. (2011, April 18). *Benchmarking of green corridors—Version 2.* A SuperGreen project report.

Ilves, I., Panagakos, G., Wahl, A. M., Georgopoulou, C., Rönkkö, S., Vanaale, E., et al. (2010, December 15). *Benchmarking of green corridors—Version 1.* A SuperGreen project report.

Kokki, P. (2013, June). *Midnordic transport study.* A Midnordic Green Transport Corridor project report, Regional Council of Central Finland (in Finnish).

Kramer, H., Sedlacek, N., Jorna, R., van der Laak, R., Bozuwa, J., Gille, J., et al. (2009, June 10). *Report on overall benchmarking framework.* A BE LOGIC project report.

Mertel, R., Sondermann, K.-U. (2007, December 6). *Final report for publication.* BRAVO project.

Moyano, H., Panagakos, G., Fozza, S., Holte, E. A., Zacharioudakis, P., & Georgopoulou, C. (2012, December 20). *Green corridors handbook—Vol. I.* A SuperGreen project document.

Öberg, M. (2013, June). *Measurement of green corridor environmental impact performance—A2A logistic corridor concept.* A Bothnian Green Logistic Corridor report.

Panagakos, G. (2012, December 17). *Green corridors handbook—Vol. II*. A SuperGreen project document.
Panagakos, G., & Psaraftis, H. (2014, April 16). *How green are the TEN-T core network corridors?* Paper presented at the TRA2014 Conference, Paris, France.
Panagakos, G., Psaraftis, H., & Holte, E. (2015). Green corridors and their possible impact on the European supply chain. In C.-Y. Lee & Q. Meng (Eds.), *Handbook of ocean container transportation logistics: Making global supply chains effective*. Singapore: Springer.
RailNetEurope. (2014). Retrieved October 18, 2014, from http://www.rne.eu/rail-freight-corridors-rfcs.html
Réseau Ferré de France. (2014). Retrieved October 18, 2014, from http://www.rff.fr/en/the-network/europe-on-rails/european-network/major-cross-border-projects
Salanne, I., Rönkkö, S., & Byring, B. (2010, July 15). *Selection of corridors*. A SuperGreen project report.
SGCI. (2012, June 8). *Green corridors criteria*. A report of the Swedish Green Corridors Initiative.
Södergren, C., Sorkina, E., Kangevall, J., Hansson, N., & Malmquist, M. (2012, December 20). *Inventory of actors, transport volumes and infrastructure in the Bothnian Green Logistic Corridor*. A BGLC project report.
Stenbæk, J., Kinhult, P., & Hæstorp Andersen, S. (2014, September). *From speed and transit to accessibility and regional development*. A Green STRING Corridor project report.
Swahn, M. (2010, June). *How to evaluate transport's environmental performance*. Article at GreenPort Journal.
TransBaltic. (2012, August 28). *Macroregional Transport Action Plan*. A TransBaltic project report.
Wålhberg, K., Jüriado, R., Sundström, M., Nylander, A., Hansson, A., & Engström R. (2012, December). *Green Corridors*. Final report of the Swedish Green Corridor Initiative.

Chapter 4
Green Corridors and Network Design

George Panagakos

Abstract The purpose of this chapter is to investigate the relation between the Trans-European Transport Network (TEN-T) and the green corridor concept. First, the need is established for a corridor governance structure that enables the close cooperation among the numerous stakeholders from both the public and private sectors engaged in all corridor related issues ranging from network design to the provision of integrated logistical solutions. The governance scheme of the recently introduced TEN-T core network corridors seems to fulfil this requirement.

Following a brief history of TEN-T development, the 2013 major overhaul of the EU transportation infrastructure policy is outlined and the basic differences with the past are pinpointed. The provisions of the new TEN-T Guidelines are scrutinized so as to check whether the TEN-T core network corridors exhibit the characteristics of a green corridor, as they have been identified in the previous chapter. Based on the results of this analysis, it is concluded that the TEN-T core network is, as far as its freight dimension is concerned, a network of green corridors.

Abbreviations

ATM	Air Traffic Management
BGLC	Bothnian Green Logistics Corridor
CEF	Connecting Europe Facility
CMS	Corridor Management System
EC	European Commission
EDI	Electronic Data Interchange
EEIG	European Economic Interest Group
EFTA	European Free Trade Association
EIB	European Investment Bank
EP&C	European Parliament & Council
ERDF	European Regional Development Fund
ERTMS	European Rail Traffic Management System

G. Panagakos (✉)
Technical University of Denmark, Lyngby, Denmark
e-mail: geopan@transport.dtu.dk

H.N. Psaraftis (ed.), *Green Transportation Logistics*, International Series in Operations Research & Management Science 226,
DOI 10.1007/978-3-319-17175-3_4

ETCS	European Train Control System
EU	European Union
EWTC	East-West Transport Corridor
ICT	Information and Communication Technology
INEA	Innovation and Networks Executive Agency
IQ-C	International Group for Improving the Quality of Rail Transport in the North-South-Corridor (Rotterdam-Genoa)
ITS	Intelligent Transport Systems
MoS	Motorways of the Sea
MoU	Memorandum of Understanding
RDS-TMC	Radio Data System—Traffic Message Channel
RIS	River Information Services
RNE	RailNetEurope
SCANDRIA	Scandinavian-Adriatic corridor for innovation and growth
SESAR	Single European Sky ATM Research
SoNorA	South-North-Axis corridor in central Europe
TEN	Trans-European Network
TEN-T	Trans-European Transport Network
TEN-T EA	Trans-European Transport Network Executive Agency
UNECE	United Nations Economic Commission for Europe
VTMIS	Vessel Traffic Monitoring and Information System

4.1 Introduction

The purpose of this chapter is to explore possibilities for green freight transportation in relation to network features. Once again, the geographical scope of the analysis is Europe and its transportation network.

This task will be performed in three parts that directly correspond to the three main sections of the chapter. The first one deals with the development and governance of green corridors. Following the description of a green corridor and its basic characteristics in the previous chapter, here we discuss issues of more practical nature pertaining to the implementation of this concept. Those who study or practice logistics know very well that cooperation among all actors involved lies at the core of this business. The international nature of green corridors makes this necessity even more critical. It follows that the governance structure should enable and facilitate the cooperation among public and private sector stakeholders who play a significant role in all facets of green freight transportation; from network design to the provision of integrated logistical solutions. An effective and enabling governance scheme is, thus, a prerequisite for a successful implementation.

The second part of the analysis looks into the Trans-European Transport Network (TEN-T). Comprising one of the basic dimensions of the Common Transport Policy in Europe, it aims to provide the infrastructure needed for the internal market to function smoothly and for the objectives of the Lisbon Agenda on growth and jobs to be achieved. It also sets out to help ensure accessibility and boost economic, social and territorial cohesion. TEN-T supports the EU citizen's right to move freely within the territory of the Member States and it integrates environmental protection requirements with a view to promoting sustainable development. Of particular importance to our work are the TEN-T Guidelines, the documents containing the EU transport infrastructure policy. Both the previous one that supported the development of the network as it exists today and the current one that places emphasis on the corridor approach will be reviewed.

The third part of the chapter investigates whether the green characteristics of a corridor, as have been identified in the previous chapter, are exhibited by the TEN-T core network corridors introduced with the new TEN-T Guidelines. Based on the results of this analysis, the chapter concludes that, as of the end of 2013, a network of green corridors has been established in Europe.

4.2 Green Corridor Governance

The purpose of this section is to present issues related to the governance and operation of green corridors. Both these issues are linked to the management of the corridor structures. The term management, of course, implies some form of control but, given the diversity of stakeholders involved, this is easier said than done. The problem is further complicated by the fact that, despite the recent establishment of numerous corridors with such a self-claimed label, in practice green corridors have not yet moved far beyond the stage of inception. In this respect, the present section handles practical matters but in a rather visionary context.

The activities of a transportation corridor involve a number of government agencies and a diverse set of transportation and logistics service providers carrying a wide variety of operations. As a result, the management of a corridor is generally performed by organizations established by government, the private sector, or jointly to plan development, disseminate information and coordinate stakeholder efforts. The appropriate structure for corridor management depends on the nature of the corridor and the specific functions to be managed.

4.2.1 Corridor Functions

Having examined a number of international transportation corridors in the framework of a World Bank project, Arnold (2006) identifies a number of general functions requiring management oversight. They can be grouped in the following categories:

- **Infrastructure and facilities**, including links and nodes along the routes, are developed and funded primarily by the public sector but increasingly constructed and maintained by the private sector. The role of management is to guide the planning and procurement of these assets. Its goal is to ensure that these assets are:
  - of sufficient capacity to meet projected demand,
  - designed to provide efficient movement of cargo along the infrastructure and through the facilities,
  - constructed and maintained so as meet required standards,
  - used efficiently, and
  - fully utilized.
- **Transportation and logistics services**. Increasingly these activities are undertaken by the private sector in a competitive market with costs recovered through user charges. The objective of the managers of individual services is to capture significant market share by offering a competitive combination of cost, time and reliability. To the extent that corridor management is responsible for overseeing these services, its objective should be to promote more efficient services, usually by encouraging competition but often by allowing vertical and horizontal integration. Addressing security concerns and encouraging the use of ICT and risk management are additional objectives.
- **Regulatory procedures** that affect the movement of goods in the corridor and the transportation and logistics providers operating in the corridor. Rarely is corridor management involved in the enforcement of the regulations or even in the enactment of these regulations. Instead it performs an advocacy role discouraging excessive regulation and reforming regulation that leads to inefficiencies. The management can encourage reform by supporting efforts to harmonize procedures across borders, to simplify documentation and procedures, and to enhance transparency. In cases involving trade and transit agreements, corridor management can be engaged in their periodic revisions and in defining the regulations ensuring their proper implementation.
- **Monitoring corridor performance**. Corridor management is the appropriate entity for monitoring and coordination efforts aiming at improving its performance. This subject has been discussed in Chap. 3 of this book.

These corridor functions require different management approaches. They can involve the public sector, the private sector or both. The first involves provision of assets in a market with limited competition and partial cost recovery, the second provision of services in a competitive market with full cost recovery, while the third deals with enforcement of laws/regulations and tax collection.

More recently, Engström (2011) reports that the Swedish Transport Administration views green corridors projects/initiatives as being divided into three main categories that interact and complement each other. These categories promote the view of logistics/transports as a system of integrated services and properties aiming

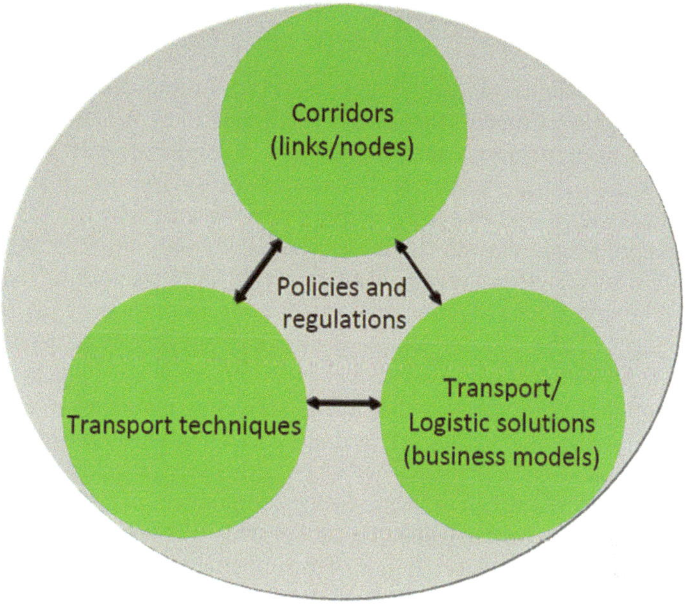

Fig. 4.1 The three pillars of green corridors. *Source*: Engström (2011)

at increased efficiency and a reduced negative ecologic impact. The three parts, shown in Fig. 4.1, are:

- **Corridors (links and nodes)**: A corridor project is a geographic subset of a designated main European Green Corridor. It is based on the needs of an efficient transportation infrastructure in a physical and/or communicative aspect. A corridor project promotes optimal use of transportation modes including transshipment nodes (hubs, cross docks etc.). It can be of either a national or international character.
- **Transport techniques**: Projects related to transportation techniques encompass features and properties of various types of equipment used in transportation operation. The main focus is on the different transportation modes, transportation/load units and transfer/reloading of goods between different modes. Examples are techniques related to trucks, trailers, railway engines, rail wagons, ships, port handling, containers, packaging, cranes, stackers etc.
- **Transport/logistics solutions**: Refers to complete solutions which integrate different partners and stakeholders mutually forming a business case that promotes efficiency and lowers environmental impact. In general terms, it is a complete freight logistic/transportation setup that meets a shipper's demand often linked to a new business model.

Although not seen as a 'pillar' in the Swedish schematic, the underlying policies and regulations are also recognized as a prerequisite for the implementation of green corridors.

4.2.2 Corridor Development Models

Based on the functions of the previous section, Arnold (2006) distinguishes between three general models that have been applied in corridor development.[1] The first is named **project coordination** and is viewed as part of a general development model. This approach is characterized by a project focus. Governments undertake improvements in the corridor infrastructure based on local requirements and problems. Growth in trade combined with liberalization of the transportation and logistics sector offers a steady improvement in the variety, quality and competitiveness of the transportation services. An evolving consensus on the concept of the corridor allows stand-alone projects to be related to the development of the corridor. This model has been most effective in providing improvements in infrastructure, but is less suitable for addressing legal or operational issues. Neither is it particularly useful for tackling bilateral and multilateral issues. Moreover, it lacks a formal corridor organization or other mechanism to identify and prioritize initiatives, as it relies on committees or similar structures.

The second is the so-called **legislative** model. This is characterized by the use of legislation to provide formal recognition of the importance of corridors, designation of specific routes, harmonization of standards, simplification of cross-border movements and funding for corridor infrastructure. Implementation is left to individual jurisdictions and government agencies. Coordination is undertaken at the regional or ministerial level and is characterized by formal meetings to review progress made by others. Development of services on the corridor is left to private sector competition. Improvements in infrastructure are undertaken by government agencies responsible for transportation. This approach is effective in targeting infrastructure funding and reducing formal impediments to movement of goods on these corridors. It is less effective for improving interconnections through modifications of regulatory constraints on cross-border and transit movements.

The third is the **consensus-building** model. This approach uses a regional institution to mobilize stakeholder support for improvements in the corridor and to push for trade facilitation reforms including improving border-crossing procedures. Its primary function is to provide information to stakeholders, including government agencies, concerning current performance, needs for improvement, and success of previous initiatives. The success of this model depends on the active participation of public and private sector stakeholders in addressing issues related to regulation, investment and quality of service. The ability to maintain a professional staff is also a critical success factor for such a model.

Bringing this taxonomy into the current European environment, one could distinguish between two models. The first is the **top-down** model that corresponds to Arnold's legislative one. It has been followed in all corridor development

[1] A fourth model relates to an institution responsible for developing public-private partnerships for improving the operation of facilities and services in the corridor. However, this model is dropped from the present analysis, as it is effective only at the domestic level.

initiatives of the European Commission, such as the RNE corridors, the ERTMS corridors, the rail freight corridors of Regulation No 913/2010 and, more recently, the TEN-T core network corridors. In a smaller scale, the Brenner corridor is a good example of a top-down model application.

The second is the **bottom-up** model, corresponding to Arnold's consensus-building one. All Scandinavian projects such as the EWTC II, SCANDRIA, TransBaltic, and BGLC corridors comprise applications of this type of model.

No European equivalent to Arnold's project coordination model is necessary, as activities such as priority setting and project identification under this model are more or less undertaken at national or local level which, nowadays in Europe, concerns only infrastructure projects of minor importance.

How do these models compare? Their distinction basically relates to the origin of the initiative. In the top-down model the initiative comes from a powerful central entity like the European Commission or a modal association. On the contrary, it is the transportation and logistics companies themselves who take the initiative in the bottom-up model.

Nevertheless, as the corridor structures mature, their success will depend on whether they exhibit features like:

- the cooperation between public and private sectors; and
- the active participation of stakeholders.

In this respect, in the long run the two models will have to converge.

If the idea of a green corridor is more popular among private businesses, the bottom-up approach should be followed. The idea is cultivated among all types of stakeholders and once sufficient support is secured, the public sector is engaged. In any event, its involvement is necessary for signing the necessary bilateral or multilateral agreements.

If, on the other hand, the idea is originated in the ministerial offices or among infrastructure managers closely related to national governments, the top-down model seems to be more appropriate. Intensive information campaigns are needed to engage the private sector in the process as early as possible.

4.2.3 Corridor Governance Structures

Regardless of the functions it serves or the development model it has followed, a corridor needs an organization engaged in the promotion and coordination of its development and operation. Where corridors have been successful, there has been strong political and market support for their development. A corridor organization provides a point of coordination for stakeholder efforts and a forum for identifying major impediments. It also provides coordination for the financing schemes. As a promoter, this organization must have the support of the private sector but be able to work closely with government agencies to improve procedures and policies. As a coordinator, it must have some form of public-private partnership as well as linkages with a regional ministerial committee that is tasked to address issues of regional harmonization.

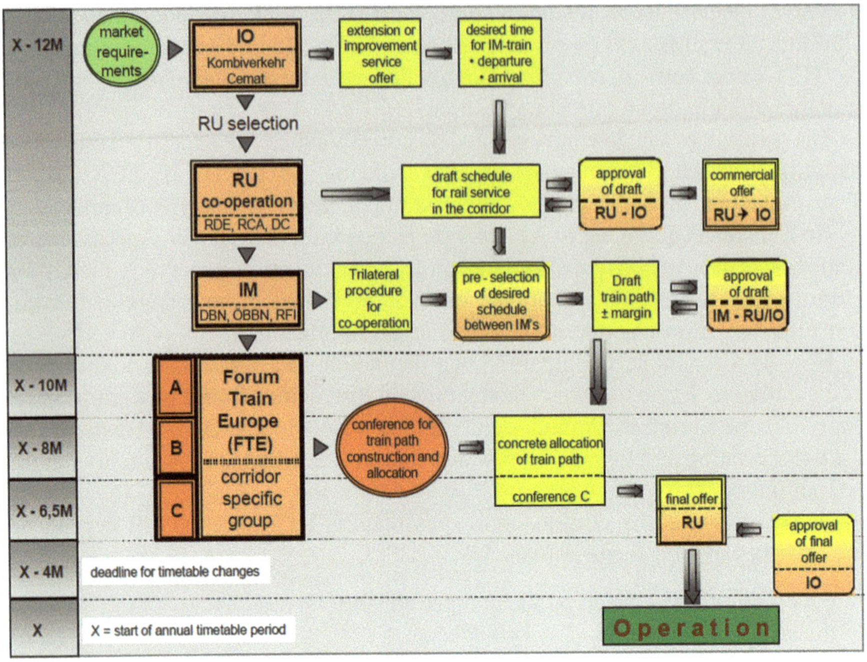

Fig. 4.2 Planning procedure of intermodal Brenner transports. *Source*: Galonske (2004)

A first attempt of the European research community to formulate an open Corridor Management System (CMS), linking the actors of an intermodal chain of transportation, was done by the BRAVO project and concerned the Brenner Corridor (Galonske, 2004).

The project succeeded in describing in detail both the role of all actors involved (Infrastructure Managers, Railway Undertakings, Intermodal Operators and Terminal Operators) and the procedures that need to be followed in order to plan an intermodal transport, which as shown in Fig. 4.2, takes about 12 months.

In relation to corridor management, the project first assessed the "Full integrator model," which gives all parties free access to all components of the CMS. After rejecting this model due to legal and institutional considerations and the existing competition between actors, the project suggested as the most suitable management structure a combination of an "open platform" integrating all actors in a non-discriminating way (e.g. guided "Round table") for the strategic and long-term tasks and a "restricted platform" for operational and commercial tasks (Fig. 4.3).

The management organization of ERTMS Corridor A (Rotterdam-Genoa) is more structured. On 9 January 2003 the transport ministers of Germany, Italy, the Netherlands and Switzerland signed a joint MoU in Lugano aimed at enhancing the quality of cross-border freight transportation by rail on the Rotterdam-Genoa corridor. The ministers entrusted the International Group for Improving the Quality

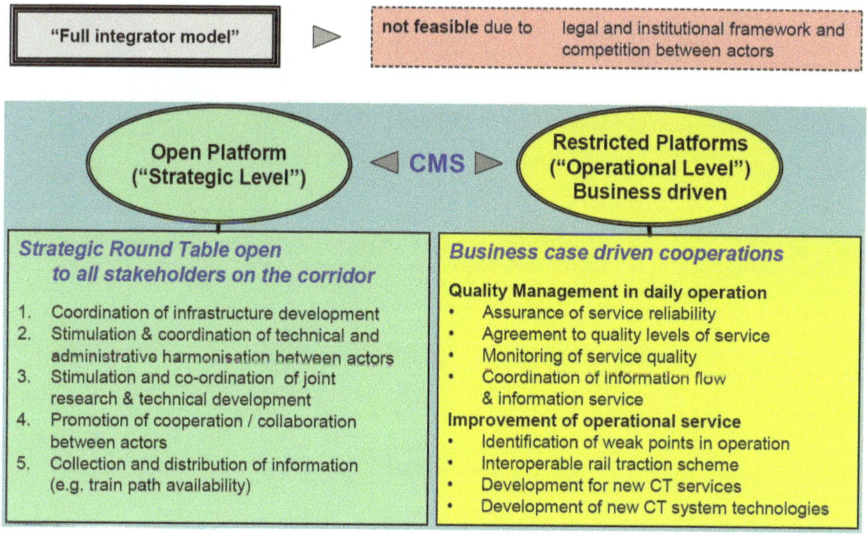

Fig. 4.3 BRAVO Corridor Management Scheme. *Source*: Mertel and Sondermann (2007)

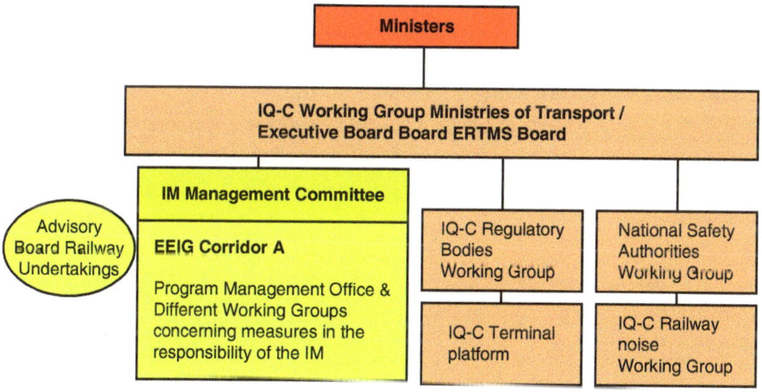

Fig. 4.4 The management structure of Corridor A. *Source*: Corridor A/IQ-C (2011)

of Rail Transportation in the North-South-Corridor or Corridor A (IQ-C) with the task of implementing a package of specific measures that were defined following a prior analysis of the main problems hampering rail freight transportation along the North-South-Corridor.

In 2006, the organization for the deployment of ERTMS/ETCS in the corridor was established. As shown in Fig. 4.4, the Infrastructure Managers set up the Management Committee to steer the overall improvement program integrating all ERTMS and other activities of IQ-C, whereas the Ministries created the Executive Board supervising the ERTMS implementation on the corridor. Since 2008, the IQ-C Working Group of the Ministries of Transport and the ERTMS Executive

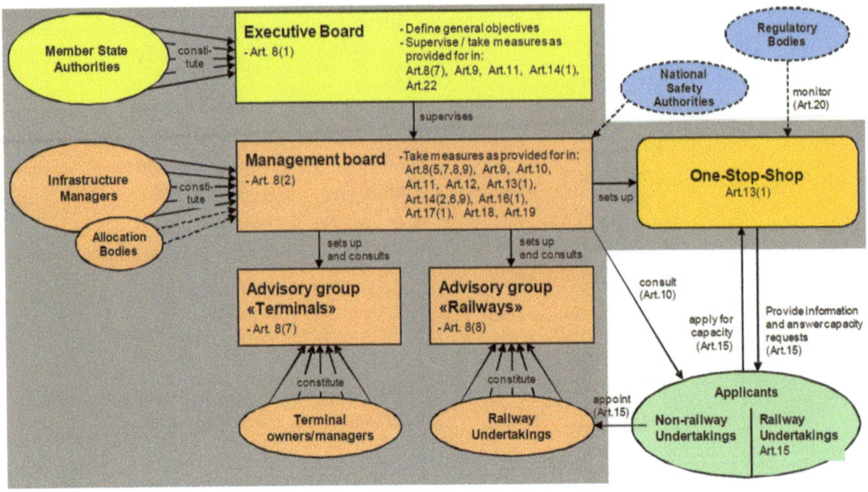

Fig. 4.5 Governance structure of a Rail Freight Corridor. *Source*: EC (2011b)

Board are working together in very close cooperation and coordinate their actions and time schedules. The same year, the Infrastructure Managers of the corridor founded the EEIG "Corridor Rotterdam-Genoa EWIV," which enabled them to act as a legal entity, financially borne by its members and associates.

On the side of Infrastructure Managers, the Program Management Office is implemented and works as one common corridor management board, which develops, steers, monitors and reports all corridor activities as an integrated action. Since 2009, the corridor organization includes a "Terminal platform" and a Working Group on Railway noise as additional parts of the organization (Corridor A/IQ-C, 2011).

This structure is basically identical to the one stipulated by Regulation EU 913/2010 establishing the Rail Freight Corridors (Fig. 4.5).

The Executive Board is composed of representatives of Member States. The Management Board is formed by the Infrastructure Managers and where relevant the Allocation Bodies. It is clearly stated that Railway Undertakings cannot be members of the Management Board, which can be an independent legal entity such as an EEIG. The Management Board has to set up two Advisory Groups, one consisting of managers and owners of the terminals of the freight corridors, the other representing Railway Undertakings using or interested in using the corridor. To simplify communication with applicants and other interested parties, the Regulation provides for the establishment of a corridor one-stop-shop. More details on the governance structure specified by the Regulation can be found in the relevant handbook (EC, 2011b).

More recently, Regulation EU 1315/2013 on the "new TEN-T guidelines" established the core network corridors. In terms of governance, this Regulation foresees European Coordinators, acting in the name and on behalf of the Commission, to facilitate the coordinated implementation of the core network corridors.

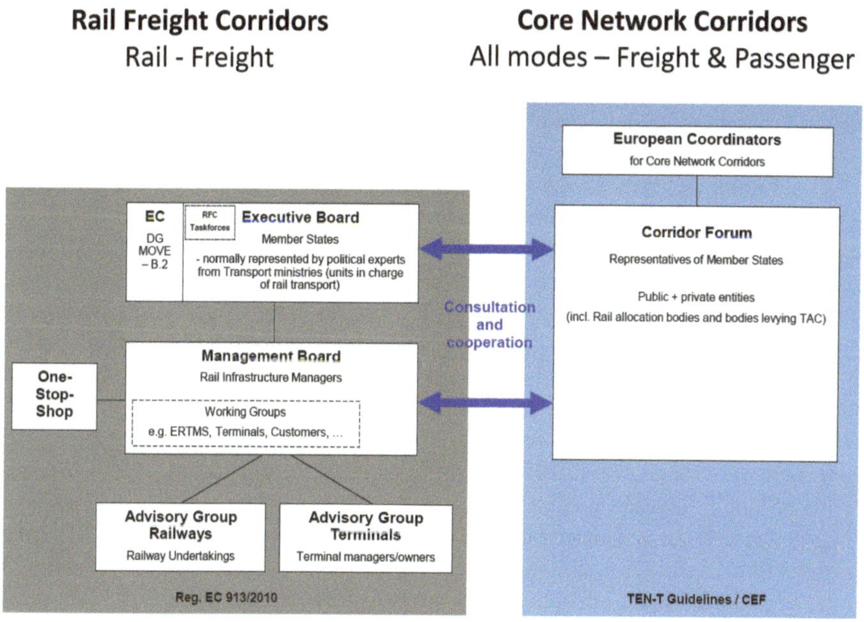

Fig. 4.6 Governance structure of the TEN-T core network corridors. *Source*: Rousseaux (2012)

Furthermore, for each core network corridor, the Member States concerned shall establish a Corridor Forum responsible for defining the general objectives of the corridor and for preparing and supervising the relevant measures. The Corridor Forum shall be composed of the Member State representatives and other appropriate public and private entities, and shall be chaired by the European Coordinator. The relation of this structure to the one foreseen for the Rail Freight Corridors appears schematically in Fig. 4.6. Although the Corridor Forum is expected to include representatives of all parties involved, its structure intentionally has been left open to be decided on a corridor level enabling consideration of corridor specific conditions.

4.3 The Trans-European Transport Network (TEN-T)

The purpose of this section is to present the TEN-T with emphasis placed on its design aspects. Following a brief piece on network development prior to the involvement of the EU, the TEN-T is presented as it looked until very recently, that is a combination of an extensive comprehensive network and a set of priority projects. The section ends with a reference to the "new TEN-T guidelines," representing a major overhaul of the European transportation infrastructure policy. The network design aspects of this attempt are given special attention.

4.3.1 The Pre-EU Era

The point of this brief heading is to underline that cross-border transportation networks were being developed for well over a century prior to the EU's influence on infrastructural integration.

The European cross-border infrastructure is being discussed since the early nineteenth century. During negotiations about the European order after the Napoleonic Wars at the Vienna Congress in 1814–1815, the French philosopher Claude-Henri de Saint Simon suggested the establishment of a European Parliament to take on matters of common European interest such as large trans-border waterway projects.

Although Saint Simon's idea of a European Parliament did not materialize for a long time, transnational networks did. Schipper and Van der Vleuten (2008) distinguish between existing and new transportation networks. Navigation and road networks were already in place. However, they were greatly improved in terms of length, density, quality and usage. Waterways were always considered as long-distance arteries. Roads, by contrast, were rediscovered as such only after the introduction of the automobile. At the dawn of the twentieth century, France was the owner of the most advanced technology in car manufacturing. A number of well-advertised road races, introducing transnational road use, were organized between Paris and other European capitals like Amsterdam, Berlin, Vienna and Madrid, before the trend was stopped abruptly following the disastrous Paris-Madrid race of 1903 (Fig. 4.7).[2]

The United Nations Economic Commission for Europe (UNECE) Declaration on Main International Traffic Arteries of 1950 was the first post-war international treaty concerning road traffic in Europe. The signatory parties "... considered it essential, in order to establish closer relations between European countries, to lay-down a coordinated plan for the construction or reconstruction of roads suitable for international traffic." So, they adopted the road network of Table 4.1 as a concerted plan, which they intended to undertake in accordance with agreed upon technical specifications within the framework of their national programs for public works or within the possibilities of international financing.

In terms of new infrastructure, railways attracted most popular attention in the nineteenth century. The highly transnational character of these networks resulted from the extensive effort of governments to re-position their countries in the European economic and military geography. Starting from the 1830s, Dutch and Belgian rail projects connected the major ports of the region to the Central-European hinterland. The Italian network was developed with the same purpose soon after the Suez Canal was opened in 1869. Alpine countries built

[2] The race was declared officially over at the end of its first leg Versailles—Bordeaux (552 km), after half of the 224 participating vehicles (170 cars and 54 motorcycles) had crashed or retired, 8 people had died (3 spectators and 5 racers) and over 100 had been wounded. No other races on public streets were allowed until 1927 (Mille Miglia).

Fig. 4.7 The 1903 Paris-Madrid race: Louis Renault (Renault) and Ferenc Szisz (mechanic). *Source*: FIAT131RACING (2015)

Table 4.1 The main international traffic arteries (*Source*: UNECE, 1950)

Number	Description[a]
E1	London-Paris-Nice-Roma-Palermo
E2	London-Lausanne-Milano-Brindisi
E3	Lisboa-Paris-Stockholm
E4	Lisboa-Bern-København-Stockholm-Helsinki
E5	London-Wien-Budapest-Beograd-Alexandroupolis-Istanbul-Ankara-Turkish/Syrian frontier
E6	Roma-Berlin-Oslo-Stjördal
E7	Roma-Wien-Warszawa
E8	London-Den Haag-Berlin-Warszawa-USSR
E9	Amsterdam-Basel Genova
E10	Paris-Bruxelles-Den Haag-Amsterdam
E11	Paris-Salzburg
E12	(Paris)-Praha-Warszawa-(Leningrad and Moskva)
E13	Lyon-Venezia
E14	Trieste-Praha-Szczecin
E15	Hamburg-Berlin-Praha-(Budapest)
E16	Bratislava-Gdynia
E17	Chagny-Salzburg
E18	Stavanger-Oslo-Stockholm
E19	(Greek/Albanian frontier)-Ioannina-Korinthos
E20	Koritza-Sofia
E21	Aosta-Torino-Savona

(continued)

Table 4.1 (continued)

Number	Description[a]
E21a	Martigny-Grand Saint Bernard-Aosta
E21b	Genève-Bonneville-Mont Blanc-Aosta
E22	Berlin-Wroclaw-Opole-Bytom-Krakow-Rzeszow-Przemysl-(USSR)
E23	Ankara-Kirsehir-Kayseri-Sivas-Erzincan-Erzurum-Agri-(Turkish/Iranian frontier)
E24	Kömürler-Gasiantep-Urfa-Mardin-Cizre-Hakkari-Bajirge-(Turkish/Iranian frontier)
E25	Burgos-Madrid-Bailén-Sevilla-Cádiz-Algeciras
E26	Barcelona-Tarragona-Castellón de la Plana-Valencia-Granada-Málaga-Algeciras

[a]Names of cities/countries appear as listed in the original document

hugely expensive railway tunnels to improve their accessibility, while Prussian and Austria-Hungarian interests promoted connection to the Balkan Peninsula and ultimately Turkey and Iraq. Russia's Trans-Siberian railway was operational by 1901 and soon travelling from the Channel to Vladivostok, an unthinkable endeavor some years ago, became a reality (Schipper & Van der Vleuten, 2008).

In the twentieth century, a fourth transportation network, aviation, further strengthened transportation across the globe, as maritime shipping had done earlier.

4.3.2 *Early EU Efforts in Network Development*

Transportation was one of the two sectors of economic activity for which a common policy is pursued by the 1957 Treaty of Rome; agriculture being the other one. In establishing the European Economic Community, the founding parties considered the creation of a common market as their primary objective. In this regard, the provisions of Title IV (Transport: Articles 74–84) were dealing exclusively with the removal of barriers to fair competition that were not uncommon in this period.

It is interesting to note that the relevant provisions were applicable only to transportation by rail, road and inland waterway (Article 84). Apparently, access to the market of international maritime services was since then much less inhibited by protective measures.

No reference to infrastructure investments was contained in the Treaty of Rome. However, the role of infrastructure in the growth of regional economies, especially in the peripheral areas, as well as in the integration of the transportation services and the EU itself soon became evident. In February 1966, the Council of Ministers introduced a consultation procedure for infrastructure investment, albeit with minimum results due to circumstantial exchange of information (Stasinopoulos, 1995).

A second consultation procedure on transportation infrastructure programs was adopted by the Council of Ministers in 1978 and a special committee was set up to coordinate national infrastructure policies. In 1979 the European Commission argued that the European transport policy would not achieve the objectives set out in the 1957 Treaty of Rome unless it related more to the infrastructure. However the Commission's early attempts to promote a European approach to investment in transportation infrastructure met with only limited success (Butcher, 2012).

In terms of financing, the European Investment Bank (EIB) had started in the 1960s to grant loans to infrastructure projects of Community interest. Since 1975, the European Regional Development Fund (ERDF) further supported transportation infrastructure in lagging regions. In general, however, the financial assistance was inadequate and the arrangements to determine Europe-wide intervention were based solely on national plans.

During the preparations for the European Single Market throughout the second half of the 1980s, the European Round Table of Industrialists expressed its concern that, because of infrastructure bottlenecks, a further development of transportation could be hampered, and that this would result in a loss of productivity gains that could otherwise be brought about by a more sophisticated division of labor (Sichelschmidt, 1999).

The European Commission (1990) expressed the view that the European infrastructure networks were still segmented and that "... the lack of interoperability between them makes it impossible for them to link up with each other beyond national frontiers and for them to be operated simultaneously or consecutively so that they offer a coherent and satisfactory service at a reasonable cost to the user. These difficulties are linked not only to the facilities and installations concerned but also to the services provided...".

The same document identified the following criteria that the Community infrastructure needs to meet:

- ability to cope with the predicted increase in the intra-Community trade unimpeded by physical, technical and, "...in the near future, tax barriers" (volume effect),
- need for existing infrastructure and services to be interconnected so that they will match the new dimensions of the market (interoperability requirement),
- taking the Community dimension into account in the design and development of future networks (dimension effect—subsidiarity principle),
- provision of adequate service quality throughout Europe (quality requirement), and
- need to draw closer all the elements of the Community space (cohesion effect).

Furthermore, the document drew the framework of an action program for Community infrastructure in the sectors of transportation, energy and telecommunications, containing indicative priority projects. In terms of transportation, only the road and rail sub-sectors are covered. The following road links are mentioned:

- Toulouse-Madrid and Bordeaux-Valencia via a tunnel under the Somport,
- Toulouse-Barcelona via the Puymorens tunnel,
- The Brenner axis,
- Road link to Ireland: A5/A55 Crewe-Holyhead link in the UK,
- Brindisi-Patras-Athens,

- Lisbon-Madrid,
- Aalborg-Frederikshaven motorway,
- Fehmarn links, and
- Athens-Evzoni-Yugoslavia.

With respect to rail transportation, the indicative projects included:

- The North high speed axis: Paris-London-Brussels-Amsterdam-Cologne,
- The South high speed axis: Seville-Madrid-Barcelona-Lyon-Turin-Milan-Venice, and hence to Tarvisio and Trieste, Oporto-Lisbon-Madrid,
- The Dublin-Holyhead-Crewe and Dublin-Belfast axes, and
- The Brenner axis.

On 7 February 1992, the Trans-European Networks (TENs) were officially introduced with the Maastricht Treaty (EC, 1992 - Article 129b):

> . . .to enable citizens of the Union, economic operators and regional and local communities to derive full benefit from the setting up of an area without internal frontiers, the Community shall contribute to the establishment and development of trans-European networks in the areas of transportation, telecommunications and energy infrastructures.

The stated objective of this Community action was "... within the framework of a system of open and competitive markets, ... to promote the interconnection and interoperability of national networks as well as access to such networks. It shall take account in particular of the need to link island, landlocked and peripheral regions with the central regions of the Community."

To meet these objectives, the Community (Article 129c):

- shall establish a series of guidelines, which would identify projects of common interest,
- shall implement any measures necessary to ensure the interoperability of the networks, in particular in the field of technical standardization,
- may support the financial efforts of the Member States for projects included in the guidelines, particularly through feasibility studies, loan guarantees, interest rate subsidies or through the Cohesion Fund to be set up no later than 31 December 1993,
- may take, in close cooperation with the Member States, any useful initiative promoting coordination among the Member States in relation to policies pursued at national level which may have a significant impact on the stated objectives, and
- may decide to cooperate with third (i.e. non-EU) countries to promote projects of mutual interest and to ensure the interoperability of networks.

The political impetus was subsequently given by the European Council (1993) in Copenhagen: it called on the Commission and the Council to speed up the adoption of master plans in the field of transportation, energy and telecommunication.

In December of the same year, the White Paper on "Growth, Competitiveness and Employment" (European Commission, 1993) was published. In this paper, the Commission presented the TENs as "...the arteries of the single market." In a true

Keynesian line of thinking, the Commission claimed that the malfunction of the networks reflected lost opportunities to create new markets and hence jobs.

An important finding of this document relates to the massive investment required for the implementation of the TENs, particularly for transportation infrastructures. Given the shortage of available public financing at both the Community and Member State level, new types of partnerships between private and public financing were needed, backed by financial engineering encompassing all different sources and types of financing. In order to effectively launch the process of this partnership, an initial list of 26 projects which were both of Community interest and had the potential to mobilize private economic operators was drawn up by the Commission. These candidate projects formed the basis on which discussions were initiated with the relevant authorities and economic circles.

In the framework of these discussions, a "group of personal representatives of the Heads of State or Government", called the "Christophersen Group" for short, was set up in December 1993 at the request of the Council to identify priority projects in transportation and energy. It proposed 35 projects to be granted priority. In its interim report to the Corfu European Council in June 1994, the Group identified a first list of 11 projects in the transportation sector as special priorities because they were either in the stage of realization or prepared for a start of realization before the end of 1996. A final report was presented to the Council meeting in Essen in December 1994 including three additional projects. This proposal (some 8,000 km of rail lines, thereof nearly 4,500 km for high speed traffic, an ample 4,000 km of motorways and 1 airport project) was finally endorsed by the European Council meeting in Essen (Sichelschmidt, 1999).

In addition to the top 14 projects, the Group produced a list of traffic management projects and a list of other projects which were important but which were not yet ready for work to begin. The Commission proposed in 2001 to add six further schemes, including the global navigation and positioning satellite system, Galileo. Finally, the High Level Group, chaired by Karel Van Miert, recommended in total 16 additional priority projects. These were added to TEN-T in 2003, bringing the total number of priority projects to 30 (Butcher, 2012).

4.3.3 *The 2010 TEN-T Guidelines*

The objectives, priorities and broad lines of measures envisaged in the area of the Trans-European Transport Network are contained in a document called "TEN-T Guidelines." The same document indicates the routes of Union importance that may be considered for EC financial support. The first set of Union guidelines for the development of the TEN-T was published in 1996. It was revised in 2001 and 2004 and recast in 2010. Although the guidelines currently in force are those of 2013 (refer to Sect. 4.3.4), the existing infrastructure of the TEN-Ts has been formed on the basis of the provisions of the 2010 Guidelines (Decision 661/2010/EU). The main features of this document are briefly presented below.

4.3.3.1 Objectives

The single objective of the EU transportation infrastructure policy is the gradual (by 2020) establishment of the TEN-T by integrating land, sea and air transportation infrastructure networks throughout the Union in accordance with a set of outline plans and specifications (EP&C, 2010).

In terms of requirements, the network must:

- ensure the sustainable mobility of persons and goods within an area without internal frontiers under the best possible social and safety conditions, while helping to achieve the Union's objectives, particularly in regard to the environment and competition, and contribute to strengthening economic and social cohesion,
- offer users high-quality infrastructure on acceptable economic terms,
- include all modes of transportation, taking account of their comparative advantages,
- allow the optimal use of existing capacities,
- be, insofar as possible, interoperable within modes of transportation and encourage intermodality,
- be, insofar as possible, economically viable,
- cover the whole territory of the Member States so as to facilitate access in general, link island, landlocked and peripheral regions to the central regions and interlink without bottlenecks the major conurbations and regions of the Union,
- be capable of connecting to the networks of the European Free Trade Association (EFTA) States, the countries of Central and Eastern Europe and the Mediterranean countries, while at the same time promoting interoperability and access to these networks, insofar as this proves to be in the Union's interest.

The comparison between these requirements and those of 1990 (refer to Sect. 4.3.2) shows that during the last 20 years the European society has become more sensitive in issues relating to the environmental and social sustainability, safety, economic viability, optimal use of existing capacities, and the external dimension of EU policies. It is believed that the European dimension (subsidiarity principle) is missing from the recent list only because it has been taken into consideration inherently in drawing up the outline plans (see below).

4.3.3.2 Scope and Priorities of the Network

In terms of scope, the trans-European network consists of transportation infrastructure, traffic management systems and positioning and navigation systems. The transportation infrastructure includes road, rail and inland waterway networks, motorways of the sea, seaports and inland waterway ports, airports and other interconnection points between modal networks.

The following priorities are mentioned in the guidelines:

- the establishment and development of the key links and interconnections needed to eliminate bottlenecks, fill in missing sections and complete the main routes, especially their cross-border sections, cross natural barriers, and improve interoperability on major routes,
- the establishment and development of infrastructure which promotes the interconnection of national networks in order to facilitate the linkage of islands, or areas similar to islands, and landlocked, peripheral and outermost regions on the one hand and the central regions of the Union on the other, in particular to reduce the high transportation costs in these areas,
- the necessary measures for the gradual achievement of an interoperable rail network, including, where feasible, routes adapted for freight transportation,
- the necessary measures to promote long-distance, short sea and inland shipping,
- the necessary measures to integrate rail and air transportation, especially through rail access to airports, whenever appropriate, and the infrastructures and installations needed,
- the optimization of the capacity and efficiency of existing and new infrastructure, promotion of intermodality and improvement of the safety and reliability of the network by establishing and improving intermodal terminals and their access infrastructure and/or by developing intelligent systems,
- the integration of safety and environmental concerns in the design and implementation of the trans-European transport network,
- the development of sustainable mobility of persons and goods in accordance with the objectives of the Union on sustainable development.

4.3.3.3 The Outline Plans and Specifications by Sector

The **road network** comprises motorways and high-quality roads, as well as infrastructure for traffic management, user information, dealing with incidents, emergencies and electronic fee collection. The network should guarantee its users a high, uniform and continuous level of services, comfort and safety.

The outline plan of the road network, as amended in 2013, appears in Fig. 4.8. In addition to those shown on the plan, projects of common interest[3] could concern:

- development of the network, and in particular:
  - widening of motorways or upgrading of roads,
  - construction or improvement of bypasses or ring roads,
  - increasing the interoperability of national networks.

[3] According to the terminology of the TEN-T Guidelines, a "project of common interest" is one that pursues the set objectives of the guidelines, corresponds to one or more of the set priorities of the guidelines, is economically viable on the basis of a socio-economic cost/benefit analysis, relates to the routes of the outline plans and meets the specifications set by the guidelines.

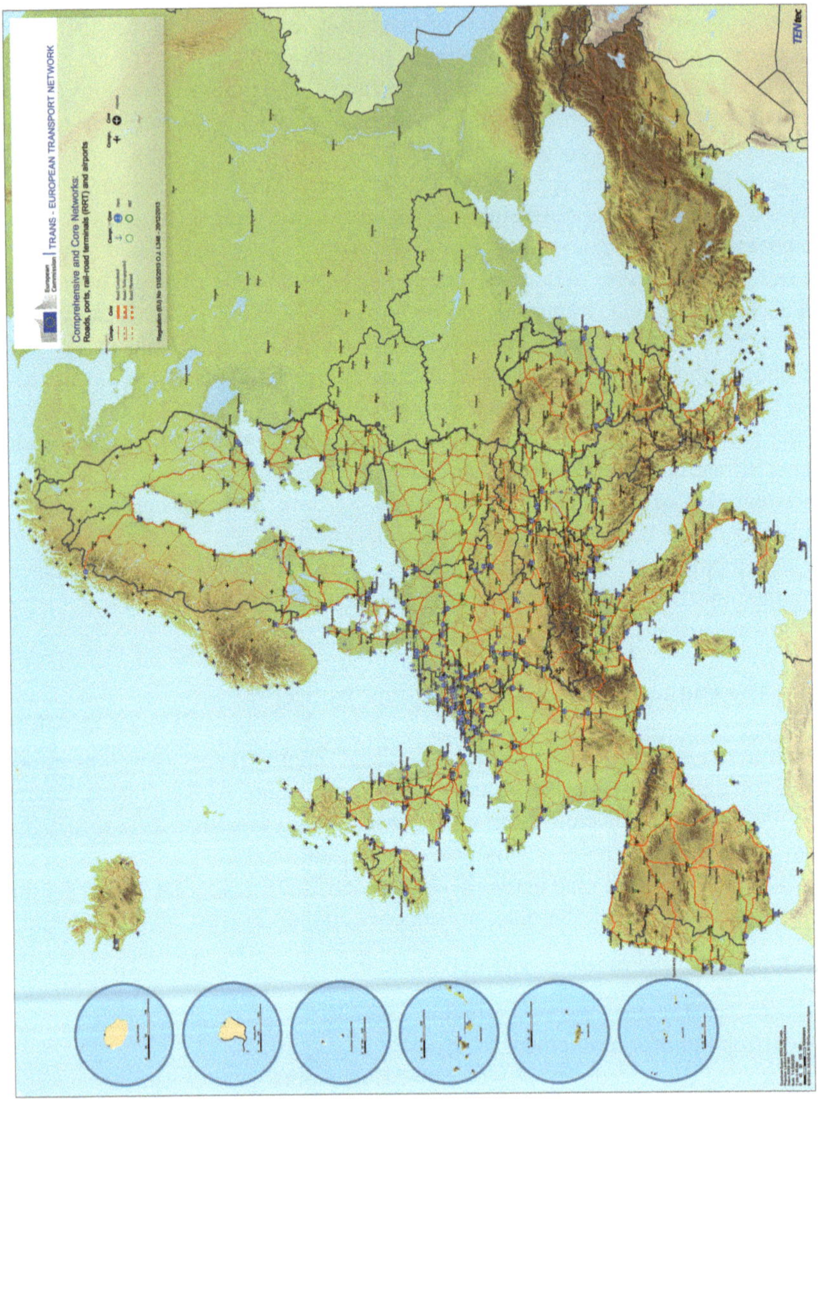

Fig. 4.8 The road, port, rail-road terminal and airport plan. *Source*: EC (2015)

- development of traffic management and user information systems, and in particular:
  - establishment of telematics infrastructures for collecting traffic data,
  - developing traffic information centers and traffic control centers, as well as exchanges of data between traffic information centers in different countries,
  - establishing road information services, in particular the RDS-TMC system,
  - technical interoperability of telematics infrastructures.

The **rail network** comprises both high-speed and conventional rail networks, as well as facilities that enable the integration of rail and road and, where appropriate, maritime and air transportation services. Technical harmonization and the gradual implementation of the ERTMS harmonized command and control system ensures the interoperability of national networks. The users should benefit from a high level of quality and safety, by virtue of its continuity and the gradual realization of its interoperability.

The outline plan of the rail network appears in Fig. 4.9. In addition to those of the plan, projects of common interest could concern:

- interoperability between trans-European railway systems,
- interconnection with networks of other modes of transportation.

The **inland waterway network** comprises rivers, canals, and inland ports. The network also includes traffic management infrastructure, and in particular an interoperable, intelligent traffic and transportation system (RIS—River Information Services), intended to optimize the existing capacity and safety of the inland waterway network as well as improve its interoperability with other modes of transportation. The minimum technical characteristics for waterways forming part of the network are those of class IV, which allows the passage of a vessel or a pushed train of craft 80–85 m long and 9.50 m wide.

The outline plan of the inland waterway network appears in Fig. 4.10. In addition to those presented on the plan, projects of common interest, which must relate solely to infrastructure open to any user on a non-discriminatory basis, could concern:

- inland ports, and in particular:
  - access to the port from waterways,
  - port infrastructure inside the port area,
  - other transportation infrastructure inside the port area,
  - other transportation infrastructures linking the port to other elements of the trans-European transport network.

- traffic management, and in particular:
  - a signaling and guidance system for vessels, in particular those carrying dangerous or polluting goods,
  - communication systems for emergencies and inland waterway safety.

The **seaport network** (refer to Fig. 4.10) permits the development of sea transportation and constitutes shipping links for islands and the points of

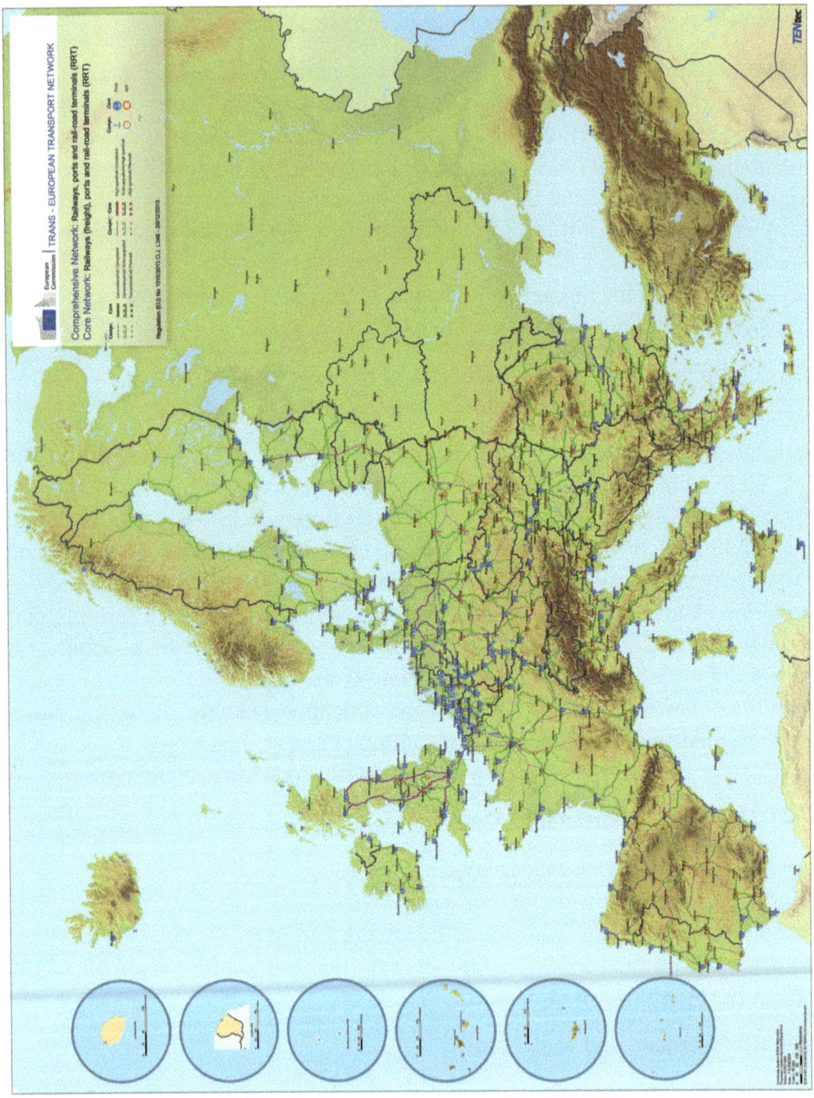

Fig. 4.9 The rail, port and rail-road terminal plan. *Source*: EC (2015)

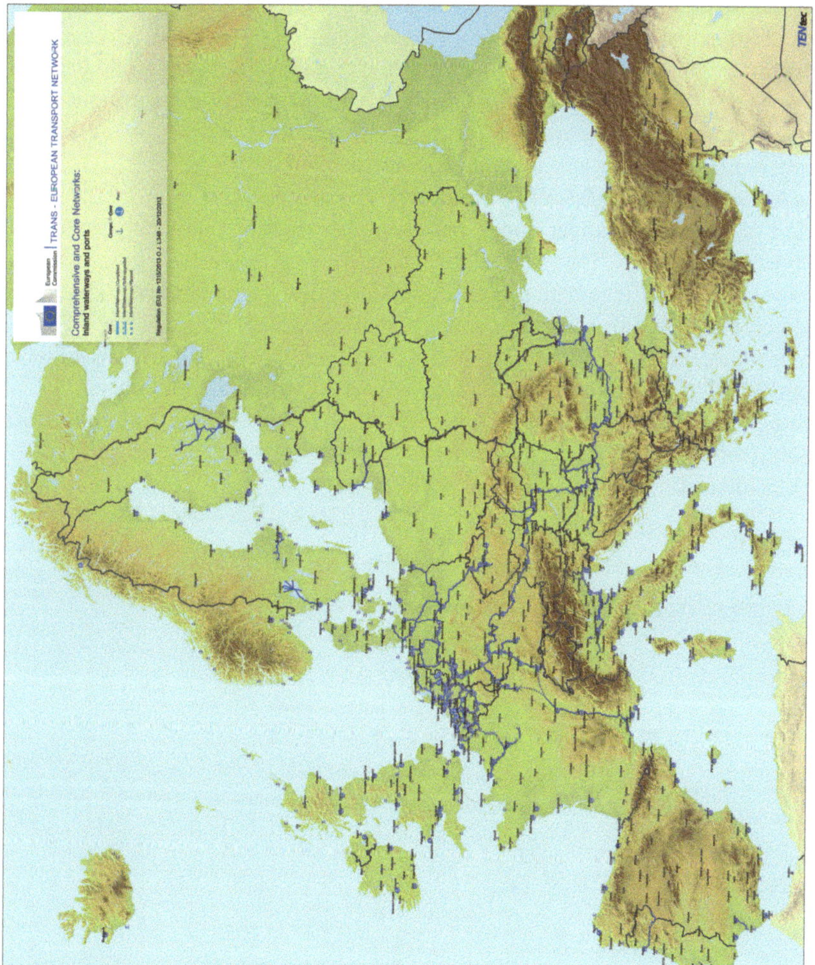

Fig. 4.10 The inland waterway and port TEN-T plan. *Source*: EC (2015)

interconnection between sea transportation and other modes of transportation. Seaports provide equipment and services to transportation operators. Their infrastructure provides a range of services for passenger and goods transportation, including ferry services and short- and long-distance shipping services, including coastal shipping, within the Union and between the latter and third countries.

The **Motorways of the Sea (MoS)** network concentrates flows of freight on sea-based logistical routes so as to improve existing maritime links and establish new viable, regular and frequent links for the transportation of goods between Member States. The concept builds on EU's goal of transforming shipping into a genuine alternative to overcrowded land transportation, and aims at introducing new intermodal maritime-based logistics chains in Europe.

The MoS network consists of facilities and infrastructure concerning at least two ports in two different Member States, one maritime operator and ideally hinterland transportation operators. The projects can include elements, such as port facilities, electronic logistics management systems, safety, security, administrative and customs procedures, as well as infrastructure for direct land and sea access, including dredging and icebreaking facilities. The projects of common interest of the MoS network should be proposed by at least two Member States and adhere to a tendering process.

Through Priority Project 21 of the TEN-T (see below), the following four corridors (refer to Fig. 4.11) have been designated for setting up projects of European interest:

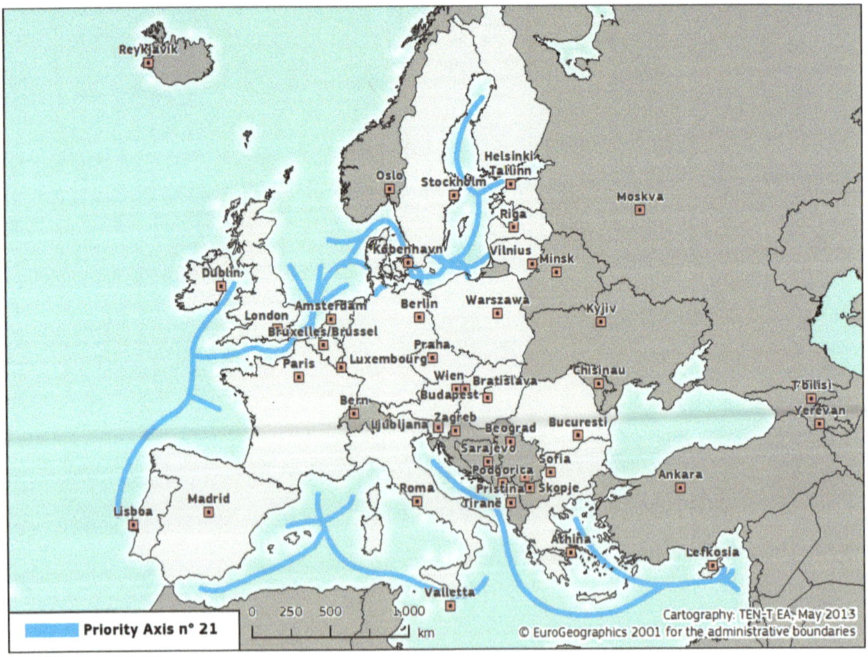

Fig. 4.11 The motorways of the sea. *Source*: INEA (2015)

- Motorway of the Baltic Sea,
- Motorway of the Sea of western Europe,
- Motorway of the Sea of south-east Europe, and
- Motorway of the Sea of south-west Europe.

The **airport network** comprises airports situated within the EU which are open to commercial air traffic (refer to Fig. 4.8). They should permit the development of air links, both within the EU and between the EU and the rest of the world, as well as the interconnection with other modes of transportation.

Airports are classified into international, Union and regional connecting points according to a set of criteria. The international and Union connecting points form the core of the TEN-T airport network. Airport projects can qualify as projects of common interest provided that they meet the following specifications:

- Optimization of existing airport capacity, and in particular:
  - Optimization of the existing capacity in terms of aircraft, passenger or freight movements, including the airport's air navigation equipment (all classes)
  - Improvement of airport security and safety (all classes)
  - Adaptation of existing infrastructures made necessary by completion of the internal market and in particular by the measures governing the free movement of persons within the Union (all classes)
- Development of new airport capacities, and in particular:
  - Development of the infrastructure and equipment which determine airport capacity in terms of aircraft, passenger or freight movements, including the airport's air navigation equipment (international and Union classes)
  - Construction of new airport to replace an existing airport or airport system which cannot be developed further on its site (international and Union classes)
- Improvement of environmental compatibility in terms of noise and the treatment of airport effluent (international and Union classes)
- Improvement or development of airport access, and in particular:
  - Improvement or development of interfaces between the airport and access infrastructures (international and Union classes)
  - Improvement and development of interconnections with other transportation networks, and more specifically the rail network (international and Union classes).

The **combined transport network** comprises railways and inland waterways which, together with the shortest possible road haulage in relation to the first and last miles, permit long-distance combined transportation of goods. It also comprises intermodal terminals equipped with installations permitting transshipment between railways, inland waterways, shipping routes and roads, as well as suitable rolling stock as required.

The outline plan of the combined rail-road terminals appears in Fig. 4.9. In addition to those shown on the plan, projects of common interest could concern:

- construction or upgrading of railway or inland waterway infrastructures in order to make the transportation of intermodal loading units technically possible and economically viable,
- construction or development of centers for transfers between inland types of transportation, including the setting up within the terminal of transshipment equipment with the corresponding infrastructure,
- adaptation of port areas, making it possible to develop or improve combined transportation between sea transportation and rail, inland waterway or road transportation,
- railway transportation equipment specially adapted to combined transportation where so required by the nature of the infrastructure.

Furthermore, the TEN-T Guidelines include the shipping management and information network, the air traffic management network and the positioning and navigation network.

4.3.3.4 The Priority Projects

The provisions of the guidelines concerning the so-called "priority projects" is the part of the document that attracts the highest attention due to special financing possibilities offered to these projects. They are projects of common interest, where examination confirms that they:

- are intended to eliminate a bottleneck or complete a missing link on a major route of the TEN-T, in particular projects which are of cross-border or cross-natural-barrier nature,
- are on such a scale that long-term planning at European level contributes significant added value,
- present, overall, potential socio-economic net benefits and other socio-economic advantages,
- significantly improve the mobility of goods and persons between Member States and thus also contribute to the interoperability of national networks,
- contribute to the territorial cohesion of the Union by integrating the networks of the new Member States and improving connections with the peripheral and island regions,
- contribute to the sustainable development of transportation by improving safety and reducing environmental damage caused by transportation, in particular by promoting a modal shift towards railways, intermodal transportation, inland waterways and maritime transportation,
- demonstrate commitment on the part of the Member States concerned to carrying out studies and evaluation procedures in time to complete the work in accordance with a date agreed in advance, based upon national plans or any other equivalent document relating to the project in question.

The Guidelines provide in annex a list of 30 priority projects, on which work was due to start before 2010, together with the agreed date of completion. This list is presented in Table 4.2 below.

But what makes these projects so special? Article 24 of the Guidelines declares these priority projects to be of "European interest," while Article 25 forces Member States to give appropriate priority to the projects declared to be of European interest when submitting their projects under the Cohesion Fund and the budget for the trans-European networks. The implications are straightforward in an environment of restricted budgets at both the Union and the Member State level.

Table 4.2 The priority projects of the 2010 TEN-T Guidelines (*Source*: EP&C, 2010)

Number	Description
PP1	Railway axis Berlin-Verona/Milan-Bologna-Naples-Messina-Palermo
PP2	High-speed railway axis Paris-Brussels-Cologne-Amsterdam-London
PP3	High-speed railway axis of south-west Europe
PP4	High-speed railway axis east
PP5	Betuwe line
PP6	Railway axis Lyon-Trieste-Divača/Koper-Divača-Ljubljana-Budapest-Ukrainian border
PP7	Motorway axis Igoumenitsa/Patra-Athens-Sofia-Budapest
PP8	Multimodal axis Portugal/Spain-rest of Europe
PP9	Railway axis Cork-Dublin-Belfast-Stranraer
PP10	Malpensa airport
PP11	Öresund fixed link
PP12	Nordic triangle railway/road axis
PP13	Road axis UK/Ireland/Benelux
PP14	West coast main line
PP15	Galileo
PP16	Freight railway axis Sines/Algeciras-Madrid-Paris
PP17	Railway axis Paris-Strasbourg-Stuttgart-Vienna-Bratislava
PP18	Inland waterway axis Rhine/Meuse-Main-Danube
PP19	High-speed rail interoperability in the Iberian peninsula
PP20	Railway axis Fehmarn Belt
PP21	Motorways of the Sea
PP22	Railway axis Athens-Sofia-Budapest-Vienna-Prague-Nuremberg/Dresden
PP23	Railway axis Gdańsk-Warsaw-Brno/Bratislava-Vienna
PP24	Railway axis Lyon/Genova-Basel-Duisburg-Rotterdam/Antwerp
PP25	Motorway axis Gdańsk-Brno/Bratislava-Vienna
PP26	Railway/road axis Ireland/United Kingdom/continental Europe
PP27	'Rail Baltica' axis Warsaw-Kaunas-Riga-Tallinn-Helsinki
PP28	'Eurocaprail' on the Brussels-Luxembourg-Strasbourg railway axis
PP29	Railway axis of the Ionian/Adriatic intermodal corridor
PP30	Inland waterway Seine-Scheldt

It is worth noticing that the selected projects show a clear tendency for a preferential treatment of the railway sector. The high share of rail projects in the TEN-T program reflects the EC's long-standing intention to bring about a modal shift in intra-European cargo transportation from road to rail (and/or sea), in order to improve the environmental effects of transportation (EP&C, 2010, preamble, recital [14]).

4.3.3.5 Implementation of the Priority Projects

The financing of infrastructure projects in the EU is supported by various instruments, including the TEN-T budget, the Structural and Cohesion Funds, and loans from the EIB. The Structural and Cohesion Funds have been a major source of finance for the investment needed to reduce imbalances in transportation endowment in lagging regions across the EU. The TEN-T budget currently co-finances projects on the TEN-T network.

Nevertheless, Community financial instruments have so far not been able to bring about a full and timely completion of all projects involved. Insufficient finance—both public and private—is probably the most important obstacle in infrastructure development. This has also been identified as one of the main reasons for delays in the implementation of certain TEN-T priority projects.

According to the 2010 annual progress report of the TEN-T Executive Agency[4] (TEN-T EA), the status of the 30 TEN-T priority projects is graphically depicted in Fig. 4.12. Five of these 30 projects have been completed:

PP5: Betuwe Line
PP9: Railway axis Cork-Dublin-Belfast-Stranraer
PP10: Malpensa airport
PP11: Öresund fixed link
PP14: West coast main line,

while significant progress has been made in some other projects. The opening of high-speed lines in Germany, Italy, Spain, France and the Benelux countries has considerably improved accessibility and brought people closer together. Rail has already captured market shares from aviation and from the passenger car. However, other projects have not been as successful: a couple of projects such as the trans-Alpine rail tunnels on Brenner and Fréjus have been designated as a 'priority' for about 20 years but they remain critical bottlenecks since then.

[4] The TEN-T EA was established in 2006 to follow the technical and financial implementation of all TEN-T projects throughout their entire lifecycle, to provide support to the beneficiaries of TEN-T financing and to coordinate with other institutional partners. It became autonomous in 2008 and was succeeded by the Innovation and Networks Executive Agency (INEA) as of 1 January 2014.

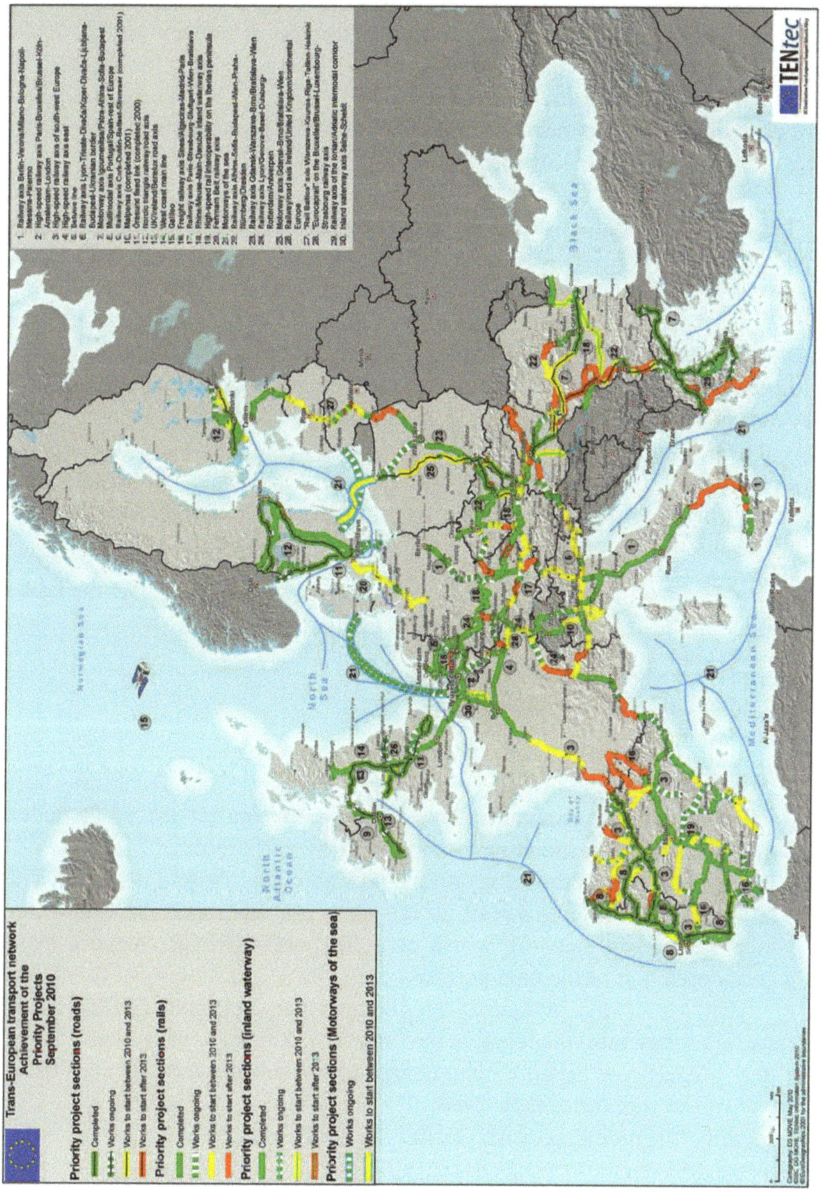

Fig. 4.12 Status of the 30 TEN-T priority projects as of September 2010. *Source*: EC (2010)

4.3.4 The Recent Reform of the TEN-T

As already mentioned in Chap. 1, in December 2013 the European Parliament and the Council adopted a legislative package defining a new policy framework for the TEN-T, which was proposed by the European Commission back in October 2011. The package includes a Regulation on the new Union guidelines for the TEN-T development with a time horizon extending to 2050 (EP&C, 2013a) and a Regulation for establishing the Connecting Europe Facility (CEF), which will govern EU funding until 2020 (EP&C, 2013b).

The TEN-T Guidelines, as the first component of the package, establish the policy basis by defining network plans including infrastructure standards, objectives and priorities for action. A dual layer network structure has been introduced, consisting of a comprehensive and a core network. The comprehensive network constitutes the basic layer of the TEN-T and is, in large part, derived from the corresponding national networks. The core network, on the other hand, overlays the comprehensive network and contains its strategically most important parts.

The core network is the result of a genuine European network planning methodology that combines geographical and economic criteria. It builds on the key nodes of political, economic, cultural and transportation-related importance and links them through all available transportation modes (EC, 2011a). More specifically, the design of the core network involved the following steps:

Step 1. Identification of the main nodes of the Core Network. These are the nodes of the highest strategic importance in the EU:

- main nodes for passengers and freight,
- main nodes for freight only,
- main nodes for passengers only.

Step 2. Identification of the links between the main nodes. Multimodal links were selected from the comprehensive network to connect the main nodes, following the corresponding (potential) main traffic flows.

Applying this methodology on inland waterways showed that almost all of them would become part of the core network. For this reason, the entire inland waterway network which complies with UNECE category IV is considered part of the core network.

The "Motorways of the Sea" are the maritime dimension of the TEN-T. As far as they fulfil the function of core network links or of sections thereof (e.g. linking core network main nodes across the sea), they are considered part of the core network, as well.

Step 3. Merging the modal network parts to the multimodal core network.

The functions of the comprehensive and the core network complement each other: whereas the purpose of the comprehensive network is to serve accessibility functions and ensure a balanced infrastructure endowment throughout the Union, the core network pioneers the development of a sustainable mobility network. It shall be completed as a priority, by 2030. The new policy basis provides more clarity with regard to the identification of a broad range of "projects of common interest" (including the closing of missing physical links, infrastructure upgrading to target standards, ITS or innovative equipment).

To facilitate implementation of the core network, the Guidelines introduce the instrument of "core network corridors"—a coordination tool aiming at coherent project implementation and at promoting technological, operational and governance-related innovation. The core network corridors also aim to strengthen a "systems" approach that links transportation infrastructure development with related transportation policy measures. Eventually, this approach seeks to promote higher resource efficiency to achieve the EU carbon emissions' reduction objectives in the transportation sector.

Due to the broad range of measures addressed with the new Guidelines, many different actors will have to contribute to their implementation. The proposed corridor governance structures (see Sect. 4.2.3) intend to foster cooperation of the various actors. Existing activities such as the rail freight corridors introduced with Regulation No 913/2010 will form an integral part of core network corridor developments.

Vis-à-vis the TEN-T guidelines, the CEF, as the financing instrument, sets out funding priorities in transportation, energy and digital broadband for the period 2014–2020, as well as the corresponding rules. Regarding transportation, it defines a geographical basis for the corridor approach and pre-identifies the most mature projects along those corridors. Annex I to the CEF Regulation lists the nine core network corridors that form the basic part of the TEN-T core network. They appear in Fig. 4.13 and, in metro format, in Fig. 1.2 of Chap. 1.

4.4 How Do Green Corridors Relate to the TEN-T?

Figure 4.14 depicts the land part of the TEN-T core network, proposed in 2011, plotted against the nine SuperGreen corridors (refer to Sect. 3.4). The geographic overlap is impressive, even after accounting for the fact that the priority projects of the TEN-T were taken into consideration, among several other criteria, when selecting the SuperGreen corridors in June 2010.

How about the conceptual relation though? Do these corridors exhibit the green characteristics as those were identified in Sect. 3.3? To refresh the reader's memory, these characteristics are:

(a) Reliance on co-modality, which in turn requires:

   - adequate transshipment facilities at strategic locations, and
   - integrated logistics concepts.

(b) Reliance on advanced technology leading to:

   - improved energy efficiency, and
   - use of alternative clean fuels.

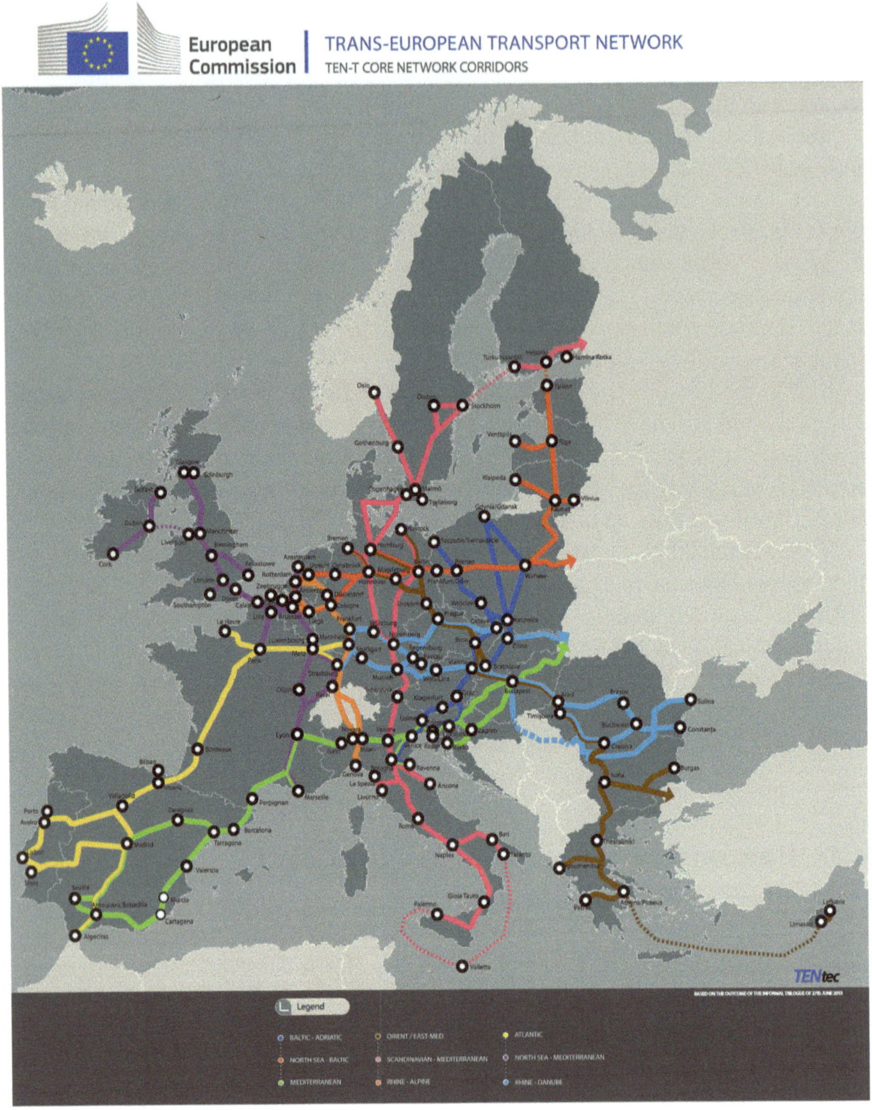

Fig. 4.13 The TEN-T core network corridors. *Source*: EC (2015b)

(c) Development and demonstration capabilities of environmentally-friendly and innovative transportation solutions, including advanced telematics applications.
(d) Collaborative business models.

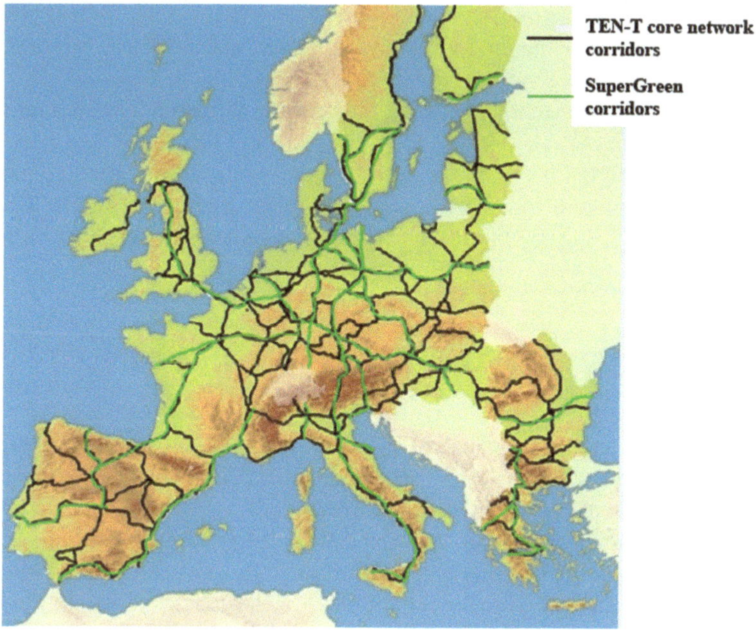

Fig. 4.14 The SuperGreen and TEN-T core network corridors. *Source*: Panagakos (2012)

The provisions of the new TEN-T Guidelines in relation to these characteristics are presented below:

4.4.1 Reliance on Co-modality

Although the term co-modality is not mentioned, the Guidelines include several references to multimodality. In fact, there is an entire section (Section 6 of EP&C, 2013a) devoted to the 'infrastructure for multimodal transport' that refers to the comprehensive network and includes logistic platforms. When it comes to the core network, Article 42 is crystal clear:

> ... In order to lead to resource-efficient multimodal transport, ... core network corridors shall be focused on modal integration, interoperability, and a coordinated development of infrastructure.

4.4.2 Adequate Transshipment Facilities

The TEN-T Guidelines provide for:

- the connection of rail freight terminals with the road infrastructure or, where possible, the inland waterway infrastructure of the comprehensive network (Article 12),
- the connection of inland ports with the road or rail infrastructure (Article 15),

- the connection of maritime ports with railway lines or roads and, where possible, inland waterways of the comprehensive network, except where physical constraints prevent such connection (Article 22),
- multimodal interconnections between airports and infrastructure of other transportation modes (Article 26),
- seamless connection between the infrastructure of the comprehensive network and the infrastructure for regional and local traffic and urban freight delivery, including logistic consolidation and distribution centers (Article 30).

4.4.3 Integrated Logistics Concepts

It is worth mentioning that the general objective of the TEN-T is to "... strengthen the social, economic and territorial cohesion of the Union and contribute to the creation of a **single European transport area** which is efficient and sustainable, increases the benefits for its users and supports inclusive growth" (Article 4).

Furthermore, one of the criteria for identifying 'projects of common interest,' which comprise the building blocks of the TEN-T, is the demonstration of 'European added value' (Article 7) which, in turn, is defined as "... the value of a project which, in addition to the potential value for the respective Member State alone, leads to a significant improvement of either transportation connections or transportation flows between the Member States which can be demonstrated by reference to improvements in efficiency, sustainability, competitiveness or cohesion ..." (Article 3).

4.4.4 Reliance on Advanced Technology

There are numerous references to advanced technology applications including ICT. The following is an indicative list:

- "[TEN-T contributes to efficiency through] ... cost-efficient application of innovative technological and operational concepts" (Article 4),
- "The TEN-T shall be planned, developed and operated in a resource-efficient way, through ... the deployment of new technologies and telematics applications, where such deployment is economically justified" (Article 5),
- "In the development of the comprehensive network, general priority shall be given to measures that are necessary for ... implementing and deploying telematics applications and promoting innovative technological development" (Article 10),
- "Telematics applications shall, for the respective transport modes, include in particular ERTMS (for railways), RIS (for inland waterways), ITS (for road transport), VTMIS and e-Maritime services (for maritime transport) and the SESAR system (for air transport)" (Article 31),

- "In order for the comprehensive network to keep up with innovative technological developments and deployments, the aim shall be in particular to support and promote the decarbonization of transport through transition to innovative and sustainable transport technologies" (Article 33),
- "The core network corridors shall support the comprehensive deployment of interoperable traffic management systems and, where appropriate, the use of innovation and new technologies" (Article 42).

4.4.5 Energy Efficiency

Relevant references include:

- "In the development of the comprehensive network, . . . particular consideration shall be given to measures that are necessary for . . . ensuring fuel security through increased energy efficiency, and promoting the use of alternative and, in particular, low or zero carbon energy sources and propulsion systems" (Article 10),
- "Member States shall pay particular attention to projects of common interest which both provide efficient freight transport services that use the infrastructure of the comprehensive network and contribute to reducing carbon dioxide emissions and other negative environmental impacts, and which aim to stimulate resource and carbon efficiency, in particular in the fields of vehicle traction, driving/steaming, systems and operations planning" (Article 32).

4.4.6 Use of Alternative Clean Fuels

The TEN-T Guidelines provide direct references to alternative fuels for all transportation modes:

- "Member States shall ensure that the railway infrastructure, save in the case of isolated networks, is fully electrified as regards line tracks and, to the extent necessary for electric train operations, as regards sidings" (Article 12),
- "Projects of common interest for motorways of the sea . . . may also include activities . . . for improving environmental performance, such as the provision of shore-side electricity . . . and alternative fuelling facilities . . ." (Article 21),
- "In order for the comprehensive network to keep up with innovative technological developments and deployments, the aim shall be in particular to make possible the decarbonization of all transport modes by stimulating energy efficiency, introduce alternative propulsion systems, including electricity supply systems, and provide corresponding infrastructure" (Article 33),

- As for the core network, Article 39 stipulates full electrification of the line tracks and selective sidings for the railways, while alternative clean fuels should be available for the road, inland waterway and maritime transport infrastructures. For air transport, the relevant requirements are reduced to the "... capacity to make alternative clean fuels available."

4.4.7 Development of Innovative Logistics Solutions

The promotion of innovative solutions is mentioned several times in the guidelines:

- "In the development of the comprehensive network, general priority shall be given to measures that are necessary for ... promoting the efficient and sustainable use of the infrastructure ..." (Article 10),
- "When developing the comprehensive network in urban nodes, Member States shall, where feasible, aim to ensure promotion of efficient low-noise and low-carbon urban freight delivery" (Article 30),
- "Member States shall pay particular attention to projects of common interest which ... aim to promote the deployment of innovative transport services ..." (Article 32),
- "Projects of common interest relate to all directly concerned stakeholders, ... [who may contribute to] ... the promotion of sustainable transport solutions, such as enhanced accessibility by public transport, telematics applications, intermodal terminals/multimodal transport chains, low-carbon and other innovative transport solutions and environmental improvements" (Article 50).

4.4.8 Collaborative Business Models

Although no direct reference to business models can be found in the guidelines, there are several ones relating to the need for enhanced cooperation among stakeholders including provision of information:

- "The ... core network corridors, is a strong means of realizing the respective potential of stakeholders, of promoting cooperation between them and of strengthening complementarity with actions by Member States" [Preamble (50)],
- "Member States shall ensure ... that freight terminals and logistic platforms, inland and maritime ports and airports handling cargo are equipped for the provision of information flows within this infrastructure and between the transport modes along the logistic chain" (Article 28),
- "Telematics applications shall be such as to enable traffic management and the exchange of information within and between transport modes for multimodal transport operations and value-added transport-related services, improvements in safety, security and environmental performance, and simplified administrative procedures" (Article 31),

- "Member States shall pay particular attention to projects of common interest which ... aim to promote the deployment of innovative transport services ... through ... the establishment of relevant governance structures" (Article 32),
- "Member States shall pay particular attention to projects of common interest which ... aim to facilitate multimodal transport service operations, including the necessary accompanying information flows, and improve cooperation between transport service providers" (Article 32).

The above references lead to the conclusion that all green characteristics of a corridor that have been identified in Sect. 3.3 are shared more or less by the TEN-T core network corridors, as they have been introduced in the new Guidelines. In conjunction with the enabling governance structure of Sect. 4.2.3, we can conclude that, through the freight dimension of the TEN-T core network, the new TEN-T Guidelines have established a network of green corridors in Europe.

Acknowledgements Work reported herein was supported in part by (a) EU project SuperGreen, (b) the Lloyd's Register Foundation in the context of the Centre of Excellence in Ship Total Energy-Emissions-Economy at the National Technical University of Athens, the author's former affiliation, and (c) an internal grant at the Technical University of Denmark. The Lloyd's Register Foundation helps to protect life and property by supporting engineering-related education, public engagement and the application of research.

References

Arnold, J. (2006, December). *Best practices in Management of International Trade Corridors*. Trade Logistics Group, Transport Papers TP-13. Washington, DC: The World Bank Group.

Butcher, L. (2012, February 7). *Trans-European Transport Networks (TEN-T)*. Library of the House of Commons, Standard Note: SN478.

Corridor A/IQ-C (2011, August). *6th Progress Report*. Executive Board ERTMS Corridor A and International Group for Improving the Quality of Rail Transport in the North-South Corridor.

Engström, R. (2011, June 16). The Swedish Green Corridor initiative—History, current situation and thoughts about the future. In *Proceedings of the 8th SoNorA University Think Tank Conference, Szczecin, Poland*.

European Commission. (1990, December 10). *Towards trans-European networks: For a community action programme*. COM(90) 585, Brussels.

European Commission. (1993, December 5). *White paper on growth, competitiveness, employment: The challenges and ways forward into the 21st century*. COM(93) 700, Brussels.

European Commission. (2010, June). *Trans-European transport network: Progress report 2010, implementation of the priority projects*.

European Commission. (2011a). *Planning methodology for the trans-European transport network (TEN-T), as used for the Commission proposal made on 19 October 2011*.

European Commission. (2011b, June 30). *Handbook on the regulation concerning a European rail network for competitive freight (Regulation EC 913/2010)*. A DG MOVE staff working document, Brussels.

European Commission. (2015a). Retrieved January 6, 2015, from http://ec.europa.eu/transport/infrastructure/tentec/tentec-portal/site/en/maps.html

European Commission. (2015b). Retrieved January 6, 2015, from http://ec.europa.eu/transport/themes/infrastructure/doc/ten-t-country-fiches/ten-t-corridor-map-2013.pdf

European Community. (1992, February 7). *The Maastricht Treaty: Provisions amending the treaty establishing the European Economic Community with a view to establishing the European Community*. Maastricht.

European Council. (1993). *Conclusions of the Presidency*. European Council in Copenhagen, 21–22 June 1993.

European Economic Community. (1957, March 25). *The treaty of Rome*.

European Parliament and Council. (2010, July 7). *Decision no 661/2010/EU of the European Parliament and of the Council of 7 July 2010 on union guidelines for the development of the Trans-European Transport Network (recast)*. Strasbourg.

European Parliament & Council. (2013a, December 11). *Regulation (EU) no 1315/2013 of the European Parliament and of the Council of 11 December 2013 on union guidelines for the development of the Trans-European Transport Network and repealing decision no 661/2010/EU*. Strasbourg.

European Parliament & Council. (2013b, December 11). *Regulation (EU) no 1316/2013 of 11 December 2013 establishing the Connecting Europe Facility, amending regulation (EU) no 913/2010 and repealing regulations (EC) no 680/2007 and (EC) no 67/2010*. Strasbourg.

FIAT131RACING. (2015). *Heroes of the past: 1903 Paris-Madrid race*. Retrieved January 6, 2015, from http://fiat131racing.canalblog.com/albums/heroes_of_the_past__heros_d_antan/photos/38908679-1903_paris_madrid___louis_renault__renault___mechanic_ferenc_szi.html

Galonske, N. (2004, December 21). *Corridor Management System (CMS)—Components and interaction, implementation and demonstration schedule*. BRAVO project.

INEA. (2015). Retrieved January 6, 2015, from http://inea.ec.europa.eu/en/ten-t/ten-t_projects/30_priority_projects/priority_project_21/priority_project_21.htm

Mertel, R., & Sondermann, K.-U. (2007, December 6). *Final report for publication*. BRAVO project.

Panagakos, G. (2012, December 17). *Green Corridors Handbook—Vol. II*. A SuperGreen project document.

Rousseaux, P. (2012, May 10). *Rail freight corridors*. Presentation in the RNE Business Conference, Frankfurt, Germany.

Schipper, F., & Van der Vleuten, E. (2008). Trans-European network development and governance in historical perspective. *Network Industries Quarterly, 10*(3), 5–7.

Sichelschmidt, H. (1999). The EU programme "trans-European networks"—A critical assessment. *Transport Policy, 6*(1999), 169–181.

Stasinopoulos, D. (1995). Common transport infrastructure policy and the development of Trans-European Networks. *Journal of Transport Economics and Policy, 29*(2), 220–222.

United Nations Economic Commission for Europe. (1950, September 16). *Declaration on the construction of main international traffic arteries*. Inland Transport Committee, Geneva.

Chapter 5
Benchmarking the SuperGreen Corridors with Green Technologies

Chara Georgopoulou, Nikolaos M.P. Kakalis, Valerio Recagno, and Sara Fozza

Abstract Among the objectives of the SuperGreen project was the establishment of a green corridor benchmarking methodology. This chapter describes the characteristics and functions of a green corridor benchmarking exercise using green technologies, starting from the SuperGreen paradigm and arriving at wider conclusions. This exercise estimates (a) the baseline performance of six of the SuperGreen corridors according to the set of KPIs established in Chap. 3; and (b) the potential improvement of the corridor performance from an assumed use of advanced technologies. Using this benchmark, we arrive at conclusions on the role of green technologies on the development of a more sustainable EU transportation system. The scope of the chapter is then extended beyond the presentation of the SuperGreen results, to point out the necessary features of an efficient corridor benchmarking methodology and analyse the technologies that could have a pivotal role in the improvement of EU transportation corridors, including alternative fuels, hybrid systems, and energy efficiency measures. Particular focus is given to the preparatory phases prior to the creation of a benchmark, including the analysis of the baseline EU corridor technologies and the investigation of green technology application areas over the corridors. It is seen that he SuperGreen benchmark may result in a set of green technology enablers that could be applied on the corridors, accompanied by estimates on their greening potential and capacity to solve bottlenecks, as well as the barriers on their deployment. The role of ICT technologies is covered in Chap. 6.

C. Georgopoulou • N.M.P. Kakalis (✉)
DNV GL Maritime Research & Development, Aitolikou 5, Piraeus 18545, Greece
e-mail: chara.georgopoulou@dnvgl.com; nikolaos.kakalis@dnvgl.com

V. Recagno • S. Fozza
D'Appolonia S.p.A., Via S. Nazaro 19, Genova 16145, Italy
e-mail: valerio.recagno@dappolonia.it; sara.fozza@dappolonia.it

H.N. Psaraftis (ed.), *Green Transportation Logistics*, International Series in Operations Research & Management Science 226,
DOI 10.1007/978-3-319-17175-3_5

Abbreviations

CNG	compressed natural gas
DSS	Deep sea shipping
EC	European Commission
GHG	Green house gas
GNSS	Global Navigation Satellite Systems
ICT	Information and Communication Technologies
ITS	Intelligent Transport Systems
IWW	Inland waterways
KPI	Key performance indicator
LNG	Liquefied natural gas
NG	Natural gas
NGV	NG-fuelled vehicles
NSB	Norwegian National Railway
OECD	Organization for Economic Co-operation and Development
PM	Particulate matter
R&D	Research and Development
RIS	River Information Services
SECU	Stora Enso Cargo Unit
SSS	Short sea shipping
TEN-T	Trans-European Transport Network
TTW	Tank-to-Wheel
WTW	Well-to-Wheel

5.1 Introduction

As per Chap. 3 of this book, transportation corridors are described as key players in international and regional economic growth. Due to their importance, their greening would eventually lead to more environmentally-friendly transportation at the international and regional levels. Evidently, the question is: how can a corridor become greener? This chapter attempts to shed light on this question, by focusing on the technologies that can be used towards that goal.

Already the development of the SuperGreen green corridor benchmarking methodology (see Chap. 3) resulted, among other things, in the selection of some Key Performance Indicators (KPIs) that can be used to assess the performance of a corridor. A question is, can these KPIs be improved by using appropriate technologies? The present chapter looks into this question and presents the findings on the possible impact of these technologies on corridor greening. We focus on technologies other than Information and Communication Technologies (ICT), such as vehicles, engines, alternative fuels, etc. This chapter can thus be viewed as complementary to Chap. 6, which focuses on the role of smart ICT in green corridors.

The structure of the rest of this chapter is as follows. Section 5.2 presents the objective and the background of the benchmarking methodology. Section 5.3 presents the results from employing the benchmarking methodology on the SuperGreen network. Section 5.4 analyses the findings on the role of key technology enablers for the greening of the corridors. Finally, Sect. 5.5 concludes with the key benchmark outcomes and suggestions for future R&D on green corridor analysis.

5.2 How to Benchmark a Green Corridor

5.2.1 Objective

One of the tasks of a green corridor benchmarking exercise is to establish a *baseline* for the comparative evaluation of a corridor's status against the hypothetical situation that market-ready technological measures are used to improve corridor operations and alleviate bottlenecks. In that sense, the benchmark quantifies the gap of the actual corridor status from the best possible conditions that could be achieved via the use of advanced technologies across the corridor's territory. In this respect, it is necessary to identify: (a) what is the most promising technology mix to improve the corridor performance, (b) what are the best practices in utilising these technologies and, (c) what is the expected change in corridor performance. Therefore, the benchmark can be used for:

- Corridor evaluation against environmental, energy efficiency and service quality criteria;
- Evaluation of the technological solutions to improve current corridor performance;
- Evaluation of the potential corridor greening.

As the use of environmental-friendly technologies is a prerequisite for greening a corridor, it is essential that the benchmark is built around a wide coverage of technologies.

5.2.2 Experience from the Literature

In the literature there is a variety of methods to analyse and quantify the performance of transportation networks, though lacking the sufficiency to address large-scale systems like corridors. While many studies assess individual modes and routes, no consistent way has been applied before to assess a transportation network of such complexity as a corridor. The SuperGreen project firstly developed a green corridor benchmark based on past research on the aforementioned topics.

An intermodal transportation chain benchmarking methodology was developed in the EC-funded project BE LOGIC[1] (2008–2011) (Kramer et al., 2009). The project aimed to benchmark intermodal transportation policies, chains and terminals, in order to assess their efficiency and, therefore, support the improvement of logistics operations. In this framework, the project developed and implemented a transport chain benchmarking tool, which served as a basis for the SuperGreen work.

Projects on innovative transportation systems were also considered the SuperGreen analysis. In the EC-funded Railenergy[2] project (2006–2009), the drastic reduction of rail energy consumption and emissions within an optimized railway system was investigated. The GHG TransporD[3] project (2009–2011) targeted to analyse emission reduction solutions for all transportation modes, with a time horizon from 2020 to 2050. The project resulted in a list of technologies related to maritime, railway and road transportation modes, covering categories such as fuels and sources of energy, engines and propulsion systems, vehicles, loading units and navigation technologies. The CREATING[4] project (2004–2007) and the PLATINA[5] technology platform assessed innovative systems for inland waterways. The EC-funded PROMIT[6] project (2006–2009) investigated intermodal technology options to increase the speed of procedures and intermodal logistics operations, including heating/cooling and cargo handling systems.

As the definition of accurate performance metrics is important, particular attention was paid to existing methods and software tools for the evaluation of transportation emissions, such as the EcoTransIT[7] tool. Platforms like the Shipping KPI (Key Performance Indicator) benchmarking system[8] and the InteGRail[9] project (2005–2008) KPI tool for railways were reviewed and considered in this work.

---

[1] BE LOCIG project, "*Benchmark Logistics for Co-modality*", www.be-logic.info

[2] Railenergy project, "*Innovative Integrated Energy Efficiency Solutions for Railway Rolling Stock, Rail Infrastructure and Train Operation*", www.railenergy.org

[3] GHG TransporD project, www.ghg-transpord.eu

[4] CREATING project, "*Concepts to Reduce Environmental impact and Attain optimal Transport performance by Inland NaviGation*", www.creating.nu

[5] www.naiades.info/platina

[6] PROMIT project, "*Promoting Innovative Intermodal Freight Transport*", www.promit-project.net

[7] www.ecotransit.org

[8] www.shipping-kpi.org

[9] InteGRail project, "*Intelligent Integration of Railway Systems*", www.integrail.info

5.2.3 The Green Corridor Benchmarking Methodology

The green corridor benchmarking methodology was structured in three phases as per Fig. 5.1.

We next describe each phase in some more detail.

5.2.3.1 Phase 1: Analysis of the Corridor Baseline

According to the methodology description of Chap. 3, the identification of the major cargo flows and the decomposition into representative transport chains was based on a questionnaire survey, which was handed to transportation operators and freight villages (a number of 15 questionnaires per corridor, on average). This included solicitation of information on typical trade routes, load factors, vehicle characteristics, time schedules, and other data (refer to Chap. 3 and to Salanne, Rönkkö, and Byring (2010), Ilves et al. (2011), Psaraftis and Panagakos (2012), Georgopoulou et al. (2012) and Georgopoulou, Kakalis et al. (2013), Aditjandra et al. (2012), Paalson et al. (2010). By compiling the survey results, an outlook of the most common corridor technologies was generated (see Table 5.1). More details are in Georgopoulou et al. (2011).

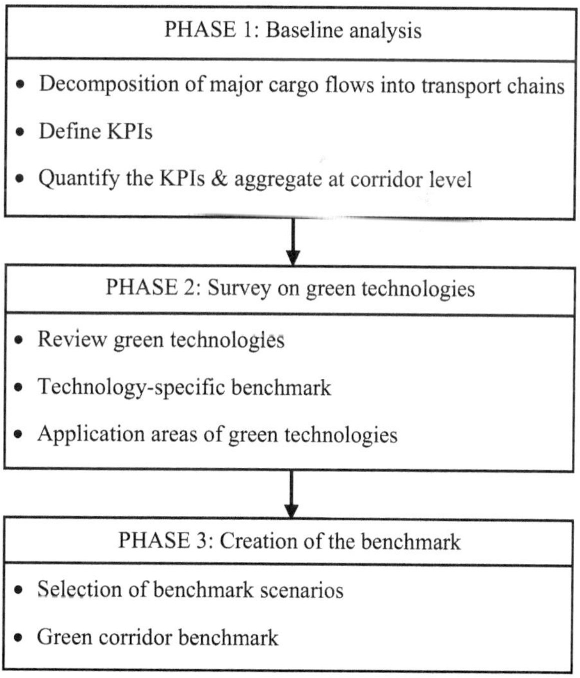

Fig. 5.1 Corridor benchmark development phases (Georgopoulou, Kakalis et al., 2013)

Table 5.1 Baseline technologies per SuperGreen corridor and technology category

	Mode	Description of the technology	Corridor coverage
Engines and propulsion systems	Road	Heavy duty diesel engines	Brenner, Edelweiss, Cloverleaf, Nureyev
	Rail	Electric traction	Brenner, Edelweiss, Cloverleaf, Nureyev
	SSS/DSS	Two/four stroke diesel engines	All maritime corridors
Fuels and energy sources	Road	Diesel oil	Nureyev, Cloverleaf
	Rail	Electricity	Brenner, Cloverleaf, Nureyev, Silk Way
	SSS/DSS	Marine heavy fuel	Mare Nostrum, Nureyev
	IWW	Distillate fuel	Strauss
Cargo handling and transfer	Intermodal	Gantry cranes	All corridors—Hubs
		Automotive lift trucks	
Heating and cooling	Intermodal	Cooling using eco-friendly refrigerants	All corridors
Loading units and their treatment	Intermodal	SECU units (Storage Enso Cargo Unit)	Nureyev
Vehicles	Road	Euro I–VI trucks	Brenner, Edelweiss, Nureyev, Cloverleaf, Mare Nostrum
	Rail	Electrified rail; capacity: 500–1,500 ton	Brenner, Silk Way
	SSS/DSS	Conventional vessel designs	Mare Nostrum, Nureyev
	IWW	All kind of vessels: pushed convoy, MCV + lighters, JOWI class vessels, etc.	Strauss
Navigation technologies	Rail	European Rail Traffic Management System	Edelweiss, Brenner, Cloverleaf
	SSS/DSS	Global Navigation Satellite Systems (GNSS)	Nureyev, Mare Nostrum, Silk Way
	IWW	River Information Services (RIS)	Strauss
Best practices	SSS/DSS	Container/Ro-Ro vessel loading schemes	Nureyev, Mare Nostrum, Silk Way
	Intermodal	Practices for proper stowage of products	All corridors

Additionally, the survey revealed bottlenecks mainly caused by the lack of infrastructure and harmonized regulations between the EU countries:

- Traffic jam and delays to the delivery of goods due to insufficiency of freight networks at urban areas and around ports.
- Ports with mixed freight and passenger traffic usually face high congestion particularly at peak seasons like summer.
- Incompatibility problems caused by different properties of the infrastructure networks between countries, such as in railways.

- Congestion at narrow passages in land (e.g. the Brenner Pass) or at sea (e.g. the Dardanelles Straits) caused by geographical barriers and limitations.
- Weather problems, such as ice in the area of the Baltic Sea or harsh weather at the Mediterranean Sea, may cause delays.
- Deficits in road infrastructure can harm vehicle and cargo condition, increasing the relative transport cost.
- Time-consuming custom procedures during border crossing cause congestion and delays.
- Infrastructural insufficiency at terminals causes delays during loading/unloading at all transportation modes (road, rail, inland waterways, maritime).
- Freight thefts cause cargo security problems at long distance road trips.

As per Chap. 3, the SuperGreen KPIs were defined after consecutive stakeholder workshops other and meetings. The final set of KPIs is as follows:

- Relative transport cost [€/ton km];
- Average speed [km/h] (or transport time, in hours);
- Reliability [% of shipments delivered within acceptable time window];
- Service frequency [no of trips per year];
- CO_2 [g/ton km];
- SO_x [g/ton km].

To that effect, and as per the methodology of Chap. 3, the transport chain KPIs were calculated, composing the SuperGreen corridor performance shown in Table 5.2. Only six of the nine SuperGreen corridors were covered, mainly due to data availability issues. The results were expressed as ranges of values, which correspond to the minimum and maximum values of the transport chain KPIs.

5.2.3.2 Phase 2: Survey of Green Technologies

In the second phase of the methodology, a *technology-specific* benchmark was developed by collecting information on green technologies and their impact on the KPIs. This phase was organized into three tasks:

Task 1. Review of green technologies, including energy efficiency and emission reduction measures, cleaner fuels and operational practices. A technology categorization is necessary to manage the amount of information gathered.

Task 2. Assessment of the potential green technology impact on the KPIs by reviewing manufacturer datasheets, scientific works and published success stories. Depending on the availability of information, the technology assessment can be either qualitative or quantitative.

Task 3. Identification of green technologies application areas over the corridors, to support or replace conventional systems and techniques.

Table 5.2 Baseline corridor benchmark (Ilves et al., 2011)

Corridor	Mode	Cost (€/ton km)	Av. speed (km/h)	Reliability (%)	Frequency (no/year)	CO_2 (g/ton km)	SO_x (g/ton km)
Brenner	Inter-modal	0.03–0.09	9–41	95–99	26–624	10.62–42.11	0.02–0.14
	Road	0.05–0.07	19–40	50–99	104–2,600	46.51–71.86	0.05–0.08
	Rail	0.05–0.80	44–98	50–100	208–572	9.49–17.61	0.04–0.09
	SSS[a]	0.04	23	100	52		0.19
Clover-leaf	Road	0.06	40–60	80–90	4.68	68.81	0.09
	Rail	0.05–0.09	45–65	90–98	156–364	13.14–18.46	0.01–0.02
Nureyev	Inter-modal	0.10–0.18	13–42	80–90	156–360	13.43–33.36	0.03–0.15
	SSS	0.05–0.06	15–28	90–99	52–360	5.65–15.60	0.07–0.14
Strauss	IWW[b]	0.02–0.44	–	–	–	9.86–22.80	0.01–0.03
Mare Nostrum	SSS	0.003–0.2	17	90–95	52–416	6.44–27.26	0.09–0.4
	DSS[c]	–	–	–	–	15.22	0.22
Silk Way	Rail	0.05	26	–	–	41	–
	DSS	0.004	20–23	–	–	12.5	–

[a]Short sea shipping
[b]Inland waterways
[c]Deep sea shipping

In Task 1, as many as 263 measures were reviewed stemming from nine technology categories, grouped as follows (for more details see Recagno, Fozza et al. (2013) and Fozza and Recagno (2012)):

- **Engines and propulsion systems**: innovative technologies concerning engines and propulsion systems in general, which can be applied to any kind of transport modes on green corridors;
- **Fuels and energy sources**: technologies related to energy production, including for instance solar panels, wind turbines and other renewable energy sources; furthermore innovative fuels will also be considered;
- **Cargo handling and transfer technologies**: technologies related to loading or unloading or cargo, transfer of loading units between different transport modes, internal handling of transport units;
- **Cargo preparation technologies**: this category is relevant to all technologies used in preparing cargo before it is transported, such as preservatives for perishable goods, packaging, sealing, etc.;
- **Heating and cooling technologies**: this category includes innovative heating or cooling technologies embedded into transport vehicles, implemented into warehouses or used during handling and transfer operations;
- **Innovative loading units and their treatment (cleaning, etc.)**: this category includes new loading units able to reduce the time required for loading/unloading and transfer operations, as well as the energy consumption and pollution emissions in case they involve heating/cooling devices. It also considers ancillary technologies needed for pre- or post-transport treatment of the loading unit;
- **Vehicles**: new vehicle concepts with the purpose of improving transport time and reducing pollution emissions shall be reported in this category;
- **Navigation technologies**: this category is referred to technologies facilitating vehicles navigation during transport, including tracking/tracing, and automatic vehicles identification (AVI);
- **Best practices of technologies integration**: this category is dedicated to the identification of best practices derived from real cases, related to the integration of innovative technologies on transport systems, with particular reference to their impact on energy and carbon footprint reduction, and their potential for exportability to different environments.

In Table 5.3 the distribution of technologies per mode of transportation is shown.

In Task 2, the green technologies collected were described by means of well-defined indicators; this allowed the analysis of their relevant characteristics and the definition of the baseline needed for identifying the most promising technologies for further analysis in the scope of the project.

A number of industry and academic works were reviewed to collect material on the technology effects on the KPIs, including studies, reports, articles and documents released by EU technology platforms on technological innovations per transport mode (see for instance DNV (2010, 2012a, 2012b), Eide, Longva,

Table 5.3 Number of green technologies per technology category and transport mode

	Engines and propulsion systems	Fuels and energy sources	Cargo handling and transfer technologies	Cargo preparation technologies	Heating and cooling technologies	Innovative loading units and their treatment	Vehicles	Navigation technologies	Best practices of technology integration	Total
Inland waterways	11	10	3	0	0	0	4	2	0	30
Maritime	11	21	29	0	1	2	3	14	2	83
Railway	8	17	4	0	0	13	12	3	11	68
Road	7	18	0	0	2	1	17	3	1	49
Multimodal	0	3	16	6	5	3	0	0	0	33
Total	37	69	52	6	8	19	36	22	14	263

Sources: Recagno, Fozza et al. (2013) and Fozza and Recagno (2012)

Hoffmann, Endresen, and Dalsøren (2011), Feo, García, and Sáez (2008), GasHighWay (2012), NESCCAF (2009), PLANCO (2007) and TOSCA (2011)).

To support this analysis, it was assumed that, if a technology has quantifiable effects on factors that affect KPIs, analogous results on the KPIs would be expected. For example, the transport cost KPI is composed by various cost factors, like fuel costs, consumables' costs, logistic activities' burden, taxes, etc. Any efficiency benefits, such as fuel savings, would be expected to reduce fuel costs and, thus, the transport cost KPI. After a series of internal SuperGreen workshops, the most important KPI factors were identified (see also Chap. 3 of this book). Regarding the KPIs of CO_2 and SO_x emissions, the objective was to collect information on both Tank-to-Wheel (TTW) and Well-to-Wheel (WTW) releases.[10] As the availability of information was limited only to some corridors, the final technology assessment was limited to TTW emissions only. In addition, a semi-qualitative assessment scheme was used, as the collected material was not of the same level of resolution (Georgopoulou et al., 2012). The qualitative ranking scheme included five ranks, with the top and bottom ranks denoting mature technologies with positive potential and negative impact to the KPIs, respectively:

- −2: Very negative effect;
- −1: Moderate negative effect;
- 0: No effect;
- +1: Moderate positive effect;
- +2: Very positive effect.

In Task 3, a second questionnaire survey collected information on: (a) which green technologies are already applied on the corridors, and (b) which green technologies have potential applicability over the corridors. After three rounds of data collection performed during the project lifetime, a map has been developed showing the potential applicability of green technologies over the segments and nodes of the corridors, covering all nine categories and all modes of transportation (road, rail, maritime and inland waterways). The analysis started from the Task 1 technology list and resulted in 202 technologies that looked promising accordingly to the SuperGreen scope and objectives. The importance of each technology was assessed using a 6-rank classification as follows:

A—Very important. These technologies are believed to have a large impact on the greening potential of cargo transportation in a transportation corridor. The technologies are mature and are considered to influence the greening potential in near future.

B—Important. These technologies are believed to have an impact on the greening potential of cargo transportation in a transportation corridor. The technologies are mostly mature and are considered to influence future greening.

---

[10] Tank-to-Wheel emissions mainly concern the energy conversion system, such as the engine of a vehicle. On the other hand, Well-to-Wheel emissions consider the whole supply chain, from the production of a fuel to its end-of-pipe use.

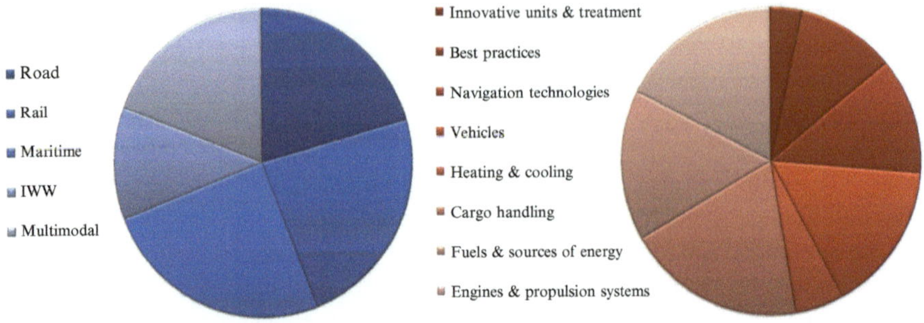

Fig. 5.2 Distribution of promising technologies by type and transport mode. *Sources*: Recagno, Fozza et al. (2013) and Fozza and Recagno (2012)

C—Low importance. These technologies are believed to have less impact on the greening potential of cargo transportation in a corridor or are less mature than those found in category A or B but are still considered valuable to the project.

D—Merged. These technologies are regarded special cases of the technologies that are placed in categories A, B or C, and are considered as valuable information to the project.

X—Need information. More information is necessary to evaluate these technologies.

Z—Not relevant. These technologies are regarded as not relevant to the SuperGreen project and are not included in the final selection.

At the end of the analysis, about 30 % of the identified technologies were selected as the most promising ones in terms of greening potential (category A + B). Their distribution in relation to type and mode appears in Fig. 5.2. The Appendix presents the final list of technologies believed to have the largest potential for the SuperGreen project (Categories A, B or C).

5.2.3.3 Phase 3: Creation of the Benchmark

The final phase of the methodology was dedicated to the green technology impact on the corridor performance, the so-called *corridor-specific* benchmark, which is accomplished by extending the technology-specific one to account for representative corridors routes. Hence, a set of scenarios was analysed, i.e. combinations of green technologies and corridor segments and nodes, which can be selected according to the criteria below:

- Relevance to the baseline trade routes;
- Solution of targeted bottlenecks;
- Technology maturity level.

The baseline analysis provided information on route features (load factor, trip duration, service frequency, vehicle details), which helped to quantify the KPI factors as percentages of the KPIs, e.g. the percentage of fuel cost to the overall cost. Via some algebraic calculations, the technology influence on the corridor KPIs was evaluated (see also Georgopoulou, Kakalis et al. (2013)).

5.3 The SuperGreen Corridor Benchmark

5.3.1 Technology-Specific Benchmark

The technology-specific benchmark can be dually used:

- To outline the technology theoretical effects on the KPIs;
- To identify the technology enablers for improving corridor sustainability footprint.

According to the SuperGreen technology-specific benchmark, green technologies could bring an average positive influence of 35 % on the corridor performance, accompanied with negative impact on the cost KPI, due to capital costs, retrofitting procedures and extra consumables (Georgopoulou et al., 2012; Georgopoulou, Kakalis et al., 2013). The results were based on semi-qualitative assessments and, thus, the remarks cannot be generalized for all application areas of the technologies.

The identified technology enablers are shown in Table 5.4. The technologies that comprise the benchmark of Table 5.4 were identified as promising for improving the corridor performance. Furthermore, the following ones were estimated to have significant impact on the corridor KPIs, for inter- and multi-modal implementation:

- Alternative fuels, such as liquefied natural gas (LNG) and compressed natural gas (CNG) are expected to play key role in evolving the transportation sector, in the future.
- New vehicle designs move to the direction of hybrid systems, where a conventional combustion engine is supported by energy storage systems to increase cycle efficiency and reduce emissions. The energy storage systems could be charged by devices that consume alternative fuels or by recovering dissipated energy, such as from the brakes (kinetic energy recovery) or the exhaust gases of a conventional engine (heat recovery).
- ICT systems are expected to increase the service quality KPI factors, such as the reduction of delays, the improvement of cargo safety and reliability.

5.3.2 Corridor-Specific Benchmark

Table 5.5 presents the SuperGreen corridor-specific benchmark, as described in Georgopoulou et al. (2012), Georgopoulou, Fozza, and Holte (2013) and Georgopoulou, Kakalis et al. (2013). The relative cost indicator considers only

Table 5.4 Technology-specific benchmark: categorization of the green technologies depending on their potential benefits on the corridor performance

KPI	Mode	Technologies with potentially positive effect on the corridor KPIs
Relative transport cost[a]	Multi-modal	Hybrid propulsion systems and locomotives (diesel-electric, batteries, etc.); fuel cells; biogas; wind energy (use in terminals); hybrid cranes (diesel-electric, super-capacitors, etc.); low emission engines for cargo handling systems; automatic-manual transmission concepts in cargo handling systems; cargo condition monitoring and control systems (temperature alarm detectors, temperature control unit, etc.); wireless internet applications (increase the efficiency of logistics operations); braking energy recovery; on board energy storage systems (super-capacitors, batteries, etc.); cardboard pallets
	Water-borne	Liquefied natural gas (LNG); dual-fuel engines; Azimuth thrusters; advanced propeller concepts (counter rotating propeller); hybrid hydraulic drive terminal tractors; sky sails systems; waste heat recovery systems; horizontal container loading practices; automated cargo handling terminal systems; River Information Services (RIS); route optimization systems
	Road	Compressed natural gas (CNG); bottoming cycle systems (waste heat recovery); electric vehicles; aerodynamic drag improvements; low rolling resistance tires
	Rail	Electric locomotives; traffic flow management; computer-aided guidance to energy efficiency driving and timetable optimization; energy settlement systems
Emissions CO_2, SO_x	Multi-modal	Hybrid propulsion systems (diesel-electric, batteries, etc.); fuel cells; exhaust abatement systems; low NO_x engines; ethanol and bio-diesel; biogas; electricity; wind energy; hybrid cranes (diesel-electric, super-capacitors, etc.); low emission engines for cargo handling systems; automatic-manual transmission concepts in cargo handling systems; braking energy recovery; on board energy storage systems (super-capacitors, batteries, etc.).
	Water-borne	Liquefied natural gas (LNG); dual-fuel engines; Azimuth thrusters; gas engines; advanced propeller concepts (counter rotating propeller); sky sails systems; waste heat recovery systems; hybrid hydraulic drive terminal tractors; horizontal container loading practices; automated cargo handling terminal systems; route optimization systems; River Information Services (RIS)
	Road	Compressed natural gas (CNG); bottoming cycle systems (waste heat recovery); electric vehicles; aerodynamic drag improvements; low rolling resistance tires
	Rail	Waggons that combine different types of cargo (lorries, buses, cars, containers, etc.) and horizontal loading; electric locomotives; computer-aided guidance to energy efficiency driving and timetable optimization; energy settlement systems

Average speed, frequency and reliability of service	Multi-modal	Cargo cassette and translifter; cargo condition monitoring and control systems (temperature alarm detectors, temperature control unit, etc.); satellite-based positioning systems
	Water-borne	Horizontal container loading practices; automated cargo handling terminal systems; Electronic Chart Display and Information System (ECDIS); route optimization systems (scheduling); River Information Services (RIS); smart docking and automated transhipment port systems
	Rail	Traffic flow management

Source: Georgopoulou, Kakalis et al. (2013)

[a]No return of investment effects on cost are considered

Table 5.5 Corridor-specific benchmark: the green technology impact on the corridor baseline performance is shown via 20 scenarios

Corridor-scenario	Technology	KPI	Impact
Mare Nostrum—Container liner service between Mediterranean ports	Waste heat recovery systems	Rel. cost [€/ton km]	1–5 %
		CO_2 [g/ton km]	1–5 %
		SO_x [g/ton km]	1–5 %
	Exhaust abatement systems	Rel. cost [€/ton km]	−4[a] to −1 %
		CO_2 [g/ton km]	−4 to −1 %
		SO_x [g/ton km]	90–96 %
	Integrated SS transport	Av. speed [km/h]	5–8 %
Nureyev—container vessel serving a port-to-port connection linking Rotterdam and Helsinki	Contra rotating propeller	CO_2 [g/ton km]	5–15 %
		SO_x [g/ton km]	5–15 %
	Mechanical Azimuth thrusters	CO_2 [g/ton km]	0–20 %
		SO_x [g/ton km]	0–20 %
	Wind propulsion—sails[a]	CO_2 [g/ton km]	0–15 %
		SO_x [g/ton km]	0–15 %
	LNG	CO_2 [g/ton km]	10–20 %
		SO_x [g/ton km]	98–100 %
	Cargo cassette trans lifter	Av. speed [km/h]	0–38 %
		Serv. freq. [no/year]	0–6 %
		Reliability [%]	0–6 %
Strauss—Inland waterways' JOWI class container vessel, serving the Rotterdam-Duisburg segment	Exhaust abatement systems	Rel. cost [€/ton km]	0–1 %
		CO_2 [g/ton km]	−5 to 8 %
	Route optimization systems	Rel. cost [€/ton km]	1 %
		CO_2 [g/ton km]	10 %
		SO_x [g/ton km]	10 %
	LNG	CO_2 [g/ton km]	10–19 %
		SO_x [g/ton km]	95–100 %
SilkWay—railway connection between China and Poland	Braking energy recovery and on board storage	CO_2 [g/ton km]	30–40 %
Brenner—roadway connecting Verona and Berlin operated by heavy duty EURO V type refrigerated trucks	Hybrid trucks	Rel. cost [€/ton km]	5–7 %
		CO_2 [g/ton km]	25 %
	Aerodynamic drag improvements	Rel. cost [€/ton km]	3–4 %
		CO_2 [g/ton km]	10–26 %
		SO_x [g/ton km]	10–26 %
	Low rolling resistance tires	Rel. cost [€/ton km]	0–1 %
		CO_2 [g/ton km]	2–4 %

(continued)

Table 5.5 (continued)

Corridor-scenario	Technology	KPI	Impact
Cloverleaf—fleet of Euro IV trucks of 24–40 ton capacity case study, serving the link between London and Duisburg	Aerodynamic drag improvements	Rel. cost [€/ton km]	2–8 %
		CO_2 [g/ton km]	10–25 %
		SO_x [g/ton km]	10–25 %
	Hybrid trucks	Rel. cost [€/ton km]	10–25 %
		CO_2 [g/ton km]	25 %
		SO_x [g/ton km]	10–25 %
Cloverleaf—electrified long train operating in Midlands-Duisburg	Energy settlement systems	Rel. cost [€/ton km]	1 %

Source: Georgopoulou, Fozza et al. (2013) and Georgopoulou, Kakalis et al. (2013)
[a]The effect on operating costs considers only the scrubber loads. The reduction of cargo space due to the installation of the system would probably increase the negative effect

operating costs, without the inclusion of capital costs, or the return of investment on green technologies. The indicator for service reliability is a semi-qualitative factor, indicating the level of improvement of the service quality in the hypothetical case that the green technology would be used.

In brief, the findings are:

- Road transportation segments could be improved up to 7 % in operating costs (excluding return of investment overheads) and 26 % in CO_2 emissions using innovative design and propulsion concepts.
- Maritime segments could improve up to: (a) 20 % in CO_2 emissions using alternative fuels, hybrid and energy efficiency technologies; (b) 38 % on the average speed with advanced cargo handling systems; and (c) 73 % in SO_x emissions using SO_x cleaning systems.
- Railway segments could be improved via electrification, use of ICT and energy settlement systems for optimal energy use.
- Alternative fuels, like LNG and CNG, can support reducing baseline emissions at road, shipping and terminal operations, but bunkering infrastructure development is necessary for their uptake.
- In baseline transport chains, there is still ample room for efficiency improvements. Energy efficiency improvements of conventional designs, such as aerodynamic drag improvements in trucks and waste heat recovery systems in ships, are promising, but the impact rate depends on the operational profile and all the parameters relevant to the operation processes, such as the driving cycle.
- Integrated technology schemes including smart ICT, such as the energy settlement system for railways, could contribute to the reduction of fuel consumption and emissions. Still, the impact depends on the utilization of the technology from the system operators.

5.3.3 Data Uncertainty

The SuperGreen project showed that a complete corridor benchmark demands extensive data collection and processing. Uniformity of data and well-defined mathematical formulae are prerequisites for a quality baseline description. However, data unavailability was determined as a major bottleneck and, hence, suggestions for future actions on corridor data collection were proposed, among which high priority should have:

- Major cargo flows;
- Operational parameters, such as driving cycle, load factors and delivery times;
- Reliability and service quality indices, such as number of accidents, cargo losses due to malicious actions, thefts, etc.

Uncertainty relevant to the baseline data may also affect the interpretation of the benchmark results. For this purpose, it is necessary to clearly define the goal and scope of the corridor benchmark, which would, then, affect the level of detail and the resolution of the analysis. For example, one goal would be to result in a tight benchmark evaluation, and a second one would be to outline the technology enablers for the future improvement of the transportation sector. In addition, it is rather difficult to quantify all the effects that a green technology may have over the corridors, as the impact may change in a per case basis. The KPI factorization offers a mapping between technologies and KPIs. However, it is necessary to have a clear definition of the factor mathematical formulae and accurate data on the technology costs. As described in Georgopoulou et al. (2012), if the overheads from the return of investment on the technologies were considered in the SuperGreen benchmark, the effect on the relative cost KPI could be different.

5.3.4 The SuperGreen Knowledge Base

The results of the SuperGreen corridor benchmark are publicly accessible through a web-based repository, the so-called "*SuperGreen Knowledge Base*", hosted at http://88.32.124.84/SuperGreen. The *SuperGreen Knowledge Base* (Fig. 5.3) is a virtual machine that presents the project results on the use of green technologies over the nine SuperGreen corridors (Recagno, Fozza et al., 2013). The objective is to demonstrate the project findings on the applicability of advanced transportation solutions over the corridors to improve performance and solve bottlenecks.

Through this virtual machine, the user can access the corridor maps, including segments, hubs and transshipment points, ports and transportation modes. The user can also select to display the applicable technologies per corridor segment and/or node, which are highlighted using a two-color code: (a) green corresponds to possible applicability; and (b) red indicates no applicability.

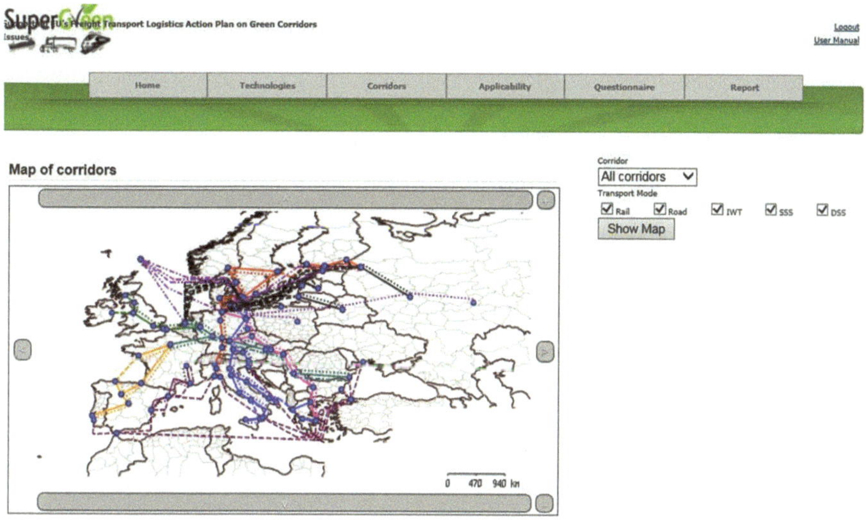

Fig. 5.3 The *SuperGreen Knowledge Base*, hosted at http://88.32.124.84/SuperGreen

The technologies considered lie in the categories of: engines and propulsion systems, fuels and energy sources, cargo handling and transfer technologies, cargo preparation technologies, heating and cooling technologies, innovative loading units and their treatment, vehicles, navigation technologies, best practices of technologies integration. For each technology, the *SuperGreen Knowledge Base* lists a set of main characteristics, including the energy source, the emissions, the efficiency, the readiness level, the time-to-market status, the applicability on the corridors and the benchmark results.

5.4 Outlook of the SuperGreen Technology Enablers

This section presents the most promising technology enablers for corridor greening in SuperGreen.

5.4.1 Cleaner Fuels: Natural gas

Natural gas (NG) is a cleaner alternative to Diesel fuel oil, offering environmental and economic benefits from the reduced emissions, price and taxation. NG has high methane concentration and close to zero sulphur and particulate matter (PM) content. It can be transported either in compressed (CNG) or liquefied (LNG) form. In the SuperGreen technology-specific benchmark, NG was assigned with the rates of Fig. 5.4; elaboration on the assessment is given in the next paragraphs, on per mode basis.

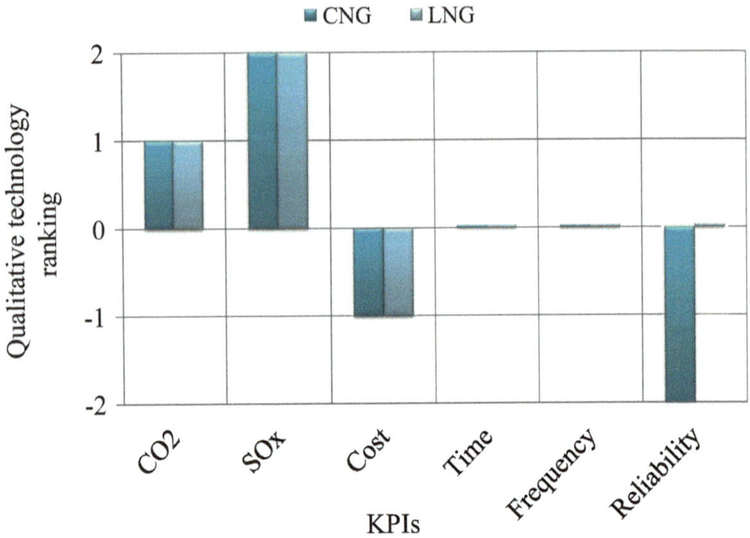

Fig. 5.4 Qualitative assessments of the LNG and CNG fuels: the top and bottom ranks denote positive and negative impact to the KPIs, respectively. *Source*: Georgopoulou, Fozza et al. (2013)

In road transportation, the first NG-fuelled vehicles (NGVs) appeared back in 1920s, but were overwhelmed by the low cost Diesel-fuelled ones (Fryczka, 2004). NGVs became an attractive option after the two oil shocks in 1974 and 1979. In the last two decades, the number of NGVs and CNG fuel stations in Europe increases, as a consequence of rising oil prices and environmental awareness. For heavy duty trucks, NG is considered to bring significant energy and environmental benefits. According to TIAX (2007), the impact on the energy efficiency is at a range of −2 to +8 % of the baseline fuel consumption. Other studies estimate that the TTW CO_2 emission reduction is about 6–7 % compared to Diesel oil, under real-world driving conditions (Krupnick, 2010). Apart from the fuel cost savings, economic benefits also depend on the investment costs, the fuel price, taxation and regulations. In general, the capital cost to buy a heavy duty NGV is higher than a conventional Diesel one, but the prices would change in accordance to the production scale of the NGVs. Also, if the environmental regulations make the green Diesel options[11] more expensive, then the NGVs could become a more attractive solution in the future. Regarding taxation, motor fuel taxes tend to be a considerable part of end-user fuel prices in OECD[12] countries (DNV, 2012a). Following this path, a general estimation of the NG impact in the relative cost is not straightforward. From a reliability aspect, NGVs are considered safe, if the safety measures are taken, as NG dissipates

[11] Vehicles that include technologies to meet the environmental regulations, like for NOx and PM reduction.

[12] Organization for Economic Co-operation and Development.

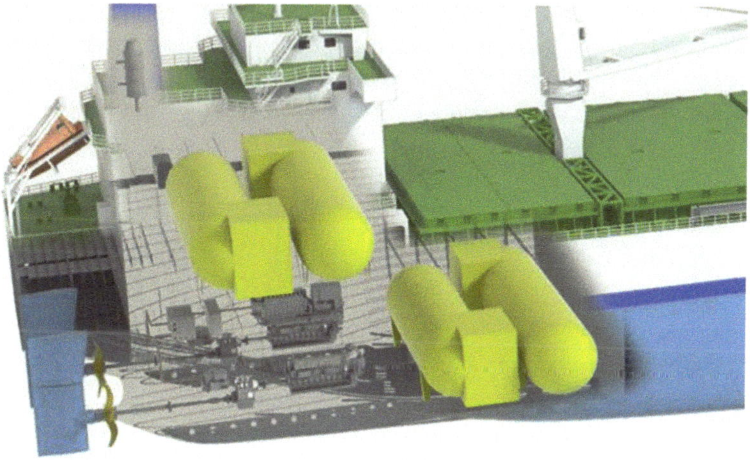

Fig. 5.5 LNG tanks location at the ECO-Ship concept developed by classification society DNV GL and Oshima Shipbuilding. *Source*: DNV (2011). Illustration: Copyright © DNV GL AS. 2014. All rights reserved

in the atmosphere, whereas diesel (or petrol) leaks on the ground, increasing the potential of fire hazard (Toy, Graham, & Hammitt, 2000).

LNG for shipping is a proven and safe technology to improve the ship's energy efficiency and environmental performance, and an alternative to after-treatment systems to reduce SO_x, NO_x and PM emissions. The interest in LNG is expected to increase due to the environmental regulations and energy efficiency mandates (MARPOL, 2011). The LNG is estimated to reduce the tank-to-propeller CO_2 emissions by 20–25 %, the NO_x emissions by 90 %, and SO_x and PM emissions by almost 100 %, compared to conventional marine fuel oil. Similar reduction effects are estimated for inland waterways (Schweighofer, Kampfer, & Seiwerth, 2006). As an example, the LNG-powered ECO-Ship is an open hatch bulk carrier concept, developed by Oshima Shipbuilding and DNV in 2011, which is expected to emit about 20 % less CO_2, 90 % less NO_x and zero PM and sulphur compared to oil-fuelled vessels (DNV, 2011). Figure 5.5 shows the ECO-Ship concept for the location of the LNG fuel storage tanks. From a technical point of view, the installation of the LNG fuel tanks may require a reduction in cargo capacity, which depends on the type of vessel and the system complexity. Of course, due to the economy of scale, large vessels could benefit more from LNG compared to small ones, assuming the availability of the respective fuel supply and bunkering infrastructure. Regarding fuel cost, currently, LNG is cheaper than oil and the price levels vary from country to country; however, there is uncertainty on the future prices (DNV, 2012a).

NG can be used for hub operations as a fuel in lift trucks and fork lifts. Apart from the energy savings, the reduction of emissions and the reduced noise levels would bring benefits to the occupational environment. Despite the rising interest for transportation of NG fuel, a drawback is the lack of sufficient bunkering infrastructure. The development of NG refuelling networks over the corridors would support the uptake of NG fuel in Europe. Therefore, it is essential that the future development of infrastructure is considered on a corridor level, allowing the use of NG at long distance international trips.

5.4.2 Hybrid Concepts: Combinations of Multiple Energy Sources

Hybrid transportation systems are characterized by the combination of a power production unit with an energy storage system (e.g. fuel cells, batteries, supercapacitors), allowing to switch or complementarily use both options for power supply, resulting in fuel economy and reduction of emissions. The SuperGreen technology assessment for hybrid systems is shown in Fig. 5.6.

In the car industry, hybrid power systems have been in use for decades, with the capital costs being a key burden in their wide application. Some hybrid truck designs recover part of the vehicle's kinetic energy during braking via a regenerative braking system and store it on board in lithium-ion batteries. In NESCCAF (2009), a review on hybrid truck concepts showed a potential reduction of baseline consumptions by 5.5–6.0 %. A hybrid truck system can cover the hotel loads during idling, bringing savings of about 7 % to the total drive cycle. In SuperGreen, hybrid trucks were estimated to reduce fuel consumption and emissions up to 25 %,

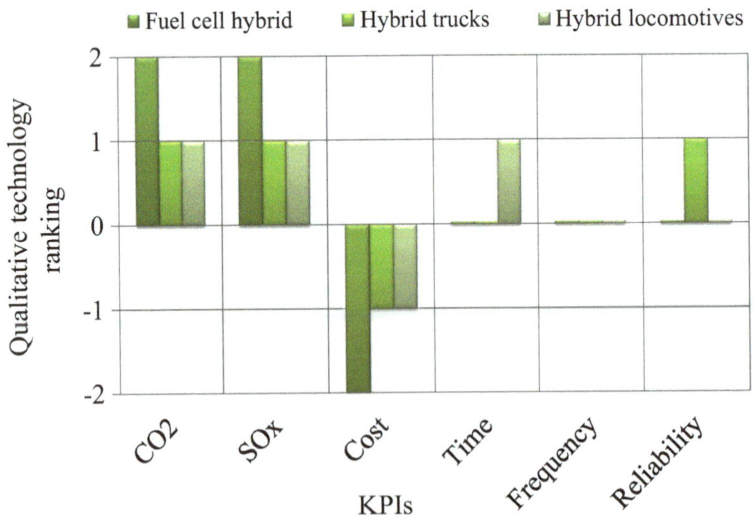

Fig. 5.6 Qualitative assessments of hybrid concepts: the top and bottom ranks denote positive and negative impact to the KPIs, respectively. *Source*: Georgopoulou, Fozza et al. (2013)

for full utilization of the technology capabilities (Georgopoulou et al., 2012, Georgopoulou, Fozza et al., 2013, Georgopoulou, Kakalis et al., 2013).

Maritime hybrid designs started to appear recently, with the objective to smartly cover the necessary power demands by optimally handling and sharing the energy from different suppliers to the on board consumers, respecting the safety constraints of the vessel. A hybrid ship example is the Viking Lady offshore supply vessel, owned by Eidesvik Offshore ASA,[13] (6,100 ton gross tonnage, 5,900 ton deadweight). Her energy system combines 4 dual-fuel LNG/Diesel engines with a molten carbonate fuel cell stack of 320 kW power capacity, and a battery pack to allow for optimal management of the power. The potential for fuel consumption and CO_2 emissions reduction is from by 20 to 30 % compared to the baseline, while the SO_x and PM emissions are reduced by almost 100 % (DNV, 2013).

5.4.3 Energy Efficiency Improvements: Extend the Capabilities of Baseline Designs

One of the ways to reduce baseline emissions is by applying energy efficiency alterations to the conventional system designs. A number of four selected energy efficiency measures for road, waterborne and rail is described as follows; their assessment is shown in Fig. 5.7.

Heavy-duty truck aerodynamic drag improvements alter the external form of the vehicle targeting to reduced drag coefficient and, therefore, less energy consumption and emissions. Such systems include shape changes, installations or retrofits, such as: reduced tractor to trailer gap, trailer side skirts and undercarriage skirts, boat tail, integrated tractor roof fairings, aerodynamic mirrors, replacement of mirrors with cameras, fuel tank fairings, bumper fairings, wheel fairings, and hidden vertical exhaust stacks. In SuperGreen, truck design improvements were considered to bear energy and emission savings of about 10–26 %.

Waste heat recovery systems (Fig. 5.8) exploit the thermal energy of the engine exhaust gases, in order to produce steam and, via a steam turbine, additional power on board. Depending on the engine size, heat recovery systems can increase the power output by 4–10 %. Their installation requires certain space, which could lead to reduced cargo capacity. For this purpose and due to the analogy between the amount of hot exhaust gases and the heat recovery potential, waste heat recovery systems are more attractive for large vessels. In a series of studies conducted by DNV GL Strategic Research & Innovation (Dimopoulos, Georgopoulou, & Kakalis, 2011; Kakalis, Dimopoulos, & Stefanatos, 2013; Stefanatos, Dimopoulos, & Kakalis, 2013; Kakalis et al. 2014; Dimopoulos et al. 2014), the optimal design of waste heat recovery systems for different types of ocean-going vessels was performed using state-of-the-art systems engineering modelling and simulation

[13] http://www.eidesvik.no/viking-lady/category253.html

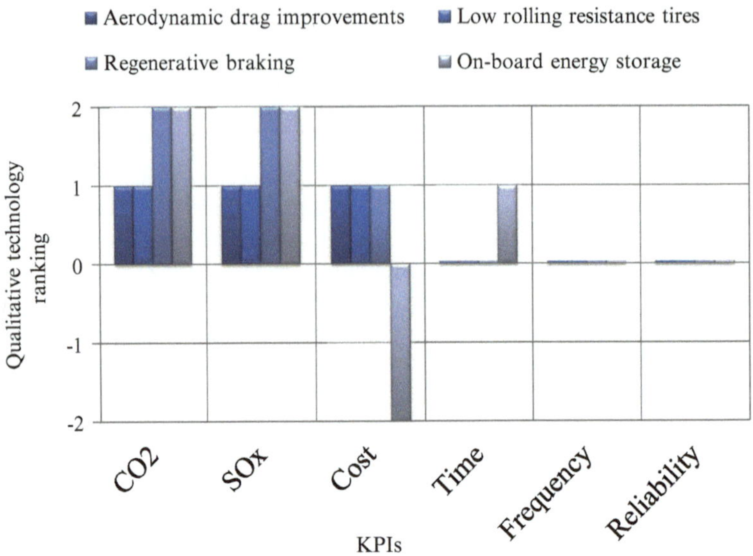

Fig. 5.7 Qualitative assessment of hybrid concepts: the top and bottom ranks denote positive and negative impact to the KPIs, respectively. *Source*: Georgopoulou, Fozza et al. (2013)

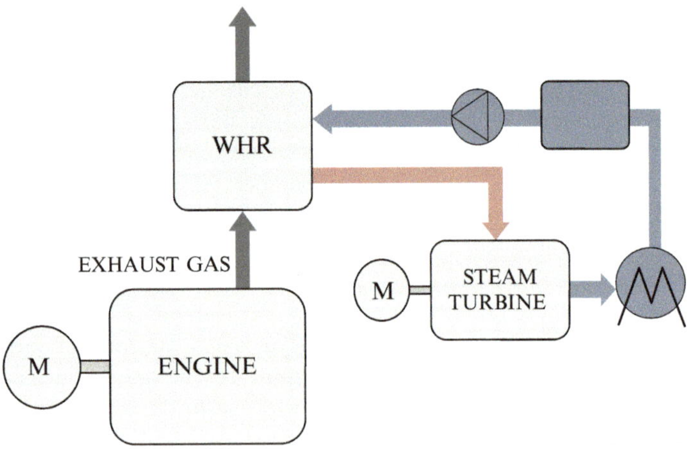

Fig. 5.8 Simplified flowchart of a waste heat recovery system

techniques. The study showed that waste heat recovery systems are attractive for the present and near-future deep sea shipping industry, while the investment payback periods depend on the vessel type and size, the system design and the annual operational profile. The overall system efficiency was estimated at about 7–10 % higher than the baseline design, with relevant CO_2 emission reduction effect.

In railways, regenerative braking is a mature and relatively standard energy efficiency technology. In conventional electric trains, the train kinetic energy is dissipated as heat by the dynamic braking systems. With the use of regenerative braking, the current in the electric motors is reversed, slowing down the train, while the motors generate electricity and return it to the power distribution system, to power up other trains or auxiliaries, like lighting. Friction brakes are still needed as backup in the case that the regenerative brakes fail. The recovered power can only be used simultaneously. Alternatively, it can be stored in energy storage systems, like supercapacitors, batteries or flywheels, and be recovered at any phase in time. In Diesel multiple units, energy storage and reuse of brake energy could lead to theoretical energy savings up to 30 and 40 % (UIC, 2002). In SuperGreen, the regenerative braking and on board storage technologies for railways were considered to have a positive impact of 30 % on the KPIs related to fuel consumption and emissions. This technology would be effective on short distance networks, as the number of full stops is high (UIC, 2002). The energy savings would, also, affect the transportation costs, as the electricity bill would benefit from the consumption reduction.

5.5 Conclusions

5.5.1 Feedback from the SuperGreen Corridor Benchmark

One of the main objectives of the SuperGreen project was to develop the first corridor benchmarking methodology in the literature and to demonstrate its application on European corridor examples. The SuperGreen project proved that such an effort is feasible, especially if the necessary information is available. Even though we had to deal with data unavailability, the derived benchmark covered the required resolution of the analysis and yield key results for the improvement of the current performance of European transport networks.

5.5.1.1 How to Develop a Representative Benchmark

From our experience in SuperGreen, a key prerequisite to the creation of a transportation corridor benchmark is the development of a representative corridor baseline, based on accurate data on trade routes, transportation technologies and practices. If such data are available, then, an accurate picture of the corridor baseline condition could be drawn. In addition, the evaluation of the baseline performance requires a clear definition of the Key Performance Indicators, including their mathematical formulae, assumptions and examples. In addition, it is important to gather information on advanced technologies and assess their effects on the factors of the KPIs. Having this mapping at hand, the potential greening of a corridor using advanced technologies can be predicted. To summarize, a representative benchmark would require reliable data on:

- The major cargo flows and their annual volumes, organized per cargo type and considering all transportation modes of the corridor;
- The description of representative long-distance routes, including the technologies used, the operating profiles and the logistic practices involved;
- The performance characteristics of green technologies, including cost indices.
- The description of typical bottlenecks that hinder the efficient use of technologies, and/or the insufficiency of supportive infrastructure, if relevant.
- The description of the regulatory framework relevant to the technology uptake over the corridor.

5.5.1.2 How to Use the Green Corridor Benchmark

The green corridor benchmark is an indicator of the "gap" between the current status of a corridor and a hypothetical condition that state-of-the art technologies are in use; the latter could be regarded as the next-generation corridor status. Theoretically, this distance could have an analogy with the investments to be made, in order the corridor to evolve according to the optimal standards of the technological frontline; although, the description of such an analogy is out of the scope of this study.

A corridor benchmark could be used to trace the "hot spots" in corridor performance and identify areas for future improvements and the solution of bottlenecks. Building a representative corridor benchmark is important for policy makers to recognize that enabling technologies and shape their decisions accordingly. In this perspective, preventative and proactive corridor planning and management is encouraged, rather than reactive end-of-pipe approaches. In a broader view, the benchmark could be used as a sustainability tool to compare between the past, current and future conditions of a corridor, suggesting relevant indicators and metrics of environmental, energy efficiency and service quality performance. Any future increase in the corridor KPIs compared to the benchmark would indicate successful decision making and/or potential for further improvements.

As a general rule, the benchmark cannot be used for comparisons between different corridors. By definition, a transportation corridor contains a number of segments and nodes, which are important for the economies at global, community, and regional levels. Therefore, a corridor cannot be disregarded, as it serves specific transportation and economy needs. In some regions, the corridor networks play crucial role in the trades, so that their substitution from other networks is impossible. In addition, the variety in network options offers advantages in the balance of regional and global markets.

Due to the lack of large statistical data on the corridor flows, the SuperGreen benchmark cannot be used in its current form in order to draw representative figures for the performance of different transportation modes, or their combinations on the corridors, and any comparison between them might generate misleading conclusions. Even in the case that all necessary information is available; the benchmark cannot be used for definite modal comparisons in its current form. On the contrary, comparisons should be made on a case-specific level, rather than based on the general picture of a corridor, but considering the particular transport chain features, such as practices, driving cycles, frequency of service, etc. At this point, it is necessary to comment that the SuperGreen benchmark does not imply any endorsement on the routes and/or the green technologies by the consortium, or the EC. Instead, it was built on the purpose to construct the methodology and support future efforts in this direction.

5.5.2 *Green Technology Enablers*

The green corridor benchmark reveals the technology enablers capable to improve the corridor performance and solve bottlenecks. In brief, the key technology enablers per transportation mode for the SuperGreen corridors are summarized hereafter:

- In road transportation, the baseline energy efficiency performance would be potentially improved from the use of advanced technologies on improving the aerodynamic design of vehicles, the main and auxiliary loads via hybrid concepts and the road-truck interaction via more efficient tires.
- In railways, the deployment of smart navigation technologies, risk management systems and energy settlement schemes would increase energy efficiency and service reliability. Existing technologies, like regenerative braking, already provide standard economic benefits and could further improve the overall energy savings, if combined with innovative schemes, like on-board energy storage by batteries or supercapacitors. More on railways can be found in Chap. 12.
- In waterborne transportation, the introduction of existing energy efficiency measures, advanced hybrid concepts and alternative fuels would bring significant reduction of emissions and increase in energy efficiency. The average speed and service reliability could be improved from the use of better cargo handling systems, automation technologies in hubs and transhipment points, and navigation technologies.

Despite the fact that there is proven technology for all modes that can offer economic, environmental and social benefits, their adoption may take long. In the SuperGreen baseline analysis, we have identified the following key barriers to innovation in transportation:

- The level of technology maturity plays important role from development to deployment of an innovative system. Apart from the technical barriers in making a new technology compatible with the conventional ones, the needs for infrastructure availability and competitive prices are crucial for the adoption of a technology. The LNG paradigm shows that the multi-modal adoption of natural gas as a fuel depends on the availability of sufficient bunkering infrastructure.

- The presence of regulatory incentives play pivotal role in the uptake of new technologies. In recent years, hybrid and electric vehicles for urban mobility became popular as a result of the enactment of supportive regulations. By ensuring an advantageous environment through policy, the confidence to invest on innovative transportation solutions is increased.

To conclude, this work can serve as a basis for a detailed investigation of green technology applications on the European corridors, which will shed more light on their greening potential and will contribute to a solid understanding of the most promising solutions on a corridor level.

Appendix: The Most Relevant Green Technologies. *Source*: Recagno, Fozza et al. (2013) and Recagno, Vio et al. (2013)

Engines and Propulsion Systems

ID	Category	Technology name	Transport mode	Description
EN02	A	Directly driven propeller	Maritime	Slow speed engine directly connected to propeller shaft, 20 year life time, running 5,500 h/a
EN03	A	Mechanically connected propeller	Maritime	Medium speed engine connected by a reduction gear to the propeller shaft, 20 year life time, running 5,500 h/a
EN07	A	Diesel-mechanic propulsion with high speed engine	Maritime	High speed engine connected by a reduction gear to the propeller shaft, 20 year life time, running 5,500 h/a
EN16	A	Full/parallel hybrid	Road	Electrical support of engine power by saving and re-use of break-energy; combination of six cylinder engine plus electrical engine
EN21	A	Exhaust abatement system	Inland waterways	Emission reduction system comprising a reactor for selective catalytic reduction of NO_x and a reactor containing a particulate matter filter for reduction of particulate matter
EN39	A	Gas engines	Inland waterways	Engines running on natural gas (different solutions available, pure gas engines, gas-diesel engines, dual fuel engines)
EN06	B	Mechanical azimuthing thrusters	Maritime	The engine runs generator. An electric motor is located inside the ship where it runs propeller shaft. Twenty year life time, running 5,500 h/a

(continued)

ID	Category	Technology name	Transport mode	Description
EN18	B	Fuel cell technology	Road	>3.5 ton transporter running on renewable fuel cell technology
EN 15	C	PG engine diesel locomotives	Railway	A propulsion system for a four-axle, standard-gauge, centre-cab locomotive using a liquefied petroleum gas (LPG) engine instead of conventional diesel
EN42	C	CCNR I engine	Inland waterways	Most existing engines comply with CCNR I standard
EN45	C	CCNR II engine	Inland waterways	Today new engines have to comply with CCNR II standard
EN48	B	CCNR III engine	Inland waterways	Still under negotiation
EN51	B	CCNR IV engine	Inland waterways	Still under negotiation
EN54	C	Kaplan propeller in nozzle	Inland waterways	Nozzle around Kaplan propeller creates additional thrust; highly effective at large propeller loads, *Source* DST
EN57	C	High screw propellers	Inland waterways	Nozzle around high skew propeller creates additional thrust; highly effective at large propeller loads, *Source* DST
EN61	C	Contra rotating propeller	Maritime	Thrust system consisting of a pair of propellers behind each other which rotates in opposite directions, so that the aft propeller recovers some of the rotational energy in the slipstream from the forward propeller
EN62	C	Diesel turbo compound	Road	Turbo compound systems can be used to affect engine operation using the energy in exhaust gas that is driving the available turbocharger. A first electrical device acts as a generator in response to turbocharger rotation. A second electrical device acts as a motor to put mechanical power into the engine, typically at the crankshaft. Apparatus, systems, steps, and methods are described to control the generator and motor operations to control the amount of power being recovered. This can control engine operation closer to desirable parameters for given engine-related operating conditions compared to actual

(continued)

ID	Category	Technology name	Transport mode	Description
EN11	B	Diesel-electric propulsion with dual fuel engine	Maritime	Medium speed engine using LNG (Liquefied Natural Gas) as primary fuel and HFO (Heavy Fuel Oil) or MDO (Marine Diesel Oil) as pilot fuel. The engine runs generator. An electric motor runs propeller shaft. Twenty year life time, running 5,500 h/a
EN24	B	Improved gas engine	Road	Integrated approach using electronic valve motion management, enhanced cylinder head cooling, near-to-valve port fuel injection system, advanced integrated control

Fuels and Sources of Energy

ID	Category	Technology name	Transport mode	Description
FU02	A	Ethanol and bio-diesel	Maritime road	Investigation about using alternative fuels
FU03	A	CGN (compressed natural gas)	Multimodal	Cleaner fuel for yard handling equipment (prime movers)
FU08	A	LNG	Multimodal	Liquefied natural gas
FU18	A	Biogas	Multimodal	Biogas is mainly produced from bio-waste, agricultural residues and residues from sewage treatment plants
FU25	A	Sky sails system	Maritime	It uses large towing kites for the propulsion of the ship. The tractive forces are transmitted to the ship via a highly tear proof, synthetic rope
FU05	B	AMP	Maritime	Alternative maritime power is a shore-side power source, that transforms the shore-side power voltage to match the vessel power system
FU06	B	Wind energy	Maritime Inland waterways	Wind turbines which will generate clean energy to power 14 container terminal quay cranes, reefer containers, repair workshops and other power consumption needs
FU13	B	Electricity	Road Railway	Electricity is today produced from fossil fuels, nuclear energy and renewable energy sources

(continued)

ID	Category	Technology name	Transport mode	Description
FU26	B	Waste heat recovery system	Maritime	It passes exhaust gases from the ship's main engine through a heat exchanger to generate steam for a turbine driven generator the electrical power generated assists ship propulsion or supplies shipboard services
FU01	C	Ultra-low sulphur diesel	Maritime Inland waterways Railway Road	Switch from industrial diesel oil (IDO 0.5 % sulphur) to ultra-low sulphur diesel (ULSD 0.005 %) for PMs and RTGs
FU04	C	Solar power network	Multimodal	A 6.600 m^2 solar panel able to generate clean energy which will reduce reliance on oil and cut electricity-related greenhouse gas emissions
FU07	C	HFO	Maritime Railway Road	Heavy fuel oil
FU14	C	Hydrogen	Road Inland waterways	Hydrogen is today mainly produced from steam reforming of fossil gas—some production from electricity and renewable sources
FU23	C	Nuclear power	Inland waterways Maritime	Nuclear power
FU30	C	Flettner rotor	Maritime	It is a vertical cylinder rotating around its axis that converts prevailing wind into propulsive energy
FU29	C	Fuel cell hybrid system	Multimodal	Develop fuel-cell systems that are capable of meeting the demands of heavy-duty transport for road, rail and marine applications. These systems will be: (a) highly efficient, above 60 %; (b) power dense; (c) powerful units of 200 kW plus; (d) durable, robust and reliable. The two FC technologies considered are Polymer Electrolyte Fuel Cell (PEFC) technology and Solid Oxide Fuel Cell (SOFC) technology. The scientific and technological approach is based on FC clustering and hybridization

Cargo Handling and Transfer

ID	Category	Technology name	Transport mode	Description
HT01	A	Diesel to electric power convertor (RTGs)	Maritime Multimodal	RTGs fitted with electrical components in place of traditional hydraulic parts. Conversion will eliminate black emissions and lower noise levels of engines
HT03	A	Hybrid hydraulic drive terminal tractors	Maritime	Storing braking energy into hydraulic system for acceleration and system
HT07	A	Low emission engines	Multimodal	Euro III/IV compliant engines burn diesel more efficiently, reducing emission of CO_2 and providing up to 5 % reduction on fuel consumption
HT10	A	Horizontal container (un) loading	Railway	Metrocargo is an innovative solution for containers cargo handling in overhead electrified railways, it's a containers horizontal movement system from an automated platform to train wagons. This technology is ready to experimentation. Metrocargo will be tested on new Maersk's Platform in Vado Ligure (SV), Italy
HT06	B	MP-RTGs	Multimodal	Mains-powered RTGs transfer the power generation from the engine of the yard crane to a far more efficient power station. Power station can be up to 40 % more efficient than equipment engine
HT11	B	Cargo cassette and translifter	Maritime	Wheel less cargo cassette is a loading platform which is used together with a translifter in a cassette system. Translifter is a steerable lifting trailer which together with cassettes replaces roll trailers in Ro-Ro and StoRo handling
HT28	B	Automatic RoRo cargo unit handling	Multimodal	The concept is based on self (un)loading of units using a roll-on/roll-off system with a special train of platform cars, called a train loader. The performance of a train loader is often limited by the operation of the stockpile and reclaim system and the capacity of the train loader surge bin. While both are separate systems, they operate in concert to achieve a given performance. Poorly designed reclaim systems, or insufficient train loader surge capacity can significantly downgrade train-loading performance

(continued)

ID	Category	Technology name	Transport mode	Description
HT24	C	FCT	Maritime	The floating container terminal collects and distributes containers originating from small calls, and bundles these currents with containers
HT08	B	ZF transmission systems	Multimodal	Installation in the new PM (prime movers) of new transmission system operating based on automatic-manual transmission concept. Reduction of fuel consumption by 10 % when compared with older existing transmission systems
HT09	B	Green schemes to improve RTGs emissions and noise	Multimodal	Addition of a super-capacitor on RTGs. When RTGs engine is running, it charges the super capacity at the same time, and when super capacitor is fully charged, it will supply
HT20	B	Barge Express (BEX)	Inland waterways	BEX is an integrated concept for large scale barge container transport aiming at automated handling at barge terminals
HT36	B	FlexiWaggon	Railway	Flexiwaggon can combine lorries, buses, cars, containers on one and the same waggon. Individual loading and unloading of waggons. Loading and unloading is done horizontally which means no consideration is necessary for overhead contact lines. The emissions will be reduced by 75 %, including carbon dioxide emissions
HT05	C	Timing device for engine start-stop	Multimodal	Applied on yard equipment (straddle carriers) to shut down the engine after a period of inactivity. This is a timing device that controls engine shutdown and start-up depending on activity level
HT32_a	C	River-Sea Push Barge System	Maritime	The river-sea push barge is a transport system in which one and the same push barge is used for the sea- and the river leg in a transport chain
HT33	C	Combined Traffic Carrier Ship/ Barge (CTCB)	Maritime	A shortsea concept based on a new type of shortsea vessel: the Trans Sea Lifter (TSL). This vessel is able to carry floating unit load carriers, in particular barges generally used in inland navigation, between inland waterways that are separated by the open sea

(continued)

ID	Category	Technology name	Transport mode	Description
HT34	C	Intermodal loading unit	Multimodal	New technical solutions for intermodal loading units including containers, dedicated adaptors and mobile internal fixtures in order to shift the main transportation route for goods from the road onto rail and inland waterways in a sustainable way. The technical activities will be focused on the development and design of large ISO containers and ISO compatible roll-off containers with the dimensions of 2,550 × 2,900 × 7,450 mm. These dimensions comply with the recommended directive of the European Commission for intermodal loading units

Cargo Preparation

ID	Category	Technology name	Transport mode	Description
CP01	C	Cardboard pallets	Multimodal	Ecological and sustainable being made of recycled materials and completely recyclable, have low weight but good strength
CP02	C	Modularized boxes	Multimodal	Containers modularized and standardized worldwide in terms of dimensions, functions and fixtures. Easy to handle, store, transport, interlock, load, unload, construct and dismantle, compose and decompose. Environment friendly materials with minimal off-service footprint
CP03	C	Passive controlled atmosphere system	Multimodal	Passive controlled atmosphere system in which the fruit itself creates the desired environment. Lower oxygen levels slow down the respiration process of the fruits
CP04	C	Cargo hold tank coatings	Multimodal	Innovative cargo hold tank coatings to reduce abrasion and corrosion
CP05	C	Software for optimal pallet configuration	Multimodal	Software for optimal pallet configuration to reduce shipping costs. The user enters primary package or box dimensions and rapidly assembles optimal pallet configurations

Heating and Cooling

ID	Category	Technology name	Transport mode	Description
HC02	B	Intelligent temperature unit	Multimodal	Current refrigerated boxcars will be built with energy efficient cooling systems, GPS (Global Positioning System) tracking, fresh air exchange and the ability to remote monitoring the systems, sometimes from thousands of km away on a network. RFID (Radio Frequency Identification) for tracking services are the main support in management systems of perishable goods
HC03	B	Temperature control units	Road	CryoTech: liquid CO_2 modules for temperature for multi temperature control (cooling/heating)
HC04	B	RFID tag antenna with temperature alarm sensor	Multimodal	RFID tag antenna with ultra-low cost temperature alarm sensors which is capable of detecting temperature violations above a critical temperature threshold
HC05	C	Natural refrigerants	Multimodal	Natural refrigerants are chemicals which occur in nature's bio-chemical processes. They do not deplete the ozone layer and make negligible contribution to global warming. Their high efficiency means they make a much lower, indirect contribution to global warming than many synthetic refrigerants
HC06	C	Systems to reduce heating costs in cold climates	Multimodal	The project will investigate two cooling approaches during the compression process. In one approach, relatively large amounts of oil are injected into the compressor to absorb heat generated throughout the compression stage. In the second approach, a mixture of liquid and vapor refrigerant from the expansion stage is injected at various points during compression to provide cooling. The added steps improve the compression process while also reducing energy losses due to friction in the expansion stage

(continued)

ID	Category	Technology name	Transport mode	Description
HC07	C	Software program QUEST	Maritime	QUEST is a CO_2 emission friendly software with focus on maintaining a constant cargo temperature. It regulates the return air temperature and allows the supply air temperature to fluctuate without exposing the cargo to chill damages
HC08	C	Truck Refrigeration Unit TDJS35HP	Road	Truck refrigeration unit enables simultaneous temperature control of two separate cargo compartments with different temperature settings entirely by heat pump

Innovative Units and Treatment

ID	Category	Technology name	Transport mode	Description
LU13	B	Braking energy recovery	Railway	Recovery of dynamic braking energy and restitution to national grid/reversible DB substation
LU14	B	Onboard energy storage systems	Railway	Supercaps, batteries, flywheels, hybrid storage; a flywheel is a mechanical device with a significant moment of inertia used as a storage device for rotational energy. Flywheel energy storage, or the rotational energy of a flywheel, and rechargeable electric traction batteries are also used as storage systems. Batteries are electrochemical energy storage systems. A supercapacitor is a tool offering very high electrical capacitance in a small package. A hybrid train is a locomotive, railcar or train that uses an onboard rechargeable energy storage system (RESS), placed between the power source (often a diesel engine prime mover) and the traction transmission system connected to the wheels

(continued)

ID	Category	Technology name	Transport mode	Description
LU11	C	APU (Auxiliary Power Unit)	Railway	An auxiliary power unit (APU) is a device on a locomotive whose purpose is to provide energy saving and to reduce the polluting emissions. Locomotive engines cannot use antifreeze in their cooling systems for technical reasons related to reactions of antifreeze chemicals on internal engine parts. Therefore, during cold weather, a locomotive engine must either be working to transport freight or idling to prevent freezing. The APU keeps the main engine warm, reducing fuel consumption and emissions while the main engine is shut down and also APU reduces railway noise levels
LU02	C	SECU unit	Multimodal	The SECU (Stora Enso Cargo Unit) is ISO certified for 93.5 gross tonnes. The dimensions are 3.6 × 3.6 × 13.8 m
LU03	C	Loading plate	Maritime	Actiw LoadPlate was developed to meet customer demands for quick loading of standard cargo space: sea containers, trailers. Solution is suitable for loading difficult cargo that is hard to containerize
LU04	C	Trailer stand	Maritime	Simple system to lash trailers
LU05	C	2.5 wide container	Multimodal	Allows two pallets to be loaded side by side

Vehicles

ID	Category	Technology name	Transport mode	Description
VE02	A	Electric locomotive	Railway	NS 999 is an entirely electric locomotive that uses a lead-acid energy storage system without the use of a diesel engine and with zero exhaust emissions
VE03	A	Hybrid truck	Road	Support engine plus auxiliary drive to operate an elevating platform of the truck; combination of six cylinder engine plus electrical engine
VE09	A	Electric vehicles	Road	Battery-electric vehicles
VE10	A	Euro VI vehicles	Road	Euro VI is compulsory for new trucks from 2013, replacing Euro V

(continued)

ID	Category	Technology name	Transport mode	Description
VE01	B	Hybrid locomotive	Railway	Hybrid locomotive was developed with the goal of creating the cleanest, most fuel-efficient high-horsepower diesel locomotive ever built
VE22	B	Road-rail cargo interchange	Railway	The Flexiwagon rail project will allow containers to be moved by road and by train by loading trucks onto railcars
VE25	B	Brake energy recovery system	Railway	Reversible DC substation for recovering of dynamic braking energy and restitution to national grid
VE29	B	Aerodynamic drag improvements	Road	Aerodynamic mirrors, cab side extenders, integrated cab roof fairings, aerodynamic front bumper, full fuel tank fairings, trailer side skirt fairings, trailer gap fairing, rear mounted trailer fairing. Ref to the "Reducing heavy -duty long haul combination truck fuel consumption and CO_2 emissions report" http://www.nescaum.org/documents/heavy-duty-truck-ghg_report_final-200910.pdf/
VE33	C	Low rolling resistance tires	Road	Tires which are designed to minimize the energy wasted as heat as the tire rolls down the road
VE35	B	Electrification of trucks on highways	Road	The eHighway concept introduces the idea of diesel-electric hybrid trucks which can work like a electric trolley when overhead electric lines are available and work as a diesel
VE04	C	Fuel cells	Road	3.5 ton F-Cell Sprinter is a transporter running on renewable fuel cell technology
VE20	C	River-Sea Push Barge System	Inland waterways	The river-sea push barge is a transport system in which one and the same push barge is used for the sea- and the river leg in a transport chain
VE21	C	Combined Traffic Carrier Ship/Barge (CTCB)	Maritime Inland waterways	A shortsea concept based on a new type of shortsea vessel: the Trans Sea Lifter (TSL). This vessel is able to carry floating unit load carriers, in particular barges generally used in inland navigation, between inland waterways that are separated by the open sea
VE31	C	Innovative bogie	Railway	New-generation of powered bogie with axles directly driven by synchronous motors is already available for light rail vehicles. Traction, running gear and braking technologies are combined in the bogie in order to form a highly integrated mechatronic system

(continued)

ID	Category	Technology name	Transport mode	Description
VE32	C	Friction control measure	Railway	Some energy expended by the train is lost to wheel-to-rail friction. Reductions in wheel-to-rail resistance can be made via improved lubrication. Efficient lubrication systems, such as top-of-rail lubrication systems, reduce wheel and rail wear and reduce fuel consumption

Navigation Technologies

ID	Category	Technology name	Transport mode	Description
NA02	A	Automatic Identification System (AIS)	Maritime	Ship-to-ship, ship-to-shore and shore-to-ship system. Main purpose is collision avoidance, ship tracking and tracing. Works on VHF (very high frequency, 30–300 MHz) radio frequency
NA15	A	WiMax	Maritime Railway Road	Worldwide interoperability for microwave access. Long range, high bandwidth wireless Internet
NA01	B	Train Control System	Railway	Train control and tracking system based on a special GPRS method
NA05	B	ECDIS	Maritime	An Electronic Chart Display and Information System (ECDIS) is a computer-based navigation information system that can be used as an alternative to paper nautical charts. Integrates position information from GPS and other navigational sensors (radar, AIS). It may also give sailing directions and fathometer
NA12	B	GEO satellites	Maritime	Geosynchronous Satellite whose orbital track on the Earth repeats regularly over points on the Earth over time. If such a satellite's orbit lies over the equator and the orbit is circular, it is called a geostationary satellite

(continued)

ID	Category	Technology name	Transport mode	Description
NA13	B	LEO satellites	Maritime	A low Earth orbit (LEO) is generally defined as an orbit within the locus extending from the Earth's surface up to an altitude of 2,000 km. Given the rapid orbital decay of objects below approximately 200 km, the commonly accepted definition for LEO is between 160 and 2,000 km (100–1,240 miles) above the Earth's surface
NA14	B	Inmarsat	Maritime	British satellite telecommunications company, offering global, mobile services. It provides telephony and data services to users worldwide, via portable or mobile terminals which communicate to ground stations through eleven geosynchronous telecommunications satellites
NA16	B	ATM	Inland waterways	The advising Tempomaat (ATM) is a computer program advising the skipper on the most economical combination of route and speed, enabling the vessel to arrive on time with a most efficient use of fuel leading to a reduction of fuel consumption and emissions
NA17	B	River Information Services (RIS)	Inland waterways	River Information Services (RIS) are customized information services for inland waterway transport and make it possible to coordinate logistical processes with actual transport situations on a constant basis. RIS play a key role in making cargo transport and passenger services on waterways more efficient leading to a reduction of fuel consumption by approximately 5 %, while at the same time increasing traffic safety
NA18	B	Predictive cruise control (PCC)	Road	The PCC assistance system uses map and satellite-based route previews and saves substantial amounts of fuel. Unlike a conventional cruise control system that tries to maintain a preset speed, regardless of how the terrain changes, the PCC system looks for its route a mile in advance and adjusts engine output to the uphill and downhill gradients ahead. Based on this information, the on-board computer calculates the optimum speed to use the momentum of the truck to maximize fuel economy

(continued)

ID	Category	Technology name	Transport mode	Description
NA07	C	Global Navigation Satellite Systems or GNSS	Maritime Railway Road	Global Navigation Satellite Systems (GNSS) is the standard generic term for satellite navigation systems ("sat nav") that provide autonomous geo-spatial positioning with global coverage. GNSS allows small electronic receivers to determine their location (longitude, latitude, and altitude) to within a few metres using time signals transmitted along a line-of-sight by radio from satellites
NA11	C	LRIT	Maritime	The Long Range Identification and Tracking (LRIT) of ships. It consists of the ship borne LRIT information transmitting equipment, Communications Service Providers (CSPs), Application Service Providers (ASPs), LRITDataCenters, the LRIT Data Distribution Plan and the International LRIT Data Exchange
NA15	B	WiMax—Worldwide Interoperability for Microwave Access	Maritime	Long range, high bandwidth wireless Internet

Best Practices

ID	Category	Technology name	Transport mode	Description
BP04	A	Traffic flow management	Railway	A system for online optimization of rail traffic flow to have minimum delays and minimum energy consumption, developed by Emkamatik on behalf of SBB
BP07	A	Carbon-free rail freight transport	Railway	DB Schenker Rail replaces the electricity required for your freight transport with regenerative energy that comes 100 % from renewable sources in Germany. This helps to avoid carbon emissions right from the outset. Even the smallest quantities can be transported in this way without carbon emissions, on a national and international scale
BP02	B	TDS	Railway	Train Control System based on a GPS application method

(continued)

ID	Category	Technology name	Transport mode	Description
BP03	B	GEKKO	Railway	A system to provide guidance to energy efficiency driving and timetable optimization, developed for Danish State Railways
BP08	B	Integrated shortsea transport	Maritime	The concept of Coaster Express (CoEx) is a short sea transport concept directed to bundling the transport flows, scaling-up the short sea facilities and standardization and automation of the transition processes
BP35	A	EREX (ERESS 2011)	Railway	The Erex system, has been designed by the European Railway Energy Saving Solution (ERESS), to help railways to save money and reduce CO_2 emissions by providing exact energy consumption data. It provides an efficient, reliable, and flexible energy settlement process, enabling railway undertakings to understand their use of energy and thereby save energy and costs. Erex has been configured with a virtual platform with almost unlimited capacity

References

Aditjandra, P., Zunder, T., Islam, D., & Vanaale, E. (2012). Investigating freight corridors towards low carbon economy: Evidence from the UK. *Procedia—Social and Behavioral Sciences, 48*, 1865–1876.

Dimopoulos, G. G., Georgopoulou, C. A., & Kakalis, N. M. P. (2011). Modelling and optimisation of an integrated marine combined cycle system. In Proceedings of the 24th International Conference on Energy, Cost, Optimization, Simulation and Environmental Impact of Energy Systems (ECOS), Novi Sad, Serbia, 4–7 July 2011, pp. 1283–1298.

Dimopoulos, G. G., Georgopoulou, C. A., Stefanatos, I. C., Zymaris, S. A., & Kakalis, N. M. P. (2014). A general-purpose process modelling framework for marine energy systems. *Energy Conversion and Management, 86*, 325–339.

DNV. (2010). *Pathways to low carbon shipping*. Det Norske Veritas (DNV) Publications.

DNV. (2011). *Eco-ship*. Det Norske Veritas (DNV) Press release.

DNV. (2012a). *Shipping 2020*. Det Norske Veritas (DNV) publications.

DNV. (2012b). *Fuel saving guidelines—For container ships*. Det Norske Veritas (DNV) publications.

DNV. (2013). *Hybrid systems*. Det Norske Veritas (DNV) Research & Innovation Position paper.

Eide, M. S., Longva, T., Hoffmann, P., Endresen, Ø., & Dalsøren, S. B. (2011). Future cost scenarios for ship CO_2 reduction. *Maritime Policy & Management, 38*(1), 11–37.

ERESS. (2011). *European partnership for railway energy settlement systems*. Annual Report.

Feo, M., García, M. L., & Sáez, C. L. (2008). Estimation of a modal choice model in the Western Europe motorway of the sea corridor: a stated preference analysis. In 2nd International Symposium on Ship Operations, Management, and Economics, SNAME Greek Section, 17–18 September, Athens, Greece.

Fozza, S., & Recagno, V. (2012). Sustainable technologies and innovation for green corridors: Survey and application. *Procedia—Social and Behavioral Sciences, 48*, 1753–1763.

Fryczka, D. (2004). *Natural gas use for on-road transport*. Diploma thesis. Faculty of Drilling, Oil and Gas, AGH University of Science and Technology in Cracow.

GasHighWay. (2012). *Route to green transportation*. Final publication.

Georgopoulou, C., Fozza, S., Aditjandra, P., Rebuffi, C., Stenvall, V., & Antikainen, T. (2011). *SuperGreen deliverable 3.3 version I—Benchmark green corridors with green technologies—Phase 1: Description of the baseline*.

Georgopoulou, C., Fozza, S., & Holte, E. (2013). *SuperGreen deliverable 3.3 version II—Benchmark green corridors with green technologies—Phase 2: Creation of the benchmark*.

Georgopoulou, C., Kakalis, N. M. P., Fozza, S., Recagno, V., Minsaas, A., & Holte, E. (2012). Assessing the sustainability potential of EU transport networks. In The Proceedings of the 40th European Transport Conference, 8–10 October 2012, Glasgow, Scotland, UK.

Georgopoulou, C., Kakalis, N. M. P., Psaraftis, H. N., Recagno, V., Fozza, S., Zacharioudakis, P., et al. (2013). Green technologies and ICT for sustainable freight transport. In U. Clausen, M. ten Hompel, R. de Souza (Eds.), Lecture notes in logistics. Efficiency and innovation in logistics—Proceedings of the International Logistics Science Conference ILSC (pp. 15–33).

Ilves, I., Holte, E. A., Georgopoulou, C., Rönkkö, S., Aditjandra, P., Vanaale, E., et al. (2011). *SuperGreen deliverable 2.4 version II—Benchmarking of green corridors*.

International Union of Railways UIC. (2002). Report. *Regenerative braking in freight trains*.

Kakalis, N., Dimopoulos, G., Stefanatos, I. (2014). A novel framework for dynamic modelling of integrated machinery system. In Proceedings of the 13th International Conference on Computer Applications and Information Technology in the Maritime Industries COMPIT 2014, 12–14 May, Redworth, UK.

Kakalis, N. M. P., Dimopoulos, G. G., & Stefanatos, I. (2013). Model-based techno-economic assessment and optimisation of marine waste heat recovery options. In 27th CIMAC World Congress, 13–16 May 2013, Shanghai, China.

Kramer, H., Sedlacek, N., Jorna, R., van der Laak, R., Bozuwa, J., Gille, J., et al. (2009). *BELOGIC deliverable 2.1—Report on overall benchmarking framework*.

Krupnick, A. (2010). *Energy, greenhouse gas and economic implications of natural gas trucks*. Resources for the Future and National Energy Policy Institute background paper.

MARPOL 73/78. International convention for the prevention of pollution from ships, 1973 as modified by the protocol of 1978. Annex VI, amendments as by 2011.

Northeast States Center for a Clean Air Future (NESCCAF), International Council on Clean Transportation (ICCT), Southwest Research Institute, TIAX LLC. (2009). *Final report on reducing heavy-duty long haul combination truck fuel consumption and* CO_2 *emissions*. USA.

Paalsson, C., Bengtsson, N., Salanne, I., Pütz, K., Tsilingiris, P., Georgopoulou, C., et al. (2010). *SuperGreen deliverable 2.2—Definition of benchmark indicators and methodology*.

PLANCO. (2007). *Economical and ecological comparison of transport modes: Road, railways, inland waterways*. PLANCO Consulting GmbH Publication.

Psaraftis, H. P., & Panagakos, G. (2012). Green corridors in European surface freight logistics and the SuperGreen project. *Procedia—Social and Behavioral Sciences, 48*, 1723–1732.

Recagno, V., Fozza, S., D'Almeida, J., Georgopoulou, C., Schoen, A. D., Varvate, M., et al. (2013). *SuperGreen deliverable 3.2—Define application areas for green technologies*.

Recagno, V., Vio, F., Fozza, S., Wahl, A. M., Holte, E. A., Georgopoulou C., et al. (2013). *SuperGreen deliverable 3.1—Identify green technologies*.

Salanne, I., Rönkkö, S., & Byring, B. (2010). *SuperGreen feliverable 2.1—Selection of corridors*.

Schweighofer, J., Kampfer, A., & Seiwerth, P. (2006). Emission reduction potentials for inland vessels. In NAV 2006, International Conference on Ship and Shipping Research, 21–23 June 2006, Genova, Italy.

Stefanatos, I., Dimopoulos, G. G., & Kakalis, N. M. P. (2013). Techno-economic analysis and optimisation of marine exhaust gas waste heat recovery systems using mathematical modelling techniques (in Greek). In ELINT (2013) Annual Meeting, Athens, Greece.

TIAX LLC. (2007). *Full fuel cycle assessment: Well-to-wheels energy inputs, emission, and water impacts*. Prepared for the California Energy Commission.

TOSCA Report. (2011). *Rail freight transport on the techno-economic analysis of energy and greenhouse gas reductions*. Technology opportunities and strategies toward climate-friendly transport (TOSCA) project publications.

Toy, E., Graham, J. D., & Hammitt, J. K. (2000). *Risk in perspective. Fueling heavy duty trucks: Diesel or natural gas? Vol. 8*(1). Harvard Centre for Risk Analysis Publications.

Chapter 6
ICT in Green Freight Logistics

Christiane Geiger

Abstract This chapter deals with the role of Information and Communication Technology (ICT) in promoting ecologically sustainable goods transportation, with a focus on Europe. After having introduced the need for acting ecologically sustainably and the particular urgency thereto within European transportation sector, a set of measures to reduce consumption of freight logistics activities is shown. Further on, the chapter takes a closer look at ICT as one of those measures. After having defined ICT and its implementing systems, their general functions and associated benefits for transportation are pointed out. In order to substantiate their application as a valuable approach towards green freight logistics, the chapter moreover deals with particular ICT systems. Based on the findings of the EU project SuperGreen the individual purpose of selected examples as well as their potential is depicted. The chapter concludes by summarizing and giving recommendations for further ICT development.

Abbreviations

AIS	Automatic identification system
CH_4	Methane
CHD	Waterway Charges and Harbour Dues service
CO_2	Carbon dioxide
EDI	Electronic data interchange
ERI	Electronic Ship Reporting
ERTMS	European Railway Traffic Management System
ETCS	European Train Control System
ETML	European Traffic Management Layer
EU	European Union
EUROPTIRAILS	European Online Optimization of International Traffic through Rail Management System

C. Geiger (✉)
Faculty of Mechanical Engineering/Institute of Transport Logistics,
TU Dortmund University, Dortmund, Germany
e-mail: geiger@itl.tu-dortmund.de

H.N. Psaraftis (ed.), *Green Transportation Logistics*, International
Series in Operations Research & Management Science 226,
DOI 10.1007/978-3-319-17175-3_6

FCD	Floating Car Data
GSM	Global System for Mobile Communications
GSM-R	Global System for Mobile Communications-Railway
GWP	Global Warming Potential
HFCs	Hydrofluorocarbons
ICT	Information and Communication Technology
IPCC	Intergovernmental Panel on Climate Change
ISA	Intelligent speed adaption
ITS	Intelligent transportation systems
KPI	Key performance indicator
N_2O	Nitrous oxide
NAIADES	European Action Programme for the promotion of inland waterway transport
NO_2	Nitrogen dioxide
PCS	Port community systems
PFCs	Perfluorocarbons
Port-MIS	Port Management Information System
PROSPER	Project for Research on Speed Adaptation Policies on European Roads
RIS	River Information Services
RWIS	Road weather information system
SF_6	Sulphur hexafluoride
SuperGreen	Supporting EU's Freight Transport Logistics Action Plan on Green Corridors Issues
SWIS	Straßenzustands- und Wetterinformationssystem
TEN-T	Trans-European Transport Network
TISA	Traveller Information Services Association
TMC	Traffic Message Channel
TRIP	Transport Research and Innovation Portal
UMTS	Universal Mobile Telecommunications System
UNECE	United Nations Economic Commission for Europe
VANET	vehicular ad-hoc networks

6.1 Introduction

Worldwide, dramatic changes in the climate system in forms of atmosphere and ocean warming, ice loss and sea level rise are observed. A main driver is the growing concentration of greenhouse gases in the atmosphere.

Greenhouse gases are gaseous substances in the atmosphere. According to the Kyoto Protocol, a climate protection treaty agreed in 1997, these include (UNFCC, 2008, p. 106):

- carbon dioxide (CO_2),
- methane (CH_4),

- nitrous oxide (N_2O),
- hydrofluorocarbons (HFCs),
- perfluorocarbons (PFCs) and
- sulphur hexafluoride (SF_6).

Greenhouse gases have the characteristic of affecting the radiation balance on earth. On the one hand, they let the solar rays pass through the atmosphere until they strike the earth. On the other hand, they absorb the heat which is reflected by the earth's surface and re-emit it in all directions. A part of the radiation is therefore reverberated to the earth's surface and warms it together with the sun light. This so-called greenhouse effect raises the global average temperature of the earth's surface from −18 to +15 °C and creates a climate that allows life to flourish. (Rahmstorf & Schellnhuber, 2012, pp. 30–33)

During the last century the greenhouse effect verifiably intensified. For the period between 1880 and 1985 a rise in global temperature was reported from 0.5 to 0.7 °C (Hansen & Lebedeff, 1987, p. 13345). The source, however, is controversial. Explanations range from usual fluctuations of the natural greenhouse effect to a largely anthropogenic cause.

The majority of climate scientists supports the latter argument. In its fifth report also the Intergovernmental Panel on Climate Change (IPCC) draws the conclusion that "it is extremely likely that human influence has been the dominant cause of the observed warming since the mid-twentieth century" (IPCC, 2013, p. 17). As one of the human activities primarily contributing to the observed warming the burning of fossil fuels (e.g., for electric power generation) is recognized, whose combustion leads to greenhouse gas emissions (IPCC, 2013, p. 11).

Of the greenhouse gases listed above CO_2 is considered to be the most important one, because it accounts for 20 % of the natural greenhouse effect. It is set as reference value for the remaining greenhouse gases by specifying them as multiples of CO_2. Basis for this approach is the gases' Global Warming Potential (GWP), which characterizes the impact of each greenhouse gas on global warming. According to the IPCC, the GWP of methane, for instance, is 25 times higher than the GWP of the same amount of CO_2 (Forster et al., 2007, p. 212). By measuring the climate impact of all greenhouse gases in terms of CO_2, it is unified by the measurement unit CO_2 equivalent, which is set as standard unit by Article 3 of the Kyoto Protocol (UNFCC, 2008, p. 3)

Initiated by the outlined development, discussions were and are still held about the issue of adopting sustainable actions and business. One focus of this debate centers on transportation since this sector, as it is shown by a sector comparison in Sect. 6.2, is a major contributor to the human greenhouse gas emissions. Against this background, Sect. 6.3 presents a wide scope of energy and emission saving measures appropriate for use in freight logistics. In the following, ICT as one of those measures is studied more closely by referring back to the results of the EU project SuperGreen. After having introduced ICT as well as the general function of its implementing systems, Sect. 6.4 deals with particular ICT systems in European freight logistics. The section treats 12 ICT solutions, explaining their individual purpose and potential to support green freight logistics. Section 6.5 picks out one of those ICT systems investigated: European Railway Traffic Management System (ERTMS). Examining its installation along a concrete European freight corridor, potentially arising benefits are estimated. Section 6.6 concludes by summarizing and giving recommendations for further ICT development.

6.2 Ecological Sustainability in the Transportation Sector

In comparison to other sectors, the transportation sector is very energy-intensive. Operating logistics processes is linked with vast energy consumption, for example, in the form of fuel for powering cars and trucks, traction current for powering railway vehicles or oil for heating the large and voluminous logistics facilities. Therefore, transportation today accounts for almost one-third of total European energy consumption (Fig. 6.1).

When covering its energy requirements the transportation sector largely reverts to fossil energy sources. According to the European Environment Agency (2007, p. 7), even 98 %, so nearly the entire energy consumption of the transportation sector consists of fossil fuels. The logically compelling consequence is that the transportation sector is among the leaders as well when greenhouse gas emissions are the criterion of comparison. Today, more than 20 % of total greenhouse gas emissions come from transportation.

The severity of the transportation sector's accountability is further aggravated by the fact that other sectors of the European industry could reduce their greenhouse gas emissions. Since 1990, as shown in Fig. 6.2, the Energy Industries, for example, could lower their emissions by 16 % and the Manufacturing and Construction Industries could even achieve a decrease by 38 %. Transportation-related emissions, however, rose by 14 % during that time.

The contrast of the transportation sector's environmental effects with other sectors indicate the necessity as well as the urgency to promote ecologically sustainable logistics, commonly referred to as green logistics. The transportation sector has to abandon its restraint and to considerably foster greenhouse gases mitigation. Transportation companies are strongly challenged to integrate an environmentally sensitive behavior into their business activities and to establish it in the long run.

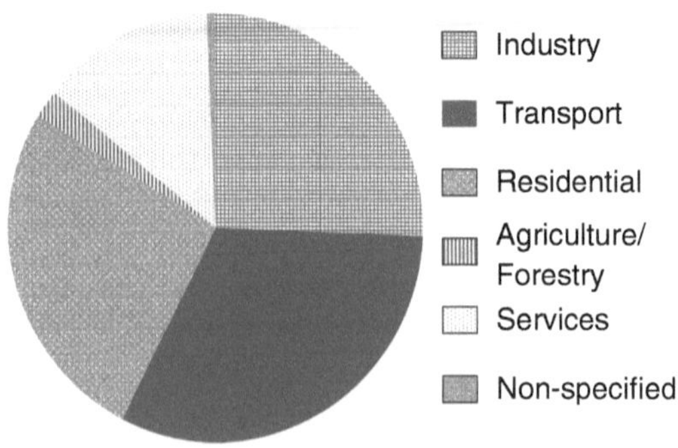

Fig. 6.1 Energy consumption of EU28 states by sector in 2012. *Source*: European Environmental Agency (2014a)

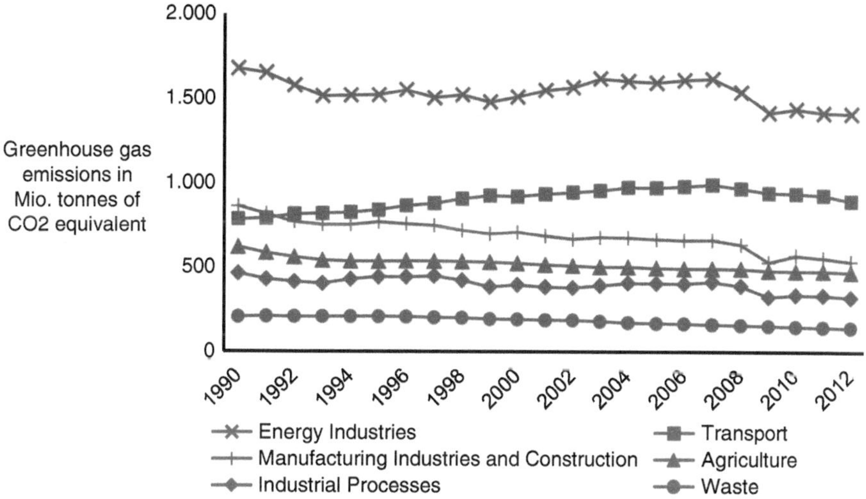

Fig. 6.2 Development of greenhouse gas emissions of EU-28 states by sector. *Source*: European Environmental Agency (2014b)

6.3 Measures to Mitigate Greenhouse Gas Emissions

Within freight transportation, the mitigation of greenhouse gas emissions can be achieved in a number of ways. As one of the most effective measures, modal shift on low-emission transportation modes has to be named. Figure 6.3 illustrates that the extent of greenhouse gas emissions differs widely between the four modes. This is particularly due to the fact that the transportation systems have varying energy requirements since they need to overcome disparate resistances or comparable resistances at different levels (Rudolph & Wagner, 2008, pp. 235–242). For example, in rail traffic, the rolling resistance between a steel wheel and the track is up to 5 times lower than the friction between the tyres of a truck and the asphalt street (Schach, Jehle, & Naumann, 2006, p. 154).

However, modal shift from road to rail or water transportation, or from air to sea borne transportation is not feasible in every case. The major requirement to mitigating measures is that they go along with the economic interest of the transportation companies. This condition unfortunately is often not fulfilled. Referring to a survey among German shipping companies and freight forwarders most important stumbling blocks are (Geiger & Schmied, 2012, p. 49):

- the extension of transport times,
- rise of price level and
- decrease of transport reliability.

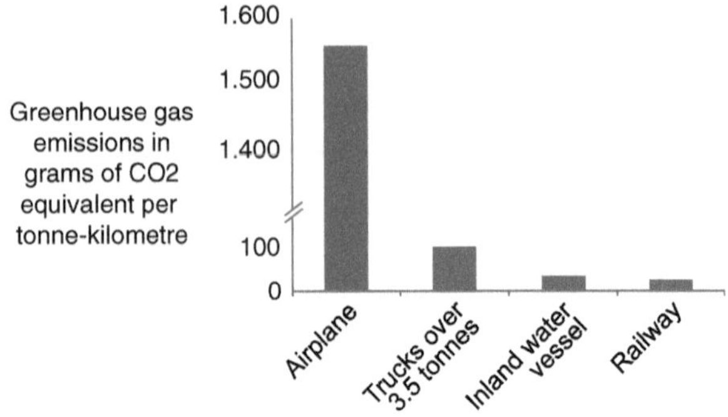

Fig. 6.3 Greenhouse gas emissions of modes. *Source*: Umweltbundesamt (2012)

Measures which open up emission reduction and cost savings at the same time have, according to the majority of respondents, to be placed when organizing the transport regardless of its mode. The scheduler should put emphasis on bundling transports, on avoiding empty runs and on abandon underutilized express transits that deny every opportunity for consolidation due to their short-term nature. (Geiger & Diekmann, 2013, p. 29; Geiger & Schmied, 2012, pp. 36–37)

Moreover, technical solutions can be employed. Chapter 5 of this book has given some examples. Within road haulage, for instance, permanent side walls and vehicle wraps can be attached that improve aerodynamics and this way lead to fuel reduction up to 7–12 % (Bode, Hermsmeier, Hocke-Anbeh, & Ziegler, 2011, p. 25). Other examples are smooth running tyres that lower the rolling friction of trucks and therefore can reduce the fuel consumption up to 6 % (Bode et al., 2011, p. 26). Although the train and ship constitute low-emission modes, there is still potential to reduce their energy demand by technical means. Examples are an aerodynamic design or the application of sky sails, which uses a towing kite to transmit the pulling force of the wind to the ship (SkySails, 2011, pp. 1–2).

Further on, behavioral measures have to be listed. Within training courses truck drives are taught to drive smoothly and with foresight. According to Swantusch (2010, p. 30) regular trainings contribute to fuel reduction by 5–8 %. An example from sea borne transportation is the operation of cargo ships at significantly less than their maximum speed, e.g., Maersk ships sailing 12 instead of 24 knots, saving 22 % of bunker fuel (Jorgensen, 2011, p. 2).

In a series of scientific papers of the last years Information and Communication technology was appreciated as an emerging measure towards energy efficiency (cf. Klunder, Malone, & Wilmink, 2010; Marchet, Perego, & PerottiMarchet et al., 2009; Perego, Perotti, & Mangiaracina, 2011). In accordance with this thesis, the present chapter focuses on ICT in the following aiming at demonstrating its value for ecologically sustainable freight logistics.

6.4 ICT in European Freight Logistics

ICT shall be defined here by regarding its compositional terms. Technology describes all applicable or actually applied working, design, production and implementation procedures of technics (Heinrich & Stelzer, 2011, p. 162). According to this definition, also future procedure developments count to technologies.

Information technologies serve for the regulated gathering and processing of information (Schulz-Spathelf, 2012, p. 236). Communication technologies are intended to transfer information between two or more partners (Klaus, Krieger, & Krupp, 2012, p. 283). In this context, partners do not only refer to individuals but also to organizations and machines.

ICT becomes manifest in systems, which according to Sjurts (2011, p. 275) comprise:

- hardware such as computer architecture elements (e.g., motherboards and processors), drives and other mass storage devices or peripheral equipment (e.g., display or sensors),
- communication networks which are spatially distributed systems for transporting messages working, for example, with radio waves as GSM or UMTS or electrical signals via wire-based transmission media as the Ethernet, and
- software in the form of suitable programs and algorithms for administrating, controlling and optimizing logistics processes.

In freight transportation logistics ICT systems record data directly at their source. After further processing, the data are transferred to successive systems in real-time and, finally, made available again (Scheer, Angeli, & Herrmann, 2001, p. 31). In general these data are associated with goods transportation. Dependent on the concrete service of the ICT system these may be vehicle or traffic data, meteorological conditions and forecasts, infrastructure data or data on shipments. In view of the mentioned combined functions of ICT systems, their application considerably creates information transparency within freight transportation logistics (Notteboom, 2013, p. 89).

ICT systems offer a monitoring of current states and activities and thus facilitate the identification of deviations from plan or target at an early stage. This in turn enables to proactively intervene and to introduce necessary compensatory measures immediately, for example, by resorting to emergency concepts to secure supply, motivating changes in behavior or to adjust original planning. Thanks to the widespread availability of data ICT systems moreover allow to eliminate some inherent components of the transportation process. This restructuring does not only provide simplification but also generates efficiency through savings in time and energy. In conclusion, ICT systems create the basis for steering material flows under the maxims of high economic performance and ecological sustainability.

The SuperGreen project has recognized this strength of ICT. Its purpose was to promote an environmentally sound development of European freight logistics. To this end, the project explored how a series of freight corridors within Europe can be arranged in a more ecologically sustainable way. In this context a corridor means a transcontinental geographic axis covering main transportation routes throughout

Europe. The series involved nine geographically and modally balanced freight corridors selected out of an initial list of 60 corridors made at the beginning of the research work (see Chap. 3 of this book and Salanne, Rönkkö, & Byring, 2010):

- Brenner,
- Cloverleaf,
- Edelweiss,
- Finis Terrae,
- Mare Nostrum,
- Nureyev,
- Silk Way,
- Strauss and
- Two Seas.

To reach this goal, the project first studied the implementation of green technologies such as novel propulsion systems, heating and cooling technologies, alternative fuels and cargo handling techniques (see Chap. 5 of this book). As second, and for the present chapter, relevant examination the usage of ICT systems within the regarded freight corridors was investigated. The motivation behind was that the above mentioned capability of ICT is still not used to the full extent. This is evident from the frequently lacking integration of ICT systems although being a key enabler for information transparency. Stakeholders in freight logistics often do not realize the benefits resulting from data accessibility. Others, however, use different ICT systems that are incompatible with each other. These systems cannot communicate leading to information gaps. In addition, a large number of systems is limited to basic and static functionalities. They do not incorporate changes over the course of time, computational algorithms or optimization routines (Zacharioudakis et al., 2012, p. 180).

In view of the painted status quo of ICT implementation, the SuperGreen project has written on its banner to promote an application of smart ICT systems within the regarded corridors and thereby to exploit their efficiency benefits for the purpose of ecologically sustainable freight logistics.

The initial step towards this aim was to draw up an inventory of smart ICT systems. Thereto, a broad collection of 54 ICT systems was set up and their individual function, purpose and scope elaborated. As this inventory formed the working background for the subsequent steps and further project objectives, the inquiry was extended over all transportation modes relevant for goods transportation within the regarded corridors (road, rail, maritime and inland waterways) as well as over intermodal transports and supply chains. Based on these results, seven clusters were defined, each of them summarizing ICT systems with common characteristics. In cases in which systems were allocable to more than one category due to their broad functionality the most prevalent cluster was chosen. Table 6.1 gives an overview of the clusters, their particular feature and those ICT systems assigned to them.

In the second step, the ICT systems were evaluated with the help of external experts. The evaluation reverted to two groups of key performance indicators (KPIs) rendered in Table 6.2. As explained above, the application of smart ICT systems must and can have an environmental and economic impact at the same time. The first group therefore consisted of efficiency KPIs that on the one hand assess the ability of each corridor to increase its environmental performance and on the other hand measure in how far the costs of the corridor usage can be reduced (Clausen, Geiger, & Behmer, 2011). However, a smart ICT system is absurd if its usability criteria are not met. To take account of this prerequisite, the second group comprised KPIs evaluating a system's availability. The evaluation was carried out on a qualitative level using a scale 1–3, with 3 points as the best score.

The result of the evaluation therefore was a filtered list of ICT solutions that are most promising for green freight logistics in Europe. Based upon this, 15 scenarios were identified within the third step, each of them comprehending an ICT/mode/corridor combination. By examining the scenarios, the importance of these particular constellations could be estimated prior to ICT installation. This way, reliable and specific recommendations could be given to the Commission in order to assist in the formulation and harmonization of policies on Green Corridors.

In the following the findings of the SuperGreen project related to the described examination are reproduced. Within the Sects. 6.4.1–6.4.12 the ICT systems'

Table 6.1 Clusters for smart ICT systems

Cluster	Cluster description	ICT systems
Expert charging systems	Appling charges to traffic	• Unified electronic toll system • Congestion charging • Toll amount depending on the pollutant category of the truck
Centralized transportation management systems	Managing traffic with a centralized decision-making scheme	• ERTMS • Traffic flow optimization • Caesar • Vessel traffic service, Vessel traffic monitoring & information systems • Electronic traffic management • River information service • Fairway information service • Information for law-enforcement • Traffic control systems • Optimar • International networking of national traffic control centers • Icebreaker net • Traffic signalling optimization
Decentralized transportation management systems	Managing traffic with a decentralized decision-making scheme	• Platooning • Intelligent speed adaption • Speed limits depending on CO_2 values

(continued)

Table 6.1 (continued)

Cluster	Cluster description	ICT systems
Broadcasting, monitoring & communication systems	Providing information to transport operators	• Conducted communication systems • Broadcasting systems (TMC, TMC Plus, TMC pro) • Mobile radio systems • Car-to-X-Communication • Electronic chart systems, Electronic chart display & information system • Broadband communication • Global navigation satellite system • Automatic identification system • Intelligent transportation systems • Long range identification and tracking, radar • SafeSeaNet • Agherra • Radio-frequency identification • Schenker smartbox • Route guidance systems, Personal navigation assistant • Head-up display • Navigation system for trucks: Map & guide professional • Green trucks • Verkehrsmanagement Audio Mobil
Safety systems	Supporting safety	• Road weather information systems • Road speed limiter • Night vision system • Distance control systems • Collision warning systems • Braking assistant systems • Lane departure warning • Lane keeping assistant • Adaptive speed limit
E-Administrative Systems	Supporting commercial functions	• Single window solutions • Janela única portuária • Freight transport information technology solutions • ShortSeaXML • Port community systems
Emissions footprint calculator systems	Calculating emissions	• Anonymized sensor data gateway • Sensors

Source: Rönkkö et al. (2012)

Table 6.2 KPIs for evaluating ICT systems

Efficiency indicators		Availability indicators	
KPIs	Description	KPIs	Description
Transport avoidance	Number of transports	Availability of ICT solution	• Available for all • Available for certain group of stakeholders • Company specific
Loading factor incl. return cargoes	Capacity utilization		
Cost efficiency of transport chains	Unit costs	Visibility and availability of information	• Open for all • Open for certain group of stakeholders • Company specific
Service quality: Transport	• Transport time		
	• Reliability		
	• Frequency of service		
	• Cargo security		
Service quality: Interface	• Cargo safety	Transport chain suitability	• All transport chains • Mostly part load transport chains • Mostly full load transport chains
Infrastructure sufficiency	• Congestion		
	• Bottlenecks		

Source: Rönkkö et al. (2012)

purpose and functioning are explained. On this basis, their approach to traffic and transportation management is deduced as the regarded ICT systems influence traffic and transportation differently. Their ways can be categorized into

- a financial approach, whereby user fees are charged,
- a decision-support approach consisting of obtaining, preparing, compressing and structuring information for human decision makers, and
- a direct time- and energy-saving approach.

Finally, the evaluation in terms of the efficiency and availability KPIs resulting from the expert survey is reflected.

Due to the vast number of diagnosed ICT systems these explanations are limited to 12 examples. Anticipating the results of the experts' evaluation, those ICT systems were selected that earned the highest ranking. For the complete results the reader is referred to Zacharioudakis et al. (2012).

Section 6.5 deals with the implementation of ERTMS in the Brenner corridor as one example of the scenarios studied in the third step.

6.4.1 Congestion Charging

Congestion charging, also named congestion pricing, is an ICT system affecting users of road transportation infrastructure. Congestion charging systems make use of variable pricing. The price is supposed to reflect the external costs, e.g., delay, air pollution, accidents and noise, each road user imposes on other users (Congress of the United States, 2009, p. 7) or on the environment (Eißel & Chu, 2014, p. 399f.). Therefore those systems charge a higher price during times or at places of peak demand and a lower price at times or places with light traffic. Congestion charging systems focusing on urban roads are currently in use in for example London, Stockholm or Milan (David-Freihsl, 2013, p. 10).

Congestion charging forces the users to pay for the negative externalities (Zacharioudakis et al., 2012, p. 51) and hence uses a financial approach to regulate and manage the use of road infrastructure. Generally, Congestion charging systems are applied to parts of road infrastructure, which face (distinctive) periods of peak demand resulting in increased external costs. By setting up an appropriate system providers, owners or operators of road infrastructure, mostly public authorities, abet the user to adjust his behavior accordingly. However, rather than the business relation between infrastructural manager and user is at the forefront of the ICT systems' networking function, congestion charging connects users by making them conscious of the costs they impose upon each other (Zacharioudakis et al., 2012, p. 51).

The implementation of such a system helps to harmonize demand of infrastructure resulting in a higher performance which becomes visible in reduced traffic congestion, shorter and more reliable travel times, decreased fuel use, less pollution and improved land use. Exceeding their cost and being incorporated in the public budget, such systems additionally generate financial benefits to society (Congress of the United States, 2009, p. 8).

This assessment goes in line with the questioned experts, who expect a reduction of infrastructural inefficiencies and emissions of greenhouse gases and other polluters and therefore awarded maximum points to these KPIs (Fig. 6.4). Considering that public authorities are responsible for the implementation of congestion charging, the system's availability and visibility of information is assessed as relatively high. Moreover, its applicability to all types of transport chains is given (Rönkkö et al., 2012, pp. 18–21).

6.4.2 Unified Electronic Toll System

The concept of a unified electronic toll system is derived from the Waterway Charges and Harbour Dues service (CHD), a component of the traffic management system River Information Services (RIS). RIS, legally specified by the EU framework directive EC/2005/44 (European Parliament & Council, 2005) and taken up in various EU policy papers (e.g., EC White paper on transport

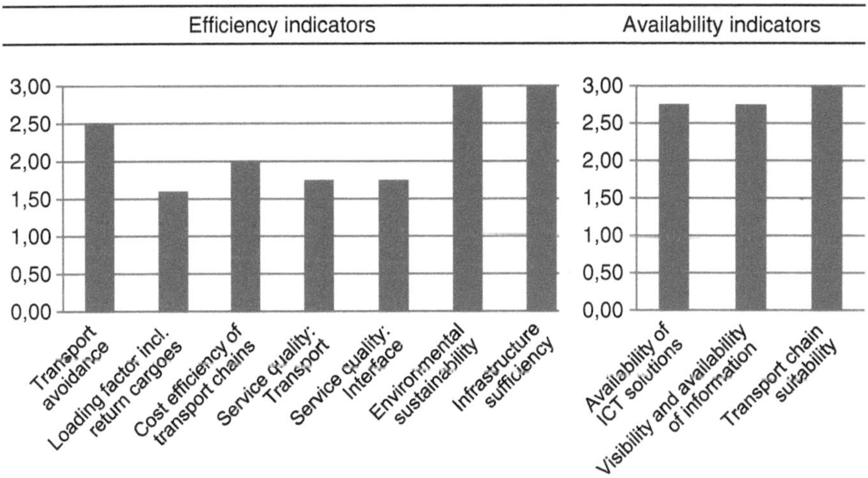

Fig. 6.4 Evaluation of Congestion charging systems. *Source*: Rönkkö et al. (2012)

(European Commission—Directorate General for Mobility & Transport, 2011)), TEN-T Guidelines (European Parliament & Council, 2013), NAIADES (via donau—Österreichische Wasserstraßen-Gesellschaft mbH, 2014) and Logistics Action Plan (Commission of the European Communities, 2007) aims at simplifying the exchange of information between inland waterway operators and users across countries' frontiers by standardization and harmonization (Vallant & Hofmann-Wellenhof, 2008, 238ff).

Its CHD service uses travel data from Electronic Ship Reporting (ERI) and Vessel Tracking and Tracing systems (Inland AIS) to automatically calculate the charge and initiate the invoicing procedure (Zacharioudakis et al., 2012, p. 152). ERI describes a set of various messages, which are used for electronic reporting in inland waterway transportation. These messages are either directed to responsible authorities (e.g., messages, which announce the cargo carried to customs, immigration, police and statistical offices) or to other parties in the transport chain (e.g., messages from shipper to the barge operator, which announce details regarding the cargo respectively containers and any further requirements for the transport) (River Information Services, 2009). Inland AIS is a standardized tracking and tracing system for the electronic exchange of nautical data with other nearby ships and between ships and shore installations (Central Commission for Navigation on the Rhine, 2008).

In Europe, a unified electronic toll system is motivated by the excessive number of different toll systems in some sections or countries. Following the CHD model, a unified electronic toll system uses standardized electronic messages and interfaces to exchange toll-relevant data. Depending on the mode, these are identification, position, load, destination or intended route. This information is exchanged on request or at fixed points and used during and after the trip to calculate according tolls (Rönkkö et al., 2012, p. 18).

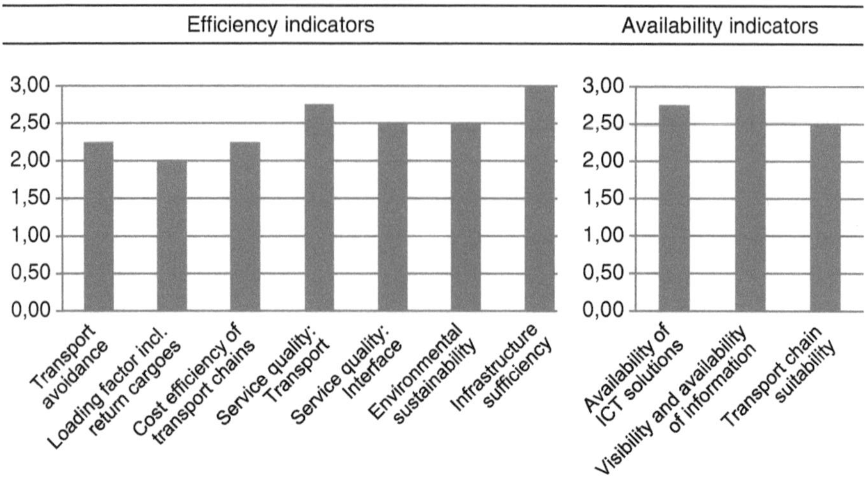

Fig. 6.5 Evaluation of a Unified electronic toll system. *Source*: Rönkkö et al. (2012)

A unified toll system is not supposed to charge additional tolls or fees. It supports the users of infrastructure to achieve shorter and more reliable travel times as well as more accurate levying of charges and hence uses a direct time- and energy-saving approach to regulate and manage the use of infrastructure.

As a general toll charging system, a unified electronic toll system can be applied to all nodes and links of all modes of transportation, which are levying charges. Predominantly, it links the instance and the party liable to pay the charges, e.g., the ship master. Moreover, a unified electronic toll system, as it is the case with CHD, can connect users and toll charging instances with governmental or regulatory authorities which are in charge of the particular infrastructure (Economic Commission for Europe, 2005, p. 21).

The benefits of such an ICT system are mirrored in a decrease in local congestion and waiting times around toll stations as well as in a reduction of many unnecessary frequent stops, both resulting in higher service quality and fewer emissions (Fig. 6.5). According to the interviewed experts, the availability of the system and visibility of information is given for all stakeholders. Its suitability, however, might be restricted to full load transport chains (Rönkkö et al., 2012, pp. 18–21).

6.4.3 *Intelligent Transportation Systems*

Intelligent transportation systems (ITS), as defined by the EU Directive 2010/40/EU, are advanced applications capable of providing innovative services relating to road transportation and its interfaces to other modes of transportation (European Parliament & Council, 2010, p. 1f.). According to this broad understanding, ITS refer to numerous applications. Examples are collision avoidance systems (e.g., by using Intelligent speed adaption), Eco-driver assistance, automatic engine start-up/shutdown, tyre pressure indicators, weather monitoring or platooning technologies. Platooning technologies refer to mechanical or electronic coupling of two or more vehicles during transport in order to decrease the distance in between, which in turn increases traffic density (Rönkkö et al., 2012, p. 40).

The technical functioning of ITS depends on the specific application, but in general relies on real-time communication (Woonsuk, Kyungmee, & Eunseok, 2011, p. 299) between vehicles (inter-vehicle communication) or between vehicles and infrastructure (vehicle-to-roadside communication) (Zacharioudakis et al., 2012, p. 37).

In the first case, the information exchange is realized with so-called vehicular ad-hoc networks (VANETs) (Bouhoute, Berrada, & El Kamili, 2014, p. 289). VANETs are automobile ad-hoc networks that are independent and self-organising with vehicles as their nodes. These vehicles collect, send and receive information, as for example about road conditions or driving situations. Due to the mobility of the nodes data transmission is wireless via radio links. (Plößl, 2009, S. 7–8)

In the latter case, infrastructure is equipped with ITS functionalities as well and therefore is able to collect and/or to send information to the vehicles. It is also common to connect these stationary facilities to superordinate traffic control centers. (Fraunhofer Institute for Communication Systems ESK, 2010, p. 2)

According to Fantechi, Flammini, & Gnesi (2012, p. 187), the impact on traffic depends on the specific ITS application and ranges from a significant contribution to improving transportation productivity, environmental performance, travel reliability as well as safety of passenger and freight transportation. As shown in Fig. 6.6, the interviewed experts agree to these comprehensive benefits of ITS.

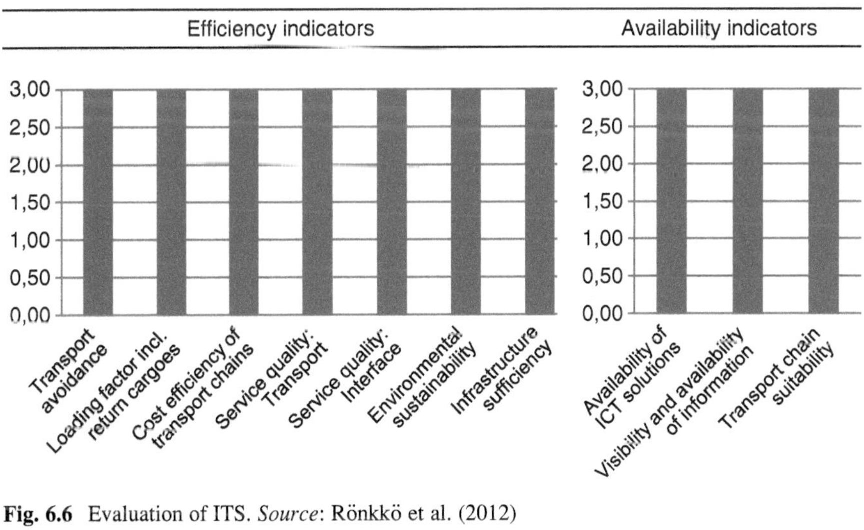

Fig. 6.6 Evaluation of ITS. *Source*: Rönkkö et al. (2012)

6.4.4 European Rail Traffic Management System (ERTMS)

Traffic flow optimization in European rail transportation faces problems derived from more than 20 different (national) control systems across the European Union (Fig. 6.7). Being stand-alone solutions these various systems are non-interoperable and therefore inhibiting information flow (Rönkkö et al., 2012, p. 30). Deviations from the timetable, however, require real-time rescheduling activities based upon continuous communication and information on the current state (Lüthi, 2008, p. 8ff). For this reason, it is almost impossible for European railway companies to reliably (re-)calculate travel times.

The ERTMS project was set up by eight members of the Association of the European Rail Industry comprising Alstom Transport, Ansaldo STS, AZD Praha, Bombardier Transportation, CAF, Mermec, Siemens Mobility and Thales—in close cooperation with the European Union, railway stakeholders and the GS—MR industry (ERTMS-The European Rail Traffic Management System, 2013). It strives to facilitate European rail transportation and enhance cross-border interoperability by creating a single Europe-wide standard.

The ERTMS comprises three basic elements: GSM-R, ETCS and ETML. To enable continuous communication and data transmission for rail purposes the Global System for Mobiles-Railway (GSM-R), based on the GSM standard, was

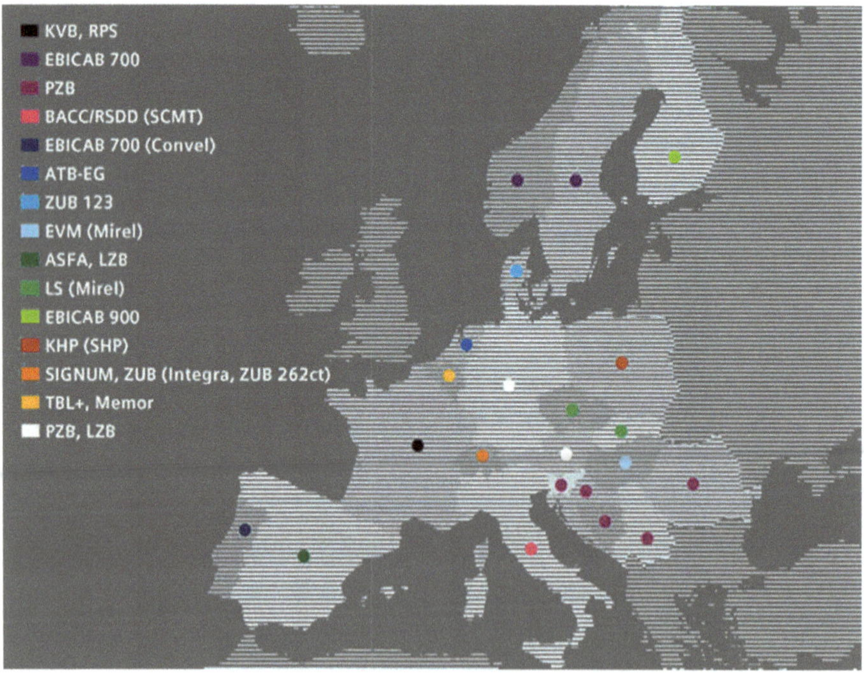

Fig. 6.7 Incompatible train protection systems in Europe (© Siemens AG). *Source*: Siemens (2014)

developed. GSM-R is the general communication technology for the other ERTMS applications. Thus it is used to install the European Train Control System (ETCS), which is the signalling, control and train protection system of ERTMS. On the operating control level, which does not apply to train safety and signalling issues, ERTMS defines the European Traffic Management Layer (ETML). ETML supports the cross-border flow of real-time information for ad-hoc rescheduling and route planning (International Union of Railways, 2013; Maschek, 2013, p. 543). First steps to implement ETML functionalities were taken within the European Online Optimization of International Traffic through Rail Management System (EUROPTIRAILS) project (Pachl, 2013, p. 79).

Aiming at an optimized traffic flow by supporting scheduling and rescheduling, ETML is assumed to have a positive effect on the overall performance of the railway network. Due to its integration into the ERTMS project, the effect of ETML singularly is hardly to number. The ERTMS project as a whole estimates that a full implementation of all ERTMS elements will result in an increased capacity on existing lines, an enhanced reliability and punctuality and improved safety for passengers and goods (ERTMS—The European Rail Traffic Management System, 2013). The interviewed experts share this view, awarding highest possible points for infrastructure sufficiency, service quality and cost efficiency (Fig. 6.8). Assessing ERTMS as suitable for all transport chains, they moreover expect a very positive impact on environmental sustainability (Rönkkö et al., 2012, p. 39).

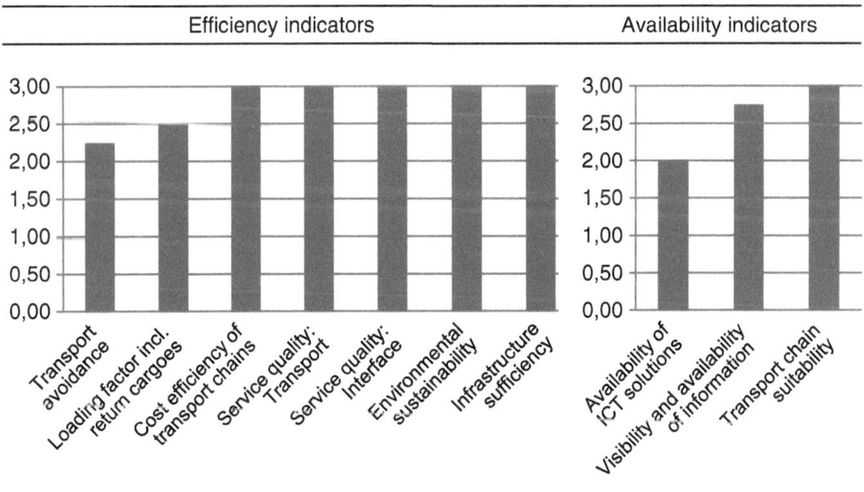

Fig. 6.8 Evaluation of ERTMS. *Source*: Rönkkö et al. (2012)

6.4.5 Traffic Control Systems

Traffic control systems inform drivers about the current traffic situation and support a relaxation in tense situations by strategic route guidance. Traffic control systems use Traffic Message Channel (TMC) which is an analogue data channel to broadcast traffic information, e.g., information on traffic flow and traffic jams, construction works and other traffic obstacles.

The information sources utilized by TMC are manifold. TMC may take into account information from police reporting, from databases (e.g., about ongoing or planned construction works or holiday traffic), from induction loops or radar sensors, which measure traffic intensity and density, as well as speed and Floating Car Data (FCD) (Zacharioudakis et al., 2012, p. 32f). FCD are data of a sample of vehicles from which the overall traffic condition is inferred (Brakatsoulas et al., 2005, p. 324). Among these data are the vehicles positions, directions and speed. By matching these data against digital infrastructural maps, the average speed and travel time is estimated. Thus, drivers are not only addressees but also collectors of traffic data.

After reception of these data, a vehicles route guidance system is able to calculate the estimated delay in the event of traffic obstructions and to suggest alternative routes. This way, Traffic control systems consult users in adapting their route to current road and traffic conditions in terms of time-saving alternatives resulting in an overall higher performance, diminished traffic congestion, shorter and more reliable travel times as well as reduced transportation costs (Fig. 6.9) (Rönkkö et al., 2012, p. 39).

TMC development was mainly shaped by the TMC forum (nowadays Traveller Information Services Association (TISA)), which, as a non-profit organization, comprises service providers, receiver manufacturers, vehicle manufacturers, map vendors, broadcasters (public and private), automobile clubs, and public authorities.

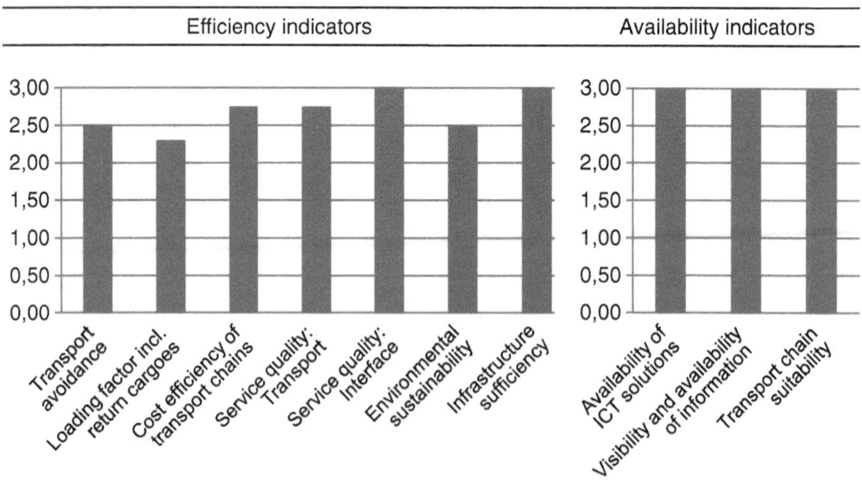

Fig. 6.9 Evaluation of Traffic control systems. *Source*: Rönkkö et al. (2012)

TMC basic functions are, as a service regulated by public law, free of charge and already widely used. The interviewed experts therefore evaluated the availability of Traffic control systems and the visibility of information as open for all (Rönkkö et al., 2012, p. 39). Enhancements of the TMC service are currently in use in Germany (TMC pro, operated by NAVTEQ®) and Austria (TMC Plus). Both use additional, automated input from stationary sensors. TMC pro for instance comprises own sensors and additionally offers a traffic forecasting service, which aims at making suggested routes more reliable regarding travel time (Zacharioudakis et al., 2012, p. 33). While the Austrian TMC Plus service remains free of charge, TMC pro as a commercial service is fee-based (Kleine-Besten, Kersken, Pöchmüller, & Schepers, 2012, p. 608).

6.4.6 Speed Limits Depending on CO_2 Values

Issuing speed limits depending on CO_2 is supported by local Traffic control systems. CO_2 emissions are measured at local scale and in temporal intervals by stationary sensors. If declared boundary values are exceeded, dynamic electronic traffic signs set lower speed limits. As soon as CO_2 concentration falls below the boundary value, the restrictions are relaxed (Rönkkö et al., 2012, p. 40).

In their current state, traffic control systems are designed for links of road networks. The principle that signalling alongside the road issues a lower speed limit if the monitored parameter exceeds a predefined threshold is called Dynamic Speed Limiting. It can be transferred to further environmental externalities (other air pollutants, noise), road conditions (severe weather (Sect. 6.4.8)) or traffic performance (density, intensity) (Garcia-Castro & Monzon, 2013, p. 117). One example for their application is Tirol, Austria, where speed limits on parts of the A 12 Inntal motorway are set depending on the NO_2 values (LGBl. Nr. 36/2011).

In terms of the latter application area, dynamic speed limits make use of the effect, that traffic intensity and thus performance is highest, if traffic flows at around 80 km/h (Schnieder, 2007, p. 193). When traffic intensity reaches a pre-set boundary value, speed therefore is limited and the road's capacity increased. Also signalling another appropriate speed limit than 80 km/h might be sufficient, since it leads to a more harmonized traffic and thus to an improved traffic performance. For example, with a speed limit of 120 km/h, 80 % of the vehicles are moving at a speed of 100–130 km/h. Only 3 % of the vehicles are driving faster than 130 km/h (Schnieder, 2007, p. 193).

By imposing speed limits on all users, these ICT systems pursue a direct time- and energy-saving approach to regulate traffic and thereby to collectively reach environmental objectives. Together with their positive impact on traffic performance, these systems therefore are regarded as strong contributors to Infrastructure sufficiency and Environmental sustainability (Fig. 6.10). Assuming that these systems are most likely installed by the public authority, the questioned experts evaluate the systems' availability as well as the visibility of information as open for all. They further evaluate these ICT systems to cover all transport chains (Rönkkö et al., 2012, pp. 40–43).

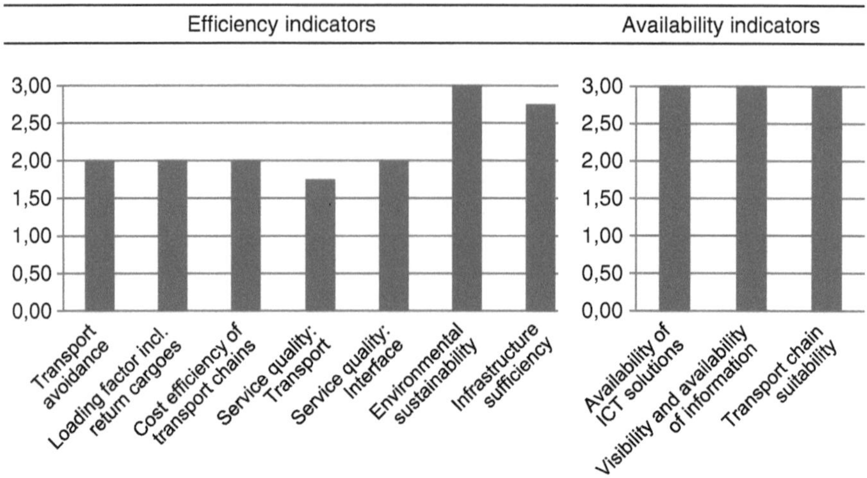

Fig. 6.10 Evaluation of Speed limits depending on CO_2 values. *Source*: Rönkkö et al. (2012)

6.4.7 Intelligent Speed Adaption

Intelligent speed adaption (ISA) is one of the more common ITS. It refers to ICT systems which provide drivers of road vehicles with support in speed controlling (Jamson, Carsten, Chorlton, & Fowkes, 2006, p. 6). ISA at first collects information about the road on which the vehicle is traveling in order to determine the correct speed. Information on the correct speed can be obtained by use of digital maps, general speed zoning information or through recognition technologies which identify speed limit signalling (Rönkkö et al., 2012, p. 41). On this basis ISA can react differently to support the driver in equalling his speed, if the prescribed limit is exceeded. In general there are three ways (Jamson et al., 2006, p. 6):

- Advisory devices (so called open ISA) simply remind the driver of the speed limit.
- Voluntary ISA (also half-open ISA) actively limit the vehicle's speed, which in turn can be overruled by the driver.
- Mandatory ISA (also closed ISA), usually in detected emergency situations, limit the vehicle's speed without allowing driver's intervention.

The regulatory approach can therefore be characterized as decision-supportive. When using a mandatory ISA, however, the system is moreover intrusive.

Besides their traffic managing function, ISA are safety systems since reduction of speed correlates with reduction of collisions. Taking into account the side effects of lesser collisions, ISA may lead to less congestion and increasing environmental sustainability (Fig. 6.11) and therefore have a positive effect on traffic in general and thus on road haulage as well (Rönkkö et al., 2012, pp. 41–43). Studies however imply that ISA only affects traffic positively, if penetration rates within the vehicle fleet reach approximately 60 % (Jamson et al., 2006, p. 15).

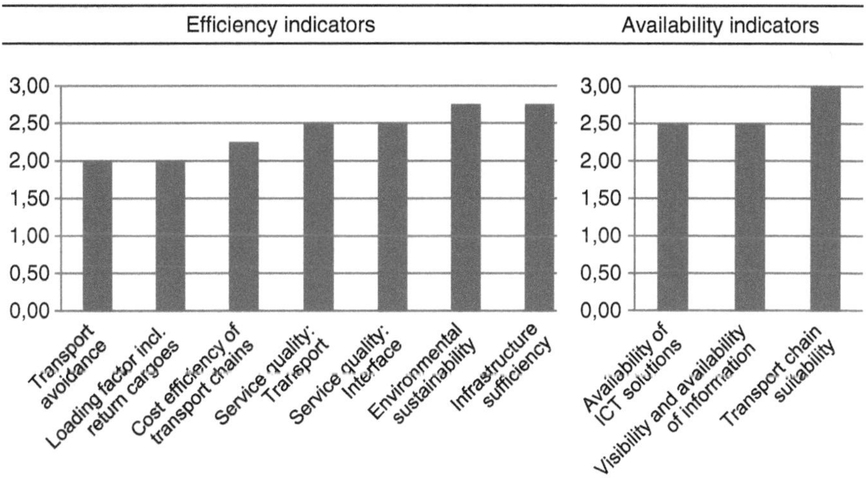

Fig. 6.11 Evaluation of ISA. *Source*: Rönkkö et al. (2012)

The penetration of such driver assistance systems emanates from the manufacturer's supply, the driver's willingness to use and technical feasibility. According to the EU Project for Research on Speed Adaptation Policies on European Roads (PROSPER) (Transport Research & Innovation Portal (TRIP), 2014) the vehicle industry and European road users are positive. Concerning the latter factor, there are, however, several barriers to be removed in order to propel the solution forward. These barriers include Europe-wide coverage of high quality speed limit information and functioning methods for cross-border applications. The current applicability of ISA therefore must be viewed as rudimentary.

6.4.8 Road Weather Information Systems

Across Europe, several Road weather information systems are in use (e.g., SWIS (Deutscher Wetterdienst, 2014), Vaisala (2014), Swedish RWIS (Vägverket—Swedish Road Administration, 2007). Although in some extent focusing on different aspects, they in general comprise sensing technologies, weather prediction systems and applications to distribute or broadcast their findings (Federal Highway Administration, 2013). Stationary sensors collect data, for instance by measuring temperature, precipitation or wind speed, which in turn are input to the prediction system.

Road weather information systems can be classified according to their users and the aspired prediction in terms of geographic dimension and type of forecast.

Narrow systems refer to defined user groups. For instance, these might be facilities that provide road maintenance services and therefore are responsible for granting proper road conditions. Their systems focus the forecast on local road surface conditions in order to support the coordination of duties such as de-icing or scattering of salt and sand.

Ample systems address all road traffic participants by alerting to severe weather conditions in general. Often, these systems offer more specific and concerted information as well, however, charging a fee for this service. Users are able, depending on the specific system, to assemble personal forecasts on road conditions regarding their route or on weather influences such as visibility or snowfall. In some cases road weather systems can, in cooperation with dynamic signalling, warn drivers alongside the road, if changing of unsafe road conditions are predicted or detected (Federal Highway Administration, 2014). Especially with regard to applying road weather information to manage traffic, these systems can be allocated to ISA (Petty, William, & Mahoney, 2007, p. 9). By analogy with ISA Road weather information systems follow a decision-support approach.

In the view of Goodwin (2003, S. 1), the impact of Road weather information systems on traffic results in fewer accidents, since they contribute to improved and safer road conditions or to more careful driving by making users aware of hazardous road conditions. As mirrored in the ratings of the interviewed experts they accordingly expect a higher service quality (Fig. 6.12), especially impelled by rising cargo safety. Concerning infrastructure sufficiency the experts reckon with only slightly less congestion since these ICT systems do not propose any deviating route out of weather-related traffic obstructions. Moreover, they conclude that unnecessary road haulage is avoided upon release of adverse weather conditions (Rönkkö et al., 2012, pp. 44–48).

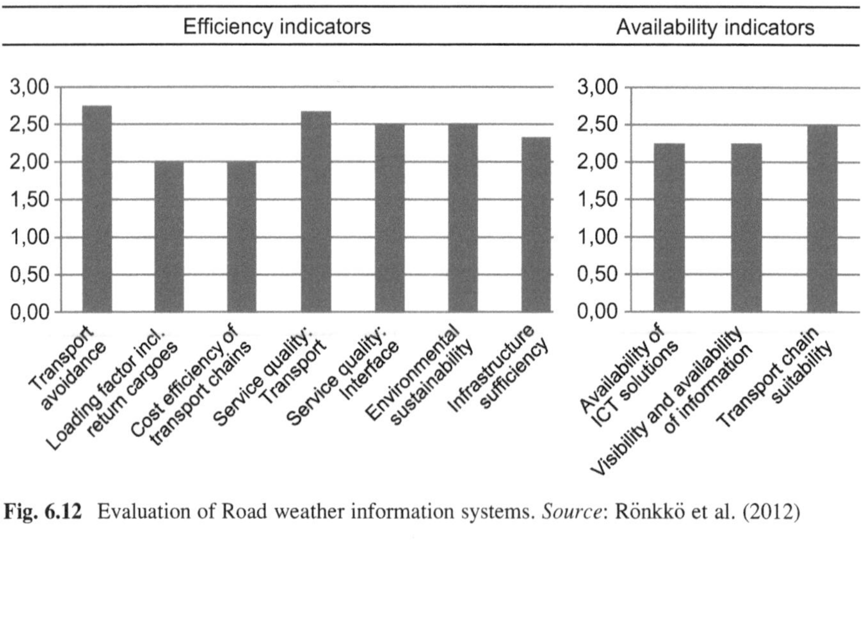

Fig. 6.12 Evaluation of Road weather information systems. *Source*: Rönkkö et al. (2012)

6.4.9 Road Speed Limiter

Road speed limiters are on-board devices which set maximum speed limits to vehicles. Their basic functioning is similar to ISA insofar as they are able to voluntarily, advisorily or mandatorily interact with the driver. In contrast to ISA their input on the speed limits is either set in general or as appropriate by the driver (Várhelyi, 2009, p. 238).

Within Europe, all vehicles over 3.5 tonnes maximum gross weight and passengers' vehicles over eight seats, depending on their national or cross-border usage and their date of first registration, have to be equipped with Road speed limiters. These devices set a maximum speed determined by the vehicle category that cannot be exceeded. (Council of the European Communities, 1992)

In general, Road speed limiters can be characterized as ICT systems following a decision-support approach for vehicle safety. Taking into account the mentioned EU-regulations, speed limiters as applications in European freight logistics use a mandatory approach imposing fines and penalties for infringements (Rönkkö et al., 2012, pp. 44–45).

According to Várhelyi (2009, p. 244) the contribution of speed limiters to road safety can be attributed to lower, more homogenous speeds and thus fewer overtaking and decreasing fuel consumption by about 5.5 %. As shown in Fig. 6.13, the interviewed experts correspondingly are confirmed that these systems lead to less emission of greenhouse gases and other polluters caused by road haulage (Rönkkö et al., 2012, p. 45).

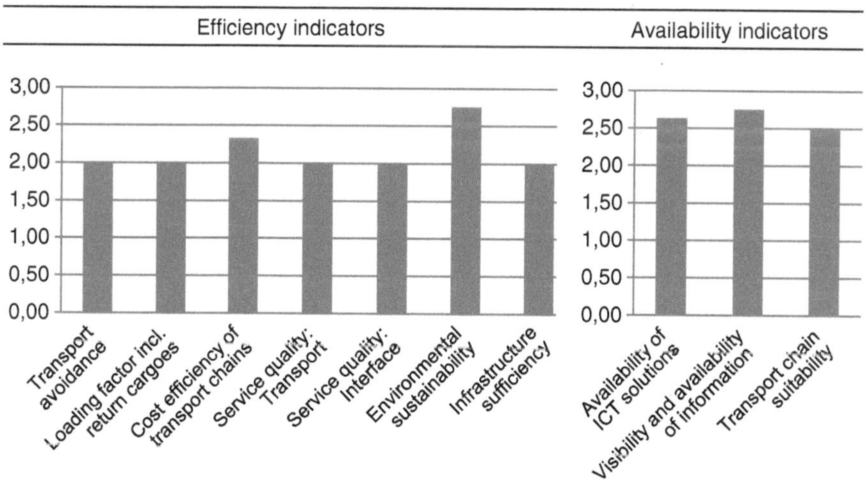

Fig. 6.13 Evaluation of Road speed limiters. *Source*: Rönkkö et al. (2012)

6.4.10 Port Community Systems and Single Window Solution

Port community systems (PCS) are neutral and open electronic platforms. Their users are public and private stakeholders with interest in an airport, seaport or inland port, and therefore making up the so-called port community (European Port Community Systems Association, 2011, p. 3).

Usually PCS are either tailored to a specific port or are designed as nationwide services (e.g., national Port-MIS system used in Korea) (Posti, 2012, p. 58). Some systems even offer a generic solution (Maguire, Ivey, Golias, & Lipinski, 2010, p. 7; Long, 2009, p. 63). This leads to differences in the offered functions ranging from basic information handling and EDI exchange through messaging with embedded databases to import/export services as customs declarations, tracking and tracing and processing of maritime and other statistics (Long, 2009, p. 63). Common to all, however, is that they realize an efficient connection of the platforms' users and thereby follow a direct time- and energy-saving approach to support transportation management.

PCS often include a single window functionality (Long, 2009, p. 63). As defined by the United Nations Economic Commission for Europe (UNECE) a Single window is "a system that allows traders to lodge information with a single body to fulfil all import- or export-related regulatory requirements" (United Nations Economic Commission for Europe, 2003, p. 2). Within a PCS, the services, which are combined by a single window, have to be defined by the involved regulatory agencies. They build up the single window environment (World Customs Organization, 2011, p. 21) and include e. g., customs, transportation, agriculture and health agencies (United Nations Economic Commission for Europe, 2003, p. 2). For traders, Single windows facilitate communication and data exchange with the involved regulatory agencies, resulting in a quicker settlement of regulatory issues.

PCS apply to nodes of air, inland waterway and sea transportation networks. These nodes include interfaces to all modes of transportation. A PCS network therefore comprises a broad range of parties, such as terminal operators, carriers (ocean, road and rail), freight forwarders, enforcement agencies (e.g., customs), port authorities, and various lobby groups (including workers' unions, environmentalists, and other policy makers) (Srour, van Oosterhout, van Baalen, & Zuidwijk, 2008, p. 2).

The main benefits of PCS are an accelerated information flow and an improved information quality and integrity. By speeding up clearance and release, delivery times and delays are reduced and costs are cut down (Fig. 6.14) (Posti, 2012, p. 58; Rönkkö et al., 2012, p. 49).

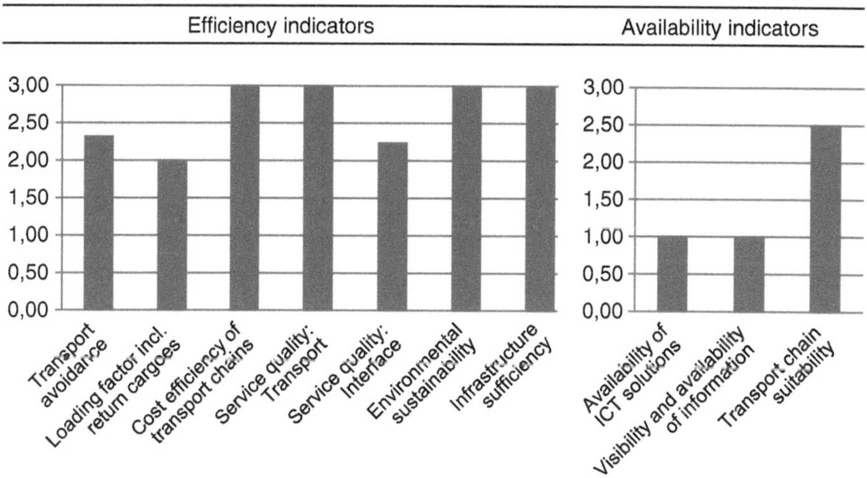

Fig. 6.14 Evaluation of PCS. *Source*: Rönkkö et al. (2012)

6.4.11 Sensors

Sensors are devices for monitoring emissions, which are imposed on the transported goods or which are imposed on the surroundings by transportation. Sensors thereto measure and record data that provide input for further managing actions. Their approach to traffic management is therefore decision-supportive.

Sensors are categorized as external Sensors, if they measure and record sound emissions of transportation to the environment (Rönkkö et al., 2012, p. 53). These sounds are either airborne or structure-borne and perceived as noise or shock (Bundesministerium für Verkehr und digitale Infrastruktur, 2014, p. 13f.). Noise and shock Sensors are applicable to all modes of transportation.

Current efforts regarding noise emissions are covered by the EU Directive 2002/49/EC and comprise the measurement of noise emissions on major roads, railways and airports (European Parliament & Council, 2002, p. 24). These data are prepared into strategic noise maps, which are basis for action plans to reduce noise exposure.

Internal Sensors, in contrast, measure shock, temperature, humidity and other climate conditions within loading equipment (e.g., Hasselmann & Lange, 2008, p. 575). They are therefore used for valuable and sensitive freight. Some Sensors function as mere data loggers only measuring the corresponding environmental impact. More integrated devices combine different Sensors and/or communication technology to transmit their records in real-time (Herzog & Schildhauer, 2009, p. 57f.).

One example of those devices is the Schenker Smartbox. Mounted on the container door, several Sensors within the box measure parameters such as door status, temperature, humidity and movement inside the container. Together with the coordinates of the container the measured data are then transmitted via mobile communication. (DB Schenker, 2013)

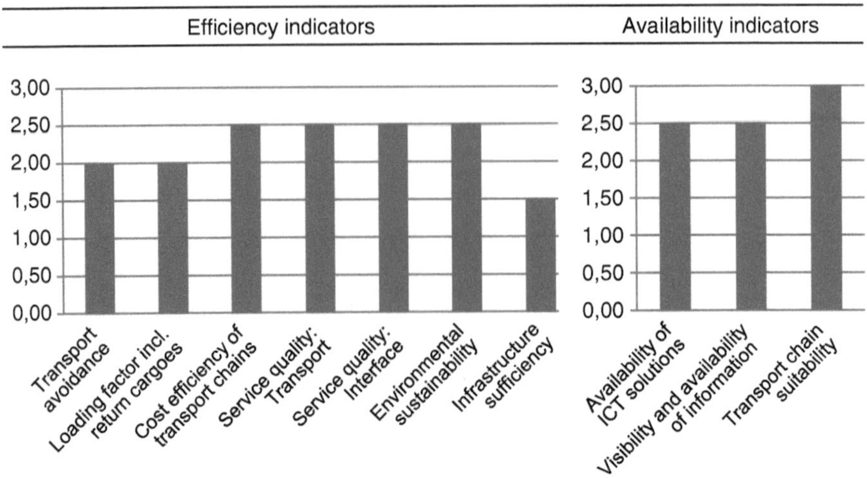

Fig. 6.15 Evaluation of Sensors. *Source*: Rönkkö et al. (2012)

Internal sensors enable operators to monitor the freight status, to detect deviations from desired conditions and to take appropriate counteractions. They therefore contribute to cost efficiency and service quality of freight transportation (Fig. 6.15) (Rönkkö et al., 2012, p. 53). Applying to the characteristics of the transported goods and not to load characteristics, internal Sensors are applicable to all types of transport chain. They are especially advisable for long-range, intermodal transports with many handling operations.

6.4.12 Green Trucks

The term green trucks denotes vehicles for road transportation, which are equipped with several ITS. Referring to the definition of ITS (Sect. 6.4.3) these are ICT systems such as Adaptive Cruise Control, Platooning, Eco-driver Assistance, automatic engine start-up/shutdown and tyre pressure indicators that are implemented into the vehicle (Rönkkö et al., 2012, p. 53). According to the basic functioning of ITS, green trucks use inter-vehicle communication and/or vehicle-to-roadside communication (Zacharioudakis et al., 2012, p. 37).

In comparison to vehicles equipped with singular ITS, green trucks benefit from various ITS and, as applicable, from interaction between the installed ITS exploiting available information on the current state and the operations of the road network more properly (Crainic, Gendreau, & Potvin, 2009, p. 542).

The impact on traffic of green trucks depends on the installed ITS. In general, to the interviewed experts, the efficiency impact of green trucks is a critical issue

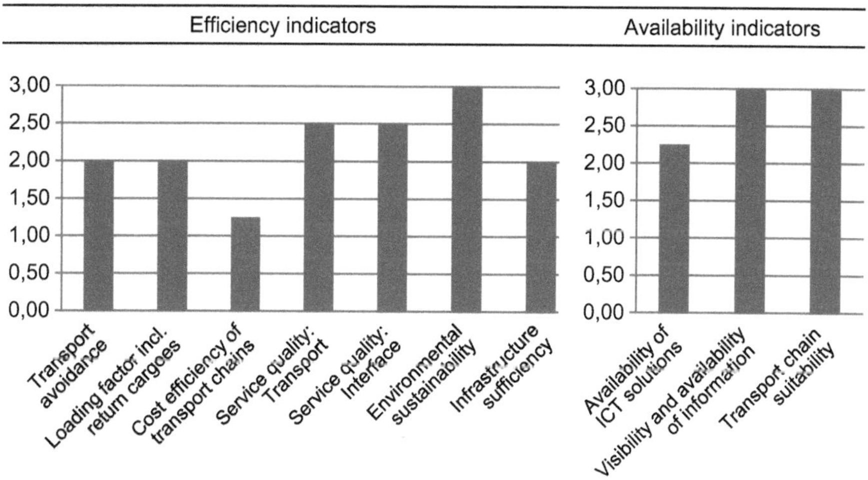

Fig. 6.16 Evaluation of Green trucks. *Source*: Rönkkö et al. (2012)

(Fig. 6.16). Going in line with their evaluation of ITS, they agree that green trucks contribute to less emission of greenhouse gases and other polluters. However, they assume that due to investments required for the systems' installations the cost of operation will increase until their amortization (Rönkkö et al., 2012, pp. 53–55).

6.5 Installation of ICT Systems in European Freight Corridors

In order to assess the benefits that possibly result from the installation of the ICT systems highlighted in Sects. 6.4.1–6.4.12 the SuperGreen project acted out 15 scenarios for concrete ICT/mode/corridor combinations. As one example of these scenarios the implementation of ERTMS in the Brenner corridor shall be presented in the following.

Thereto, the status quo of ERTMS installation on the European rail network is introduced by describing the related plans and their actual status. Further, the cost data are reported as they are crucial for assessing the financial feasibility of installation. Based upon this, the Brenner corridor is portrayed and potential benefits after its equipment with ERTMS are deduced by transferring the general evaluation results of Sect. 6.4.4 to the regarded scenario and this way concretizing them for the Brenner corridor.

6.5.1 Status Quo of ERTMS Installation

Although ERTMS's history started already in 1989 with first technical drafts, it is not adopted widely yet (Zacharioudakis et al., 2013, p. 46). However, the deployment of ERTMS as a unique signalling standard is expected to bring considerable benefits as increased capacity, reliability and speed as well as reduced maintenance costs. This way, ERTMS will at least sharpen railways' competitiveness along international freight corridors. (UNIFE, 2012a)

To ensure cross-border interoperability, special attention has to be paid to care coordination when implementing ERTMS. For instance, neighboring countries have to simultaneously install ERTMS on their connecting lines. In collaboration with railway stakeholders, the European Commission therefore worked out a dedicated European deployment plan which was adopted in July 2009. This plan pursues a corridor approach, containing a list of six corridors whose equipment is followed up with priority (Fig. 6.17) (UNIFE, 2012b, 2012c):

Fig. 6.17 ERTMS deployment plan. *Source*: Own representation based on European Commission—Directorate General for Transport and Energy (2010)

- Corridor A: Rotterdam-Genoa;
- Corridor B: Stockholm-Naples;
- Corridor C: Antwerp-Basel;
- Corridor D: Budapest-Valencia;
- Corridor E: Dresden-Constanta;
- Corridor F: Aachen-Terespol.

According to Commission's Working Document on the state of play of the implementation of the ERTMS Deployment Plan (2014, p. 5) the deployment on corridors is behind schedule. Major delays occurred in:

- Corridor A, which will be on time (2015) in the Netherlands and Switzerland, operational in 2018 in Germany (3 years delay) and 2020 (5 years delay) in Italy;
- Corridor B that should be equipped by 2020. Apart from the Austrian section which is ready, Germany has announced a delay of 10 years (2030);
- Corridor C, which will be operational in France in 2020–2023 (5–8 years delay). Belgium and Luxemburg sections will be equipped on time (2015);
- Corridor D which will be operational in France in 2021 (delay of 6 years), Italy will be on time, apart from the section between the French border and Torino (5 years delay); Slovenia and Bulgaria will be on time. The Spanish section expected by 2015 will be on time;
- Corridor E, whose delays vary from 0 to 5 years;
- Corridor F, concerning which Germany has announced the date of 2027 and Poland has announced the date of 2020 for some parts.

Furthermore, as mentioned in Chap. 1, ERTMS has been prescribed as a requirement for all TEN-T Core Network Corridors, which have been introduced in 2013 and to a large extent supersede the ERTMS corridors.

6.5.2 Installation Costs

The investment costs of ERTMS contain on the one hand development cost that incurred during the deployment phase and were borne by the Railway Undertaktings and Infrastructure Managers (Zacharioudakis et al., 2013, p. 51).

On the other hand, the investment costs concern the installation of on-board and trackside modules. The cost level of the on-board module is determined by the type of locomotives or train sets. For new equipment, the level is around 100,000 €, for existing equipment prices vary between 200,000 and 300,000 €. The adaption of existing rolling stock is linked to the problem of finding adequate space to add new antennae on the trains or a new screen in the driver's cab. In this course, the compatibility of different systems has to be inquired to avoid any interference, e.g., questions of electromagnetic compatibility have to be clarified. As the fitting of ETCS can require the complete renovation of the line, the range of infrastructure cost is rather wide with estimates varying between 30,000 and 300,000 € per kilometer. Exact cost statements can only be obtained on the basis of a line-specific analysis. (European Commission-Directorate General for Mobility & Transport, 2006, pp. 1–2)

6.5.3 Benefits of ERTMS Installation

Linking Sweden with Italy the Brenner corridor (Fig. 6.18) is running across the cities Malmö, Trelleborg, Rostock/Sassnitz, Berlin, Munich, Salzburg, Verona, Bologna, Naples, Messina to Palermo, with the two branches Salzburg, Villach and Trieste (Tauern Axis) as well as Bologna, Ancona/Bari/Brindisi, Igoumenitsa/ Patras and Athens. On this axis road and rail are used as main transportation modes. Parts are further handled by short sea shipping, such as Naples to Palermo and Patras/Igoumenitsa to Ancona/Bari/Brindisi. Of these modes, rail operations are regarded within this scenario. (Ilves et al., 2011, p. 27)

Being installed along the Brenner Corridor ERTMS is expected to considerably facilitate cross-border traffic movements, e. g., between Austria and Italy, because locomotives will not need to be exchanged at each border. Thus, train speed will rise significantly. Estimates suggest that transportation time can be decreased to an

Fig. 6.18 Brenner Corridor. *Source*: Ilves et al. (2011)

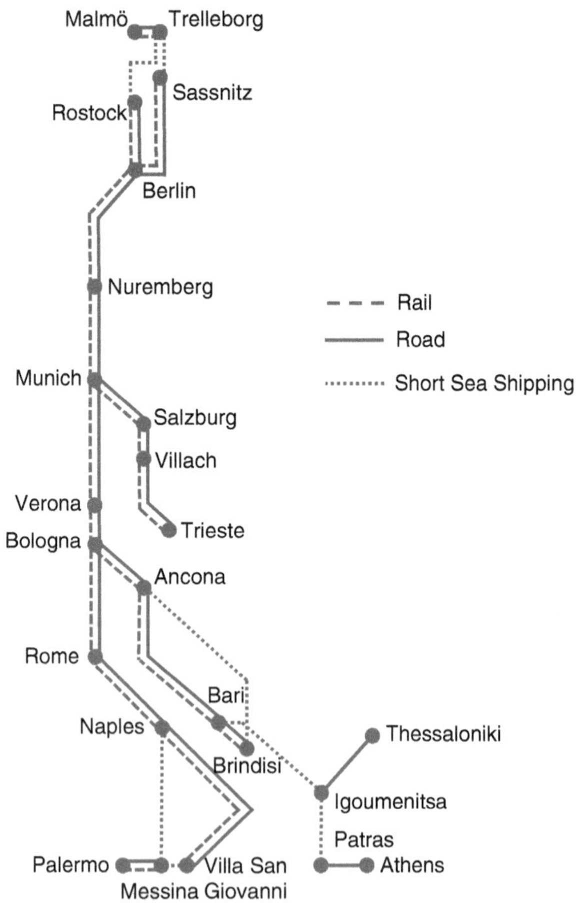

average of 60 %. Goods which so far are transported by airplanes and trucks on this axis can be transferred using the high speed rails, resulting in a tremendous modal shift from air and road to rail transportation and GHG emission savings (Zacharioudakis et al., 2013, pp. 55–56).

Moreover, locomotives must not be equipped with multiple signalling systems as PZB and BACC (cp. Fig. 6.7). Rail freight operators need to purchase only the on-board equipment of ETCS leading to reduced investment and maintenance costs. Since ERTMS enables to reduce the headway between trains, the capacity of existing lines will increase up to 40 %. Consequently, there will be further cost cutting due to decreased needs for new infrastructure. (Zacharioudakis et al., 2013, p. 56)

6.6 Conclusion

Only with accurate, full and actual information, business activities of the transportation sector can be carried out efficiently and with foresight. Information transparency therefore is the key to both economic and ecologically sustainable freight logistics. Due to their combined functions of gathering transportation related data, processing them to valuable traffic and transportation management information and providing them again to relevant transportation actors, ICT systems assist in creating thorough transparency.

Within the SuperGreen project concrete ICT systems applicable in freight logistics were examined, generally evaluated by an expert survey as well as deeply analysed in ICT/mode/corridor-scenarios. According to the gained results, the application of ICT systems as Congestion charging systems, Unified electronic toll systems and PCS, can remove obstructions and facilitate traffic flows on land, sea and in the air. Moreover, ICT systems as ERTMS, Traffic control systems, ISA, Road weather information systems and Sensors, enable real time monitoring of e.g., vehicle location and speed, traffic performance, weather condition and shipment status and thereby provide the ability to respond to events as they occur (Panagakos et al., 2012, p. 104).

In conclusion, for a large number of ICT systems it was demonstrated that they are instrumental in realizing green logistics. The SuperGreen project therefore strongly opted for advancement of smart ICT solutions for transportation. When developing new systems, particular attention, however, has to be paid to their European-wide applicability and interoperability in order to achieve and exploit synergies. Moreover, in view of the in many respects pluralistic European society, human aspects shall be considered. Systems are needed that provide individual information and convey those individually correctly.

References

Brakatsoulas, S., Pfoser, D., & Tryfona, N. (2005). Practical data management techniques for vehicle tracking data. In Proceedings of the 21st International Conference on Data Engineering (pp. 324–325).

Bode, W., Hermsmeier, J., Hocke-Anbeh, A., Ziegler, A. (2011). *Praxisleitfaden "Grüne Logistik"*. Retrieved September 29, 2014, from http://www.ris-logis.net/Gruene_Logistik/pdf/Praxisleitfaden.pdf

Bouhoute, A., Berrada, I., & El Kamili, M. (2014). A formal driving behavior model for intelligent transportation systems. In G. Noubir & M. Raynal (Eds.), *Networked systems—second international conference—revised selected papers* (pp. 289–312). Cham, Switzerland: Springer.

Bundesministerium für Verkehr und digitale Infrastruktur. (2014). *Lärmschutz im Schienenverkehr—Alles über Schallpegel, innovative Technik und Lärmschutz an der Quelle*. Retrieved September 23, 2014, from http://www.bmvi.de/SharedDocs/DE/Publikationen/VerkehrUndMobilitaet/Schiene/laermschutz-im-schienenverkehr-broschuere.pdf?__blob=publicationFile.

Central Commission for Navigation on the Rhine. (2008). *Leaflet Inland AIS—Vessel Tracking and Tracing Standard for Inland Navigation*. Retrieved August 19, 2014, from http://www.ris.eu/docs/File/393/ccr_draft_leaflet_vtt_2008.doc.

Clausen, U., Geiger, C., & Behmer, C. (2011). Green corridors by means of ICT applications. In transport research arena—Europe 2012. *Procedia—Social and Behavioral Sciences, 48*, 1877–1886.

Commission of the European Communities. (2007). *Freight transport logistics action plan*. Retrieved September 19, 2014, from http://eur-lex.europa.eu/legal-content/EN/TXT/PDF/?uri=CELEX:52007DC0607&from=EN.

Congress of the United States. (2009). *A CBO study—using pricing to reduce traffic congestion*. Retrieved September 23, 2014, from http://www.cbo.gov/sites/default/files/cbofiles/ftpdocs/97xx/doc9750/03-11-congestionpricing.pdf.

Council of the European Communities. (1992). Directive 92/24/EEC of 31 March 1992 relating to speed limitation devices or similar speed limitation on-board systems of certain categories of motor vehicles—Amended by: Directive 2004/11/EC of the European Parliament and of the Council of 11 February 2004. *Official Journal of the European Community, L129*, 154–175. Retrieved September 23, 2014, from http://eur-lex.europa.eu/LexUriServ/LexUriServ.do?uri=CONSLEG:1992L0024:20040217:EN:PDF.

Crainic, T. G., Gendreau, M., & Potvin, J.-Y. (2009). Intelligent freight-transportation systems: Assessment and the contribution of operations research. *Transportation Research, 17*(6), 541–557.

David-Freihsl, R. (2013). Die Londoner Citymaut als Mutprobe. Der Standard, p. 10.

Deutscher Wetterdienst. (2014). *Website of the road weather information service SWIS*. Retrieved September 22, 2014, from http://www.dwd.de/SWIS.

Economic Commission for Europe. (2005). *Guidellines and recommendations for river information services—resolution no. 57*. New York: United Nations.

Eißel, D., & Chu, C. P. (2014). The future of sustainable transport system for Europe. *AI & SOCIETY, 29*(3), 387–402.

ERTMS—The European Rail Traffic Management System. (2013). *ERTMS benefits*. Retrieved September 19, 2014, from http://www.ertms.net/

European Commission—Directorate General for Mobility and Transport. (2006). *ERTMS in 10 questions*. Retrieved November 17, 2014, from http://ec.europa.eu/transport/modes/rail/interoperability/ertms/doc/ertms_10_questions_en.pdf

European Commission—Directorate General for Mobility and Transport. (2011). *White paper on transport—roadmap to a single European transport area—towards a competitive and resource-efficient transport system*. Retrieved September 24, 2014, from http://ec.europa.eu/transport/themes/strategies/doc/2011_white_paper/white_paper_com%282011%29_144_en.pdf

European Commission—Directorate General for Transport and Energy. (2010). *European deployment plan for ERTMS*. Retrieved February 13, 2014, from http://ec.europa.eu/transport/modes/rail/interoperability/ertms/doc/edp/ertms_map.pdf

European Environmental Agency. (2007). *Fact sheet term 2006 02*. Retrieved September 25, 2014, from http://www.eea.europa.eu/data-and-maps/indicators/transport-emissions-of-greenhouse-gases-6/term_2006_02___transport_emission_of_greenhouse_gases_final_version.pdf

European Environmental Agency. (2014a). *Greenhouse gas emissions by sector*. Retrieved September 24, 2014, from http://epp.eurostat.ec.europa.eu/tgm/table.do?tab=table&init=1&language=en&pcode=tsdcc210&plugin=1

European Environmental Agency. (2014b). *Final energy consumption by sector*. Retrieved September 24, 2014, from http://epp.eurostat.ec.europa.eu/tgm/table.do?tab=table&init=1&language=en&pcode=tsdpc320&plugin=1

European Parliament and Council. (2002). Directive 2002/49/EC of the European Parliament and the Council of 25 June 2002 relating to the assessment and management of environmental noise. *Journal of the European Communities, 18*, 12–25.

European Parliament and Council. (2005). Directive 2005/44/EC of the European Parliament and of the Council of 7 September 2005 on harmonised river information services (RIS) on inland waterways in the community. *Official Journal of the European Union, 255*, 152–159.

European Parliament and Council. (2010). Directive 2010/40/EU of the European Parliament and of the Council of 7 July 2010 on the framework for the deployment of Intelligent Transport Systems in the field of road transport and for interfaces with other modes of transport. *Official Journal of the European Union, L207*, 1–13.

European Parliament and Council. (2013). Regulation (EU) no 1315/2013 of the European Parliament and of the Council of 11 December 2013 on Union guidelines for the development of the trans-European transport network and repealing Decision No 661/2010/EU. *Official Journal of the European Union, 56*, 1–128.

European Port Community Systems Association. (2011). *How to develop a port community system*. Retrieved September 17, 2014, from http://www.unece.org/fileadmin/DAM/trade/Trade_Facilitation_Forum/BkgrdDocs/HowToDevelopPortCommunitySystem-EPCSAGuide.pdf

Fantechi, A., Flammini, F., & Gnesi, S. (2012). Formal methods for intelligent transportation systems. In T. Margaria & B. Steffen (Eds.), *Leveraging applications of formal methods, verification and validation—applications and case studies—proceedings, Part II* (pp. 187–189). Berlin, Germany: Springer.

Federal Highway Administration. (2013). *Road weather management program*. Retrieved September 22, 2014, from http://ops.fhwa.dot.gov/weather/faq.htm

Federal Highway Administration. (2014). *Active traffic management*. Retrieved September 22, 2014, from http://ops.fhwa.dot.gov/atdm/approaches/atm.htm

Forster, P., Ramaswamy, V., Artaxo, P., Berntsen, T., Betts, R., Fahey, D. W., et al. (2007). *Changes in atmospheric constituents and in radiative forcing*. In Climate change 2007—The physical science basis. Contribution of working group I to the fourth assessment report of the Intergovernmental Panel on Climate Change (IPCC). Retrieved September 29, 2014, from http://www.ipcc.ch/pdf/assessment-report/ar4/wg1/ar4-wg1-chapter2.pdf

Fraunhofer Institute for Communication Systems ESK. (2010). *Car-to-Roadside Communication*. Retrieved September 19, 2014, from http://www.esk.fraunhofer.de/content/dam/esk/en/documents/PDB_C2R_en.pdf

Garcia-Castro, A., Monzon, A. (2013). *Variable speed limits—review and development of an aggregate indicator based on floating car data*. In Transport and Telecommunication Institute, Proceedings of the 13th International Conference "Reliability and Statistics in Transportation and Communication", 117–127.

Geiger, C., & Diekmann, D. (2013). Logistik und Umwelt. In U. Clausen & C. Geiger (Eds.), *Verkehrs- und Transportlogistik* (pp. 21–32). Berlin, Germany: Springer.

Geiger, C., Schmied, M. (2012). *Carbon footprint—Teilgutachten "Abbau von Hemmnissen zur Emissionsminderung"*. In Umweltbundesamt (Ed.), TEXTE 30/2012.

Goodwin, L. C. (2003). *Best practices for road weather management*. Retrieved September 22, 2014, from http://www.velocidade.prp.pt/Sites/velocidade/CentroDocumentacao/InvestigacaoEstudos/75%20-%20Best%20Practices%20-%20CaseStudiesFINALv2-RPT.pdf

Hansen, J., & Lebedeff, S. (1987). Global trends of measured surface air temperature. *Journal of Geophysical Research, 92*, 13345–13372.

Hasselmann, G., & Lange, V. (2008). Temperaturgeführte transporte. In D. Arnold, H. Isermann, A. Kuhn, H. Tempelmeier, & K. Furmans (Eds.), *Handbuch logistik* (pp. 570–580). Berlin, Germany: Springer.

Heinrich, L., & Stelzer, D. (2011). *Informationsmanagement—Grundlagen, Aufgaben, Methoden*. München, Germany: Oldenbourg Wissenschaftsverlag.

Herzog, O., & Schildhauer, T. (2009). Beispiele Intelligenter Objekte. In O. Herzog & T. Schildhauer (Eds.), *Intelligente Objekte—Technische Gestaltung, Wirtschaftliche Verwertung, Gesellschaftliche Entwicklung* (pp. 35–62). Berlin, Germany: Springer.

Ilves, I., Holte, E. A., Georgopoulou, C., Rönkkö, S., Aditjandra, P., Vanaale, E., et al. (2011). *SuperGreen—Supporting EU's Freight Transport Logistics Action Plan on Green Corridors Issues—Deliverable D2.4—Benchmarking of Green Corridors—Version 2*. Retrieved June 29, 2014, from http://www.supergreenproject.eu/docs/public/D2.4-2%20REVISED%20v.3.1%20FINAL%20PUBLIC.pdf

International Union of Railways. (2013). *What is ERTMS?* Retrieved September 19, 2014, from http://www.uic.org/spip.php?article381

IPCC—Intergovernmental Panel on Climate Change. (2013). Summary for policymakers. In T. F., Stocker, D. Qin, G.-K. Plattner, M. Tignor, S. K. Allen, J. Boschung, et al. (Eds.), *Contribution of working group I to the fifth assessment report of the intergovernmental panel on climate change, climate change 2013: The physical science basis*. Cambridge: Cambridge University Press.

Jamson, S., Carsten, O., Chorlton, K., & Fowkes, M. (2006). *Intelligent speed adoption—literature review and scoping study*. Leeds, England: University of Leeds, MIRA Ltd and Transport for London.

Jorgensen, R. (2011). *Slow steaming—the full story*. Retrieved September 29, 2014, from https://www.yumpu.com/en/document/view/7803715/slow-steaming-the-full-story-maersk

Klaus, P., Krieger, W., & Krupp, M. (Eds.). (2012). *Gabler Lexikon Logistik—Management logistischer Netzwerke und Flüsse*. Wiesbaden, Germany: Springer.

Kleine-Besten, T., Kersken, U., Pöchmüller, W., & Schepers, H. (2012). Navigation und Telematik. In H. Winner, S. Hakuli, & G. Wolf (Eds.), *Handbuch Fahrerassistenzsysteme—Grundlagen, Komponenten und Systeme für aktive Sicherheit und Komfort*. Wiesbaden, Germany: Vieweg + Teubner.

Klunder, G., Malone, K., Wilmink, I. (2010). Impact of information and communication technologies on energy efficiency in road transport. In *Transport Research Arena Europe 2010 (TRA2010)*, Brussels, Belgium.

Long, A. (2009). Port community systems. *World Customs Journal, 3*(1), 63–67.

Lüthi, M. (2008). *Optimizing traffic flow in heavily used railway networks: influence factors and potential strategies*. Retrieved September 18, 2014, from http://www.strc.ch/conferences/2008/2008_Luethi_Rescheduling.pdf

Maguire, A., Ivey, S., Golias, M. M., Lipinski, M. E. (2010). *Relieving congestion at intermodal marine container terminals: Review of tactical/operational strategies*. Retrieved September 22, 2014, from http://www.trforum.org/forum/downloads/2010_161_Relieving_Congestion_Marine_Terminals_Strategies.pdf

Marchet, G., Perego, A., & Perotti, S. (2009). An exploratory study of ICT adoption in the Italian freight transportation industry. *International Journal of Physical Distribution & Logistics Management, 39*(9), 785–812.

Maschek, U. (2013). Leit- und Sicherungstechnik. In L. Fendrich & W. Fengler (Eds.), *Handbuch Eisenbahninfrastruktur* (pp. 495–551). Berlin, Germany: Springer.

Notteboom, T. (2013). Maritim transportation and seaports. In J.-P. Rodrigue, T. Notteboom, & J. Shaw (Eds.), *The SAGE handbook of transportation studies* (pp. 83–102). Los Angeles, CA: Sage.

Pachl, J. (2013). *Systemtechnik des Schienenverkehrs—Bahnbetrieb planen, steuern und sichern*. Berlin, Germany: Springer.

Panagakos, G., Moyano, H., Psaraftis, H., Lyridis, D., Mpaliami, T., Webb, S., et al. (2012). *SuperGreen—supporting EU's freight transport logistics action plan on green corridors issues—deliverable D6.2—policy recommendations—version 2*. Retrieved June 06, 2014, from http://www.supergreenproject.eu/docs/public/D6.2-2.pdf

Perego, A., Perotti, S., & Mangiaracina, R. (2011). ICT for logistics and freight transportation: A literature review and research agenda. *International Journal of Physical Distribution & Logistics Management, 41*(5), 457–483.

Petty, K. R., William, P., & Mahoney, W. P. (2007). *Weather applications and products enabled through vehicle infrastructure integration (VII)—feasibility and concept development study*. Boulder, CO: National Center for Atmospheric Research. Retrieved September 19, 2014, from http://www.ops.fhwa.dot.gov/publications/viirpt/viirpt.pdf.

Plößl, K. (2009). Mehrseitig sichere Ad-hoc-Vernetzung von Fahrzeugen. In A. Pfitzmann, H. Reimer, K. Rihaczek, & A. Roßnagel (Eds.), *DuD-Fachbeiträge*. Wiesbaden, Germany: Gabler.

Posti, A. (2012). *E-PORT—improving the efficiency of Finnish ports with intelligent systems—final report of the mobile port project*. Turku, Finland: The Centre for Maritime Studies, University of Turku. Retrieved September 19, 2014, from http://mkkdok.utu.fi/pub/A58-E-PORT%20improving%20the%20efficiency%20.pdf.

Rahmstorf, S., & Schellnhuber, H.-J. (2012). *Der Klimawandel*. München, Germany: Verlag C.H. Beck.

River Information Services. (2009). *Electronic reporting international—ERI messages*. Retrieved September 18, 2014, from http://www.ris.eu/docs/File/417/eri_message__clarification_and_explanation_eri.pdf

Rönkkö, S., Salanne, I., Valli, R., Geiger, C., Psaraftis, H., Zacharioudakis, P., et al. (2012). *SuperGreen—supporting EU's freight transport logistics action plan on green corridors issues—deliverable D4.3—potential for green supply chain management*. Retrieved September 23, 2014, from http://www.supergreenproject.eu/docs/public/D4.3%20PUBLIC.pdf

Rudolph, M., & Wagner, U. (2008). *Energieanwendungstechnik—Wege und Techniken zur effizienteren Energienutzung*. Berlin, Germany: Springer.

Salanne, I., Rönkkö, S., & Byring, B. (2010). *SuperGreen—supporting EU's freight transport logistics action plan on green corridors issues—deliverable D2.1—selection of corridors*. Retrieved June 29, 2014, from http://www.supergreenproject.eu/docs/public/FINAL%20PUBLIC%20Deliverable%20D2%201_WP%202_Super%20Green.pdf

Schach, R., Jehle, P., & Naumann, R. (2006). *Transrapid und Rad-Schiene-Hochgeschwindigkeitsbahn*. Berlin, Germany: Springer.

Scheer, A.-W., Angeli, R., & Herrmann, K. (2001). Informations- und Kommunikationstechnologie als Treiber der Logistik. In H.-C. Pfohl (Ed.), *Jahrhundert der Logistik—customer related, glocal, e-based* (pp. 29–58). Berlin, Germany: Erich Schmidt.

Schenker, D. B. (2013). *Transparenz entlang der Transportkette*. Retrieved September 23, 2014, from http://www.dbschenker.com/file/6053042/data/smartbox-broschuere.pdf

Schnieder, E. (2007). *Verkehrsleittechnik—Automatisierung des Straßen- und Schienenverkehrs*. Berlin, Germany: Springer.

Schulz-Spathelf, J. (2012). Informationssysteme, integrierte logistische. In P. Klaus, W. Krieger, & M. Krupp (Eds.), *Gabler Lexikon Logistik—Management logistischer Netzwerke und Flüsse* (pp. 236–243). Wiesbaden, Germany: Springer.

Siemens, A. G. (2014). *Current situation: More than 20 incompatible train protection systems*. Retrieved November 13, 2014, from http://www.mobility.siemens.com/mobility/global/en/interurban-mobility/rail-solutions/locomotives/vectron/technology/modular-locomotive-concept/modular-train-protection-concept/Pages/modular-train-protection-concept.aspx# Current_20situation_20More_20than_2020_20incompatible_20train_20protection_20systems

Sjurts, I. (Ed.). (2011). *Gabler Lexikon Medienwirtschaft*. Wiesbaden, Germany: Gabler.

SkySails. (2011). *Neue Energie für die Schifffahrt—Pressemappe*. Retrieved September 29, 2014, from http://www.skysails.info/fileadmin/user_upload/Presselounge/Dokumente/deutsch/DE_Pressemappe_SkySails.pdf

Srour, F. J., van Oosterhout, M., van Baalen, P., Zuidwijk, R. (2008). *Port community system implementation: Lessons learned from international scan*. In Transportation Research Board 87th Annual Meeting, Washington, DC. Retrieved September 29, 2014, from http://www.researchgate.net/publication/236278299_Port_Community_System_Implementation_Lessons_Learned_from_International_Scan/file/3deec52a96410ec2fa.pdf

Swantusch, R. (2010). *Stop & Go*. In Verkehrs Rundschau, Heft 07/2010, 30.

Transport Research and Innovation Portal (TRIP). (2014). *Website of the EU PROSPER Project*. Retrieved September 19, 2014, from http://www.transport-research.info/web/projects/project_details.cfm?ID = 36783

Umweltbundesamt. (2012). *Daten zum Verkehr, Ausgabe 2012*. Retrieved September 29, 2014, from http://www.umweltbundesamt.de/sites/default/files/medien/publikation/long/4364.pdf

UNFCC. (2008). *Kyoto protocol reference manual*. Retrieved September 25, 2014, from http://unfccc.int/resource/docs/convkp/kpeng.pdf

UNIFE. (2012a). *Factsheet no. 1. from trucks to trains*. Retrieved November 17, 2014, from http://www.unife.org/uploads/2012_ERTMS_Facts_sheets_1_-_18.pdf

UNIFE. (2012b). *Factsheet no. 9. A unique signaling system for Europe*. Retrieved November 17, 2014, from http://www.unife.org/uploads/2012_ERTMS_Facts_sheets_1_-_18.pdf

UNIFE. (2012c). *Factsheet no. 15. International freight corridors equipped with ERTMS*. Retrieved November 17, 2014, from http://www.unife.org/uploads/2012_ERTMS_Facts_sheets_1_-_18.pdf

United Nations Economic Commission for Europe. (2003). *The single window concept—enhancing the efficient exchange of information between trade and government*. Geneva, Switzerland: United Nations.

Vägverket—Swedish Road Administration. (2007). *Road weather information system*. Retrieved September 22, 2014, from http://publikationswebbutik.vv.se/upload/4213/89052_road_weather_information_system.pdf

Vaisala. (2014). *Website of the road weather information service VAISALA*. Retrieved September 22, 2014, from http://www.vaisala.com/en/roads/Pages/default.aspx

Vallant, J., & Hofmann-Wellenhof, B. (2008). River information services. *Elektrotechnik & Informationstechnik, 6*, 238–243.

Várhelyi, A. (2009). Speed management via in-car devices: Effects, implications, perspectives. *Transportation, 29*, 237–252.

via donau—Österreichische Wasserstraßen-Gesellschaft mbH. (2014). *Website of the EU NAIADES initiative*. Retrieved September 19, 2014, from http://www.naiades.info/

Woonsuk, S., Kyungmee, C., & Eunseok, L. (2011). Towards reliable intelligent transportation systems for e-government. In K. N. Andersen, E. Francesconi, Å. Grönlund, & T. M. van Engers (Eds.), *Electronic government and the information systems perspective—Second International Conference—2011 proceedings* (pp. 299–314). Berlin, Germany: Springer.

World Customs Organization. (2011). *WCO compendium—how to build a single window environment—volume 1: The executive guide*. Retrieved September 19, 2014, from http://www.wcoomd.org/en/topics/facilitation/activities-and-programmes/single-window/~/media/252D1B-F37A814526BF5BFFEAB7F13692.ashx

Zacharioudakis, P. G., Mpaliami, T., Kapetanis, G., Ventikos, N., Kontovas, C., Livaniou, S., et al. (2013). *SuperGreen—supporting EU's freight transport logistics action plan on green *corridors issues—deliverable D4.4- benchmarking green corridors with smart ICT version 2.* Retrieved September 23, 2014, from http://www.supergreenproject.eu/docs/public/D4.4v2.PDF

Zacharioudakis, P. G., Mpaliami, T., Lyridis, D., Panagakos, G., Kvamstad, B., Geiger, C., et al. (2012). *SuperGreen—supporting EU's freight transport logistics action plan on green corridors issues—deliverable D4.1- identify smart ICT and information flows.* Retrieved September 23, 2014, from http://www.supergreenproject.eu/docs/public/D4.1v3.pdf

Chapter 7
Green Vehicle Routing

Tolga Bektaş, Emrah Demir, and Gilbert Laporte

Abstract *Green vehicle routing* is a branch of *green logistics* which refers to vehicle routing problems where externalities of using vehicles, such as carbon dioxide-equivalents emissions, are explicitly taken into account so that they are reduced through better planning. This chapter presents an overview of some of the recent developments in the area, including the description of some vehicle emission models and their applications in road freight transportation.

Acronyms and Abbreviations

Notation	*Description*
ALNS	Adaptive large neighborhood search algorithm
COPERT	Computer programme to calculate emissions from road transportation
CMEM	Comprehensive modal emission model
DSOP	Departure time and speed optimization procedure
EMVRP	Energy-minimizing vehicle routing problem
EVRP	Emissions vehicle routing problem
FCVRP	Vehicle routing problem with fuel consumption rate
GHGs	Greenhouse gases
GVR	Green vehicle routing
MEET	Methodology for calculating transportation emissions and energy consumption
NAEI	National atmospheric emissions inventory

T. Bektaş (✉)
University of Southampton, Southampton, Highfield SO17 1BJ, UK
e-mail: T.Bektas@soton.ac.uk

E. Demir
Eindhoven University of Technology, 5600 MB Eindhoven, The Netherlands
e-mail: E.Demir@tue.nl

G. Laporte
HEC Montréal, 3000 chemin de la Côte-Sainte-Catherine, Montréal, QC, Canada, H3T 2A7
e-mail: Gilbert.Laporte@cirrelt.ca

H.N. Psaraftis (ed.), *Green Transportation Logistics*, International Series in Operations Research & Management Science 226,
DOI 10.1007/978-3-319-17175-3_7

PRP	Pollution-routing problem
SOP	Speed optimization procedure
VRP	Vehicle routing problem

7.1 Introduction

The Vehicle Routing Problem (VRP) is a central problem in freight transportation. Introduced by Dantzig and Ramser (1959), it consists of designing optimal delivery or collection routes for a set of vehicles from a central depot to a set of geographically scattered customers. The VRP often includes a number of constraints, such as those related to vehicle capacity, route length, time windows, precedence relations between customers, etc. (Laporte 2007). Since it was introduced more than 50 years ago, the VRP has been mostly studied in terms of minimization of routing costs. However, several researchers have reexamined it in the broader context of green logistics (see, e.g., Palmer 2007; Demir 2012; Qian 2012). Recent developments in green logistics have also been discussed in several papers in the framework of operations research (see, e.g., Dekker et al. 2012; Lin et al. 2013; Demir et al. 2014a).

Externalities in freight transportation are multi-faceted. The more prominent ones include accidents, noise and greenhouse gas emissions (Forkenbrock 1999) of which CO_2 emissions are an increasing source of concern, due to their negative effects on human health and on the environment. Freight transportation also generates significant amounts of other types of pollutants, including particulate matter (small particles of dust, soot, and organic matter suspended in the atmosphere), carbon monoxide (a colorless, odorless, poisonous gas produced when carbon-containing fuel is not burned completely), ozone (formed when emissions of nitrogen oxides (NO_x) and volatile organic compounds (VOCs) chemically react in the presence of sunlight) and hazardous air pollutants, also referred to as air toxics (chemicals emitted into the atmosphere that cause or are suspected to cause cancer or other severe health effects). Wright et al. (2011) suggest that a significant proportion of emissions can be captured through the measurement of the two most prominent anthropogenic greenhouse gases (GHGs); CO_2 and CH_4. The carbon dioxide equivalent (CO_2e) measures how much global warming a given type and amount of GHG may cause, using the functionally equivalent amount or concentration of CO_2 as the reference.

Road transportation is by far the most dominant mode of freight transportation. In particular, road haulage accounted for 76 % of the total inland freight movements in the 28 EU member states in 2011 (Sanchez Vicente 2013). Significant efforts are being made to reduce emissions from transportation, as a result of which the year 2011 saw a slight reduction of 0.6 % in overall GHG emissions, but these were still above the 1990 levels by 25 %. CO_2 emissions are directly proportional to the amount of fuel consumed by a vehicle. CH_4 emissions, on the other hand, are a function of many complex aspects of combustion dynamics and of the type of emission control systems used. A good understanding of these models allows for a more accurate modeling of emissions at the operational planning level.

This chapter provides an overview of the recent research in green vehicle routing, an area of research that aims to reduce environmental externalities of transportation by better planning. It focuses on freight as the type of transport, and on emissions and fuel consumption as the environmental externality. This is where research on green vehicle routing has shown most growth. The aim of the chapter is to present an overview of the area, including some well-known vehicle emission models, algorithms and applications within the context of green routing. The exposition is limited to approaches that make use of the current infrastructure and traditional means of freight transportation (e.g., diesel trucks) as opposed to those making use of new technologies such as electric vehicles, although some discussion will be devoted to the latter in the concluding remarks. The chapter assumes a basic knowledge of the classical vehicle routing problem, for which the reader is referred to Golden et al. (2008) and to Toth and Vigo (2014), for example.

The remainder of the chapter is organized as follows. Sections 7.2 and 7.3 review emissions modeling and pollution-routing problems, respectively. The models of Sect. 7.2 are descriptive whereas those of Sect. 7.3 are prescriptive. Section 7.4 presents a review of case studies in green routing. Conclusions are stated in Sect. 7.5.

7.2 Fuel Consumption Models

This section reviews the major vehicle emissions models introduced in the mechanical and transport engineering literature. We categorize fuel consumption models into two main groups of increasing levels of complexity: macroscopic and microscopic models. Macroscopic models use average aggregate network parameters to estimate network-wide emissions. In contrast, microscopic models estimate the instantaneous vehicle fuel consumption and emission rates at a more detailed level.

Most vehicle emissions models focus on vehicle, traffic, and environmental influences, but do not capture driver related issues which are relatively difficult to measure. More specifically, the driver controls vehicle speed, acceleration rate, brake usage, shifting technique, trailer gap setting, idle time, tire inflation pressure, and more. According to Demir et al. (2014a), the difference in vehicle emissions between the best and worst drivers can be as much as 25 % over a 100 km road segment.

There exist more emission models than those presented here, but the aim of this chapter is to provide examples as opposed to an exhaustive coverage. A detailed survey of vehicle emissions models is presented by Demir et al. (2014a).

7.2.1 Macroscopic Models

We first present three macroscopic emissions models that can be used in a wide-area emission assessment.

7.2.1.1 Methodology for Calculating Transportation Emissions and Energy Consumption

A publication of the European Commission by Hickman et al. (1999) on emission factors for road transportation (INFRAS 1995) describes a methodology for calculating transportation emissions and energy consumption, named MEET. It covers several vehicle technologies for different classes of vehicles. For vehicles weighing less than 3.5 tons, the rate of CO_2 emissions per km is estimated using a speed-dependent regression function of the form $e(v) = 0.0617v^2 - 7.8227v + 429.51$. For other classes, MEET suggests the use a function the form $e(v) = K + av + bv^2 + cv^3 + d/v + e/v^2 + f/v^3$, where $e(v)$ is the rate of CO_2 emissions (g/km) for an unloaded goods vehicle on a road with a zero gradient. The parameters K and a to f are predefined coefficients whose values can be found in Table 7.1.

The emission factors and functions of MEET refer to standard testing conditions (i.e., zero road gradient, empty vehicle, etc.) and are typically calculated as a function of the average vehicle speed. Depending on the vehicle type, a number of corrections may be needed to account for the effects of road gradient and vehicle load on the emissions once a rough estimate has been produced. The following road gradient correction factor is used to take the effect of road gradient into account: $GC(v) = A_6v^6 + A_5v^5 + A_4v^4 + A_3v^3 + A_2v^2 + A_1v + A_0$, where the values of the coefficients A_0 to A_6 can be found in Table 7.2.

The following correction factor is used to take the load factor into account: $LC(v) = k + nv + pv^2 + qv^3 + r/v + s/v^2 + t/v^3 + u/v$, where the values of the coefficients k and n to u can be found in Table 7.3.

MEET suggests estimating the total CO_2 emissions (g) for a given speed v and distance D as

$$E(v, D) = e(v) \cdot GC(v) \cdot LC(v) \cdot D. \tag{7.1}$$

We note that the parameters of MEET model were calibrated in 1999 and updates to these parameters would be needed as new engine technologies and aerodynamic designs of vehicles are introduced.

Table 7.1 Emission parameters used in MEET

Weight class	K	a	b	c	d	e	f
$3.5 < \text{Weight} \leq 7.5$	110	0	0	0.000375	8,702	0	0
$7.5 < \text{Weight} \leq 16$	871	−16.0	0.143	0	0	32,031	0
$16 < \text{Weight} \leq 32$	765	−7.04	0	0.000632	8,334	0	0
$\text{Weight} > 32$	1,576	−17.6	0	0.00117	0	36,067	0

Table 7.2 Road gradient factors for MEET

Weight class	A_6	A_5	A_4	A_3	A_2	A_1	A_0	Slope
Weight ≤ 7.5	0	−3.01E−09	5.73E−07	−4.13E−05	1.13E−03	8.13E−03	9.14E−01	[0, 4]
Weight ≤ 7.5	0	−1.39E−10	5.03E−08	−4.18E−06	1.95E−05	3.68E−03	9.69E−01	[−4, 0]
7.5 < Weight ≤ 16	0	−9.78E−10	−2.01E−09	1.91E−05	−1.63E−03	5.91E−02	7.70E−01	[0, 4]
7.5 < Weight ≤ 16	0	−6.04E−11	−2.36E−08	7.76E−06	−6.83E−04	1.79E−02	6.12E−01	[−4, 0]
16 < Weight ≤ 32	0	−5.25E−09	9.93E−07	−6.74E−05	2.06E−03	−1.96E−02	1.45E+00	[0, 4]
16 < Weight ≤ 32	0	−8.24E−11	2.91E−08	−2.58E−06	5.76E−05	−4.74E−03	8.55E−01	[−4, 0]
Weight > 32	0	−2.04E−09	4.35E−07	−3.69E−05	1.69E−03	−3.16E−02	1.77E+00	[0, 4]
Weight > 32	0	−1.10E−09	2.69E−07	−2.38E−05	9.51E−04	−2.24E−02	9.16E−01	[−4, 0]

Table 7.3 Load correction factors for MEET

Weight class	*k*	*n*	*p*	*q*	*r*	*s*	*t*	*u*
Weight ≤ 7.5	1.27	0.0614	0	−0.00110	−0.00235	0	0	−1.33
7.5 < Weight ≤ 16	1.26	0.0790	0	−0.00109	0	0	−2.03E−7	−1.14
16 < Weight ≤ 32	1.27	0.0882	0	−0.00101	0	0	0	−0.483
Weight > 32	1.43	0.121	0	−0.00125	0	0	0	−0.916

Table 7.4 Set of sample data for COPERT

Payload (%)	Gradient (%)	*a*	*b*	*c*	*d*	*e*
0	0	530.707	0.0634	2,704.528	0.512	157.588
0	−2	546.477	0.064	9,599.652	0.766	61.960
0	+2	1,051.552	0.424	−67.668	0.084	0
50	0	505.770	0.051	4,762.796	0.609	180.436
50	−2	479.620	0.047	7,858.071	0.677	40.246
50	+2	2,074.874	1.008	−0.534	0	0
100	0	502.941	0.041	9,343.090	0.729	195.202
100	−2	1,144.824	0.981	−0.400	0	0
100	+2	1,882.813	1.006	−0.422	0	0

7.2.1.2 Computer Programme to Calculate Emissions from Road Transportation

The computer programme to calculate emissions from road transportation (COPERT) is an European Economic Area funded emissions model (Kouridis et al. 2010), which estimates vehicle emissions for a range of vehicles by engine classification and vehicle type. It is driven by a database of emissions as a function of vehicle class, engine technology and speed. It estimates emissions for all major air pollutants, as well as GHGs produced by different vehicle categories. Similar to MEET, COPERT uses a number of regression functions to estimate fuel consumption, which are specific to vehicles of different weights. An example of total fuel consumption function (g) with different load and gradient factor is given by

$$F(v,D) = (e + (a\,exp(-b\,v)) + (c\,exp(-d\,v))) \cdot D, \qquad (7.2)$$

where a to e are the output coefficients. Table 7.4 shows a set of sample data for a 20–26 ton rigid truck running on Euro-5 diesel.

7.2.1.3 National Atmospheric Emissions Inventory

The national atmospheric emissions inventory (NAEI) model was developed for use in a large range of sectors including agriculture, domestic activity, industry and transport (NAEI 2012). Emissions from road transportation are calculated either

Table 7.5 Sample parameter sets for NAEI

Fuel type	k	a	b	c	d	e	f	g
Pre Euro I	0.037	13,690.286	390.990	54.374	−2.587	0.048	0	0
Euro I	0.037	13,661.548	−220.208	85.387	−3.401	0.060	0	0
Euro II	0.037	4,747.573	1,341.727	−23.117	0.103	0.002	0	0
Euro III	0.037	9,584.751	464.510	42.172	−2.069	0.038	0	0
Euro IV	0.037	10,297.467	210.752	53.532	−2.367	0.043	0	0
Euro V	0.037	10,537.515	220.217	54.175	−2.404	0.043	0	0
Euro VI	0.037	10,537.515	220.217	54.175	−2.404	0.043	0	0

from a combination of total fuel consumption data and fuel properties, or from a combination of driving-related emission factors and road traffic data. For each vehicle category, the fuel consumption with average speed functions for hot exhaust in the NAEI can be calculated as

$$F(v) = k(a + bv + cv^2 + dv^3 + ev^4 + fv^5 + gv^6)/v, \tag{7.3}$$

where k and a to g are the output coefficients and $F(v)$ is in L/100 km. Table 7.5 shows a setting of these coefficients for a 20–26 ton rigid truck. More information can be found in NAEI (2012).

7.2.2 *Microscopic Models*

We now review two microscopic emissions models to estimate hot-stabilized vehicle emissions. The models are used to calculate emissions at a given point t in time and require, as input, instantaneous vehicle kinematic parameters, such as speed and acceleration, or in more aggregated settings, parameters such as time spent in each traffic mode, cruise, acceleration and deceleration.

7.2.2.1 An Instantaneous Fuel Consumption Model

Bowyer et al. (1985) described an energy-related emissions estimation model which uses vehicle characteristics such as mass, energy, efficiency parameters, drag force and fuel consumption components associated with aerodynamic drag and rolling resistance, and approximates the fuel consumption per second. According to this model, the fuel consumption of a vehicle can be calculated as

$$f(t) = \begin{cases} \alpha + \beta_1 R(t)v + (\beta_2 M a^2 v/1000) & \text{for} \quad R_t > 0 \\ \alpha & \text{for} \quad R_t \leq 0, \end{cases}$$

where $f(t)$ is the fuel consumption per unit time (mL/s), $R(t)$ is the total tractive force (kN) required to move the vehicle and calculated as the sum of drag force, inertia force and grade force as $R(t) = b_1 + b_2v^2 + Ma/1000 + gM\omega/100000$. Furthermore, α is the constant idle fuel rate (in mL/s), β_1 is the fuel consumption per unit of energy (in mL/kJ), β_2 is the fuel consumption per unit of energy-acceleration (in mL/(kJ$\cdot$ m/s^2)), b_1 is the rolling drag force (in kN), and b_2 is the rolling aerodynamic force (in kN/(m/s^2)). The total amount of fuel consumption $F(T)$ (mL) for a journey of duration T can be calculated as

$$F(T) = \int_0^T f(t)dt. \tag{7.4}$$

The model works best at a micro-scale level and is better suited for short trip emission estimations.

7.2.2.2 A Comprehensive Modal Emission Model

A comprehensive modal emission model (CMEM) for heavy-goods vehicles was developed and presented by Scora and Barth (2006), Barth et al. (2005) and Barth and Boriboonsomsin (2008). It is based on second-by-second tailpipe emissions. The CMEM needs detailed vehicle specific parameters for the estimations such as the engine friction coefficient, and the vehicle engine speed. The CMEM follows, to some extent, the model of Ross (1994) and is composed of three modules, namely engine power, engine speed and fuel rate, which are detailed below.

- The engine power module: The power demand function for a vehicle is obtained from the total tractive power requirements P_{tract} (kW) placed on the vehicle at the wheels:

  $$\mathrm{P}_{tract}(t) = (Ma + Mg\sin\omega(\theta) + 0.5C_d\rho Av^2 + MgC_r\cos\omega(\theta))v/1000. \tag{7.5}$$

  To translate the tractive requirement into engine power requirement, the following relationship is used:

  $$P(t) = P_{tract}(t)/\eta_{tf} + P_{acc},$$

  where P is the second-by-second engine power output (kW), and η_{tf} is the vehicle drive train efficiency.
- The engine speed module: Engine speed is approximated in terms of vehicle speed as

  $$N(v) = S(R(L)/R(L_g))v,$$

where $N(v)$ is the engine speed (in rpm), S is the engine-speed/vehicle-speed ratio in top gear L_g, $R(L)$ is the gear ratio in gear $L = 1, \ldots, L_g$, and η is the efficiency parameter for diesel engines.

- The fuel rate module: The fuel rate (g/s) is given by the expression

$$f_{cm}(t) = \xi(kN(v)V + P(t)/\eta)/43.2. \tag{7.6}$$

The total fuel consumption (g) can be calculated as

$$F_{cm}(T) = \int_0^T f_{cm}(t)dt. \tag{7.7}$$

The CMEM can be seen as a state-of-the-art microscopic emission model because of its ease of applicability.

7.3 Pollution-Routing Problems

Incorporating the fuel consumption models reviewed in Sect. 7.2 into the existing VRP models is a way of explicitly accounting for fuel consumption, and consequently emissions, into the route planning process. This section presents an overview of this class of problems, called Pollution-Routing Problems (PRPs). The name Pollution-Routing Problem was coined by Bektaş and Laporte (2011) who defined the PRP as an extension of the classical VRPTW, where the aim is to route a number of vehicles to serve a set of customers *and* to determine their speed on each route segment, so as to minimize a function comprising fuel, emission and driver costs. In the PRP, it is assumed that in a vehicle trip all parameters remain constant on a given arc, but the load and speed may change from one arc to another. We provide a more detailed description of the PRP below.

7.3.1 Mathematical Model for the PRP

The PRP is defined on a complete directed graph $\mathcal{G} = (\mathcal{N}, \mathcal{A})$ where $\mathcal{N} = \{0, \ldots, n\}$ is the set of nodes, 0 is a depot and $\mathcal{A} = \{(i, j) : i, j \in \mathcal{N}$ and $i \neq j\}$ is the set of arcs. The distance from i to j is denoted by d_{ij}. A fixed-size fleet of vehicles denoted by the set $\mathcal{K} = \{1, \ldots, m\}$ is available, and each vehicle has capacity Q. The set $\mathcal{N}_0 = \mathcal{N} \setminus \{0\}$ is a customer set, and each customer $i \in \mathcal{N}_0$ has a non-negative demand q_i as well as a time interval $[a_i, b_i]$ for service. Early arrivals at the nodes are permitted but the vehicle has to wait until time a_i before service can start. The service time of customer i is denoted by t_i. An integer programming formulation for the PRP was first presented by Bektaş and Laporte (2011), and was subsequently extended by Demir et al. (2012) to allow for low travel speeds. The formulation works with a discretized speed function defined by R non-decreasing speed levels $\overline{v}^r \quad (r = 1, \ldots, R)$. Binary variables x_{ij} are equal to 1 if and only if arc (i, j) appears in solution. Continuous variables f_{ij} represent the total amount of flow on each arc $(i, j) \in \mathcal{A}$. Continuous variables y_j represent the time at which service starts at node $j \in \mathcal{N}_0$. Moreover, s_j represents the total time spent on a route that has a node $j \in \mathcal{N}_0$ as last visited before returning to the depot. Finally, binary variables z_{ij}^r indicate whether or not arc $(i, j) \in \mathcal{A}$ is traversed at a speed level r. An integer linear programming formulation of the PRP is shown below:

$$\text{Minimize} \sum_{(i,j)\in\mathcal{A}} f_c k N V \lambda d_{ij} \sum_{r=1}^{R} z_{ij}^r / \overline{v}^r \tag{7.8}$$

$$+ \sum_{(i,j)\in\mathcal{A}} f_c w \gamma \lambda \alpha_{ij} d_{ij} x_{ij} \tag{7.9}$$

$$+ \sum_{(i,j)\in\mathcal{A}} f_c \gamma \lambda \alpha_{ij} d_{ij} f_{ij} \tag{7.10}$$

$$+ \sum_{(i,j)\in\mathcal{A}} f_c \beta \gamma \lambda d_{ij} \sum_{r=1}^{R} z_{ij}^r (\overline{v}^r)^2 \tag{7.11}$$

$$+ \sum_{j\in\mathcal{N}_0} f_d s_j \tag{7.12}$$

subject to

$$\sum_{j\in\mathcal{N}} x_{0j} = m \tag{7.13}$$

$$\sum_{j\in\mathcal{N}} x_{ij} = 1 \qquad \forall i \in \mathcal{N}_0 \tag{7.14}$$

$$\sum_{i\in\mathcal{N}} x_{ij} = 1 \qquad \forall j \in \mathcal{N}_0 \tag{7.15}$$

$$\sum_{j\in\mathcal{N}} f_{ji} - \sum_{j\in\mathcal{N}} f_{ij} = q_i \qquad \forall i \in \mathcal{N}_0 \tag{7.16}$$

$$q_j x_{ij} \leq f_{ij} \leq (Q - q_i) x_{ij} \qquad \forall (i,j) \in \mathscr{A} \tag{7.17}$$

$$y_i - y_j + t_i + \sum_{r \in \mathscr{R}} d_{ij} z_{ij}^r / \overline{v}^r \leq K_{ij}(1 - x_{ij}) \qquad \forall i \in \mathscr{N}, j \in \mathscr{N}_0, i \neq j \tag{7.18}$$

$$a_i \leq y_i \leq b_i \qquad \forall i \in \mathscr{N}_0 \tag{7.19}$$

$$y_j + t_j - s_j + \sum_{r \in \mathscr{R}} d_{j0} z_{j0}^r / \overline{v}^r \leq L(1 - x_{j0}) \qquad \forall j \in \mathscr{N}_0 \tag{7.20}$$

$$\sum_{r=1}^{R} z_{ij}^r = x_{ij} \qquad \forall (i,j) \in \mathscr{A} \tag{7.21}$$

$$x_{ij} \in \{0,1\} \qquad \forall (i,j) \subset \mathscr{A} \tag{7.22}$$

$$f_{ij} \geq 0 \qquad \forall (i,j) \in \mathscr{A} \tag{7.23}$$

$$y_i \geq 0 \qquad \forall i \in \mathscr{N}_0 \tag{7.24}$$

$$z_{ij}^r \in \{0,1\} \qquad \forall (i,j) \in \mathscr{A}, r = 1, \ldots, R. \tag{7.25}$$

The objective function (7.8)–(7.11) is derived from (7.7). The terms (7.9) and (7.10) calculate the cost incurred by the vehicle curb weight and payload. Finally, the term (7.12) measures the total driver wages. Constraints (7.13) state that each vehicle must leave the depot. Constraints (7.14) and (7.15) are the degree constraints which ensure that each customer is visited exactly once. Constraints (7.16) and (7.17) define the arc flows. Constraints (7.18)–(7.20), where $K_{ij} = \max\{0, b_i + t_i + d_{ij}/v^l - a_j\}$ and L is a large number, enforce the time window restrictions. Constraints (7.21) ensure that only one speed level r is selected for each arc and $z_{ij}^r = 1$ if $x_{ij} = 1$. Constraints (7.22)–(7.25) define the domains of decision variables. Table 7.6 presents the typical values of the parameters used in the PRP.

7.3.2 ALNS Metaheuristic for the PRP

The formulation presented in Sect. 7.3.1 only allows to solve small-scale instances to optimality, as was empirically shown by Bektaş and Laporte (2011). In order to

Table 7.6 Required parameters for the PRP

Notation	Description	Typical values
w	Curb-weight (kg)	6,350
ξ	Fuel-to-air mass ratio	1
k	Engine friction factor (kJ/rev/L)	0.2
N	Engine speed (rev/s)	33
V	Engine displacement (L)	5
g	Gravitational constant (m/s^2)	9.81
C_d	Coefficient of aerodynamic drag	0.7
ρ	Air density (kg/m^3)	1.2041
A	Frontal surface area (m^2)	3.912
C_r	Coefficient of rolling resistance	0.01
n_{tf}	Vehicle drive train efficiency	0.4
η	Efficiency parameter for diesel engines	0.9
f_c	Fuel and CO_2e emissions cost per liter (€)	1.4
f_d	Driver wage per (€/s)	0.0022
κ	Heating value of a typical diesel fuel (kJ/g)	43.2
ψ	Conversion factor (g/s to L/s)	737
υ^l	Lower speed limit (m/s)	5.5 (or 20 km/h)
υ^u	Upper speed limit (m/s)	25 (or 90 km/h)

solve larger scale instances, Demir et al. (2012) developed an algorithm which iterates between the solving a vehicle routing problem with time windows (VRPTW) and a speed optimization problem. The former problem was solved through adaptive large neighborhood search (ALNS) (Ropke and Pisinger 2006) and the latter was solved through a polynomial time speed optimization procedure (SOP) to be discussed later in this chapter. The authors introduced new removal and insertion operators aimed at improving solution quality in terms of CO_2 emissions. Computational results were presented on benchmark instances generated by randomly sampling cities from the United Kingdom and using real geographical distances. All instances can be downloaded from http://www.apollo.management.soton.ac.uk/prplib.htm. Demir et al. (2012) show that, through the approach presented, CO_2 emissions can be reduced by 10 % on average.

7.3.3 *The Bi-objective PRP*

Most real-world problems involve multiple objectives. This is also the case of the PRP where two important objectives, namely minimization of fuel consumption and the total driving time, come to the fore. Fuel consumption depends on the energy required to move a vehicle from one point to another, and is proportional to the amount of emissions. As discussed in Demir et al. (2012), for each vehicle there exists an optimal speed yielding a minimum fuel consumption. However, this speed is generally lower than the speed preferred by drivers in practice. Another important indicator in road transportation is time, particularly when it comes to measuring service quality. In freight transportation, time is money and it is essential for firms to perform timely deliveries in order to establish and keep a good reputation. Reduction in time spent on a route can be achieved by traveling at a higher speed, but this, in turn, increases fuel costs and emissions. Since the two objectives of minimizing fuel and time are conflicting, the problem requires the use of multi-objective optimization to allow an evaluation of the possible trade-offs.

Demir et al. (2014b) have investigated the trade-offs between fuel consumption and driving time with the help of CMEM. They showed that trucking companies need not compromise greatly in terms of driving time in order to achieve a significant reduction in fuel consumption and CO_2 emissions. The converse of this insight also holds, i.e., considerable reductions in driving time are achievable if one is willing to increase fuel consumption only slightly.

In order to solve the bi-objective PRP, Demir et al. (2014b) used an enhanced version of the ALNS algorithm as a search engine to find a set of non-dominated solutions. This was done through four *a posteriori* methods, namely the weighting method, the weighting method with normalization, the epsilon-constraint method and a new hybrid method. More information on these methods can be found in Demir et al. (2014b).

We analyze the Pareto solutions identified by the ALNS algorithm for speeds between 20 and 100 km/h. To this end, Fig. 7.1 illustrates a sample instance for comparing the four methods. The values on the x-axis represent the driving time objective and the values on the y-axis show the fuel consumption.

Figure 7.1a–d exhibit the Pareto solutions found by each method. A recurring finding in the results summarized in Fig. 7.1d is that driving time can be decreased from about 39 h to 38 h, depending on the instance tested, without much change in the fuel consumption. Conversely, fuel consumption can be brought down quite significantly, from around 278 L to 275 L with only a slight increase in driving time.

7.3.4 Speed Optimization

One prominent variable in the models reviewed in Sect. 7.2 is vehicle speed, which has significant effect on the amount of fuel consumed by a vehicle. This section presents a class of problems to optimize speeds on given routes to minimize the fuel consumption, broadly called speed optimization. We review two procedures to solve such problems in the following.

The first type is the *speed optimization procedure (SOP)*, a polynomial time exact algorithm (Hvattum et al. 2013) developed the context of maritime routing, and one used by Demir et al. (2012) for the PRP. Given a vehicle route as a sequence of nodes, the SOP consists of finding the optimal speed on each arc of the route between successive nodes so as to minimize an objective function comprising fuel consumption costs and driver wages. The objective of SOP is

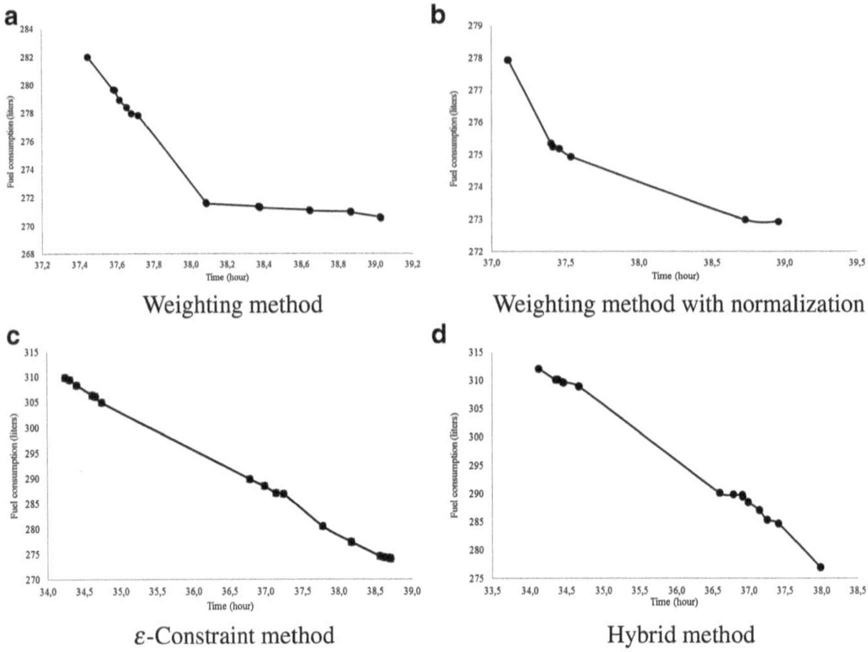

Fig. 7.1 Pareto optimal solutions found for a sample instance (Demir et al. 2014b). (**a**) Weighting method; (**b**) weighting method with normalization; (**c**) ε-constraint method; (**d**) hybrid method

non-linear due to the function used to estimate the fuel consumption of a vehicle. The SOP is defined on a feasible path $(0,\ldots,n+1)$ of the nodes served by a single vehicle, where 0 and $n+1$ are two copies of the depot. The model uses the variable w_i to denote the waiting time at each node i, the variable v_i to represent the speed at which a vehicle travels between nodes i and $i+1$, and the variable e_i for the arrival time at node i. The vehicle has a minimum and maximum speed, represented by υ_i^l and υ_i^u, between nodes i and $i+1$. The formulation of SOP is as follows:

$$\text{Minimize} \sum_{r=1}^{n} f_c F_i(v_i) + f_d e_{n+1} \tag{7.26}$$

subject to

$$e_{i+1} = e_i + w_i + t_i + d_i/\upsilon_i \qquad i = 0, \ldots, n \tag{7.27}$$

$$a_i \leq e_i + w_i \leq b_i \qquad i = 1, \ldots, n \tag{7.28}$$

$$\upsilon_i^l \leq \upsilon_i \leq \upsilon_i^u \qquad i = 0, \ldots, n \tag{7.29}$$

$$w_i \geq 0 \qquad i = 1, \ldots, n \tag{7.30}$$

$$e_i \geq 0 \qquad i = 1, \ldots, n+1 \tag{7.31}$$

$$v_i \geq 0 \qquad i = 1, \ldots, n \tag{7.32}$$

$$w_0 = e_0 = t_0 = 0, \tag{7.33}$$

where $F_i(v)$ is the total fuel consumption as derived in (7.7) for each $i = 0, \ldots, n$.

The objective function (7.26) minimizes the total cost of fuel consumption and driver wages. Furthermore, f_c is the fuel and CO_2e emissions cost per liter and f_d is the driver wage per second. Constraints (7.27) ensure that the arrival time at node $i + 1$ is the sum of the arrival time at node i, the waiting time at node i, the service time at node i and the travel time to node i. Constraints (7.28) guarantee that service at node i should start between a_i and b_i. Constraints (7.29) define upper and lower limits for speed. Constraints (7.30)–(7.32) impose the non-negativity restrictions on the variables.

The second type of problem reviewed here is the so-called *departure time and speed optimization procedure (DSOP)*, an extension of SOP introduced by Franceschetti et al. (2013) to not only optimize the speeds, but also to calculate departure times from the depot and from each customer. The closed formulation of DSOP is as follows:

$$\text{Minimize} \qquad f_c F(w, v) + f_d W(w, v) \tag{7.34}$$

subject to

$$T(w, v) + w \leq b_{n+1}, \tag{7.35}$$

$$v^l \leq v \leq v^u \tag{7.36}$$

$$w \geq \varepsilon \tag{7.37}$$

where $F(w, v)$ is the fuel consumption function for the departure time w and the free-flow speed v, and $W(w, v)$ is the time for which the driver is paid. Moreover, $T(w, v)$ denotes the travel time spent by the vehicle depending on its departure time from the depot, and the free-flow speed. Furthermore, b_{n+1} is the upper time window bound of the depot, and ε is the earliest time at which the vehicle can leave the depot. Franceschetti et al. (2013) present an algorithm to solve this problem which yielded optimal solutions for all instances tested, although whether this algorithm is exact is still an open question.

7.3.5 Other Types of PRPs

This section presents other types of PRPs introduced in the green routing literature.

7.3.5.1 The Homogeneous and Heterogeneous Emissions VRP

Figliozzi (2010) defined and introduced the emissions vehicle routing problem (EVRP) which concerns minimization of emissions and fuel consumption using of MEET (Eq. (7.1)). According to this model, The volume of emissions generated by traveling from node i to node j and departing at time b_i is calculated as

$$v_{ij}(b_i) = \sum_{l=0}^{l=p} (\alpha_0 + \alpha_1 s_{ij}^l + \alpha_2 (s_{ij}^l)^3 + \alpha_3 / (s_{ij}^l)^2) d_{ij}^l, \quad (7.38)$$

where α_0–α_3 are the emission coefficients, s_{ij}^l represents the traveling speed between node i and j at time interval l and d_{ij}^l is the distance between node i and j at time interval l.

The EVRP is an extension of the time-dependent vehicle routing problem (TDVRP). Its objective is the minimization of the emission cost, which is proportional to the amount of GHG emitted, which in turn is a function of travel speed and distance traveled. Figliozzi described a formulation and a solution algorithm for the EVRP. In the proposed algorithm, a partial EVRP is first solved to minimize the number of vehicles by means of a TDVRP algorithm, following which and emissions are optimized subject to a fleet size constraint. The departure times are also optimized for every pair of customers using the proposed algorithm. The author worked with three traffic conditions: uncongested, somewhat congested, and congested. The results presented on the Solomon (1987) instances suggest that uncongested travel speeds tend to reduce emissions on average; however, this is not always the case, and the opposite trend is sometimes observed. The author suggests that a 20 % reduction is possible by optimizing departure times.

The emission minimization vehicle routing problem using different types of vehicle was introduced by Kopfer et al. (2013). The model minimizes fuel consumption instead of driving distance by offering the possibility of using a heterogeneous fleet of vehicles. The authors compared their model to that of the traditional VRP. They replaced the distance minimizing objective function with an affine and piecewise linear fuel consumption function which considers payload. According to their results, a significant amount of reduction is possible through the use of a heterogeneous fleet of vehicles.

7.3.5.2 The Energy-Minimizing VRP

The energy-minimizing vehicle routing problem was introduced and formulated by Kara et al. (2007). The objective of the EMVRP is to minimize a weighted load function as a way of estimating fuel consumption. This function is based on a physics rule stating that on a flat surface work equals force times distance. The work done by a vehicle over an arc (i,j) is calculated as

$$W_{ij} = q_{ij}d_{ij}, \tag{7.39}$$

where q_{ij} is the weight of the load between node i and j and d_{ij} is the distance traveled between node i and j.

The integer linear programming model proposed for the EMVRP is based on that of the capacitated VRP. Since the model minimizes the total work done on the road, the authors argue that this leads to minimizing the total energy requirements, at least in terms of total fuel consumption. They study the differences between distance-minimizing and energy-minimizing solutions on benchmark capacitated VRP instances from the literature and find that energy usage increases as total distance decreases. The authors conclude that there is considerable difference between solutions that minimize energy and distance, and that the cost of the resulting routes minimizing total distance may be up to 13 % less than those minimizing energy.

7.3.5.3 The VRP with Fuel Consumption Rate

Xiao et al. (2012) incorporated fuel consumption into the capacitated VRP by using a regression model based on statistical data proposed by the Ministry of Land, Infrastructure, Transport and Tourism of Japan. According to this model, the fuel consumption between node i and j for a given load q is calculated as

$$p_{ij}(q) = p_0 + (p^* - p_0)q/Q, \tag{7.40}$$

where p_0 is the fuel consumption rate for an empty vehicle, p^* is the fuel consumption rate for a fully loaded vehicle, and Q is the maximum weight that a vehicle can carry. Xiao et al. (2012) presented a mathematical model for this problem and describe a solution algorithm based on simulated annealing. Their computational experiments show that the algorithm is both effective and efficient for solving the FCVRP.

7.3.5.4 Other Related Work

An application of CMEM within vehicle routing is presented by Ramos et al. (2012) who looked at the service areas and routes that minimize the CO_2 emissions of a transportation system with multiple products and depots. The authors proposed a decomposition-based solution algorithm and applied it to a case study in order to reshape the current system and to create a more environmental-friendly routes. Their results suggest that, in comparison to the CO_2 emissions of the current system, a decrease of 23 % can be achieved if the company reshapes both its service areas and vehicle routes. The overall potential savings can be up to 20 % if the company keeps its current service areas.

Jabali et al. (2012) developed a model that considers travel time, fuel consumption and CO_2 emission costs in a vehicle scheduling problem with time-dependent travel times between customers. Through the application of tabu search procedure, the authors showed that reducing emissions leads to reducing costs. In particular, limiting vehicle speed is desirable from a total cost perspective.

Franceschetti et al. (2013) extended the PRP to a time-dependent setting in order to take into account traffic congestion, in which vehicles have very low travel speeds. The authors considered a two-period planning horizon in which the first period corresponds to that of congestion and the latter is for free-flow traffic. The authors identified the conditions under which it is optimal to wait idly at certain locations in order to avoid congestion and to reduce the cost of emissions. Their results suggest that a 20 % reduction in the overall cost is achievable based on the instances generated by Demir et al. (2012).

7.4 Case Studies

This section presents an overview some of the case studies in green vehicle routing where the emphasis is on reducing CO_2e emissions. The classification of the reviewed studies is geographical.

- *The United States*: Christie and Satir (2006) discuss the estimation of emission reduction benefits and the potential energy savings that can be achieved through optimization. The authors aim to quantify the benefits and the potential efficiency gains in terms of emissions reduction using a computerized vehicle routing and scheduling optimization (CVRSO) method, wherein fuel consumption is estimated using a simple function of distance. Their results suggest that reductions of up to 40 % in energy consumption and GHG emissions can be achieved by implementing the CVRSO in the trucking industry, compared with manual solution techniques.
- *Cape Verde*: Tavares et al. (2008) looked at the optimization of routing networks for waste transportation. The authors proposed the use of geographic information systems (GIS) 3D route modeling to optimize the route with the aim of minimizing fuel consumption in different municipalities of the island of Santo Antao of Cape Verde. Their model takes into account both the road angle and the vehicle load. Their findings indicate that optimization of fuel consumption yields savings of up to 52 % in fuel consumption when compared to routes

with the shortest distance, even if this implies increasing the travel distance by 34 %. Another work by Tavares et al. (2009) relates to the optimization of municipal solid waste collection routes to minimis fuel consumption using 3D GIS modeling. The authors make use of the COPERT. For the case of the city of Praia, their approach reduces the traveled distance by 29 % and fuel consumption by 16 %. For the case of the Santiago island, the saving in fuel consumption is found to be 12 %.

- *Turkey*: Another real-life application was presented by Apaydin and Gonullu (2008). These authors attempted to control emissions in the context of route optimization of solid waste in Trabzon, Turkey, with a constant emission factor to estimate fuel consumption. The aim was to minimize the distance traveled by the trucks. Results suggest that the route distance and time can be decreased by 24.6 % using the proposed approach, with CO_2 emission reductions of 831.4 g on each route.
- *United Kingdom*: Maden et al. (2010) proposed a heuristic algorithm to minimize the total travel time. The proposed algorithm also considers the current driving legislation by inserting breaks for a driver when it is necessity in the context of the VRPTW; time-dependent travel times are taken into account. The approach is applied to schedule a fleet of delivery vehicles operating in the South-West of the UK using the NAEI model. Preliminary experiments are conducted on the Solomon benchmark instances. Their results suggest that the total savings in CO_2 may be up to 7 %.
- *Spain*: The last case study reviewed here is by Ubeda et al. (2011), who investigated the environmental effects of routing in Eroski, Spain. The authors compared four different approaches, namely the current approach, rescheduling (CVRP), backhauling (VRPB) and green VRP. They used a matrix of emissions based on the estimation of CO_2 emitted between each link as described by Palmer (2007). Their results suggest that the implementation of the green routing approach yields benefits from the economic and ecological perspectives. Their results suggest that savings of 13.06 % in distance and of 13.15 % in emissions can be achieved using the green VRP approach.

7.5 Conclusions

This chapter provided an overview of studies aiming to reduce the environmental effects of road freight transportation within the context of vehicle route planning, with an emphasis on the minimization of CO_2e emissions. This is a relatively new area of research, but one that is fast growing, due to the significant impacts of such externalities on human health and the environment. Our review has shown the need to well understand the ways in which these externalities can be quantified, or modeled, which allows for a more accurate representation in the planning process. To this end, we have reviewed a number of emission models extracted from the mechanical and transport engineering literature, and we have shown a number of studies where they are embedded within the classical models and algorithms available for the VRP. Our review has also identified the following gaps in the literature, where more research is needed.

- The basic assumption made in most studies is to use average speeds without an explicit consideration of possible traffic direction. There are, however, some notable exceptions. Maden et al. (2010), for example, were among the first researchers to consider congestion within route planning. They have shown that avoiding congestion at the expense of increased traveled distance is preferred when emissions are of concern. Figliozzi (2010) has also looked at settings with congestion. It is known that CO_2e emissions are particularly high under very low travel speeds, as is the case in congestion. This area is still in need of further investigation, particularly in the context of route planning. The studies by Jabali et al. (2012) and by Franceschetti et al. (2013) constitute interesting steps in this direction.
- Most studies address the problem of green route planning assuming that all parameters are known with certainty. Dynamic or real-time problems, where problem parameters are subject to foreseeable changes (e.g., rush-hour traffic) or random events (e.g., accidents, breakdowns) would be interesting directions to pursue from the perspective of emissions minimization. Real-time vehicle routing is another promising area of research where look-ahead policies are needed to avoid congestion and to prescribe alternative routes that reduce fuel consumption and emissions in the presence of incidents. Some contributions have already appeared on such issues (see, e.g., Ehmke et al. 2013), but the relevant literature is still young.
- The actual fuel consumption depends on a multitude of factors, as discussed in Demir et al. (2011, 2014a). It is very difficult to quantify some parameters, which is one reason why none of the available models provides a complete solution for the estimation of fuel consumption. It remains to be seen how some of the more qualitative parameters, such as driver behavior, can be incorporated into the classical routing algorithms and which extent this can be used to reduce emissions.
- Our review has shown that a majority of the existing studies focus only on the routing aspect of green logistics. Other problems which can be linked to routing may offer former reductions in emissions, such as location-routing problems. For example, relocating a depot or using alternative facilities may reduce the overall emissions.
- A growing area of research within green routing is related to the use of alternative types of vehicles, and in particular electric vehicles, which make use of new technologies. The use of electric vehicles opens up a completely new line of research within this field, including problems ranging from locating charging stations (Wang and Lin 2013) to working with limited battery capacities and options of recharging (Schneider et al. 2012). This line of research is beyond the scope of this chapter, but one which we believe needs to be pursued more deeply, particularly in what concerns the use of electric vehicles and their impact on energy savings and GHG emissions.

Acknowledgements The authors gratefully acknowledge the funding by the University of Southampton under the Faculty Strategic Research Fund, by the Eindhoven University of Technology, and by the Canadian Natural Sciences and Engineering Research Council under grant 39692-10. Thanks are due to Professor Harilaos N. Psaraftis for his valuable comments.

References

Apaydin, O., & Gonullu, M. T. (2008). Emission control with route optimization in solid waste collection process: A case study. *Sadhana, 33*(2), 71–82.

Barth, M., & Boriboonsomsin, K. (2008). Real-world CO_2 impacts of traffic congestion. *Transportation Research Record: Journal of the Transportation Research Board, 2058*(1), 163–171.

Barth, M., Younglove, T., & Scora, G. (2005). Development of a heavy-duty diesel modal emissions and fuel consumption model. Technical report. UC Berkeley, CA, USA: California Partners for Advanced Transit and Highways (PATH). Available at http://www.path.berkeley.edu/PATH/Publications/PDF/PRR/2005/PRR-2005-01.pdf (1.2.2014).

Bektaş, T., & Laporte, G. (2011). The pollution-routing problem. *Transportation Research Part B: Methodological, 45*(8), 1232–1250.

Bowyer, D. P., Biggs, D. C., & Akçelik, R. (1985). *Guide to fuel consumption analysis for urban traffic management* (no. 32). Vermont South, Australia: Australian Road Research Board Transport Research.

Christie, J. S., & Satir, S. (2006). Saving our energy sources and meeting Kyoto emission reduction targets while minimizing costs with application of vehicle logistics optimization. In *2006 Annual Conference and Exhibition of the Transportation Association of Canada: Transportation Without Boundaries*, Charlottetown, PE.

Dantzig, G. B., & Ramser, J. H. (1959). The truck dispatching problem. *Management Science, 6*(1), 80–91.

Dekker, R., Bloemhof, J., & Mallidis, I. (2012). Operations research for green logistics—an overview of aspects, issues, contributions and challenges. *European Journal of Operational Research, 219*(3), 671–679.

Demir, E. (2012). *Models and algorithms for the pollution-routing problem and its variations.* Ph.D. thesis, University of Southampton, Southampton, England.

Demir, E., Bektaş, T., & Laporte, G. (2011). A comparative analysis of several vehicle emission models for road freight transportation. *Transportation Research Part D: Transport and Environment, 6*(5), 347–357.

Demir, E., Bektaş, T., & Laporte, G. (2012). An adaptive large neighborhood search heuristic for the pollution-routing problem. *European Journal of Operational Research, 223*(2), 346–359.

Demir, E., Bektaş, T., & Laporte, G. (2014a). A review of recent research on green road freight transportation. *European Journal of Operational Research, 237*(3), 775–793.

Demir, E., Bektaş, T., & Laporte, G. (2014b). The bi-objective pollution-routing problem. *European Journal of Operational Research, 232*(3), 464–478.

Ehmke, J. F., Campbell, A. M., & Thomas, B. W. (2013). *A sampling approach for minimizing emissions in urban logistics.* Presented at the INFORMS Annual Meeting 2013, Minneapolis, MN.

Figliozzi, M. A. (2010). Vehicle routing problem for emissions minimization. *Transportation Research Record: Journal of the Transportation Research Board, 2197*(1), 1–7.

Forkenbrock, D. J. (1999). External costs of intercity truck freight transportation. *Transportation Research Part A: Policy and Practice, 33*(7), 505–526.

Franceschetti, A., Honhon, D., Van Woensel, T., Bektaş, T., & Laporte, G. (2013). The time-dependent pollution routing problem. *Transportation Research Part B: Methodological, 56*, 265–293.

Golden, B. L., Raghavan, S., & Wasil, E. A. (2008). *The vehicle routing problem: Latest advances and new challenges* (Vol. 43). New York: Springer.

Hickman, J., Hassel, D., Joumard, R., Samaras, Z., & Sorenson, S. (1999). *MEET-Methodology for calculating transport emissions and energy consumption. European Commission, DG VII*. Technical report. Retrieved February 1, 2014 from http://www.transport-research.info/Upload/Documents/200310/meet.pdf.

Hvattum, L. M., Norstad, I., Fagerholt, K., & Laporte, G. (2013). Analysis of an exact algorithm for the vessel speed optimization problem. *Networks, 62*(2), 132–135.

INFRAS. (1995). Workbook on emission factors for road transport: Explanatory notes. Technical report. Bern, Switzerland: INFRAS. Available at http://www.inrets.fr/ur/lte/cost319/M22.pdf (1.2.2014).

Jabali, O., Van Woensel, T., & de Kok, A. G. (2012). Analysis of travel times and CO_2 emissions in time-dependent vehicle routing. *Production and Operations Management, 21*(6), 1060–1074.

Kara, I., Kara, B. Y., & Yetiş, M. K. (2007). Energy minimizing vehicle routing problem. In A. Dress, Y. Xu, & B. Zhu (Eds.), *Combinatorial optimization and applications. Lecture notes in computer science* (Vol. 4616). Berlin/Heidelberg: Springer.

Kopfer, H. W., Schönberger, J., & Kopfer, H. (2013). Reducing greenhouse gas emissions of a heterogeneous vehicle fleet. *Flexible Services and Manufacturing, 26*(1–2), 221–248.

Kouridis, C., Gkatzoflias, D., Kioutsioukis, I., Ntziachristos, L., Pastorello, C., & Dilara, P. (2010). Uncertainty estimates and guidance for road transport emission calculations. Technical report, European Commission Joint Research Centre Institute for Environment and Sustainability. Retrieved February 1, 2014 from http://publications.jrc.ec.europa.eu/repository/handle/111111111/14202.

Laporte, G. (2007). What you should know about the vehicle routing problem. *Naval Research Logistics, 54*(8), 811–819.

Lin, C., Choy, K. L., Ho, G. T. S., Chung, S. H., & Lam, H. Y. (2013). Survey of green vehicle routing problem: Past and future trends. *Expert Systems with Applications 41*(4), 1118–1138.

Maden, W., Eglese, R. W., & Black, D. (2010). Vehicle routing and scheduling with time-varying data: A case study. *Journal of the Operational Research Society, 61*(3), 515–522.

NAEI (2012). *Emission factors*. Technical report, National Atmospheric Emissions Inventory, England. Retrieved February 1, 2014 from http://naei.defra.gov.uk/data/emission-factors.

Palmer, A. (2007). *The development of an integrated routing and carbon dioxide emissions model for goods vehicles*. Ph.D. thesis, Cranfield University, Bedford, England.

Qian, J. (2012). *Fuel emission optimization in vehicle routing problems with time-varying speeds*. Ph.D. thesis, Lancaster University, Lancaster, England.

Ramos, T. R. P., Gomes, M. I., & Barbosa-Póvoa, A. P. (2012). Minimizing CO_2 emissions in a recyclable waste collection system with multiple depots. In *EUROMA/POMS Joint Conference*, Amsterdam.

Ropke, S., & Pisinger, D. (2006). An adaptive large neighborhood search heuristic for the pickup and delivery problem with time windows. *Transportation Science, 40*(4), 455–472.

Ross, M. (1994). Automobile fuel consumption and emissions: Effects of vehicle and driving characteristics. *Annual Review of Energy and the Environment, 19*(1), 75–112.

Sanchez Vicente, A. (2013). A closer look at urban transport. In *TERM 2013: Transport indicators tracking progress towards environmental targets in Europe*. Copenhagen: European Environment Agency.

Schneider, M., Stenger, A., & Goeke, D. (2014). The electric vehicle routing problem with time windows and recharging stations. *Transportation Science*, 48(4), 500–520.

Scora, G., & Barth, M. (2006). *Comprehensive modal emission model (CMEM), version 3.01: User's guide*. Technical Report, University of California, Berkeley, CA.

Solomon, M. M. (1987). Algorithms for the vehicle routing and scheduling problems with time window constraints. *Operations Research, 35*(2), 254–265.

Tavares, G., Zsigraiova, Z., Semiao, V., & Carvalho, M. G. (2008). A case study of fuel savings through optimisation of MSW transportation routes. *Management of Environmental Quality, 19*(4), 444–454.

Tavares, G., Zsigraiova, Z., Semiao, V., & Carvalho, M. G. (2009). Optimisation of MSW collection routes for minimum fuel consumption using 3D GIS modelling. *Waste Management, 29*(3), 1176–1185.

Toth, P., & Vigo, D. (2014). *Vehicle routing: Problems, methods, and applications. MOS-SIAM series on optimization* (Vol. 18, 2nd ed.). Philadelphia: SIAM.

Ubeda, S., Arcelus, F. J., & Faulin, J. (2011). Green logistics at Eroski: A case study. *International Journal of Production Economics, 131*(1), 44–51.

Wang, Y. W., & Lin, C. C. (2013). Locating multiple types of recharging stations for battery-powered electric vehicle transport. *Transportation Research Part E: Logistics and Transportation Review, 58*, 76–87.

Wright, L. A., Kemp, S., & Williams, I. (2011). Carbon footprinting: Towards a universally accepted definition. *Carbon, 2*(1), 61–72.

Xiao, Y., Zhao, Q., Kaku, I., & Xu, Y. (2012). Development of a fuel consumption optimization model for the capacitated vehicle routing problem. *Computers & Operations Research, 39*(7), 1419–1431.

Chapter 8
Green Maritime Transportation: Market Based Measures

Harilaos N. Psaraftis

Abstract The purpose of this chapter is to introduce the concept of Market Based Measures (MBMs) to reduce Green House Gas (GHG) emissions from ships, and review several distinct MBM proposals that have been under consideration by the International Maritime Organization (IMO). The chapter discusses the mechanisms used by MBMs, and explores how the concept of the Marginal Abatement Cost (MAC) can be linked to MBMs. It also attempts to discuss the pros and cons of the submitted proposals.

Abbreviations

BAU	Business As Usual
CBDR	Common But Differentiated Responsibilities
CBO	Congressional Budget Office
CH_4	Methane
CO_2	Carbon dioxide
DNV	Det Norske Veritas
DWT	Deadweight
EC	European Commission
ECA	Emissions Control Area
EEDI	Energy Efficiency Design Index
EEOI	Energy Efficiency Operational Indicator
EIS	Efficiency Incentive Scheme
ETS	Emissions Trading Scheme
EU	European Union
FOE	Friends Of the Earth
GHG	Green House Gas
HFO	Heavy Fuel Oil
IMAREST	Institute of Marine Engineering, Science and Technology

H.N. Psaraftis (✉)
Department of Transport, Technical University of Denmark, Bygningstorvet 1
Building 115, Room 018 2800, Kgs. Lyngby, Denmark
e-mail: hnpsar@transport.dtu.dk

H.N. Psaraftis (ed.), *Green Transportation Logistics*, International Series in Operations Research & Management Science 226,
DOI 10.1007/978-3-319-17175-3_8

IMO	International Maritime Organization
IOPCF	International Oil Pollution Compensation Fund
IPTA	International Parcel Tanker Association
IUCN	International Union for the Conservation of Nature
LDC	Lesser Developed Country
LIS	Leveraged Incentive Scheme
LNG	Liquefied Natural Gas
MAC	Marginal Abatement Cost
MBM	Market-Based Measure
MCR	Maximum Continuous Rating
MDO	Marine Diesel Oil
MEPC	Marine Environment Protection Committee
MRV	Monitoring, Reporting and Verification
N_2O	Nitrous oxide
NO_x	Nitrogen oxides
PM	Particulate Matter
RM	Rebate Mechanism
SECA	Sulphur Emissions Control Area
SECT	Ship Efficiency and Credit Trading
SIDS	Small Island Developing State
SO_x	Sulphur oxides
UK	United Kingdom
UNFCCC	United Nations Framework Conference on Climate Change
US	United States
VES	Vessel Efficiency System
VLCC	Very Large Crude Carrier
VOC	Volatile Organic Compound
WSC	World Shipping Council
WTO	World Trade Organization

8.1 Introduction

8.1.1 Background

Gases emitted from ships can be classified into several categories. Green House Gases (GHGs) include carbon dioxide (CO_2), methane (CH_4), and nitrous oxide (N_2O), among others. Non-GHGs include mainly sulphur oxides (SO_x) and nitrogen oxides (NO_x). Various other pollutants, such as particulate matter (PM), volatile organic compounds (VOC), black carbon, and others, are also emitted. The effects of all of the above gases on global climate are diverse and most are considered negative if not kept under control. Among other effects, GHGs contribute to global warming, SO_x cause acid rain and deforestation, and NO_x cause undesirable health effects.

As early as in 1997 in Kyoto, the United Nations Framework Conference on Climate Change (UNFCCC) has designated the International Maritime Organization (IMO), the United Nations specialized agency with responsibility for the safety and security of shipping and the prevention of marine pollution by ships, as the body responsible for regulating maritime air emissions. However, progress on that front has generally been slow. In 2008, the Marine Environment Protection Committee (MEPC) of the IMO adopted amendments to the MARPOL Annex VI regulations that deal with SO_x and NO_x emissions. But on the GHG front, and in spite of much discussion, shipping is still not being included in the UNFCCC global emissions reduction target for CO_2 and other GHGs, and in fact until very recently, shipping was the only mode of transport for which GHG emissions were not regulated. The era of non-regulation for shipping GHGs officially came to an end in 2011, when the MEPC adopted the Energy Efficiency Design Index (EEDI) for new ships (see Sect. 1.6.1 of Chap. 1 of this book for a discussion). Even so, further measures to curb future GHG growth in shipping are being sought with a high sense of urgency.

For non GHG emissions, already in 2008 the IMO had designated the Baltic Sea, the North Sea and the English Channel as 'Sulphur Emissions Control Areas' (SECAs), with the purpose of limiting SO_x emissions, and in 2010 the IMO designated the entire US-Canadian and US-Caribbean coastal zones as an 'Emissions Control Area' (ECA), with 2012 as kick off year and ambitious goals to reduce SO_x, NO_x, and PM emissions. More on ECAs and SO_x emissions can be found in Chap. 10 of this book.

8.1.2 Three Classes of Measures to Reduce Maritime Emissions

It has been customary to break down the spectrum of measures to reduce maritime emissions (GHG and others) into basically three major classes.

First, technological measures include more efficient (energy-saving) engines, more efficient ship hulls and designs, more efficient propellers, cleaner fuels (low carbon content, LNG), alternative fuels (fuel cells, biofuels, etc.), devices to trap exhaust emissions (scrubbers, etc.), energy recuperation devices (exhaust heat recovery systems, etc.), "cold ironing" in ports, various kites, and others. Compliance with EEDI, which is a design index, will mainly induce technological measures. Also Chap. 5 gives a flavor of such technologies for all surface modes, including maritime.

Second, logistics-based (tactical and operational) measures include speed optimization, optimized weather routing, optimal fleet management and deployment, efficient supply chain management, and others that impact the logistical operation. Chapter 9 of this book provides more insights on such measures.

Third, we have what we call market-based measures or MBMs. These include Emissions Trading Schemes (ETS), an International Fund based on a contribution imposed on fuel, and a variety of others, as will be explained later.

This chapter focuses on the third category of measures, the MBMs, even though it will also touch upon the other two categories whenever warranted.

We note that the partition into the above three categories is, in many respects, artificial. This is so because an MBM may induce the ship owner to adopt:

- Logistics-based measures in the short run, and
- Technological measures in the long run.

Both sets of measures would result in emissions reductions.

8.1.3 Possible Role of MBMs

To obtain some insights into the possible role of MBMs, consider the practice of slow steaming, widely applied in recent times mainly to reduce fuel costs and help sustain a fragile market by absorbing excess shipping capacity. From basic naval architecture, the dependency of fuel consumption on ship speed is at least cubic. GHG emissions being directly proportional to fuel consumed, a simple way to reduce these emissions, perhaps drastically, is for a ship to slow down.

Slowing down can be done at two levels. The first level is logistics-based (operational), that is, have an existing ship go slower than its design speed. The second level is technological (strategic), that is, build future ships with a reduced installed horsepower so that they cannot go more than a prescribed speed.

Even though slow steaming has been practised historically without any MBM being applied (see also Chap. 9 of this book), it is clear that an MBM can affect the extent of this practice. At the operational level, paying more for bunker fuel via an imposed charge on fuel may, depending on the state of the market, induce more slow steaming. At the strategic level, the same measure may make a ship owner purchase a ship that is more fuel efficient, or just cannot go very fast. Either way, less GHGs would be produced. These emissions reductions are known as 'in sector' reductions.

By making a ship owner pay for his ship's CO_2 emissions, an MBM is an instrument that implements the 'polluter pays' principle. In that sense, it helps internalize the external costs of these emissions (see also Chap. 2). In addition, monies raised by an MBM can be used to reduce CO_2 emissions outside the marine sector, for instance by purchasing what are known as 'offsets'. Such offsets could be used to invest in projects such as for instance a wind farm in New Zealand, a solar cell farm in Indonesia, or others, and so contribute to GHG reduction outside the marine sector. These are known as 'out of sector' reductions.

The rest of this chapter is organized as follows. Section 8.2 discusses some basic concepts including that of the Marginal Abatement Cost curve. Section 8.3 outlines the MBM proposals currently on the table at the IMO. Section 8.4 comments on the modeling effort to evaluate these proposals. Section 8.5 reviews the MBM proposals in more detail. Finally Sect. 8.6 discusses the way ahead, both for MBMs and for some parallel developments on the subject, including the concept of Monitoring, Reporting and Verification (MRV) of emissions, and the 2014 update of the IMO GHG study.

8.2 Basic Concepts

Before we proceed with MBMs, some basic concepts are in order.

8.2.1 Carbon Coefficients

There is a linear relationship between fuel burned and CO_2 produced, with the proportionality constant being known as the 'carbon coefficient'. The IMO GHG study of 2000 used a coefficient of 3.17 (tonnes of CO_2 per tonne of fuel) independent of fuel type, but its 2009 update (Buhaug et al., 2009) used slightly lower coefficients, which ranged from 3.021 for Heavy Fuel Oil (HFO) to 3.082 for Marine Diesel Oil (MDO). The factor that IMO uses for EEDI reference line calculations is 3.11, for both main engine and auxiliaries. For alternative fuels such as LNG, the carbon coefficient can range from 2.6 to 2.8. This feature makes LNG more attractive than fossil fuels for propulsion, among other advantages, such as lack of sulphur and other substances and producing more energy per unit weight than fossil fuels. However, a disadvantage of LNG is the so-called 'methane slip', as some CH_4 is released by LNG use. CH_4 is a GHG that is some 20 times more potent than CO_2.

8.2.2 CO_2 Produced by International Shipping

According to the 2009 GHG study by the IMO (Buhaug, et al., 2009, see also Fig. 1 of the Preface to this book), international shipping contributed 2.7 % of the CO_2 emitted globally (2007 baseline year). The top CO_2 producer was electricity and heat production (35 %) and the top CO_2 transport mode was road (21.3 %). According to the same study, the total quantity of CO_2 emitted was 870 million tonnes for international shipping and 1,050 million tonnes for all shipping, including domestic and fishing. Figure 8.1 shows amounts of CO_2 produced per transportation mode (2005 data).

As mentioned in Chap. 2, in 2014 the IMO updated the GHG study. The new study (Smith et al., 2014) provided updated estimates of CO_2 emissions from international shipping from 2007 to 2012. The 2012 More on this in Section 8.6.2 of this chapter.

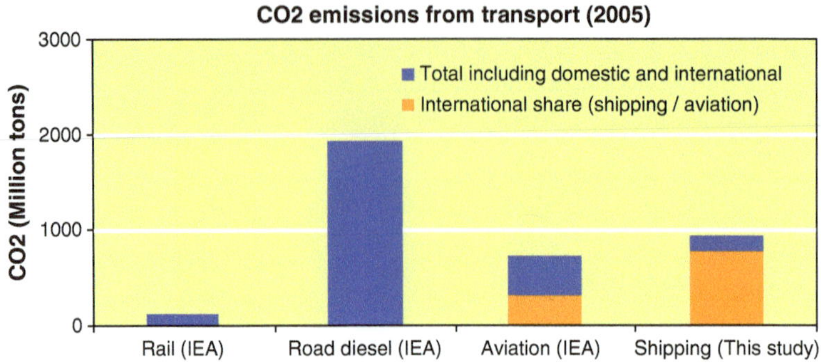

Fig. 8.1 2005 CO_2 emissions per transportation mode. *Source*: Buhaug, et al. (2009)

8.2.3 Marginal Abatement Costs

The concept of the Marginal Abatement Cost (MAC) has already been introduced in Chap. 2, but we revisit this concept here in the context of MBMs. Let us call A a well-defined available technological measure to avert CO_2. For instance, measure A can be a more efficient hull form, a more fuel-efficient engine, a more efficient propeller, a device such as a kite that can save energy, or other. Suppose that we implement measure A on a given ship and we compare the ship without measure A to the ship with measure A, ceteris paribus.

Define also:

- $\Delta GCOST(A)$: the total gross cost in implementing measure A, defined as the gross difference in annualized costs of the ship with measure A, minus those costs without measure A, gross meaning excluding fuel costs.
- $\Delta FUEL(A)$: the total annual fuel consumption averted by implementing measure A, for the same ship.
- $\Delta NCOST(A)$: the total net cost in implementing measure A, defined as the net difference in annualized costs of the ship attributed to measure A, net meaning including fuel costs.
- $\Delta CO_2\ (A)$: the total tonnes of CO_2 averted by measure A.
- *PFUEL*: the average price of fuel over a year, and
- *F*: the carbon coefficient (between 3.02 and 3.11).

Then the Marginal Abatement Cost (MAC) of measure A is defined as follows:

$$MAC(A) = \Delta NCOST(A)/\Delta CO_2(A). \tag{8.1}$$

Given that

$\Delta NCOST(A) = \Delta GCOST(A) - \Delta FUEL(A)*PFUEL$ and that
$\Delta CO_2\ (A) = \Delta FUEL(A)*F$

it follows that

$$MAC(A) = \Delta GCOST(A)/\Delta CO_2(A) - PFUEL/F. \tag{8.2}$$

The negative term in the right-hand side of (8.2) reflects the savings in costs (per tonne of CO_2 averted) due to fuel consumption reduction. This also means that for any measure A, *MAC(A)* can be negative if the price of fuel is high enough. Measures for which the ratio $\Delta GCOST(A)/\Delta CO_2(A)$ is low are also more likely to have a negative *MAC(A)* than other measures for which these costs are high.

A negative MAC means that the ship owner would have an economic incentive to implement the respective measure. Doing so would increase his profits, and, as an important side-effect, would also reduce CO_2. It would be a win-win proposition, and would not need a regulation mandating the measure.[1] Conversely, if the MAC of a measure is positive, then the ship owner would have no incentive to adopt it. The measure would have to be mandated in order to be implemented.

8.2.4 MAC Curves

If one examines a set of feasible measures to reduce CO_2 and compute the MACs for such measures applied to the world fleet, one comes up with what is known as the MAC curves. Several attempts to construct MAC curves are known, see for instance DNV (2009), Eide et al. (2010), and IMAREST (2011). MAC curves are supposed to be constructed for the entire set of possible measures to reduce CO_2. The horizontal axis of a MAC curve measures the total amount of CO_2 averted and the vertical axis measures the corresponding MAC.

Figure 8.2 is taken from a report of Norwegian classification society Det Norske Veritas, DNV (2009) and shows a typical MAC curve for the projected world shipping fleet in 2030, together with a corresponding set of measures to reduce CO_2.

The first observation is that the MAC curve is a supply curve, in the sense that measures are rank-ordered by non-decreasing order of MAC. This means that before a certain measure is implemented, all other measures with a MAC lower than the MAC of this measure have been implemented. In practice this may not necessarily be the case, as some measures may be mandated and therefore given priority over others, and in these situations a MAC curve may not be monotonically increasing. Also the MAC curve assumes no interdependencies among measures which also may not be the case. In the figure, each measure is represented by a rectangular box, the horizontal dimension of which is the amount of CO_2 averted by the measure and the vertical dimension is the corresponding MAC.

[1] This would also assume the absence of barriers that would make the adoption of the measure difficult or impossible. See Chap. 2 for a discussion of barriers vis-à-vis negative MACs.

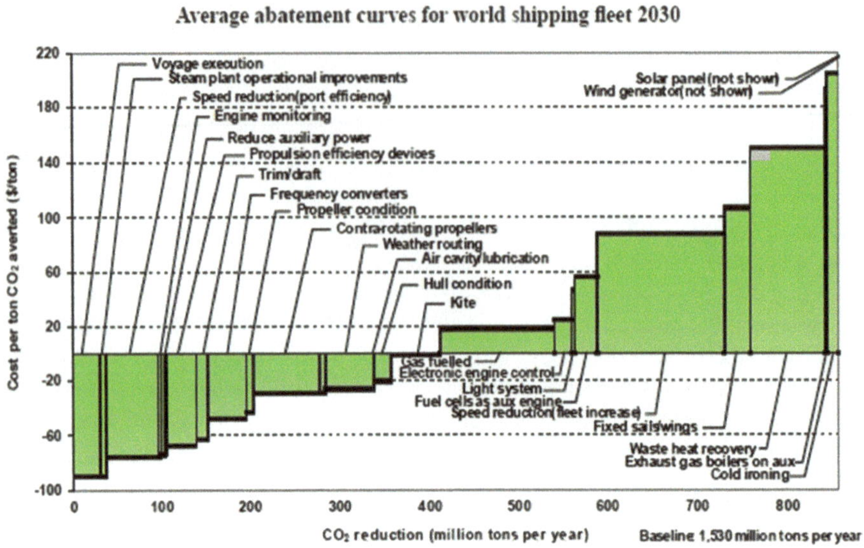

Fig. 8.2 A MAC curve. *Source*: DNV (2009)

A second observation comes from looking at the specific set of measures to reduce CO_2. Among the various technological measures, we also see an operational measure in the graph, speed reduction. In fact we see this measure twice. In one case its MAC is negative and in the other it is positive. The first case, with a negative MAC, is labelled "speed reduction (port efficiency)", and the second case, with a positive MAC, is labelled "speed reduction (fleet increase)". Presumably the first case concerns the case in which speed reduction is offset by an increase in port efficiency so that overall transit time is not increased, and the second case speed reduction necessitates more ships to cover demand throughput. A problem with such formulation is that speed reduction (see also Chap. 9 of this book) is typically a response of the ship owner or the charterer to the bunker price and freight rate under which they are called to operate. In that sense, it cannot really be considered as an independent measure to reduce emissions.

Figure 8.3 shows another such MAC curve, taken from the Expert Group report on MBMs (IMO, 2010b) and carried out again by DNV, which was commissioned by the IMO for the task. The data used for this analysis was not made publicly available. One can see here that some of the MAC curves are not monotonic, meaning that some measures may take precedence over other measures even though their MAC is higher.

It is also important to realize that the MAC curves directly depend on the projected price of fuel, as eq. (8.2) above stipulates. The MAC curve will shift up and down depending on what PFUEL will be. It will shift down by about \$100/ton for each increase of \$300/ton in the price of fuel. In DNV (2009), DNV assumed a fuel price of 350 \$/ton for a standard bunker oil and 500 \$/ton for a high quality

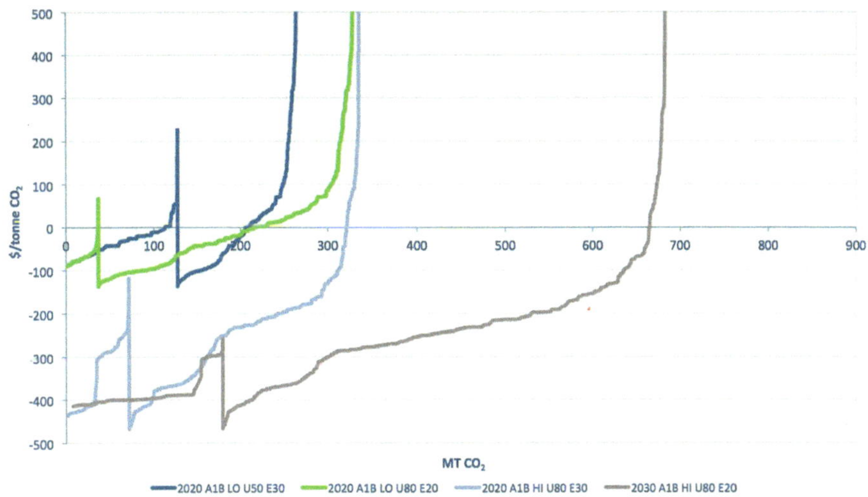

Fig. 8.3 Sample MAC curves by DNV. *Source*: IMO (2010b)

bunker oil for 2030. If these prices change, the MAC curves will change. In IMO (2010b), DNV and the MBM Expert Group examined a variety of scenarios on projected future parameters including fuel prices. More on this in Sect. 8.4.

8.2.5 *Effect of a Bunker Levy on MAC Curves*

The MAC curve can be useful if one wants to evaluate the effect of a bunker levy (or tax) on the amount of CO_2 emissions. Figure 8.4 below shows how. The figure shows two MAC curves. The one on top is before a levy is applied, and the one below is after the levy. Applying a levy equal to *LEVY* means that the price of fuel will increase from *PFUEL* to *PFUEL* + *LEVY*.

Assuming that $LEVY > 0$ and that the gross costs of each the various measures to reduce emissions ($\Delta GCOST$) do not change as a result of the levy, the MAC curve will uniformly go down by an amount equal to *LEVY*/*F*. Note that this is an first order approximation, as the new fuel price may result in speed reduction and a corresponding reduction of fuel consumption and hence emissions. So a second order effect may move the MAC curve also to the left and not only down.

Ignoring for the moment this second order effect, a first order approximation of the CO_2 reduction ΔCO_2 can be estimated if we assume that for every fuel price/levy the equilibrium CO_2 emissions will be at the point where $MAC = 0$.

We can even make a crude estimate of ΔCO_2 as equal to $LEVY/(F^*s)$, where s is the slope of the MAC curve at $MAC = 0$. However one has to be careful as from Fig. 8.3 one can see that this slope can vary widely.

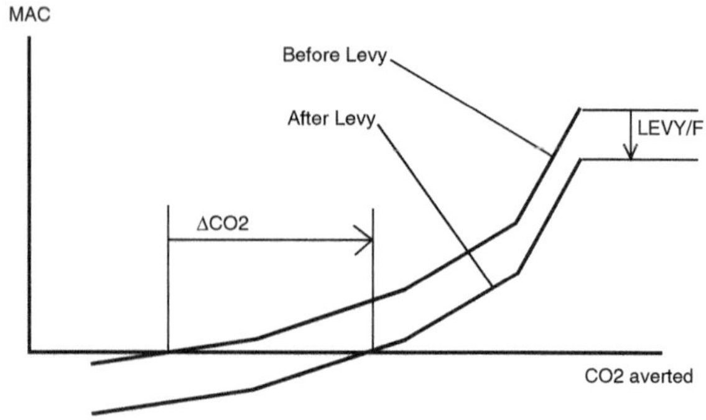

Fig. 8.4 Using MAC curves to determine the effect of a bunker levy

8.3 MBM Proposals Considered by the IMO

Following the update of the IMO GHG study in 2009 (Buhaug, et al., 2009), IMO activity on GHGs has been largely on two "parallel" tracks. The first track mainly concerns EEDI (see Sect. 1.6.1 of Chap. 1). The second track concerns MBMs. It is interesting that discussion on these two tracks has been conducted thus far with no apparent connection between the two, even though both tracks concern the same objective (reduce GHG emissions from ships). It will also be seen later that in reality these tracks are not disjoint, as some of the proposed MBMs embedded EEDI in their formulation.

For MBMs, an Expert Group was appointed by the IMO's Secretary General after solicitation of member states and was tasked to evaluate as many as ten separate MBM proposals, submitted by various member states and other organizations. All submitted MBM proposals describe programs and procedures that would target GHG reductions through either 'in-sector' emissions reductions from shipping, or 'out-of-sector' reductions via the collection of funds to be used for mitigation activities in other sectors that would contribute towards global reduction of GHG emissions.

The IMO formulated nine criteria for evaluation of GHG reduction measures, including MBMs, the following:

1) Environmental effectiveness
2) Cost-effectiveness and potential impact on trade and sustainable development
3) The potential to provide incentives to technological change and innovation
4) Practical feasibility of implementing MBM
5) The need for technology transfer to and capacity building within developing countries, in particular the least developed countries (LDCs) and the small island development states (SIDS)
6) The relation with other relevant conventions (UNFCCC, Kyoto Protocol and WTO) and the compatibility with customary international law
7) The potential additional administrative burden and the legal aspects for National Administrations to implement and enforce MBM

8) The potential additional workload, economic burden and operational impact for individual ships, the shipping industry and the maritime sector as a whole, of implementing MBM
9) The compatibility with the existing enforcement and control provisions under the IMO legal framework.

Brief descriptions of each of the ten original IMO MBM proposals are as follows (see IMO (2010b) for more details):

1. The International Fund for Greenhouse Gas emissions from ships (GHG Fund) originally proposed by Cyprus, Denmark, the Marshall Islands, Nigeria, and the International Parcel Tanker Association-IPTA (Denmark, 2010).

Liberia and the Republic of Korea were later added as co-sponsors of this MBM. This Fund would establish a global reduction target for international shipping, set by either the UNFCCC or the IMO. Emissions above the target line would be offset largely by purchasing approved emission reduction credits. The offsetting activities would be financed by what the proposers called a 'contribution' paid by ships on every tonne of bunker fuel purchased. It is envisaged that contributions would be collected through bunker fuel suppliers (Option 1) or via direct payment from ship owners (Option 2). The contribution rate would be adjusted at regular intervals to ensure that sufficient funds are available to purchase project credits to achieve the agreed target line. Any additional funds remaining would be available for adaptation and mitigation activities via the UNFCCC and R&D and technical co-operation within the IMO framework.

2. The Leveraged Incentive Scheme (LIS) to improve the energy efficiency of ships based on the International GHG Fund proposed by Japan (Japan, 2010)

This resembles the aforementioned Fund scheme with an important difference: The concept of the Leveraged Incentive Scheme is that a part of the GHG Fund contributions, which are collected on marine bunker is refunded to ships meeting or exceeding agreed efficiency benchmarks and labelled as "good performance ships". To that effect, the EEDI index is the main instrument in determining which ships are efficient and should receive the refunds. In that sense, this MBM is a hybrid one, as it includes EEDI as part of its formulation.

3. Achieving reduction in greenhouse gas emissions from ships through Port State arrangements utilizing the ship traffic, energy and environment model, STEEM (PSL) proposal by Jamaica (Jamaica, 2010)

Under this MBM, Member States participate in levying a uniform emissions charge on all vessels calling at their respective ports based on the amount of fuel consumed by the respective vessel on its voyage to that port (not bunker suppliers). The proposal is directly aimed at reducing maritime emissions of CO_2 without regard to design, operations, or energy source. The Port State Levy would be structured to achieve the global reduction targets for GHG and could be leveraged in a manner as proposed by Japan to reward vessels exceeding efficiency targets.

4. The United States proposal to reduce greenhouse gas emissions from international shipping, the Ship Efficiency and Credit Trading (SECT) (USA, 2010)

This MBM is designed to focus emission reduction activities just in the shipping sector. Under SECT, all ships, including those in the existing fleet, would be subject to mandatory energy efficiency standards, rather than a cap on emissions or a surcharge on fuel. As one means of complying with the standard, SECT would establish an efficiency-credit trading programme. The stringency level of these efficiency standards would be based on energy efficiency technology and methods available to ships in the fleet. These standards would become more stringent over time, as new technology and methods are introduced. Similar to the EEDI, these efficiency standards would be based on a reduction from an established baseline and would establish efficiency standards for both new and existing ships. As the LIS MBM by Japan, the SECT MBM by the US is a hybrid MBM, as it embeds EEDI within its formulation. However, the mechanism is different. Under SECT, ships would trade on EEDI. A 'good EEDI' ship would sell credits to a 'bad EEDI' ship.

5. Vessel Efficiency System (VES) proposal by World Shipping Council (WSC, 2010)

VES would establish mandatory efficiency standards for both new and existing ships. Each vessel would be judged against a requirement to improve its efficiency by X% below the average efficiency (the baseline) for the specific vessel class and size. Standards would be tiered over time with increasing stringency. Both new build and existing ships would be covered. New builds must meet the specified standards or they may not operate. Existing ships may comply by improving their efficiency scores through technical modifications that have been inspected and certified by the Administration or recognized organizations. Existing ships failing to meet the required standard through technical modifications would be subject to a fee applied to each tonne of fuel consumed. The total fee applied (non-compliant ships only) would vary depending upon how far the vessel's efficiency (as measured by the EEDI) falls short of the applicable standard. This is another example of a hybrid MBM, as it embeds EEDI within its formulation.

6. The Global Emission Trading System (ETS) for international shipping proposal by Norway (Norway, 2010)

This MBM would set a sector-wide cap on net emissions from international shipping and establish a trading mechanism to facilitate the necessary emission reductions, be they in-sector or out-of-sector. The use of out-of-sector credits allows for further growth of the shipping sector beyond the cap. In addition the auction revenue would be used to provide for adaptation and mitigation (additional emission reductions) through UNFCCC processes and R&D of clean technologies within the maritime sector. A number of allowances (Ship Emission Units) corresponding to the cap would be released into the market each year. It is proposed that the units would be released via a global auctioning process. Ships would be required to surrender one Ship Emission Unit, or one recognized out-of-sector allowance or one recognized out-of-sector project credit, for each tonne of CO_2 they emit. The Norwegian ETS would apply to all CO_2 emissions from the use of fossil fuels by ships engaged in international trade above a certain size threshold. The proposal also indicates that limited exemptions could be provided for specific voyages to Small Island Developing States.

7. Global Emissions Trading System (ETS) for international shipping proposal by the United Kingdom (UK, 2010)

This is very similar in most respects to the global ETS proposal by Norway. Two aspects of the UK proposal that differ from the Norwegian ETS proposal are the method of allocating emissions allowances and the approach for setting the emissions cap.

8. Further elements for the development of an Emissions Trading System (ETS) for International Shipping proposal by France (France, 2010)

This MBM sets out additional detail on auction design under a shipping ETS. In all other aspect the proposal is similar to the Norwegian proposal for an international ETS.

9. Market-Based Instruments: a penalty on trade and development proposal by the Bahamas (Bahamas, 2010)

This MBM does not set explicit standards or reductions to be achieved in the shipping sector or out-of-sector for GHG reductions. The proposal clearly sets forth that the imposition of any costs should be proportionate to the contribution by international shipping to global CO_2 emissions. Bahamas has indicated that it is assuming that mandatory technical and operational measures would be implemented such as the EEDI.

10. A Rebate Mechanism (RM) for a market-based instrument for international shipping proposal by IUCN (IUCN, 2010)

This MBM focuses on a Rebate Mechanism to compensate developing countries for the financial impact of a MBM. This Rebate Mechanism would 'piggy back' any of the other MBMs. A developing country's rebate would be calculated on the basis of their share of global costs of the MBM, using readily available data on a developing country's share of global imports by value as a proxy for that share (or another metric such as value-distance if data becomes available).

In addition to the above, the following developments took place after the above ten MBMs were submitted:

a) A German ETS proposal (Germany, 2010) that was not included in the original MBM list for administrative reasons was reinstated as part of the MBM roster.
b) The LIS and VES proposals were combined into what was relabeled the Efficiency Incentive Scheme (EIS) (Japan, 2011)

c) The Bahamas submitted what they called an evolution of their original proposal (Bahamas, 2011), which they subsequently withdrew altogether.
d) The US MBM proposal was radically restructured.

8.4 Modeling to Evaluate the MBM Proposals

After considerable discussion, a 300+ page report (IMO, 2010b) evaluating the MBM proposals was prepared by the MBM Expert Group and was presented and discussed at MEPC 61 (September 2010). The report went at length in assessing each MBM according to the evaluation criteria, in modeling future scenarios and in assessing the impact of MBMs on trade and developing countries. However, the report contained no horizontal comparison of MBMs and no recommendation as to which MBMs should be further pursued.

The Expert Group's modeling effort, which also involved the work of external consultants, was to develop and apply a model to make quantitative estimates of emissions reductions, revenues generated, costs and other attributes of each MBM proposal. Modeling scenarios included:

- two growth rates (1.65 and 2.8 %)
- three targets lines /caps for GHG Fund and ETS (0, 10 and 20 % below 2007 level)
- 28 % revenue used for mitigation for Rebate Mechanism and 25, 50, and 75 % revenue refunded for LIS
- low, medium and high stringency standards for VES and SECT
- two carbon price scenarios (medium and high) and two fuel price scenarios (reference and high).

For instance, the following modeling scenario (labelled A1B) assumed:

- 2.8 % growth;
- a target of 10 % below 2007 GHG emission levels (as per Second IMO GHG study 2009) for the GHG Fund, and ETS MBM proposals, with an additional 10 % contribution assumed under the GHG Fund for adaptation and R&D purposes (shown as remaining proceeds).
- 28 % of revenues are used for mitigation under the IUCN proposal and 50 % of revenues are refunded to "good performing ships" under the LIS proposal
- medium level stringency efficiency index standards for the SECT and VES proposals; and
- a medium carbon price and reference fuel price scenario.

These inputs produced results shown in Table 8.1.

Projections of emissions and remaining proceeds were also made. Figure 8.5 shows the results for the GHG Fund and for ETS. Full details can be found in IMO (2010b).

Table 8.1 results of the A1B modeling scenario

	GHG fund	LIS	PSL	SECT	VES	ETS Norway and France	ETS UK	Bahamas	IUCN
MBM in sector reductions (Mt)	11	76	64	142	30	60	60	0	64
MBM out-of-sector reductions (Mt)	501	0	0	0	0	452	452	0	345
Cost of MBM out-of-sector reductions at a carbon price of $40/tonne ($billion)	20	0	0	0	0	18	18	0	15
Gross costs ($billion)	25	49	49	Unknown	7	49	49	0	49
Remaining proceeds ($billion)	5	24	49	0	7	31	31	0	36[6]
Potential for purchase of supplementary out-of-sector reductions using remaining proceeds (Mt)	135	610	1,232	0	184	783	0	0	900

Source: IMO (2010b)

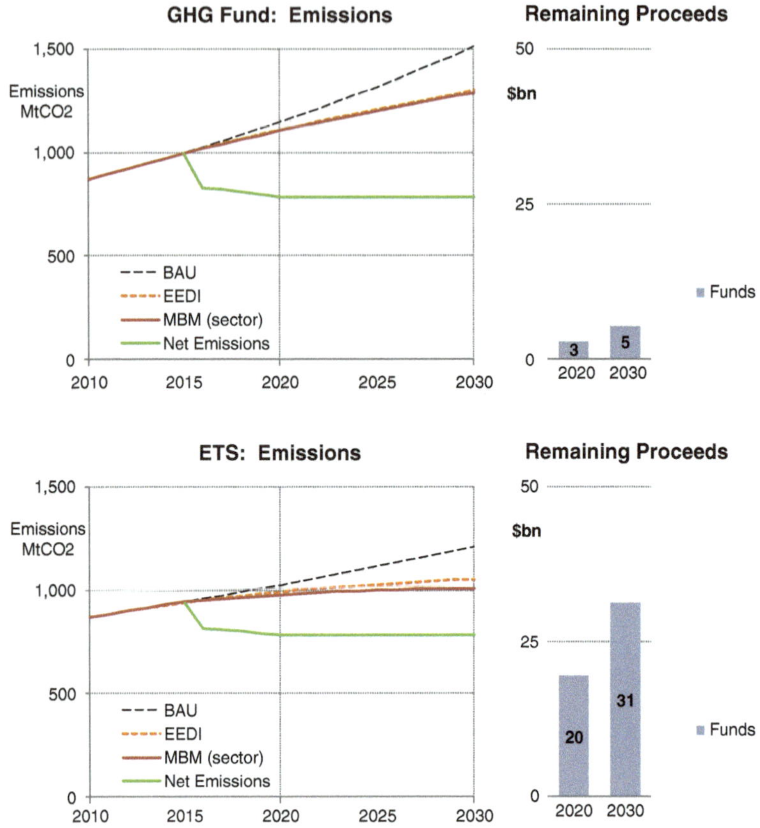

Fig. 8.5 Projections of emissions and remaining proceeds for GHG Fund and ETS. *Source*: IMO (2010b)

Even though this effort was certainly worthy of note, reservations are expressed herewith for some of the modeling assumptions. As an example (which is one of several), a key assumption was made that an increase in fuel prices of 100 % over the long-term will result in a 4 % reduction in emissions below the so-called "Business As Usual" (BAU) scenario. However, this percentage (4 % or other) critically depends on the slope of the Marginal Abatement Cost (MAC) curve at the point it crosses the x-axis. As illustrated by the DNV MAC curves for the 72 scenarios examined (see Fig. 8.3), that slope can vary widely from very low to very high, projected future fuel price being the main determinant. Therefore the fixed 4 % assumption is not necessarily correct. In that sense, strong reservations are expressed on all the numerical results of this model, which are sometimes difficult to follow and, at a minimum, should be interpreted with caution.

The same applies to the numerical results that pertain to a variety of estimates for each MBM proposal, such as in-sector and out-of-sector emissions reductions, revenues generated, costs, and a variety of others. Even estimates of CO_2 reductions with or without EEDI enacted were produced. All of these results critically depend on a variety of assumptions, which cannot be fully substantiated.

It should also be pointed out that the data and models on the MAC curves supplied to the IMO MBM Expert Group were not fully available to scrutiny, and this significantly limits their usefulness. Both the data and the models are subject to confidentiality clauses. The problem here is, if the models are not made available for scrutiny by the experts or anybody else (remaining virtually a black box) then obviously the correctness of their results cannot be confirmed.

Another reservation concerns the apparent conclusion of the report that the GHG Fund proposal is a weak driver for uptake of in-sector technological measures to reduce emissions, whereas the various ETS proposals are strong drivers. This is not so. In fact, the GHG Fund will result in a much larger reduction in CO_2 emissions than an ETS with the same average permit price.

To achieve the same amount of CO_2 reduction, if we assume equal efficiency for both systems (which is not necessarily the case), the GHG Fund 'contribution' and the ETS carbon price must be the same. Policy-makers get to choose either the target reduction (for the ETS proposal) or the target contribution (for the Fund proposal). Either can be high or low. The target reduction and the target contribution being the policy-maker's choice should aim at the same result, i.e., either the same target contribution or the same target reduction. If one goes for a modest target reduction, the carbon price will be low, in fact close to zero according to the Marginal Abatement Cost (MAC) curves submitted by DNV. However, a contribution can be fixed for a longer period (a year or more). Permit prices by their nature are both volatile and unpredictable. Owners who are unsure of the carbon price they will be facing great difficulty justifying expensive investments in carbon reduction technology.

The same applies as regards revenues generated, which are (again erroneously) postulated in the report to be higher for ETS than for the Fund (see also Fig. 8.5). If carbon price and CO_2 reductions are the same in both schemes, revenues will also be about the same after accounting for efficiency. But as will be argued later, ETS will be more expensive to maintain, and (in that sense) less efficient. This means that in-sector CO_2 reductions for the GHG Fund proposal can be much higher than those shown in the IMO report.

8.5 Review of the MBM Proposals

In the absence of credible modeling results that would lend themselves to a possible horizontal comparison among the MBM proposals, any further discussion of these proposals by necessity will be qualitative and (to some extent) subjective. The rest of this chapter makes an attempt at such a discussion.

We structure the discussion as follows: We start by commenting on the Bahamas proposal, the first version of which was essentially a 'do nothing' proposal. Then we comment on MBM proposals that are hybrid, such as those of the US, Japan and

WSC. We follow with the proposals of Jamaica and IUCN. Last but not least, we focus on the GHG Fund and ETS proposals. The discussion draws from Psaraftis (2012).

8.5.1 The Bahamas Proposal

The Bahamas original proposal (Bahamas, 2010) was a non-proposal, that is, advocated adopting no MBM, arguing that this would be an obstacle to trade. According to what was presented in Sect. 8.2, a 'do-nothing' proposal does not imply zero CO_2 reductions, as any measures that have a negative MAC would be implemented and those measures would entail a CO_2 reduction. As stated before, fuel price is a key driver to such an outcome.

The Bahamas subsequently submitted an updated MBM proposal (Bahamas, 2011), which was labelled an evolution of their former proposal. In their updated submission, the Bahamas argued that only through operational and technical measures CO_2 emissions can be cut.

The new proposal, among other things, envisioned collecting CO_2 statistics through either the collection of EEOI (Energy Efficiency Operational Indicator) data, or simply by recording emissions from the funnel using a suitable sensor. EEOI is another energy efficiency index discussed at the IMO, to be used on a voluntary basis. It is based on operational as opposed to design considerations, but resembles EEDI in many respects. The ship would be required to submit emission records to the flag State or recognized organization for annual verification. The statistics collected would then show how much emissions were actually emitted over the data collection period. The problem here is, it is actually impossible to establish a reliable EEOI for any ship with just 2–3 years of data, especially in the tramp/bulk sector. Also it is impossible to establish EEOI baselines. In addition, sister ships can have vastly different EEOIs.

Interestingly enough, and following considerable debate among IMO delegations, the Bahamas withdrew their MBM proposal in 2012.

8.5.2 Hybrid MBM Proposals

There have been three (and subsequently two) MBM proposals that are hybrid in the sense that they all include a ship's Energy Efficiency Design Index (EEDI) as part of their formulation. These are the original US SECT proposal (USA, 2010), Japan's LIS proposal (Japan, 2010), and WSC's VES proposal (WSC, 2010). The last two proposals have been subsequently merged under the name of Efficiency Incentive Scheme (EIS) (Japan, 2011).

All of the above MBM proposals use (each in a different way) the idea of rewarding ships that are supposed to be good environmentally, and in all three

EEDI is proposed as a way to measure good environmental performance. A problem may occur if a ship with a low EEDI is not the ship with the lowest CO_2. It may emit more CO_2 than another ship whose EEDI is higher. A low EEDI may mean an underpowered ship, which, in its attempt to maintain speed in bad weather, may emit more than a ship with a larger engine (Greece, 2010). Another issue is that even though EEDI is supposed to be an index that is intended to be used to assess how future ships can reduce emissions by having good hull forms, efficient engines propellers, etc., if any of these hybrid proposals is adopted, will make it applicable to existing ships as well, via the MBM mechanism. Note that there has been no discussion on applying EEDI to existing ships (directly or indirectly), nor is there a plan to have such a discussion. Applying EEDI on existing ships would likely necessitate sea trials and would not be straightforward.

Thus, one would think that keeping these hybrid proposals on the table would necessitate reopening the discussion on EEDI, and specifically how EEDI can be applied to existing ships. There is currently no provision for such a discussion at the IMO, neither in the context of the discussion on EEDI, nor in the context of the discussion on MBMs. In fact, the adoption of EEDI by the IMO in 2011 rules out the application of EEDI to existing ships.

8.5.3 Jamaica's Proposal

Although in principle Jamaica's approach (Jamaica, 2010) has merit in that it aims to internalize the external costs of CO_2 emissions, important questions regarding its practical implementation can be raised. Monitoring actual emissions is very difficult, even though it is certainly feasible technologically. The idea of measuring emissions produced for each segment of a ship's journey, plus those in port, for all of the world fleet, is a laudable one, but in our opinion we are far away from being able to implement it in a cost-effective manner taking into consideration all the pertinent parameters. Another significant problem is that, member states that choose not to participate in this system, or member states that lack proper monitoring and enforcement mechanisms, run the risk of evolving into "mega hubs" of shipping traffic, for the sole purpose of catering to the needs of those who want to evade the scheme.

8.5.4 The IUCN Proposal

The main focus of the IUCN proposal (IUCN, 2010) is a rebate mechanism which its authors claim is compatible with the Common But Differentiated Responsibilities (CBDR) principle. The principle of CBDR has been a widely accepted principle that underlines such international agreements as the Kyoto Protocol. The essence of the CBDR has two aspects. The first is common responsibility, which is raised from the concept of common heritage and common concern of humankind and reflects the duty of countries to equally share the burden of environmental protection for common resources; the second is differentiated responsibility, which addresses different social and economic situations across countries.

The IUCN proposal in principle can be applied to any type of MBM. In that sense, it is a proposal that can 'piggy back' on any other MBM proposal. The examples given however, have some common elements with the GHG Fund proposal, with upper and lower bounds on prices. The rebate mechanism uses a country's share of global imports as a key without specifying which imports.

In principle such a system might work, provided the implementation of the rebate is carried out in a fair way. If the GHG Fund is used as the MBM, a potential problem concerns the fluctuations of the carbon price, even though these are constrained by the upper and lower bounds on price. In that respect, the GHG Fund scheme provides higher investor certainty (unless of course upper and lower bounds are very close or coincide). The administrative costs will of course be higher. They will be those of the MBM system chosen, plus those of administering the rebate system.

8.5.5 The GHG Fund and ETS Proposals

This section discusses the GHG Fund and ETS proposals. These two classes of MBMs are in many respects mirrors of each other and, in our opinion, if any short list of MBM proposals should be developed, it should include these two classes of MBMs.

8.5.5.1 General Considerations

We start by noting that the proposers of the GHG Fund MBM (submitted by Cyprus, Denmark, Liberia, Nigeria, the Marshall Islands, the Republic of Korea and IPTA) did not call their proposal a 'levy' proposal, much less a 'tax' proposal, and they actually call funds collected a 'contribution'. Essentially however, and to the extent that this contribution is not voluntary and will have to be imposed to ships if this MBM is adopted, this proposal essentially involves a levy on fuel. Still, we shall keep the name given to this MBM by its proposers when referring to it.

We also note that one of the basic functions of the GHG Fund proposal is to use monies collected so as to purchase "offsets" that can be used to reduce "out-of-sector" CO_2 emissions. By doing so, the proposers may run the risk of giving the impression that "in-sector" CO_2 reductions are of lesser importance. This is not necessarily the case however, since if the contribution is high enough, significant in-sector reductions can be achieved.

We have identified a number of documents that review the levy and ETS systems in both a general context and in specific applications. Below is a brief and non-exhaustive discussion.

On the pro-ETS side we cite the book by Ellerman and Joskow (2008) describing experience with the EU ETS. This scheme began operation on 1/1/2005 and now covers more than 10,000 installations and approximately half of the EU's CO_2 emissions, being the world's largest company-level "cap-and-trade" system for trading in emissions of CO_2. All EU Member States participate fully in the scheme as well as Norway, Iceland and Liechtenstein. The system currently covers CO_2 emissions from large emitters in the power and heat generation industries and in selected energy-intensive industrial sectors. The authors of the report believe that although some 'glitches' need to be fixed, the EU ETS is basically sound and can become a prototype for a global climate policy regime.

On the pro-levy side, the US Congressional Budget Office document "Policy Options for reducing CO_2 emissions" (CBO, 2008) compares cap-and-trade with a levy system. The CBO paper compares the efficiency and CO_2 reduction potential of ETS vs. levy and concludes that a levy on emissions would be the most efficient incentive-based option for reducing emissions and could be relatively easy to implement. Further analysis in said document shows that the CO_2 reductions would be nearly double with a levy scheme than a cap and trade scheme. In other words, according to CBO a levy can bring the same environmental result (alternatively: can have the same environmental impact) at half the price of ETS or even less.

Two Friends of the Earth (FOE) reports adopt a similar stance. FOE (2009) identifies six central problems with carbon trading, namely that it is ineffective at driving emissions reductions, it fails to drive technological innovation, it leads to lock-in of high-carbon infrastructure, it allows for, and relies on, offsetting, it creates a risk of sub-prime carbon, and it provides a smokescreen for lack of action on climate finance by the developed world. FOE (2010) outlines why carbon trading is not the solution to climate change and sets out some of the real solutions for cutting greenhouse gas emissions and delivering climate finance. It calls on national governments to urgently dedicate time and resources to develop and implement these and other more viable, equitable and effective solutions to the climate crisis.

PE (2010) puts forward the case that a carbon tax is the most cost-effective measure to reduce CO_2 emissions. Exploring the relative theoretical and practical merits of carbon permit trading and carbon taxation, the report makes a strong argument for taxation, given the likely shape of the damage and cost functions associated with climate change—and the possibility that, in choosing a quantity-based approach we might easily pick the wrong quantity. A tax may enable a more long-term, credible carbon price to be established.

8.5.5.2 Certainty in Cap vs. Certainty in Price

One of the main 'selling points' of the various ETS proposals at the IMO is what is claimed as "full certainty on the emission reductions achieved by the mechanism", that is, if one sets a cap on emissions, that cap will absolutely be met. This stems from the fact that CO_2 emission allowances (or permits) will be auctioned (at a price that is established by the market) and if no more such allowances exist, a ship would not be able to legally emit CO_2. Let us assume for the moment the claim is correct and the cap is enforceable.[2]

A question then is, what should be the cap. For somebody to select it, he or she will have to know what the costs and benefits will be for that particular selection, so that this selection is better than another selection. In a sense, the correct level of reduction is the level at which the costs of further reduction are larger than the benefits of that additional reduction. Unfortunately, no one knows what that level is. So, absolute precision in meeting a cap, if it is unclear what the cap should be, is a problem, and in our opinion makes that feature less credible. And if a cap is set wrong, it is difficult to change.

On the other hand, even though we may reach the cap we selected, the carbon price that will be established will be completely unknown, being a function of future supply and demand for carbon. All carbon forecasting reports are full of many assumptions and caveats and still nobody has seen a previous forecast prove accurate. In a University of Cambridge study for the IMO (IMO, 2010a), the ETS price starts at $177 per tonne and then skyrockets to $3,200 per tonne. It can go the other way too. EU ETS carbon prices have dropped precipitously as a result of the recent economic crisis and (perhaps) as a result of too many allowances being issued.

Even though in general we may not know exactly what the impact of a levy on CO_2 may be, under certain scenarios one can make some pretty good estimates. Devanney (2010) estimates that with a base bunker fuel oil price of $465/tonne, a $50/tonne bunker levy will achieve a 6 % reduction in total VLCC emissions over their life cycle. A reasonable estimate of the reduction for a $150/tonne levy is 11.5 %.

In short, the levy and ETS systems mirror each other. The levy system targets the price and the ETS system targets the quantity of emissions. With ETS we gain cap certainty (with all the previous caveats) and lose price certainty. With a levy scheme, we gain price certainty and can always alter the price to achieve the cap, at least approximately.

[2] This may be a big assumption. It is suspected that the costs of enforcement will be the high side. Also, as the cap is being reached, carbon price may skyrocket, even years before the cap is reached. Market fears and expectations may skyrocket prices, which may in turn collapse as was the case with the EU ETS.

From an investor's perspective, facing a predictable price and basing one's investment in green technologies on that price is far less risky than facing an unpredictable price. Investors typically respond to price, not a cap on emissions.

Coming back to shipping, it is also important to note that whereas the effect of a levy on slow steaming is automatic (the owner or whoever pays for the fuel simply responds to the increased price he faces—see also Chap. 9 of this book), with an ETS things are more complicated, for it is nearly impossible to connect a carbon price paid at a certain time to purchase emissions allowances to slow steaming decisions at a later point in time.

8.5.5.3 Administrative Burden

Regarding the GHG Fund MBM, a question is which, among the two options proposed is better from an administrative viewpoint. Recall that Option 1 collects the money at the bunker supplier level and Option 2 collects it at the ship level. At first glance it would seem that Option 1 has a lower administrative effort, as involving a lower number of transactions than Option 2. In fact one might consider yet another option, Option 3, to collect the money at the refinery level. Theoretically, the higher one goes up the fuel chain, the easier it would be to administer it due to the reduced number of transactions. However, it is not yet clear how each of these options could work in practice, not only from an administrative burden viewpoint, but also in terms of enforcement and evasion avoidance. Some actually believe that none of the above options is viable, and instead propose the money to be collected via Option 4, by direct measurement of CO_2 emissions, via a suitable ultra-sound device within a ship's stack (Devanney, 2011). However, ship owner circles have raised questions on the reliability of such systems.

The administrative costs for ETS include all those administrative costs associated with Option 2 of the GHG Fund proposal (the one which is ship based), plus, many more additional costs associated with issuing the allowances, trading, monitoring compliance, avoiding fraud, and others. Therefore among these two systems, ETS is definitely heavier administrative-wise. Figure 8.6 depicts the administrative procedures of the Norwegian ETS.

To grasp some of the effects of the administrative complexity of ETS, one can examine the bareboat or term charterer issue. While a ship is on bareboat or term charter, the charterer is the effective owner. He decides where the ship goes and at what speed. Legally, he is the disponent owner. This is recognized in the charter party which puts fuel expense to the charterer's account. If an ETS is going to impact the charterers' decision on what speed the ship will go, it has to do the same. This means a shipping ETS not only has to do all of the above, but it also has to keep track of whether or not the ship was on charter and, if so, who the charterer was when the fuel was purchased. Also a chartered ship can be sub-chartered, and so on.

If alternatively the ETS ignores the ship's charter status and requires permits from the owner for all the fuel consumed on his ship regardless of what the charterer does, looking to the owner to recover the permit cost from the charterer, this would

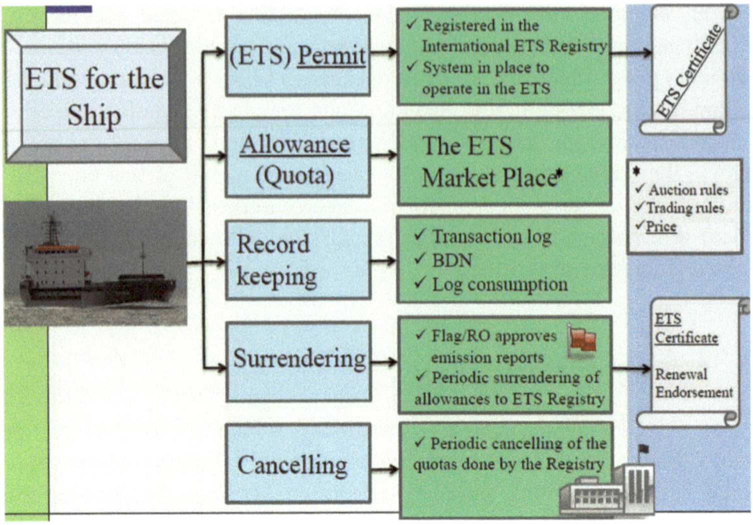

Fig. 8.6 Administrative procedures, Norwegian ETS proposal

put the owner in an untenable position. He would be responsible for emissions from bunkers which are not his, and permit expenses over which he has no control, and which in many cases are not known until well after the charter is complete. In this way the uncertain price of the permits would not influence the charterer to reduce speed in order to reduce fuel consumption. All of this is a non-issue for the GHG Fund scheme. Whoever pays for the fuel also pays the Fund's 'contribution'. Incidentally, the GHG Fund MBM (or, in general, a Levy MBM) is the only MBM that can handle slow steaming automatically, by directly impacting the speed decision of either the ship owner (in case of a spot charter) or the charterer (in case of time or bareboat charter). More on slow steaming and speed optimization in Chap. 9 of this book.

8.5.6 Carbon Leakage, Evasion and Fraud

For ETS, a way to keep the administrative burden from skyrocketing to a high level would be to place limits on coverage, e.g., limit the scheme to ships above a certain size. In fact, this is precisely the reason these limits are suggested in the ETS proposal. If all ships are included the scheme would be unmanageable. Note that according to the IMO 2009 GHG study, if the limit is set at 10,000 GRT, it would amount to 16,000 ships covering some 67 % of total CO_2 emissions. Thus, side-effects of any limit would be that a percentage of the fleet would be exempted and hence produce CO_2. One may see additional side-effects like many ships of 9,900 GRT being built if the limit is 10,000 GRT.

Avoidance of carbon leakage is likely to be problematic in ETS. One reason is the high number of exemptions built into the scheme. Already mentioned is the problem associated with the ship size cut-off. See also the exemption of cargoes associated with small developing island states (SIDS). This could result in traffic being diverted to these countries which could develop into mega transshipment hubs, just for the purpose of emissions exemptions.

Thus, under ETS the potential for evasion is substantial. Also, already several fraud cases have been reported within the EU ETS and elsewhere.

8.5.7 Experience from Other ETS Contexts

Even though similarity of ETS with other trading systems is claimed, these other systems concern really different industries, mostly land-based, which do not operate on an international basis. There is nothing directly comparable to international shipping.

An ETS system that does operate on an international basis is the EU ETS. However, the structure, economics, legal regime and role of the industries covered by the EU ETS are very different from the equivalent attributes of international shipping. A coke oven, a steel plant or a paper mill cannot change flag and relocate if they do not like the stipulations of the ETS (or for any commercial reason, for that matter). The concept of cross-trading, very much prevalent in international shipping, is nowhere to be seen in the sectors covered by the EU ETS, including international aviation, which was included in the EU ETS as of 1/1/2012. That alone might render any allocation scheme unfair and subject to misuse. It is fair to say that the differences among the two sets of sectors are much more than any of the conceivable similarities.

The inclusion of air transport within the EU ETS is an experiment that, at least for the time being, does not seem to proceed very smoothly. In that sense, it is premature to consider aviation as a successful ETS paradigm, some elements of which can be applied to international shipping. Chapter 13 of this book discusses related matters in more detail.

8.5.8 Comparison Between GHG Fund and ETS

In Table 8.2 below we present an horizontal comparison of the GHG Fund and the ETS proposal in terms of the evaluation criteria. All ETS proposals have been combined in the table. Comments on criteria 6 (compatibility to UNFCCC and other international laws) and 9 (compatibility with existing IMO framework) have been omitted as they are covered by the Expert Group report (IMO, 2010b). We also include some additional criteria.

Table 8.2 Comparison of GHG fund and ETS proposals

Main criterion	GHG fund (Denmark 2010)	ETS (Norway, UK, France, Germany)
1. Environmental effectiveness (how certain is MBM to achieve a specific reduction target)	There is less certainty of CO_2 reductions than ETS, but MAC curves of DNV can give an estimate. If price is same, CO_2 reductions are same with ETS[a]. Offsets can contribute to meeting a cap. See also criterion 2 below	There is higher certainty of CO_2 reduction, but reduction target is arbitrary (or very difficult to determine). Plus, enforcing the cap can be difficult and carbon price may skyrocket if we are close to the cap
		Significant carbon leakage risks exist (for instance, if not all ships are covered, some countries like LDCs excluded, etc.)
2. Cost effectiveness	High. Costs are known as price is known. Simplest scheme (except Bahamas). According to several studies, Levy is most efficient way to reduce emissions	Low. High administrative costs, very unpredictable carbon prices
3. Incentives to technological change	High. Investors will respond to known price	Low. Investors will not know what future prices they will encounter and will pay high administrative costs
4. Practical feasibility	Reasonable. Can be modeled from IOPCF	Questionable. All GHG Fund (Option 2) processes, plus auction permits, monitor allowance market, enforce compliance, identify fraud, etc.
5. Impact on LDCs and SIDS	Neutral. From a revenue perspective, if prices are same, revenue is same as ETS	Distortions likely, as traffic to LDCs-SIDS countries is exempted, which may lead to traffic being diverted through these countries
6. Compatibility to UNFCCC and other international laws	See IMO (2010b)	See IMO (2010b)
7. National administrative burden	Reasonable. Tracking bunkers is not trivial but burden is lower than all other schemes (except Bahamas)	Significant. High administrative costs to track, monitor, enforce, avoid evasion and fraud, etc. If all ships in the scheme, impossible to implement
8. Administrative burden on industry	Same as above	Same as above
9. Compatibility to IMO framework	See IMO (2010b)	See IMO (2010b)

(continued)

Table 8.2 (continued)

Main criterion	GHG fund (Denmark 2010)	ETS (Norway, UK, France, Germany)
Other criteria		
Impact in slow steaming	Taken care of automatically	Difficult to impossible to implement or assess
Impact on safety	Neutral	Neutral
Risk of fraud	Average. Low at refinery level.	High-documented cases in EU ETS and elsewhere
Money collected	Limited to in-sector contributions. Depends on level of Levy	If GHG Fund Levy and ETS carbon price are same, amount of money collected for ETS is same as GHG Fund minus difference in administrative costs

Adapted from Psaraftis (2012)
[a]Assuming equal cost-effectiveness which not the case

8.6 Way Ahead

We believe that this chapter supports the conclusion that appropriately formulated and implemented MBMs can potentially play an important role in reducing CO_2 emissions from shipping. However, to this date, and in spite on much analysis and debate, how exactly to proceed on this front remains wide open. Below we discuss the way ahead, both for MBMs and for some other parallel developments.

8.6.1 Fate of MBMs

Providing more detail in the analysis of the previous section, Psaraftis (2012) included all ten MBMs in an horizontal assessment according to the evaluation criteria. This has been, to our knowledge, the only comparison of these proposals to date. The Expert Group report (IMO, 2010b) contained no recommendation on which MBM should be chosen, and discussion on MBMs at the IMO level after 2010 was pretty non-productive. In March 2011, an Intersessional Meeting of the Working Group on GHG emissions from ships took place, with a view to making progress toward the ultimate selection of an MBM, but did not arrive at a similar recommendation either. The same was the case at MEPC 62 (July 2011), as discussion there was entirely devoted to EEDI. The period immediately after the adoption of EEDI was focused on practical matters involving its implementation and there was little discussion on MBMs. A proposal by Greece in 2012 (who had submitted no MBM proposal of its own) for the IMO to decide on a short-list of MBMs was rejected. The same happened to a proposal by the Chairman of the

MEPC in 2012 to conduct an impact assessment study, as political considerations and lack of agreement between developed and developing countries prevented a decision on the matter.

In fact, reception of the proposed MBMs at the IMO has been mixed at best. In addition to the lack of consensus among MBM proposers, the group of developing countries, such as China, India, Brazil and others, were as much against any MBM as they were against EEDI. This was mainly on the ground that MBMs are not compatible with the principle of CBDR. Among others, an issue of importance but also disagreement has been how monies collected by the MBM would be used for the benefit of developing countries (capacity building, technology transfer, etc.). Among industrial stakeholders, the International Chamber of Shipping, BIMCO and several ship owners associations have come out against an ETS, on the ground that it would be unworkable for the shipping industry. Interestingly enough, these include the German and Norwegian ship owners associations, even though their national maritime administrations are for ETS.

Then in May of 2013 the MEPC decided to suspend discussion on MBMs, at least for the time being. This reflected a channeling of the discussion towards the subject of Monitoring, Reporting and Verification (MRV) of emissions.

8.6.2 Monitoring, Reporting and Verification (MRV)

As noted in Sect. 1.6.3 of Chap. 1 of this book, the purpose of MRV is to monitor the energy efficiency and greenhouse gas emissions of the world merchant fleet. In order to document and track global energy efficiency gains, data from ships must be collected and a robust data collection and reporting system must be established. Already the European Commission has proposed a Regulation on MRV (EC, 2013), which is currently under consultation. At the IMO, several proposals have been made, but the discussion is ongoing and many issues are currently wide open.

Establishing an efficient and effective international framework on MRV is of paramount importance, but not so straight forward. It is clear that an MRV by itself may not lower emissions, but can be the first and necessary step for subsequent measures to reduce them. In that sense, the discussion on possible MBMs, suspended at IMO in 2013, can only resume if an efficient and effective global MRV system is established. The same is the case for any other emissions reduction measures that may be implemented at the operational level, or even to see what retrofit measures are the most effective. This means that any MRV system will have to be designed with a longer term view on what will be the next step, after the MRV is established.

In general, any energy efficiency data collection system should be consistent, transparent, objective, documented, and as simple as possible while still meeting its objectives. The administrative burden for both industry and flag and port states should be minimized. As a practical matter, the system should build wherever possible on existing MARPOL Annex VI instruments, guidelines and practices. Furthermore, data collection requirements for ships need to be flag-neutral to create a level playing field and to minimize distortion of competition.

Challenges in an MRV system are various and include (list is not exhaustive):

- The usefulness of reporting information relating to operational aspects like 'cargo carried' and 'transport work' is questionable, given that they seem irrelevant to measuring fuel consumption or CO_2 emissions, notwithstanding that they may be commercially sensitive.
- Any future energy efficiency measure should take into consideration that, while ship operators have a commercial interest in maximising utilisation rates, they have no control over imbalances in trade flows between different regions or factors such as sea and weather conditions or port congestion.
- Monitoring on a per-voyage basis is not practical for vessels operating in short sea trades, particularly for those vessels performing multi-voyages per day, as it would create a substantial administrative burden.

It is also interesting to note the difference between maritime transport and other transport modes (road in particular). Whereas in road transport whatever emissions reduction measures are imposed on the manufacturer and in fact on a fleet level, in maritime transport it is the operator who is the main responsible player and in fact this has to be on an individual ship level. The reason for this difference is probably political, but it may have ramifications as in some instances shipping may compete with land based modes and in fact may be losing at the competitiveness front, a possible result being more CO_2 overall (see also Chap. 10 of this book on this subject).

In a parallel but related development, and as also mentioned in Chap. 2 of this book, the IMO commissioned an update of the 2009 GHG study. The so-called Third IMO GHG study 2014 (Smith et al., 2014) provided updated estimates of CO_2 emissions from international shipping from 2007 to 2012. The 2012 figure, estimated by a 'bottom up' method, was 796 million tonnes, down from 885 million (updated figure) in 2007, or 2.2 % of global CO_2 emissions. CO_2 from all shipping was estimated at 940 million tonnes, down from 1,100 tonnes in 2007. The reduction was mainly attributed to slow steaming. Figure 8.7 shows the CO_2 distribution among major ship types. More on slow steaming in Chap. 9 that follows.

To sum up, MBMs can play an important role in reducing CO_2 emissions from shipping. However, to this date, and in spite on much analysis and debate on the subject, how exactly to proceed on this front remains wide open.

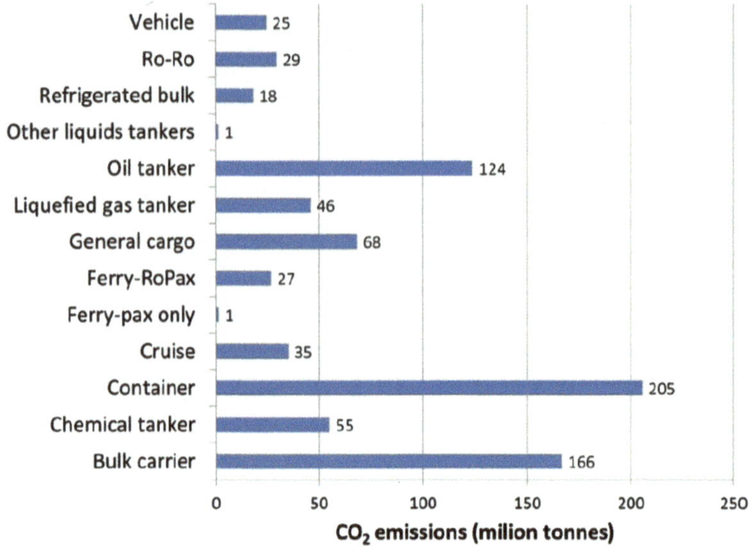

Fig. 8.7 Bottom-up CO_2 emissions from international shipping, 2012. *Source*: Smith et al. (2014)

Acknowledgments The author has been a member of the IMO/MEPC Expert Group on Market Based Measures. Opinions in this chapter are the author's alone. He would like to thank Jack Devanney, George Gratsos, Christos Kontovas and Panos Zachariadis for their assistance for some enlightening discussion and debate on MBM and GHG-related topics.

References

Bahamas. (2010). *Market-based instruments: A penalty on trade and development proposal by the Bahamas*. IMO doc. MEPC 60/4/10.

Bahamas. (2011). *How technical and operational measures are the only direct and effective means to deliver cuts in CO_2 emissions*. IMO doc. GHG WG 3/2.

Buhaug, Ø., Corbett, J. J., Endresen, Ø., Eyring, V., Faber, J., Hanayama, S., et al. (2009). *Second IMO GHG study 2009*. IMO document MEPC59/INF.10.

CBO. (2008, February). *Policy options for reducing CO_2 emissions*. The Congress of the United States, Congressional Budget Office.

Denmark. (2010). *The international fund for greenhouse gas emissions from ships (GHG Fund) proposed by Cyprus, Denmark, the Marshall islands, Nigeria and IPTA*. IMO doc. MEPC 60/4/8.

Devanney, J. W. (2010). *The impact of EEDI on VLCC design and CO_2 emissions*. Center for Tankship Excellence, USA. Retrieved from www.c4tx.org

Devanney, J. (2011, May 4). Direct taxation is the best way to curb CO_2 emissions. *Lloyds List*.

DNV. (2009). Pathways to low carbon shipping. *DNV report*.

EC. (2013). Proposal for a regulation of the European Parliament and of the Council on the monitoring, reporting and verification of carbon dioxide emissions from maritime transport and amending regulation (EU) no 525/2013, COM(2013) 480 final, Brussels, 28 June 2013.
Eide et al. (2010). Cost-effectiveness assessment of CO_2 reducing measures in shipping. *Maritime Policy and Management, 36*(4), 367–384.
Ellerman, A. D., & Joskow, P. L. (2008). *The European Union's emissions trading scheme in perspective*. Cambridge, MA: Massachusetts Institute of Technology.
FOE. (2009). *A dangerous obsession: The evidence against carbon trading and for real solutions to avoid a climate crunch*. Friends of the Earth report. Retrieved from www.foe.co.uk/resource/reports/dangerous_obsession.pdf
FOE. (2010). *Clearing the air: Moving on from carbon trading to real climate solutions*. Friends of the Earth report. Retrieved from www.foe.co.uk/resource/reports/ clearing_air_summ.pdf
France. (2010). *Further elements for the development of an emissions trading system (ETS) for international shipping proposal by France*. IMO doc. MEPC 60/4/41.
Germany. (2010). *Impact assessment of an emissions trading scheme with a particular view on developing countries*. IMO doc. MEPC 60/4/54.
Greece. (2010). *The energy efficiency design index (EEDI) and underpowered ships*. IMO doc. MEPC 60/4/17.
IMAREST. (2011). *Marginal abatement costs and cost effectiveness of energy-efficiency measures*. IMO doc. MEPC 62/INF.7.
IMO. (2010a). *Scientific study on international shipping and market-based instruments*. IMO doc. MEPC 60/INF.21.
IMO. (2010b). *Full report of the work undertaken by the expert group on feasibility study and impact assessment of possible market-based measures*. IMO doc. MEPC 61/INF.2.
IUCN. (2010). *A rebate mechanism (RM) for a market-based instrument for international shipping proposal by IUCN*. IMO doc. MEPC 60/4/55.
Jamaica. (2010). *Achieving reduction in greenhouse gas emissions from ships through Port State arrangements utilizing the ship traffic, energy and environment model, STEEM (PSL) proposal by Jamaica*. IMO doc. MEPC 60/4/40.
Japan. (2010). *The leveraged incentive scheme (LIS) to improve the energy efficiency of ships based on the international GHG fund proposed by Japan*. IMO doc. MEPC 60/4/37.
Japan and WSC. (2011). *Consolidated proposal of "efficiency incentive scheme" based on the leveraged incentive scheme and the vessel efficiency system*. IMO doc. GHG WG 3/3/2.
Norway. (2010). *The global emission trading system (ETS) for international shipping proposal by Norway*. IMO doc. MEPC 61/4/22.
PE. (2010). *Greener, cheaper*. Contributing authors Dr Robert McIlveen and Professor Dieter Helm, University of Oxford, Policy Exchange.
Psaraftis, H. N. (2012). Market based measures for green house gas emissions from ships: A review. *WMU Journal of Maritime Affairs, 11*, 211–232.
Smith, T. W. P., Jalkanen, J. P., Anderson, B. A., Corbett, J. J., Faber, J., Hanayama, S., et al. (2014, June). *Third IMO GHG study 2014*. London: International Maritime Organization (IMO).
UK. (2010). *Global emissions trading system (ETS) for international shipping proposal by the United Kingdom*. IMO doc. MEPC 60/4/26.
USA. (2010). *The United States proposal to reduce greenhouse gas emissions from international shipping, the ship efficiency and credit trading (SECT)*. IMO doc. MEPC 60/4/12.
WSC. (2010). *Vessel efficiency system (VES) proposal by World Shipping Council*. IMO doc. MEPC 60/4/39.

Chapter 9
Green Maritime Transportation: Speed and Route Optimization

Harilaos N. Psaraftis and Christos A. Kontovas

Abstract Among the spectrum of logistics-based measures for green maritime transportation, this chapter focuses on speed optimization. This involves the selection of an appropriate speed by the vessel, so as to optimize a certain objective. As ship speed is not fixed, depressed shipping markets and/or high fuel prices induce slow steaming which is being practised in many sectors of the shipping industry. In recent years the environmental dimension of slow steaming has also become important, as ship emissions are directly proportional to fuel burned. Win-win solutions are sought, but they will not necessarily be possible. The chapter presents some basics, discusses the main trade-offs and also examines combined speed and route optimization problems. Some examples are finally presented so as to highlight the main issues that are at play.

Abbreviations

AIS	Automatic Identification System
CEO	Chief Executive Officer
CIF	Cost Insurance Freight
CO_2	Carbon dioxide
COA	Contract Of Affreightment
DWT	Deadweight Ton
GHG	Green House Gas
GPCI	Global Ports Congestion Index
HFO	Heavy Fuel Oil
IMO	International Maritime Organization
LNG	Liquefied Natural Gas
LPG	Liquefied Petroleum Gas
MBM	Market Based Measure
MCR	Maximum Continuous Rating

H.N. Psaraftis (✉) • C.A. Kontovas
Department of Transport, Technical University of Denmark,
Bygningstorvet 1 Building 115, Room 018 2800 Kgs, Lyngby, Denmark
e-mail: hnpsar@transport.dtu.dk; kontova@transport.dtu.dk

H.N. Psaraftis (ed.), *Green Transportation Logistics*, International Series in Operations Research & Management Science 226,
DOI 10.1007/978-3-319-17175-3_9

MEPC	Marine Environment Protection Committee
MSC	Mediterranean Shipping Company
NTUA	National Technical University of Athens
OPEX	Operating Expenses
OR/MS	Operations Research/Management Science
Ro/Pax	Ro/Ro Passenger
Ro/Ro	Roll On Roll Off
SECA	Sulphur Emissions Control Area
SO_x	Sulphur oxides
TEU	Twenty ft Equivalent Unit
VLCC	Very Large Crude Carrier
WS	World Scale (index)

9.1 Introduction

As mentioned in Chap. 8, the classical breakdown of measures to reduce maritime emissions divides such measures into the following three major classes:

- *Technological* measures include more efficient engines, ship hulls and propellers, cleaner fuels, alternative fuels, devices to trap exhaust emissions, energy recuperation devices, "cold ironing" in ports, various kites, and others. Chapter 5 gave a flavor of such technologies for all surface transportation modes, including maritime.
- *Logistics-based* (tactical or operational) measures include speed optimization, optimized weather routing, optimal fleet management and deployment, efficient supply chain management, and others that impact the logistical operation.
- *Market-based measures* (MBMs). These were examined in Chap. 8, but we will see them again in this chapter from another angle.

We note again the remark made in Chap. 8: the above taxonomy is, in many respects, artificial. Indeed, MBMs can induce logistics-based measures in the short run and technological measures in the long run. With this proviso, the purpose of this chapter is to deal with logistics-based, or tactical/operational measures. In particular, we shall focus on the important tool of *speed optimization*, including its interface with *ship routing*.

Before we focus on speed optimization, we note that the spectrum of logistics-based problems in maritime transportation is very broad and can be broken down in the categories broadly shown in Table 9.1 below. Some related references are also shown in the table (neither list is encyclopedic).

It is important to note that, in much of the OR/MS maritime literature, environmental criteria such as emissions reduction are scarce, traditional economic criteria such as cost reduction being the norm. Sometimes such economic criteria map directly into environmental criteria: if for instance *fuel cost* is the criterion, as it is

Table 9.1 Sample of logistics-based problems in maritime transportation and sample of related references

Problem category	Related references
Ship routing and scheduling (general)	Christiansen, Fagerholt, Nygreen, and Ronen (2007), Christiansen, Fagerholt, and Ronen (2004, 2013), Ronen (2011)
Ship routing and scheduling (tramp)	Andersson, Duesund, and Fagerholt (2011), Fagerholt et al. (2010), Jetlund and Karimi (2004), Lin and Liu (2011)
Ship routing and scheduling (offshore supply)	Aas, Gribkovskaia, Halskau, and Shlopak (2007), Halvorsen-Weare and Fagerholt (2011, 2013)
Fleet deployment (liner)	Andersson et al. (2014), Meng and Wang (2011), Powell and Perakis (1997)
Fleet size and mix (liner)	Alvarez, Tsilingiris, Engebrethsen, and Kakalis (2011), Zeng and Yang (2007)
Speed optimization (general)	Devanney (2007), Fagerholt and Ronen (2013), Gkonis and Psaraftis (2012), Hvattum et al. (2013), Norstad et al. (2011), Psaraftis and Kontovas (2013, 2014, 2015).
Network design (liner)	Agarwal and Ergun (2008), Brouer, Alvarez, Plum, Pisinger, and Sigurd (2013), Imai et al. (2009), Meng, Wang, Andersson, and Thun (2013), Reinhardt and Pisinger (2012).
Weather routing (general)	Lo and McCord (1998), Perakis and Papadakis (1989).
Transshipment (liner)	Hsu and Hsieh (2005), Wang and Meng (2012a, 2012b)
Terminal management	Du et al. (2011), Goodchild and Daganzo (2007), Moccia, Cordeau, Gaudioso, and Laporte (2006), Stahlbock and Voß (2008).

directly proportional to *emissions*, if fuel cost is to be minimized as an objective, so will emissions, and the solution is win-win.

However, for other objective functions this direct relationship may cease to exist and one would need to look at environmental criteria in their own right. Even though such criteria were not very common in the past, the body of knowledge that includes such criteria is growing in recent years. Among the set of maritime logistics problems which are important as regards both economic and environmental criteria, perhaps *speed optimization* is the most important.

The importance of ship speed on ship emissions can be seen in Fig. 9.1, which breaks down CO_2 emissions from the world commercial fleet by ship type-size combination (Psaraftis and Kontovas, 2009a). The data of Fig. 9.1 is from the IHS Fairplay database and the base year is 2007 (45,620 commercial ships accounted for).

According to this analysis, containerships are the top CO_2 emitters in the world fleet. This is perhaps something to be expected, given the relatively high design speeds of these vessels (20–26 knots) as opposed to those carrying bulk cargoes (13–15 knots) and given the nonlinear relationship between speed and fuel consumption and hence emissions. This is also in line with later results. See for instance Fig. 8.7 of this book, taken from the Third IMO GHG study (2014), in which the

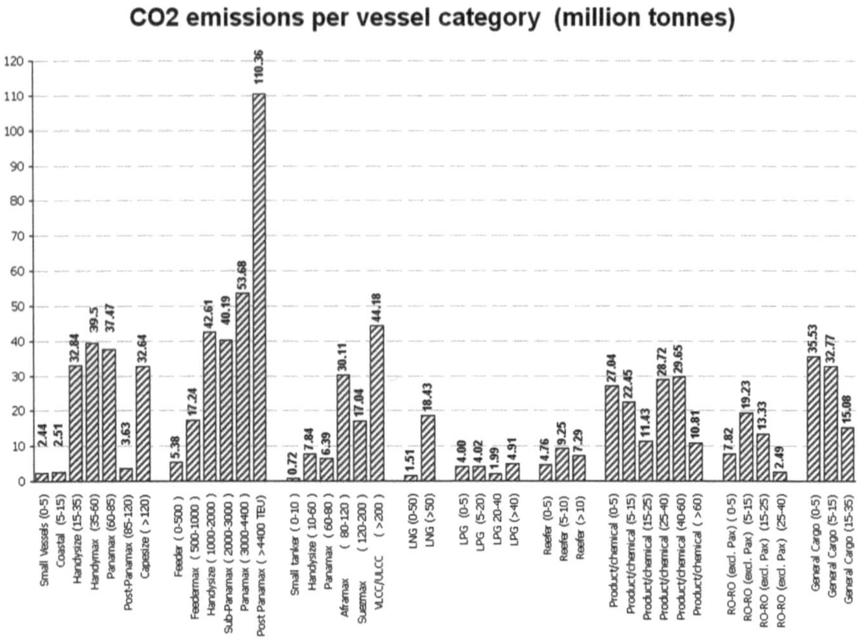

Fig. 9.1 CO_2 emissions, world fleet, 2007. *Source*: Psaraftis and Kontovas (2009a)

baseline year was 2012, and where this class of vessels was identified as the top CO_2 emitter of the world fleet.

What is perhaps not so obvious to expect and can be seen in Fig. 9.1 is that just the top tier category of container vessels (712 vessels of 4,400 TEU and above) are seen to produce 110.36 million tonnes of CO_2 emissions, which is higher than the 106 million tonnes produced by the entire crude oil tanker fleet (2,028 vessels). This means that if ship speed were to be reduced, perhaps uniformly across the board, or even selectively for some categories of vessels, emissions would be reduced too, perhaps drastically. Reducing speed could also have important side benefits: cost reduction is one, and helping a depressed market in which shipping overcapacity is the norm these days is another. In that sense, reducing ship speed may conceivably be a 'win-win' proposition.

Even though ships travel slower than the other transportation modes, a basic premise has always been that there is value in ship speed. The FAST series of conferences, held every 2 years, have been the world's leading technical conferences addressing fast sea transportation issues. As long-distance trips may typically last 1–2 months, the benefits of a higher speed may be significant: they mainly entail the economic added value of faster delivery of goods, lower inventory costs and increased trade throughput per unit time.

The need for higher speeds in shipping was mainly spurred by strong growth in world trade and development, and in turn was made possible by significant technological advances in maritime transportation in a broad spectrum of areas, including hull design, hydrodynamic performance of vessels, engine and propulsion efficiency, to name just a few. By extension, developments in cargo handling systems and supply chain management and operation have also contributed significantly to fast door-to-door transportation.

However, the above basic premise is being challenged whenever shipping markets are depressed and whenever fuel prices are on the increase. In such situations, ships tend to slow down.

Perhaps the most significant factor that is making a difference in recent years is fact that a ship has to be environmentally friendly as regards air emissions. Previous chapters have examined this issue from various angles (for instance, the technology angle in Chap. 5 and the market-based angle in Chap. 8). Because of the non-linear relationship between speed and fuel consumption, it is obvious that a ship that goes slower will emit much less than the same ship going faster.

If one starts with the simple way to reduce fuel costs (and by extension emissions) by reducing speed, this can be done at two levels. One level is the technological one, that is, build future ships with reduced installed horsepower so that they cannot sail faster than a prescribed speed. The first cellular containerships of the late 1960s and early 1970s that went up to 33 knots in the late 1960s when fuel was cheap are gone forever. Maersk's new flagship 'Triple-E' fleet of 18,000-TEU containerships (see Fig. 9.2 next page) have a design speed of 17.8 knots, down from the 20–26 knots range that has been the industry's norm, and will emit 20 % less CO_2 per container moved as compared to the *Emma Maersk*, previously the world's largest container vessel, and 50 % less than the industry average on the Asia-Europe trade lane (Maersk, 2013).[1] Triple-E stands for Economy of scale, Energy efficiency and Environmentally improved performance. Perhaps as an extreme example of how far speed reduction can go, EU-funded research project "Ulysses," whose logo is, conveniently enough, a snail, aims at designing tankers and bulk carriers that can sail as slow as 5 knots (Ulysses, 2012).

The other level of speed reduction is the logistics-based (tactical/operational) one. At that level, an existing ship can sail slower than its design speed. In shipping parlance this is known as "slow steaming" and may involve just slowing down or even 'derating' a ship's engine, that is, reconfiguring the engine so that a lower power output is achieved, so that even slower speeds can be attained. Such a reconfiguration may involve dropping a cylinder from the main engine or other measures. Depending on engine technology, 'slow steaming kits' are provided by engine manufacturers so that ships can smoothly reduce speed at any desired level. In case speed is drastically reduced, the practice is known as "super slow steaming".

In practice, super slow steaming has been pioneered by Maersk Line after it initiated trials involving 110 vessels beginning in 2007. Maersk Line North Asia

[1] The 18,000 TEU yardstick as the world's largest containership size was fated to be surpassed. As this chapter was being completed, the baton was being held by the 19,224 TEU *MSC Oscar*, of the Mediterranean Shipping Company (MSC).

Fig. 9.2 The *Majestic Maersk*, one of Maersk's Triple-E container ships, at dock in Copenhagen harbor. Photo courtesy H. N. Psaraftis

Region CEO Tim Smith said that the trials showed it was safe to reduce the engine load to as low as 10 %, compared with the traditional policy of reducing the load to no less than 40–60 % (TradeWinds, 2009). Given the non-linear relationship between speed and power, for a containership a 10 % engine load means sailing at about half of the design speed. Furthermore, China Ocean Shipping (Group) and its partners in the CKYH alliance (K Line, Yang Ming Marine and Hanjin Shipping) were also reported to introduce super-slow steaming on certain routes (Lloyds List, 2009).

Slow steaming is not only practiced in the container market, although it may seem to make more sense there due to the higher speeds of containerships. Slow steaming is reported in every market. In December 2010, Maersk Tankers was reported to have their Very Large Crude Carriers (VLCCs) sailing at half their speed. The speed of 16 knots (design speed) was reduced to less than 10 knots on almost one third of its ballast legs and between 11 and 13 knots on over one third of its operating days. For example, a typical voyage from the Persian Gulf to Asia normally takes 42 days (at 15 knots laden and 16 knots in ballast). Maersk Tankers decreased speed to 8.5 knots on the ballast leg, thus increasing roundtrip time to 55 days and saving nearly $400,000 off the voyage's bunker bill (TradeWinds, 2010).

Slow steaming has also an important role on absorbing fleet overcapacity. Since early 2009, the total containership capacity absorbed due to the longer duration of total roundtrip time for long haul services has reached 1.27 million TEU in October 2013 (taking early 2009 as a starting point), based on Alphaliner's latest estimates (Alphaliner, 2013). The average duration of Far East-North Europe strings had increased from 8 weeks in 2006 to 9 weeks in 2009 when slow steaming was first adopted. The application of even lower speeds has pushed the figure to 11 weeks currently as carriers continue to seek further cost reductions by adopting slower sailing speeds. The same phenomenon has been observed on Far East-Med strings, where the average duration has risen to 10 weeks, compared to only 7 weeks in 2006. As a record number of deliveries of new vessels is continuing to hamper the supply and demand momentum, analysts expect that slow steaming is here to stay. As a record number of vessels were scrapped in 2013; the idle fleet averaged 595,000 TEUs in 2013 compared to 651,000 TEUs in 2012. The lay-up of surplus box ships has been the worst and has lasted for the longest period since early 2009. The twin impact of extra slow steaming and longer port stays has helped to absorb much of capacity but it seems that sailing at even slower speeds is not an option. A similar situation pertains to bulk carriers and tankers. Thus, slow steaming is here to stay for the foreseeable future.

At the same time, and even though win-win solutions may look as natural consequences of speed reduction, the practice may have other ramifications which may not be beneficial. For instance, in the long run more ships will be needed to produce the same transport throughput, and this will entail some costs, some of them financial and some environmental, such as lifecycle emissions due to shipbuilding and recycling (see Chap. 11 of this book for a discussion on this topic).

Also, in-transit inventory costs will generally increase, due to the increased transit time of the cargo. These inventory costs are proportional to the value of the cargo, so if a ship hauls high-value goods, sailing at a lower speed may entail significant costs to the shipper (we shall come back to this point later in the chapter).

Yet another side effect of speed reduction is that in the short run, freight rates will go up once the overall transport supply shrinks because of slower speeds. Reducing speed may help a depressed market, but it is the shippers who will suffer and in fact they will do so in two ways: they will pay more, and receive their cargo later. For a discussion how tanker spot rates may be impacted as a result of slow steaming see Devanney (2007).

Psaraftis and Kontovas (2009b) investigated, among other things, the option to slow down in Sulphur Emissions Control Areas (SECAs) to reduce the quantity of SO_x produced. It was shown however that if the ship speeds up outside the SECA to make up for lost time within the SECA, more emissions will be produced overall, including SO_x. Fagerholt et al. (2015) examine route-speed alternatives in the context of SECAs. More on SECAs can be found in Chap. 10 of this book.

Last but not least, another possible side effect concerns effects that speed reduction may have on other modes of transportation, to the extent these are alternatives to sea transportation. This is the situation mostly as regards short-sea trades, in Europe but also in North America. If ships are made to go slower, shippers may be induced to prefer land-based transportation alternatives, mostly road, and that may increase overall GHG emissions. Even in long-haul scenarios such as the Far East to Europe trade, some cargoes may tempted to use the rail alternative (via the Trans-Siberian railway) if the speed of vessels is low enough (see Psaraftis and Kontovas (2010) for a discussion).

An important point we would like to stress is connected to an issue also brought up in Chap. 8 in the context of the Marginal Abatement Cost curves. Even though speed reduction can be considered as *a measure* to reduce emissions, this measure is typically manifested as *a response* of whoever pays for the fuel, to exogenous market parameters such as the fuel price and the freight rate. Unless speed limits are mandated by legislators (and this has been the objective of various lobbying groups—thus far without success), speed reduction should be considered not as an independent measure but as a reaction of shipping operators to such external parameters. This is of course within the feasible space dictated by ship and engine technology, as well as by the contractual arrangements between the ship owner and the charterer.

This chapter will examine ship speed optimization from various angles. In that context, some basics will be outlined, the main trade-offs will be analysed, and some decision models will be presented, including combined speed and routing scenarios. The examples to be presented will highlight the main issues that are at play. Material of this chapter is mainly taken from various papers and other work by the authors and their colleagues. These include Gkonis and Psaraftis (2012), Kapetanis, Gkonis, and Psaraftis (2014) and Psaraftis and Kontovas (2013, 2014).

The rest of this chapter is organized as follows. Section 9.2 presents some basics of speed optimization. Section 9.3 discusses factors that may impact fuel consumption. Section 9.4 discusses the possible impact of inventory costs. Section 9.5 summarizes results for tankers and bulk carriers. Section 9.6 discusses speed vis-à-vis mixed chartering scenarios. Section 9.7 presents combined speed-route optimization model. Last but not least, Sect. 9.8 presents the chapter's conclusions and discusses some extensions.

9.2 Ship Speed Optimization Basics

Before we see how ship speed can be optimized, we present some basics. We do this so as to clear possible misconceptions and highlight some issues which we find important.

The first basic is that *ships do not trade at fixed or predetermined speeds.*

In the charter (tramp) market, those who pay for the fuel, that is, the ship owner whose ship trades on the spot market, or the charterer if the ship is on time or bareboat charter, will typically choose ship speed as a function of two main input parameters: (a) the fuel price and (b) the market freight rate. In periods of depressed market conditions, as is the typical situation in recent years, ships tend to slow steam. The same is the case if bunker prices are high. Conversely, in boom periods or in case fuel prices are low, ships tend to sail faster.

An exception to the above is in case the ship is *on spot charter* (rental of the ship for a single voyage) and its speed is prescribed in the charter party contract, either explicitly (speed is, say, 15 knots) or implicitly (cargo pickup and delivery dates are prescribed). In spot charters the fuel is paid for by the ship owner. Agreeing on a prescribed speed in the charter party involves in most cases only the laden part of the trip, with the owner free to choose his speed on the ballast return leg. The speed that is agreed upon for the laden leg may or may not be the speed that the ship owner would have freely chosen if no explicit agreement were in place. If it is higher, the ship owner may ask for a higher rate than the prevailing market spot rate, understanding of course that in this case he may lose the customer to a competitor ship, with whom the charterer can obtain more favorable terms. For a discussion of possible distortions and additional emissions that can be caused by charter party speed agreements, see Devanney (2011).

A similar situation plays out in the liner market. Container and Ro/Ro operators typically operate a mixed fleet of vessels, some of which are owned vessels and some are chartered from independent owners who are not engaged in liner logistics. In either case, fuel is paid for by the liner operator. The operator receives income from the multitude of shippers whose cargoes are carried on the ship and the rates charged to these shippers can be high or low depending on the state of the market. As in the charter market, high fuel prices and/or depressed market conditions imply lower speeds for the fleet.

In spite of the above, many of the models found in the OR/MS maritime literature assume fixed and known ship speeds. See for instance Agarwal and Ergun (2008), Hwang, Visoldilokpun, and Rosenberger (2008), Grønhaug, Christiansen, Desaulniers, and Desrosiers (2010), Rana and Vickson (1991) and Song and Xu (2012), among others. In these models, ship speed is typically considered not as a decision variable but as a *fixed input* to the problem. Most of the time this input is *implicit*, in the sense that it is used to compute various other *explicit inputs* that depend on speed, such as sailing times, due dates for cargo pickup and delivery, and ship operating costs, of which fuel cost is an important component.

Assuming fixed ship speeds is typically also the case for models that compute shipping emissions worldwide, even though these do not belong to the OR/MS literature. See for instance the 2009 IMO GHG study (IMO, 2009) and Psaraftis and Kontovas (2009a), among others. In their calculations, these models typically take as input *design speeds* extracted from commercially available ship databases, such as those maintained by IHS Fairplay, among others. Such information may be inaccurate and does not necessarily represent actual ship speeds. The 2014 IMO GHG study (IMO, 2014) is more advanced in that it uses actual ship speeds in its calculations. Actual ship speeds were taken from ship Automatic Identification System (AIS) data.

Coming back to maritime transportation OR/MS models, it is clear that *not* considering speed as a decision variable may render solutions suboptimal. This is because doing so ignores the economic trade-off between (a) the lower voyage and cargo inventory costs associated with a higher speed and (b) the higher fuel costs associated with such higher speed. Assuming a fixed speed precludes the balancing of such trade-offs.

A speed that is assumed fixed may also in some cases remove flexibility in the overall decision making process. For problems that include port capacity constraints, berth occupancy constraints, time window constraints or other constraints that preclude the simultaneous service of more than a given number of vessels (see, for instance, Cordeau, Laporte, Legato, and Moccia (2005) and Halvorsen-Weare and Fagerholt (2013), among others), satisfying such constraints would conceivably be easier to meet were it not for the assumed constancy in ship speed. The same is the case for problems that analyze disruptions of service due to weather or other unpredictable events. It is clear that removing the flexibility to adjust ship speed in such scenarios would render any response to the disruption suboptimal.

Still, dealing with speed is not new in the maritime transportation literature and this body of knowledge is rapidly growing. In Psaraftis and Kontovas (2013) some 42 relevant papers were reviewed and a taxonomy of these papers according to various criteria was developed. Several additional papers dealing with ship speed appeared after the above paper was published. Its Google Scholar citations in April 2015 stood at 48, of which there was even a related paper in *Meat Science* (Mills, Donnison, & Brightwell, 2014). This indicates a growing interest of researchers in this topic.

We have amended the Psaraftis and Kontovas (2013) taxonomy and enlarged it to include 51 papers, including some of the most recent ones. The full table is available in the Appendix at the end of this chapter.

Another basic property of optimal speeds is not immediately obvious. It applies mainly to the charter (tramp) market and compares, for a specific ship and a specific route, the speed optimization problem of its ship owner and that of a time charterer who may charter the same ship. The ship owner wants to maximize average profit per day and the charterer wants to minimize average cost per day. Even though these two optimization problems appear at first glance different, the optimal ship speed for both problems turns out to be the same. For a proof in a rudimentary tramp scenario see Devanney (2010).

In Psaraftis and Kontovas (2013) it was further shown that both the above problems reduce to the following formulation:

$$\min_v \{\rho f(v) - Qv/L\}$$

where
v = sailing speed (nautical miles per day[2])
$\rho = P_{FUEL}/s$
P_{FUEL} = fuel price ($/tonne) and
s = spot rate received by the owner ($/tonne)
$f(v)$ = ship's daily fuel consumption (tonnes/day)
Q = ship's cargo capacity (tonnes)
L = roundtrip distance (nautical miles)

[2] This is 24 times the ship speed in knots. We use this unit to avoid carrying the number 24 through the calculations. One knot is one nautical mile per hour (1.852 km per hour) and is the typical unit of ship speed.

In fact from the above it can be seen that a key determinant parameter of the speed optimization problem is ρ, *the non-dimensional ratio of the fuel price divided by the market spot rate* (both expressed in \$/tonne). Higher ρ ratios will generally induce lower speeds than lower ratios. This corresponds to the typical behavior of shipping lines, which tend to slow steam in periods of depressed market conditions and/or high fuel prices and go faster if the opposite is the case.

For the simple case of a cubic fuel consumption function, $f(v) = kv^3$ and no constraints on speed, the optimal solution to the above problem is $v^* = (Q/3k\rho L)^{1/2}$, confirming the basic dependency of the optimal speed to the ρ ratio. Later we will see more realistic fuel consumption functions.

All OR/MS models that include speed that we have reviewed incorporate fuel prices as part of their input. However, in many models such inclusion is only *implicit*, whereas in others it is *explicit*. An implicit formulation is a fuel cost function $FC(v)$ instead of the product $P_{FUEL}f(v)$ and means that P_{FUEL} is not explicitly a part of the problem's input. An implicit formulation has the drawback of not allowing someone to directly analyze the functional dependency between P_{FUEL} and the optimal value of v, which can be very important.

Together with fuel price, another important input parameter is the state of the shipping market and in particular the freight rate (spot rate or other). Yet, a typical modeling assumption that is reflected in many OR/MS models that deal with ship speed is to *not* include the state of the market as part of their formulation. In most of these models it is assumed (at least implicitly) that fuel costs are being borne by the ship owner. In the tramp shipping market (served by tankers, dry bulk carriers, product carriers, and gas carriers) this is the case if the ship is on spot charter. It is known that the predominance of charter party contracts are time charters, in which fuel costs are borne by the charterer. Even though most models assume the ship owner as the party that bears the costs, including fuel, the related optimization problem is typically cost minimization rather than profit maximization. This is tantamount to assuming that revenue for the service is fixed. This is not the case however if the ship speeds up to make more profit-earning trips per unit time. Thus, some of the OR/MS models that optimize speed do not capture the trade-off between a higher speed to make more trips per unit time and the impact of such higher speed on costs (mainly on fuel).

Figure 9.3 is adapted from Gkonis and Psaraftis (2012) and captures the impact of both freight rate and bunker price on optimal speed for a specific Very Large Crude Carrier (VLCC) trading from the Persian Gulf to Japan. Optimal here means maximize average per day profit for the ship owner, and speeds are optimized in both laden and ballast conditions. Two market conditions are shown for the spot rate, one at Worldscale (WS) 60 and one at WS120.[3] Bunker prices (HFO, Heavy Fuel Oil) range from \$400 to \$1,000 per tonne. It can be observed that the impact of

[3] WS is a nondimensional index measuring the spot rate and is exclusively used in the tanker market. For a specific route, WS is proportional to the spot rate on that route (in \$/tonne) and is normalized by the 'base rate' on that route. See Stopford (2009) for a detailed definition.

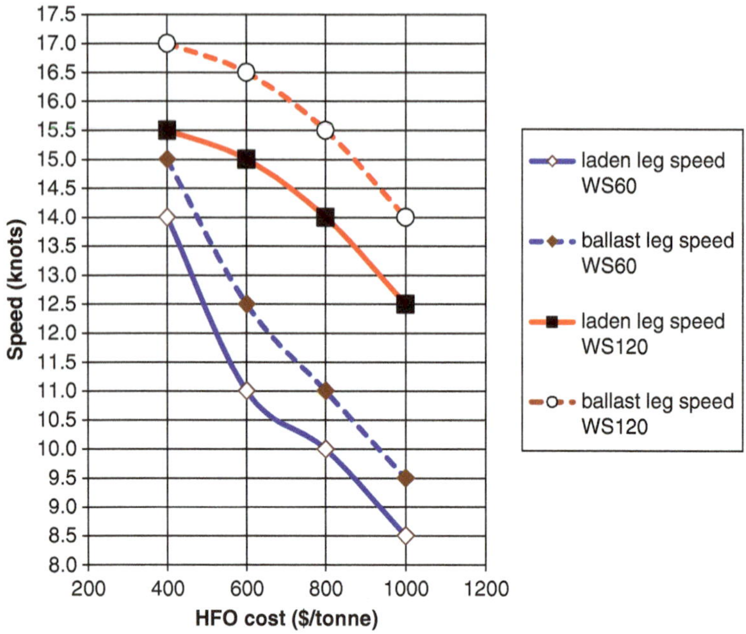

Fig. 9.3 Optimal VLCC speed as a function of spot rate and bunker price. Adapted from Gkonis and Psaraftis (2012)

both freight rate and bunker price on optimal speed can be quite dramatic, and that the range of optimal speeds can be very broad, depending on the combination of values of these two input parameters. It can be also observed that ballast speeds are typically higher than laden speeds by 1.0 knot in the lower rate scenario and by 1.5 knots in the higher rate scenario

In general there is wide variation on the input parameters that are assumed by the various models in the literature. In general these are problem-dependent. The taxonomy of the Appendix sheds some light on this issue, among others, for each of the papers in the taxonomy.

Other input parameters and model assumptions may be important in speed optimization. The following may or may not be true as regards assumptions that are used in a speed model:

(a) fuel consumption is a function of payload,
(b) fuel price is an input (explicit or implicit),
(c) freight rate is an input, and
(d) in-transit cargo inventory cost is considered.

Table 9.2 below lists a limited sample of papers of the taxonomy of the Appendix, including some of the most recent ones, and lists whether or not each of (a) to (d) above is true.

As argued throughout this chapter, inclusion of (a) to (d) above within a speed model's formulation can be important.

Table 9.2 Sample of papers and whether certain parameters are included in the model

Papers (listed chronologically)	Shipping market, logistical context	Fuel consumption a function of payload?	Fuel price explicit or implicit input?	Freight rate an input?	In-transit cargo inventory cost considered?
Ronen (1982)	Tramp, fixed route	No	Explicit	Yes	No
Perakis and Jaramillo (1991)	Liner, fleet deployment	No	Explicit	Yes	No
Devanney (2007)	World petroleum network	Only for laden and ballast conditions	Explicit	Equilibrium spot rate computed	Yes
Notteboom and Vernimmen (2010)	Container, fixed route	No	Explicit	No	No
Norstad et al. (2011)	Tramp, pickup and delivery	No	Implicit	No	No
Cariou and Cheaitou (2012)	Container, fixed route	No	Explicit	No	Yes
Gkonis and Psaraftis (2012)	Tanker, fixed route	Only for laden and ballast conditions	Explicit	Yes	Yes
Hvattum et al. (2013)	General, fixed route	No	Implicit	No	No
Fagerholt and Ronen (2013)	Tramp, pickup and delivery	No	Implicit	Only for spot cargoes[a]	No
Psaraftis and Kontovas (2014)	General, fixed or flexible route	For any loading condition	Explicit	Yes	Yes
Andersson et al. (2014)	Ro/Ro, fleet deployment	Only for laden and ballast conditions	Implicit	No	No
Doudnikoff and Lacoste (2014)	Liner, fixed route in SECAs	No	Explicit	No	No
Wang et al. (2014)	Container, schedule design	No	Explicit	No	Yes

[a]A known revenue is assumed for each of the spot cargoes, implying a freight rate for them

9.3 Factors that Affect Fuel Consumption

It is known from basic naval architecture that fuel consumption depends non-linearly on both ship sailing speed and ship payload. Many papers assume that fuel consumption per day is a *cubic function* of ship speed. The cubic approximation is reasonable for some ship types, such as tankers, bulk carriers, or ships of small size, but may not be realistic at slow or near-zero speeds and for some other ship types such as high-speed large container vessels. Even at zero speed the ship consumes some fuel, as its auxiliary engines are typically on to produce electricity. An exception is if electricity is provided to the ship by shore-side supply (also known as 'cold ironing'), but this is currently an exception rather than the rule. In-port fuel consumption is proportional to overall total port residence time.

A more serious assumption in many related models is that no dependency between fuel consumption and ship payload is considered. This assumption is reasonable in case ship payload is constant or does not change much. Cruise vessels, passenger vessels, and sometimes Ro/Ro carriers and Ro/Pax vessels belong to this category. However, if this assumption is not valid, it can cause serious under- or over-estimation of fuel costs. Ship resistance and hence fuel consumption at a given speed can be drastically different if the ship is full, empty or at an intermediate loading condition.

In tankers and bulk carriers we have a 'binary' situation, as the ship is typically either full or empty, and the difference in fuel consumption between these two extreme conditions can be quite substantial. In container vessels the ship is typically intermediately laden most of the time, but ships in some trunk routes (e.g. Far East to Europe) are mostly full in one direction and mostly empty in the opposite. This can come close to a binary situation and one would expect non-trivial differences in fuel consumption as a result.

In general, if a ship's loading condition varies along the legs of a ship's route (which is typical in pickup and delivery scenarios in which the ship is not fully laden all of the time), it is important that the dependency between ship load and fuel consumption along that route be realistically modeled. In an optimization setting, it would not make sense to claim solutions within, say, 1, 2 or 5 % from the optimal solution, or even solutions at the exact optimum, if the fuel consumption function, and hence fuel costs, are misrepresented by 10, 20 or 30 %.

In order to capture this dependency, it is useful to extend the previous formulation of the daily 'at-sea' fuel consumption of the ship $f(v)$ and assume that it is a known function $f(v,w)$ of both v, the ship's speed, and w, the ship's payload, which may actually vary along the ship's route. Function $f(v,w)$ depends on the ship, and essentially on the hull geometry-engine-propeller configuration. It can even be defined for $v = 0$ (ship in port) and $w = 0$ (ship going on ballast), and it need not be assumed in closed form, but could be given as a point/wise function, as a table, or even as the output of a relevant subroutine. Strictly speaking, f must also take into account the reduction of the ship's total displacement due to fuel being consumed along the ship's route. However, since displacement would not change much as a

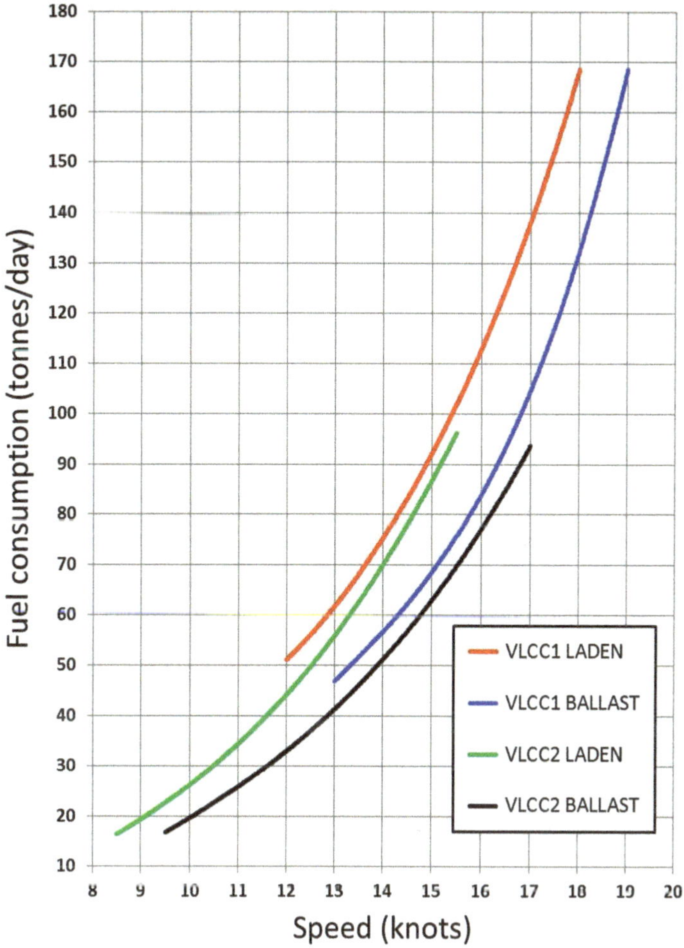

Fig. 9.4 Typical fuel consumption functions for two VLCCs for both the laden and ballast conditions. *Sources*: undisclosed

result of that consumption, one can practically assume *f* independent of en-route fuel consumption.

Figure 9.4 shows fuel consumption curves for two distinct Very Large Crude Carriers (VLCCs), for both the laden and ballast conditions. Relevant data was solicited and obtained for these ships under confidentiality conditions. It can be seen that the difference between laden and ballast fuel consumption at the same speed is on the order of 25–30 %.

In Fig. 9.4 it can be seen that ship speeds have upper and lower bounds. Both bounds are dictated by the maximum power and technology of the engine, and by the ship's payload. The upper bound exists because of limits in the ship's power and the lower bound exists because it is simply impossible for a ship engine to run slower than a certain power. Modern, electronically controlled engines can run slower than older, camshaft controlled engines.

A realistic closed-form approximation of f that takes both v and w into account is $f(v,w) = k(p + v^q)(w + A)^{2/3}$ with k, p and q constants such as $k > 0$, $p \geq 0$ and $q \geq 3$. A is the 'lightship weight', that is, the weight of the ship if empty including fuel and other consumables (modified admiralty formula). The rationale for such a formulation is that fuel consumption is proportional to the wetted surface of the ship, which is crudely proportional to the displacement of the ship $\Delta = w + A$, raised to the power of 2/3; see also Barrass (2005).

As said earlier, most papers in the literature assume a cubic function, that is, $p = 0$ and $q = 3$ and no dependency on payload.

The fuel consumption function also depends on the prevailing weather conditions along the route, which may actually vary in time and space. The way weather conditions are treated in the literature ranges from non-treatment (implying that the *average* weather conditions the ship expects along its route are implicitly factored into the function *f*, perhaps by a '*sea margin*' coefficient), to more sophisticated approaches in which *f* depends on the specific weather conditions along the ship's route, including wave height, wave direction, wind speed, wind direction, sea currents, and possibly others. These factors, most of which may be stochastic, can significantly influence both wave and wind resistance and hence fuel consumption and cost. Weather routing models typically take the more sophisticated approach, whereas all other OR/MS models including ship routing and scheduling, fleet deployment, and other models typically follow the simpler one.

Hull condition can also be an important factor that influences the frictional resistance of a ship, and, as a result, its fuel consumption. A foul hull from seaweed and other sea organisms would entail a higher resistance (and hence fuel consumption) than a clean hull, and efforts are being made (via anti-fouling paints and hull cleaning at regular intervals) to maintain a clean hull. To our knowledge, no OR/MS model takes into account such factor, all assuming an average hull condition.

9.4 Impact of In-Transit Cargo Inventory Costs

Many of the reviewed speed models do not include *in-transit cargo inventory costs* as part of the cost function. These are inventory costs that accrue while the ship is in transit, and they can be a non-trivial component of the cost that the owner of the cargo bears if the ship will sail at a reduced speed. They can be important if timely delivery of the cargo is significant. They can also be important if the voyage time and/or the quantities to be transported are non-trivial. This can be the case in long-haul problems.

It is clear that in-transit inventory costs are important for the charterer, assuming that he is the owner of the cargo. These costs are also important for the ship owner, as a charterer will prefer a ship that delivers his cargo earlier than another ship that sails slower. Thus, if the owner of the slower ship would like to attract that cargo, he may have to rebate to the charterer the loss due to delayed delivery of cargo. In that sense, the in-transit inventory cost is very much relevant in the ship owner's profit equation, as much as it is relevant in the charterer's cost equation.

The same is the case if the charterer does not move his own cargo but uses the ship to move somebody else's cargo. This is a typical situation in liner trades, where a significant part of a liner company's fleet consists of chartered ships, owned by independent ship owners but operated by the liner company. As the cargo owner will prefer a ship that moves his cargo faster, his in-transit inventory costs are again very much part of the chartered ship's cost equation.

If we call β the per day and per tonne in transit inventory cost of the cargo, it is straightforward to see that β is equal to $PR/365$, where P is the CIF value of the cargo (value of cargo at destination) and R the cargo owner's cost of capital. This represents the revenue that is lost due to a delayed delivery of 1 tonne of the cargo by 1 day. This means (as expected) that expensive cargoes are more costly than cheaper cargoes in terms of inventory cost. This also explains why expensive cargoes tend to get hauled by liner ships that go faster, whereas cheaper (bulk) cargoes go by tramp ships that go slower. Conversely, it also means that in periods of low interest rates this cost component is less important.

Cargo inventory costs can be important in the liner business which involves trades of higher valued goods than those in bulk trades. The unit value of the top 20 containerized imports at the Los Angeles and Long Beach Ports in 2004 varied from about \$14,000/tonne for furniture and bedding to \$95,000/tonne for optic, photographic and medical instruments (CBO, 2006). Delaying 1 tonne of the latter category of cargo by 1 week because of reduced speed would cost some \$91 if the cost of capital is 5 %. For a 80,000 tonne payload this would amount to some \$7.25 million. This may or may not be greater than the economic benefit of a reduced speed.

Psaraftis and Kontovas (2009b) assumed a hypothetical string of 100 identical Panamax container vessels, each with a payload of 50,000 tonne. If the fleet baseline speed is 21 knots (both ways) and the fuel consumption at that speed is 115 tonne/day, then for a fuel price of \$600/tonne (corresponding to a period of high fuel prices, before the slump of 2008), the daily fuel bill would be \$69,000 per ship. Running the same type of ship at a reduced speed of 20 knots (one knot down), and assuming a cube law the fuel consumption would drop to 99.34 tonnes/day and the daily fuel bill would drop to \$59,605 per ship.

Assume these 100 ships go back and forth a distance of 2,100 miles (each way) and are 100 % full in both directions. This is not necessarily a realistic operational scenario, as containerships visit many ports and as capacity utilizations are typically lower both ways, depending on the trade route. However, a generalization of this analysis to many ports and different capacity utilizations in each leg of the trip should be straightforward. For simplicity, also assume 365 operating days per year and zero port loading and unloading times. For non-zero port times and less than 365 days, the analysis will be more involved but will lead to similar results.

Then we will have two cases:

Case A: 100 Ships Going 21 Knots
Total fuel burned/year/ship: 115 tonnes/day*365 = 41,975 tonnes
For 100 ships = 4,197,500 tonnes
Transit time (one way) = 100 days
Total annual fuel cost (X$600) = $2,518,500,000.

Case B: 105 Ships Going 20 Knots
To reach the same throughput of cargo per year, we will need 105 ships.
Total fuel burned/year/ship: 99.34 tonnes/day*365 = 36,259 tonnes
For 105 ships = 3,807,256 tonnes
Transit time (one way) = 105 days
Total annual fuel cost for 105 ships (X$600) = $2,284,353,741 (reduced vis-à-vis Case A).
Reduction of CO_2 emissions (per year), vis-à-vis Case A: 1,237,073 tonnes.

Fuel cost difference (per year) = $234,146,259 for five more ships, that is, $46,829,252 per additional ship. Dividing by 365, this difference is $128,299 per day. This means that if the sum of additional cargo inventory costs plus other additional operational costs of these ships (including the time charter) is less than $128,299 a day, then case B is overall cheaper. One would initially think that such a threshold would be enough. But it turns out that this is not necessarily the case if in-transit inventory costs are factored in.

To compute in-transit inventory costs for this case, we hypothetically assume that cargo carried by these vessels consists of high value, industrial products (e.g., computers, luxury cars, or similar), whose average value at the destination (CIF price) is $20,000/tonne. We also assume the cost of capital being 4 %. This means that 1 day of delay of 1 tonne of cargo would entail an inventory cost of 20,000*0.04/365 = $2.19. This may not seem like a significant figure, but it is. Computing the in-transit inventory costs for this case gives a total annual difference of $200,000,000 ($4,200,000,000−$4,000,000,000) in favor of case A, which moves cargo faster. This figure is significant, of the same order of magnitude as the fuel cost differential.

Assuming also a time charter rate of $25,000 per day (typical charter rate for a Panamax containership in 2007), the total other operational costs of the reduced speed scenario are $958,125,000 per year for 105 ships, versus $912,500,000 for 100 ships going full speed. Tallying up we find a net differential of $11,478,741 per year in favor of Case 1, meaning that in-transit inventory and other operational costs offset the positive difference in fuel costs.

Of course, other scenarios may yield different results, and the reduced speed scenario may still prevail in terms of overall cost, under different circumstances. For instance, if the average value of the cargo is $10,000/tonne, and everything else is the same, then the difference in annual inventory costs drops to $100,000,000, rendering the reduced speed scenario a profitable proposition (with a total cost reduction of $88,521,259 per year). Actually, speed reduction remains profitable if

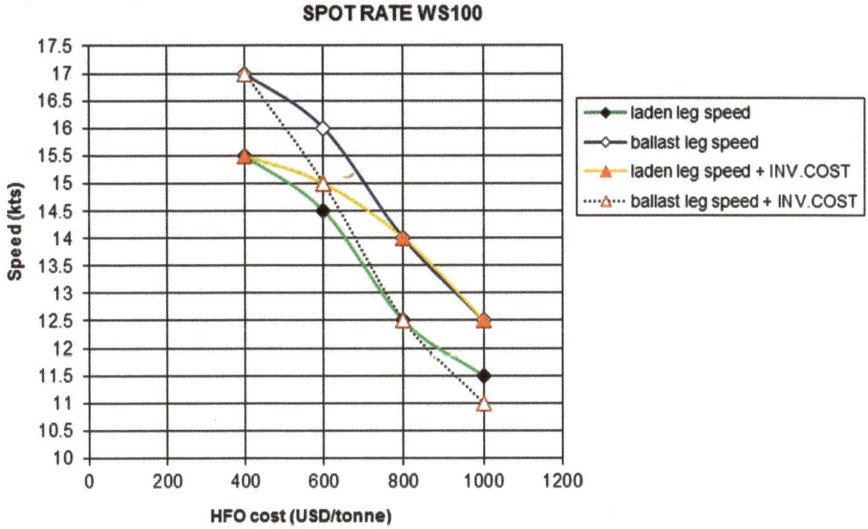

Fig. 9.5 Optimal VLCC speeds with and without inventory costs. *Source*: Gkonis and Psaraftis (2012)

the value of the cargo is no more than about $18,800/tonne (which can be considered as a break-even CIF price).

But the liner market is not the only one in which inventory costs are important. Figure 9.5 shows that optimal VLCC speeds vary, depending on whether or not in-transit inventory costs are taken into account. Including that cost component would generally increase the optimal speed. The example is for a given VLCC and assumes that the market spot rate is WS 100. Per earlier considerations, if no inventory costs are factored in, ballast speeds are uniformly above laden speeds, by 1.0–1.5 knots. But in case inventory costs are factored in, this is not necessarily the case. In the example, one can observe that if fuel (HFO) prices are higher than about $600/tonne (a break-even price), optimal laden speeds are higher than the equivalent ballast speeds.

For crude oil tankers, and in working with curves such as in Fig. 9.5, another effect should also be taken into account, which is not immediately obvious. If fuel prices change, the same in general will happen to the value of the crude oil that is carried, although there may not be a direct mapping between the two values. So in-transit inventory costs if HFO price is 800 $/tonne may be higher than those if HFO price is 400 $/tonne. We found that a higher value of the cargo while at the same time fuel prices increase will make the ship sail slightly faster (no more than 0.5 knots) than if the value of the cargo is kept fixed.

9.5 Speed Optimization in Mixed Chartering Scenarios

There are a number of papers in the OR/MS literature that examine 'mixed' chartering scenarios (see for instance Norstad, Fagerholt, and Laporte (2011) and Fagerholt and Ronen (2013), among others). In such scenarios, a number of the cargoes are labeled '*mandatory*' cargoes, to be moved under COA (contract of affreightment) terms, and the rest are '*optional*' cargoes, to be moved under spot charter terms. A COA contract is a voyage charter in which the ship owner undertakes the obligation to carry specified cargoes between specified ports at some point in the future, without naming the ships which will undertake the assignment. This gives flexibility to the ship owner in fulfilling his obligation, and in fact at time of COA signature he may not even have control of the ships that will be eventually used. He can use his own ships, or he can use ships from the charter market for that purpose.

Optional (spot) cargoes in such a mixed scenario mean that the owner has the option whether or not to embed these cargoes within his service plan, in addition to the COA cargoes which are considered mandatory. Adding spot cargoes would generate *additional revenue* for the ship owner, whereas the revenue of the mandatory cargoes is considered fixed and therefore does not impact the optimization problem. Given this additional potential revenue from the optional spot cargoes, it may make sense for some ships in the fleet to pick up some of these cargoes along their route. It may also make sense for the ships to sail at a higher speed in order to accommodate these optional spot cargoes and not violate time window constraints or other contractual obligations that may exist for the mandatory cargoes.

Such scenarios are encountered in the OR/MS maritime logistics literature but may actually be more complex than a first glance would suggest. A first observation is that the distinction between mandatory and optional cargoes may be, in many ways, artificial. Any cargo, including the COA cargoes, is optional until the ship owner and the charterer decide to enter into a mutually binding contract for that cargo. After COA contract signature, serving the COA cargo in question becomes mandatory. The same is true for the spot cargoes: they are only optional until an agreement to carry them is reached, and if so they become mandatory as well. What really distinguishes these two types of cargo is the timing of contract signature, which is presumably different: signature of the COA precedes that of the spot cargoes. The period during which the COA cargoes are mandatory and the spot cargoes are optional is the period between the signatures of the two contracts, and hence is only a transient period.

It is known that a COA does not require a ship owner to name a ship at the time of contract signature. But as the contract will specify the size of the shipments, the size of the ships that will carry them under the COA is likely to be more or less indirectly implied, usually leaving little or no extra space for spot cargoes, if the latter would have to be on the ship together with the COA cargoes. Fulfilling the COA with ships larger than required so as to allow space for potential spot cargoes is always an option. But assigning larger ships for the COA *before* potential spot cargoes are known may entail a financial risk and a potential loss to the ship owner. Even though several papers in this area also deal with uncertainty, we have seen no models that try to capture this specific risk as part of their formulation. One way to avoid such risk altogether is if the decision which spot cargoes to serve and at what speed is made *simultaneously* with the decision which ships to assign to fulfill the COA obligations. Another way is if the spot cargoes are served separately from the COA cargoes, for instance on the return leg of the COA route. However, this is likely to involve delays for the COA cargoes.

Another issue that is not often mentioned but may further complicate things is that the owner who has signed a COA may have to obtain the permission of the COA charterer to serve the additional spot cargoes. This may involve amending the COA itself, not necessarily an easy proposition, unless commitment to both the COA and spot cargoes happens at roughly the same time (something that would eliminate the distinction between mandatory and optional cargoes). The permission of the COA charterer may be warranted because adding spot cargoes may imply delayed delivery of the COA cargoes and also may involve liability issues, for instance in case the spot cargoes are dangerous or semi-dangerous and the ship (together with the COA cargoes) is lost or damaged.

Irrespective of the above, in mixed scenarios such as the above, ship payload will generally vary along the ship's route. Note however that, to our knowledge, all papers in the literature that deal with such mixed scenarios assume fuel consumption functions that are independent of ship payload. For reasons outlined earlier, this may misrepresent the fuel costs along the route and hence may lead to suboptimal solutions.

9.6 Selected Results for Tankers and Bulk Carriers (Fixed Route)

In addition to VLCCs (tankers over 250,000 DWT), Gkonis and Psaraftis (2012) developed speed optimization models for several other tanker classes, such as Suezmax (120,000–200,000 DWT), Aframax (80,000–120,000 DWT), Panamax (65,000–80,000 DWT) & product tankers (above 10,000 DWT), as well as Liquefied Natural Gas (LNG) and Liquefied Petroleum Gas (LPG) ships. Also Kapetanis et al. (2014) developed similar models for Handymax bulk carriers (54,000 DWT). In this section we reproduce some of the results of these models, referring the reader to the above papers for more details.

Figure 9.6 depicts the effect of varying freight rates and fuel prices on annual CO_2 emissions for a specific VLCC running the route Ras Tanura-Yokohama. It can be seen that as the freight rate level decreases from WS120 to WS60, emissions decrease by 29–64 %, depending on the fuel price (higher reductions for higher fuel prices). This sharp reduction in emissions is of course due to speed reduction.

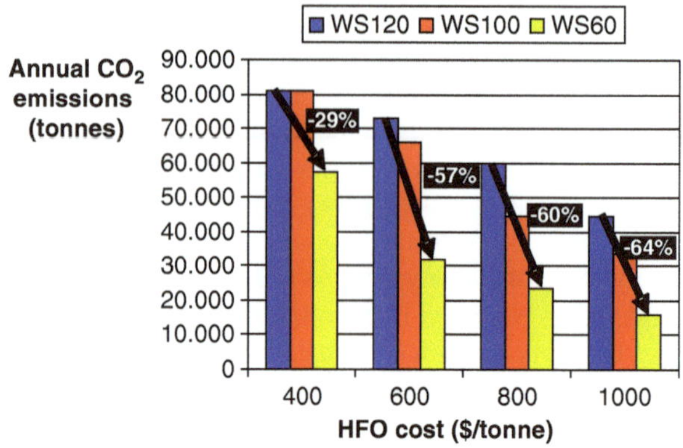

Fig. 9.6 Reduction in annual CO_2 emissions (single VLCC) for a 50 % reduction in freight rates. *Source*: Gkonis and Psaraftis (2012)

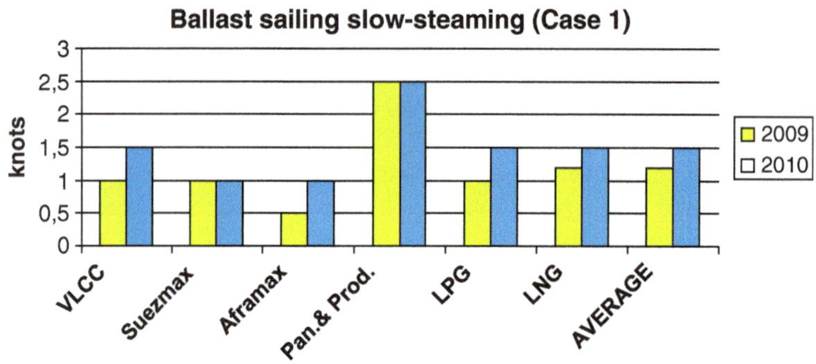

Fig. 9.7 Difference between laden and ballast speeds in Case 1 for several tanker types. *Source*: Gkonis and Psaraftis (2012)

Figure 9.6 can also be used to assess the effect of a levy (or tax) on ship emissions. If the fuel price is (say) 600 $/tonne and a levy of 200 $/tonne is applied on bunker fuel, the reduction of annual CO_2 emissions can be computed from the figure (for WS120 the drop is from approximately 73,000 to 60,000 tonnes a year, or some 18 %). This can attest to the usefulness a Market Based Measure (MBM) can have on emissions reduction (more on MBMs in Chap. 8).

Figure 9.7 shows for each tanker type examined in Gkonis and Psaraftis (2012) and for the reference years 2009 and 2010, how much slower is the ship's ballast speed versus its laden speed, assuming that laden speed is (per chartering agreement) constrained to about 90 % of MCR speed ± 1 knot, and that only the ballast speed is allowed to be free (this is denoted as Case 1). The resulting slower-steaming in ballast is on average of the order of 1.5 knots, but can vary from 0.5 to 2.5 knots.

In addition to Case 1, Case 2 is the scenario when both speeds, laden and ballast, are free to be optimized. In Fig. 9.8, we see (again for each tanker category and for both 2009 and 2010), the difference between the Case 2 laden speed and the Case 1 laden speed. One can see that if laden speed is allowed to be free (Case 2), that

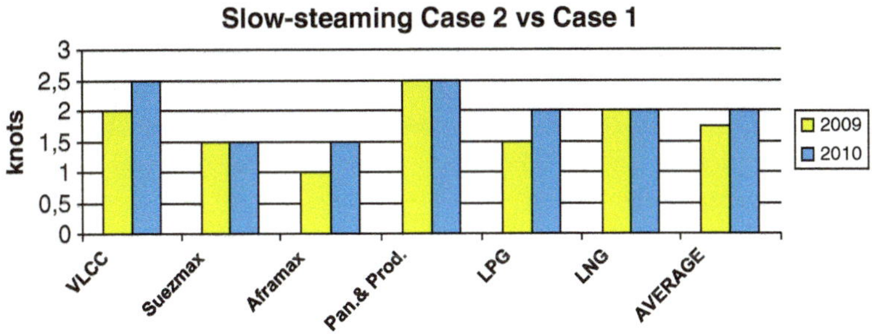

Fig. 9.8 Difference between Case 2 vs. Case 1 laden speeds for several tanker types. *Source*: Gkonis and Psaraftis (2012)

speed is generally lower than the laden speed of Case 1, and the difference is of the order of 2 knots on the average.

Kapetanis et al. (2014) performed a similar analysis for Handymax bulk carriers and investigated the effect of a bunker levy on optimal speed and emissions. As per Chap. 8, one of the effects of a bunker levy would be speed reduction. Among other results, they found that a \$200/tonne levy on bunker fuel would reduce laden speed from 14 to 12 knots and CO_2 emissions by 9.1 % for the Handymax world fleet (2,119 ships in 2010), even though 214 more ships would be needed to produce the same amount of transport work.

9.7 Combining Speed and Routing Decisions

9.7.1 General Considerations

Speed optimization can be extended into combined ship routing and speed scenarios. A number of papers in the literature have looked at such combined scenarios, see for instance Hvattum, Norstad, Fagerholt, and Laporte (2013) and Fagerholt and Ronen (2013), among others. The considerations of Chap. 7 of this book on green vehicle routing can be considered as a parallel here, although obviously the cost functions in a maritime setting are very different from those in a road setting.

In the following we examine combined single-ship scenarios in which the fuel consumption function depends on both ship speed and payload and in which fuel price, charter rate and inventory costs are also taken onboard. By increasing order of complexity, these scenarios include (see Psaraftis and Kontovas (2014) for more details):

- Fixed-route scenarios: A ship going from port A to port B, or even on a multiple leg route in which the sequence of port visits is already determined at a higher level, but ship payload varies along the route.

- Feeder scenarios: A feeder ship collecting cargoes from several ports and bringing them into a hub port or vice versa.
- Combined pickup and delivery scenarios: A ship picking up cargoes from distinct origins and delivering them to distinct destinations. The route and sequence of pickups and deliveries has to be determined, along with the ship speed at each leg of the route.

The latter scenario is actually a generalized version of the feeder scenario and includes several sub-scenarios itself, depending on whether each port has one or multiple pickup cargoes, to be delivered to one or several delivery ports.

Whatever the scenario, assume we are given a set of ports $N = \{0, 1, 2, \ldots, n\}$. Inter-port distances are known and equal to s_{ij} $(i \in N, j \in N)$, in nautical miles. Also we are given an origin/destination (O/D) matrix $[d_{ij}]$, representing the weight of cargo that has to go from port i to port j $(i \in N\backslash 0, j \in N\backslash 0, i \neq j)$, in thousands of tonnes. This matrix is not necessarily symmetric. We assume that the set of cargoes is fixed and that each cargo is considered a distinct commodity and cannot be split.

In all scenarios the ship is assumed to be initially located at port *0* (home port), and has to: (a) pick up from each port the cargoes destined to other ports, (b) deliver to each port the cargoes originating from other ports, and, optionally, depending on the scenario, (c) return to port 0. Ship capacity is Q and cannot be exceeded. It is assumed that $Q \geq max_{(i,\ j)}\ d_{ij}$, otherwise the problem is infeasible.

In all scenarios we need to decide on the appropriate sailing speeds for each leg of the route, as well as the route itself, if the latter is not fixed.

The chartering context assumed is that of a *time charter*, and the assumption is that the charterer of the ship is also the cargo owner. The charterer would like to minimize the total cost of the trip, which has the following three components: (a) fuel cost, (b) time charter cost, and (c) cargo inventory cost, as further elaborated below.

Fuel cost: Since in a time charter the charterer pays for the fuel, a basic tradeoff for the charterer is whether he should complete the trip as soon as possible, so as to reduce the charter paid to the ship owner (see below), or go slower so as to reduce fuel cost. Fuel is assumed to be purchased at a known fuel price of P_{FUEL} ($/tonne). The default scenario ignores port-related costs to be borne by the charterer, even though including these costs is a straightforward extension (see Psaraftis and Kontovas (2014)).

The daily at sea fuel consumption of the ship sailing from i to j is equal to $f(v,w)$ (tonnes/day), which is assumed a known function of the ship's speed v and payload w from i to j $(0 \leq w \leq Q)$. In-port fuel costs are assumed proportional to overall total port residence time, but as the latter is a constant proportional to total cargo moved, they can be ignored. In general, different speeds can be chosen for different legs of the route, so long as they are within the speed window $[v_{LB}(w), v_{UB}(w)]$, where $v_{LB}(w)$ and $v_{UB}(w)$ are lower and upper bounds (respectively) on the speed.

Time charter cost: In a time charter, the charterer pays to the ship owner a known freight rate of F ($/day), with F being an exogenous variable mainly determined by market conditions. It can be high in boom periods or low in depressed market periods. It is assumed that the time charter ends with the termination of the route and that the value of F is independent of charter duration and is agreed upon before the voyage commences.[4]

Cargo inventory cost: The third component of the cost that we assume the charterer bears is the inventory cost of the cargo. Per earlier considerations, we consider this cost irrespective of whether or not the charterer is the cargo owner. In addition to the per unit volume and per unit time in-transit inventory cost of β, as defined earlier, we assume that the per unit volume and per unit time cargo inventory cost of the cargo awaiting to be picked up at the port of origin is equal to α (cost accrues from time 0 until cargo is on the ship). Both α and β are known constants ($/tonne/day), and both are non-negative.

Coefficient α may be different from β for various reasons. For instance, the case $\alpha = 0$ assumes that cargo is available at the loading port in a 'just-in-time' fashion and related waiting or delay costs are zero. Also, these costs would generally depend on whether the cargo is at the origin's warehouse or inside the ship. The case $\alpha = \beta = 0$ means that inventory costs are insignificant or are ignored altogether.

It can be seen that even for each of the scenarios described earlier, several variants of the problem may exist, depending on the objective function. It turns out that these variants can be defined by an appropriate choice of the inputs.

The **minimum trip time problem** is tantamount to setting $P_{FUEL} = \alpha = \beta = 0$ and leaving F as the only nonzero cost coefficient.

At the other extreme, the **minimum emissions problem** is tantamount to setting $\alpha = \beta = F = 0$ and leaving P_{FUEL} as the only nonzero cost coefficient.

It is important to realize that different objective functions will generally produce different solutions, as will be seen in some examples that will be presented in the sections that follow.

9.8 Decomposition Property

Whatever the scenario, be it fixed route or flexible route, a property of the optimal solution is that the speed decision at each route leg can be decomposed from speed and (if applicable) routing decisions at subsequent route legs. Looking at an individual leg of the route, and assuming the ship is at port i and wants to sail to the next port j, the total cost on leg (i, j) is equal to

$$COST(i,j) = (P_{FUEL} f(v,w) + \alpha u + \beta w + F) \cdot \frac{s_{ij}}{v}$$

[4] The assumption that F is independent of charter duration is valid if the charter duration is within a reasonably narrow range. For large variations of time charter duration (e.g. a few months versus a multi-year charter), we expect that F will generally vary with charter duration.

with

v: ship speed during leg
w: ship payload during leg
u: total weight of cargo not yet picked up while ship sails on leg

This cost can be minimized with respect to speed v. As we can factor out the leg distance s_{ij}, the leg's optimal speed is the solution of the following problem:

$$C^* = \min_{v \in V} \left\{ \frac{P_{FUEL} f(v,w) + au + \beta w + F}{v} \right\}$$

with

$$V = \{v : v_{LB}(w) \leq v \leq v_{UB}(w)\}$$

C^* is the minimum per mile cost of the leg. We note that it is independent of leg distance s_{ij}. In addition, and other than the fact that it depends on variables w and u, which depend on the ship's entire route history up to i, this minimum per mile cost is also independent of either i or j, that is, is independent of which is the route leg under consideration.

It is also important to realize that, in the absence of time windows, the speed decision on leg (i, j) has no repercussion on subsequent routing or speed decisions, as it does not affect the subsequent values of the parameters w or u at port j. The speed decision on this leg depends on the values of w and u at port i, on the fuel price P_{FUEL} and the charter freight rate F. It also depends on coefficients α and β, as well as the function f. Generally, high values for F, α and β and/or low values for P_{FUEL} would induce higher speeds than if the opposite is the case.

This means that, provided we know the parameters w and u at port i, the speed optimization problem for any specific leg can be solved in a stand-alone mode, the solution method depending on the form of function f. If a general form is given, the problem can be solved by complete enumeration, perhaps over a finite set of discrete speed values. As these results are not leg-specific, all of these calculations can be carried out in advance only once, and the results can even be tabulated in an appropriate parametric form and be ready for subsequent use. We note that this is true independent of the algorithm that is used for the routing part, be that exact or heuristic.

We can also explore special cases if a mathematical form is given for function f. Then a closed-form solution can be given.

As an example, assume that $f(v,w) = g(p + v^q)(w + A)^{2/3}$
with g, p, q and A known constants.

Then we want to minimize with respect to v the function

$$H(v) = \frac{P_{FUEL}\left(g(p+v^q)(w+A)^{2/3}\right) + \alpha u + \beta w + F}{v} = k_1 v^{q-1} + \frac{k_2}{v}$$

with $k_1 = P_{FUEL} g(w+A)^{2/3}$
and $k_2 = P_{FUEL} g p(w+A)^{2/3} + \alpha u + \beta w + F$
Let $v^* = \left(\frac{k_2}{k_1(q-1)}\right)^{1/q}$
(this is the speed that sets the first derivative of $H(v)$ equal to zero)

Note that

$$\frac{k_2}{k_1} = \frac{P_{FUEL} g p(w+A)^{2/3} + \alpha u + \beta w + F}{P_{FUEL} g(w+A)^{2/3}} = p + \frac{\alpha u + \beta w + F}{P_{FUEL} g(w+A)^{2/3}} \tag{9.5}$$

Then if $v_{LB}(w) \leq v^* \leq v_{UB}(w)$, $v_{OPT} = v^*$
If $v^* < v_{LB}(w)$, $v_{OPT} = v_{LB}(w)$
If $v^* > v_{UB}(w)$, $v_{OPT} = v_{UB}(w)$

For the minimum emissions (or minimum fuel consumption) speed, one can set $\alpha = \beta = F = 0$ and $H(v)$ is as follows:

$$\begin{aligned} H(v) &= \frac{CARB \cdot P_{FUEL}\left(g(p+v^q)(w+A)^{2/3}\right)}{v} \\ &= CARB \cdot P_{FUEL}\left[g\left(\frac{p}{v} + v^{q-1}\right)(w+A)^{2/3}\right] \end{aligned} \tag{9.6}$$

with *CARB* being the "carbon coefficient", tonnes of emissions per tonne of fuel burned. For CO_2 emissions and fossil fuels, *CARB* is between 3.02 and 3.11.

In this case $\frac{k_2}{k_1} = p$ (=0 if f is cubic) and $v^* = (p(q-1))^{1/q}$

For minimum emissions, quite likely $v_{OPT} = v_{LB}(w)$ (this is surely so if f is cubic).

From the above it can be seen that F and other input parameters such as P_{FUEL}, α and β can influence the speed decision at each leg. In particular, the optimal speed is a non-decreasing function of α, β and F, and a non-increasing function of P_{FUEL}. High rates, expensive cargoes and cheap fuels will induce higher speeds than low rates, cheaper cargoes and more expensive fuels.

As a parenthesis we note that such a property is also valid in a multiple ship setting. If which ship serves which set of cargoes and which route is known, and if the objective is the same as above for the fleet as a whole, a similar speed selection rationale should be applied for each of the ships in the fleet.

Table 9.3 Interport distances (nautical miles)

i\j	0	1	2	3
0	–	200	180	360
1	200	–	160	180
2	180	160	–	200
3	360	180	200	–

9.8.1 Freight Rate and Other Input Parameters May Influence the Routing Decision

What is less obvious is that input parameters such as the above may also influence the routing decision. This is indeed the case and it can be shown by a rudimentary example as follows.

A cargo ship of lightship weight equal to $A=5$ and capacity equal to $Q=11$,[5] loads two cargoes of sizes 10 and 1 (all sizes in 1,000 tonnes) at hypothetical port 0, and has to deliver them to hypothetical ports 1 and 2 respectively, and then proceed to port 3 on ballast. Interport distances are given by Table 9.3.

For simplicity assume port dwell times are zero. Note that the route of the ship in this example is an open path as the ship does not return to port 0, but this causes no loss of generality as the path and tour problems are reducible to one another (or one could assume that the ship after visiting port 3 returns to port 0).

Assume that daily fuel consumption (in tonnes) is equal to $FC = kv^3(w+A)^{2/3}$, where v is the ship speed, w is the payload and k is a constant such that at full capacity and at a speed of 14 knots fuel consumption is 30 tonnes/day. For simplicity also assume that the ship's maximum and minimum speeds are 14 and 8 knots respectively, and are independent of payload. Assume finally that $P_{FUEL}=\$600$/tonne and that $\alpha=\beta=0$ (ignore cargo inventory costs).

In case the ship wants to minimize total emissions (or equivalently minimize total fuel consumed or total fuel cost), it is straightforward to see that all legs should be sailed at the minimum speed (8 knots) and that the optimal route is 0-1-2-3. This is so even though total distance sailed (560 nautical miles) is longer than that of the alternative route 0-2-1-3 (520 nautical miles). The reason that 0-1-2-3 is better than 0-2-1-3 is because in 0-1-2-3 the heavier cargo is delivered first, which makes the ship consume less fuel in subsequent legs (and in total). Table 9.4 shows these calculations.

However, if the objective is to minimize total cost, including cost paid for chartering the ship at a rate of F (\$/day), then if F is high enough the ship would follow the shorter route 0-2-1-3, even though in this case the heavier cargo

[5] In terms of ship size, this corresponds roughly to a feeder containership of about 1,000 TEU capacity. It could also be a product carrier or a small bulk carrier.

Table 9.4 Minimum emissions solution (optimal speed = 8 knots)

	Fuel consumption (tonnes)	Fuel cost ($)	Trip time (days)
Route 0-1-2-3			
Leg 0–1 (200 nm)	5.83	3,499	1.04
Leg 1–2 (160 nm)	2.43	1,455	0.83
Leg 2–3 (200 nm)	2.69	1,611	1.04
Total (560 nm)	10.94	6,565	2.92
Total CO_2 emissions (tonnes)	34.02		
Route 0-2-1-3			
Leg 0–2 (180 nm)	5.25	3,149	0.94
Leg 2–1 (160 nm)	4.47	2,681	0.83
Leg 1–3 (180 nm)	2.42	1,450	0.94
Total (520 nm)	12.13	7,280	2.71
Total CO_2 emissions (tonnes)	37.72		

Table 9.5 Minimum total cost solution

	Optimal speed (knots)	Fuel consumption (tonnes)	Fuel cost ($)	Chartering cost ($)	Total cost ($)	Trip time (days)
Route 0-1-2-3						
Leg 0–1 (200 nm)	10.46	9.96	5,977	11,954	17,931	0.80
Leg 1–2 (160 nm)	13.00	6.41	3,845	7,690	11,536	0.51
Leg 2–3 (200 nm)	13.54	7.69	4,616	9,231	13,847	0.62
Total (560 nm)		24.06	14,438	28,876	43,314	1.93
Total CO_2 emissions (tonnes)		74.83				
Route 0-2-1-3						
Leg 0–2 (180 nm)	10.46	8.97	5,379	10,759	16,138	0.72
Leg 2–1 (160 nm)	10.61	7.86	4,714	9,427	14,141	0.63
Leg 1–3 (180 nm)	13.54	6.92	4,154	8,308	12,462	0.55
Total (520 nm)		23.75	14,247	28,494	42,741	1.90
Total CO_2 emissions (tonnes)		73.86				

would be delivered last. Table 9.5 summarizes the cost components for both routes if $F = \$15,000$/day. The table also shows the optimal ship speed in all legs of the route.

9.8.2 *Multiple Optimal Speeds*

If ship payload varies along the ship's route, optimizing ship speed at each leg of the route is better than finding a single optimal speed, the same for all legs. This is of

Table 9.6 Results of the variable speed scenario

Leg	Distance (nm)	Payload (000 tonnes)	Speed (knots)	Fuel cost ($)	Charter cost ($)	Total cost ($)	Trip time (days)
0–1	396	0	13.54	9,139	18,278	27,417	1.22
1–2	165	5	11.61	4,442	8,884	13,326	0.59
2–3	191	6	11.36	5,252	10,504	15,756	0.70
3–4	201	8	10.95	5,736	11,472	17,208	0.76
4–5	508	11	10.46	15,182	30,364	45,545	2.02
Total	1,461			39,751	79,502	119,253	5.30

Adapted from Psaraftis and Kontovas (2014)

course to be expected, as the feasible solution space of the single speed case is a subset of the feasible solution space of the multiple speed case.

Consider a fixed-route situation with the ship of the previous scenario assumed to be visiting, *in this order*, ports 0, 1, 2, 3, 4 and 5.

Assume the ship starts empty at port 0 and has to collect cargo shipments of sizes 5,000, 1,000, 2,000 and 3,000 tonnes at ports 1, 2, 3, and 4 respectively and deliver all of them to port 5. Inter-port distances for legs (0,1), (1,2), (2,3), (3,4) and (4,5) are respectively 396, 165, 191, 201 and 508 nautical miles. As before, assume that P_{FUEL} is $600/tonne, that F is $15,000/day and that port dwell times and inventory costs can be ignored ($\alpha = \beta = 0$).

Note that in this scenario, if the fuel consumption function was assumed independent of ship payload, the ship's computed optimal speed would be the same on each leg of the route. However, with a payload-dependent fuel consumption function, different speeds on each leg would generally be warranted.

Table 9.6 shows the results of the variable speed scenario.

A pertinent question is, with the same fuel consumption function, if for whatever reason the ship is to keep the same speed along the route, can we at least find the common speed that minimizes total cost? It turns out that this speed is 11.375 knots, as shown in Fig. 9.9.

Table 9.7 shows detailed results of this scenario.

In comparing Tables 9.6 and 9.7, and as expected, the results of Table 9.6 are better for the objective to be optimized (total cost in this case). But it happens that in this instance better results are also obtained with respect to total fuel cost, total charter cost and total trip time. Note also that the single optimal speed (11.375 knots) is lower than the average speed of the multiple optimal speed scenario (11.48 knots). The difference in total costs depends on the scenario. For this one, it is not that pronounced.

If the charter rate *F* is higher, optimal speeds will tend to increase, and this is true for both the single optimal speed and for the multiple optimal speed scenarios. Figure 9.10 shows such a behavior, by plotting the optimal (single) speed as a function of the charter rate.

Figure 9.10 captures a typical market behavior in shipping: ships tend to speed up when the market is up, and slow down when the market is down. Also it can be

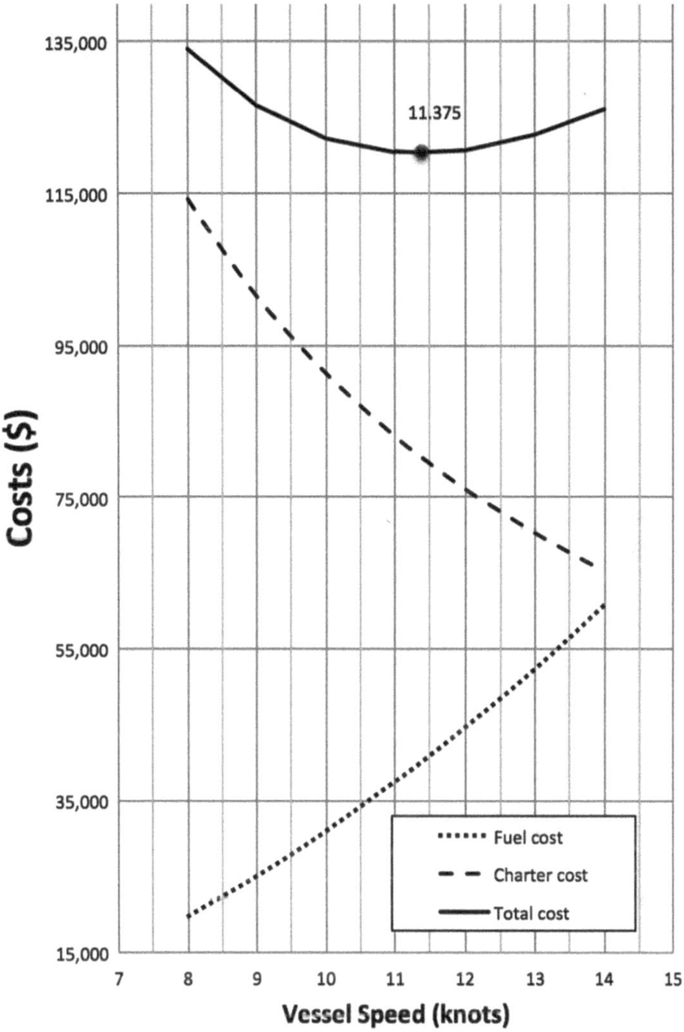

Fig. 9.9 Fuel, Charter and Total costs as functions of vessel speed. *Source*: Psaraftis and Kontovas (2014)

seen that above or below certain charter rates, the speed hits its upper or lower bounds respectively.

A similar behavior also pertains to variations in fuel price. For this particular form of fuel consumption function, keeping the ratio of fuel price to charter rate constant would result in the same speeds.

Table 9.7 Results of the fixed speed scenario

Leg	Distance (nm)	Payload (000 tonnes)	Speed (knots)	Fuel cost ($)	Charter cost ($)	Total cost ($)	Trip time (days)
0–1	396	0	11.375	6,449	21,758	28,207	1.45
1–2	165	5	11.375	4,266	9,066	13,332	0.60
2–3	191	6	11.375	5,262	10,495	15,756	0.70
3–4	201	8	11.375	6,190	11,044	17,233	0.74
4–5	508	11	11.375	17,966	27,912	45,878	1.86
Total	1,461			40,132	80,275	120,407	5.35

Adapted from Psaraftis and Kontovas (2014)

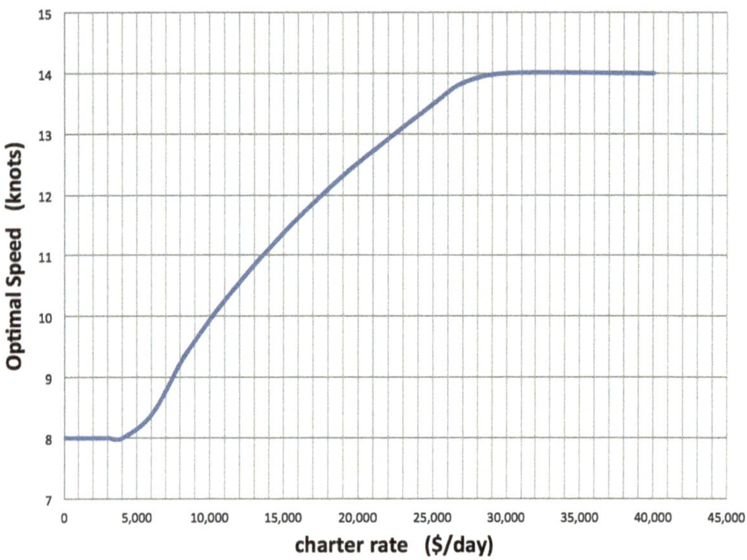

Fig. 9.10 Optimal speed as a function of the charter rate. *Source*: Psaraftis and Kontovas (2014)

The above results also tend to confirm those of Gkonis and Psaraftis (2012), to the effect that, in the absence of constraints on laden speed, if one wishes to reduce fuel costs and by extension emissions, it is better to apply speed reduction in the laden condition than in the ballast condition. Extending this argument, the more loaded the ship is, the lower its speed should be. In practice however, the opposite is often the case, as many ships maintain a constant speed and tend to sail faster when laden than in ballast. This behavior can be explained if there are contractual obligations or other constraints that force the ship to sail at the laden condition faster than in the ballast condition, or if cargo inventory costs are taken into account. This point is further elaborated next.

Table 9.8 Variation of optimal speed with value of cargo

Value of cargo ($/tonne)		Payload (000 tonnes)	0	5,000	10,000	15,000	20,000	25,000
			Speed (knots)					
Leg	0-1	0	13.54	13.54	13.54	13.54	13.54	13.54
	1-2	5	11.61	12.12	12.58	13.02	13.43	13.81
	2-3	6	11.36	11.96	12.49	12.99	13.45	13.88
	3-4	8	10.95	11.70	12.36	12.96	13.51	14.00
	4-5	11	10.46	11.42	12.24	12.96	13.61	14.00
Fuel cost ($)			39,751	44,433	48,808	52,945	56,890	59,854
Charter cost ($)			79,502	75,324	72,136	69,580	67,461	65,996
Inventory cost ($)			0	13,542	25,480	36,310	46,318	56,189
Total cost ($)			119,253	133,299	146,424	158,835	170,669	182,039
CO_2 emitted (tonnes)			206.04	230.31	252.99	274.43	294.88	310.24
Trip time (days)			5,30	5,02	4,81	4,64	4,50	4,40

Adapted from Psaraftis and Kontovas (2014)

9.8.3 Expensive Cargoes Sail Faster and Induce More CO_2

If we take cargo inventory costs into account in the previous example, optimal per leg speeds will change. The more expensive the cargo, the higher the optimal speed will be. If we assume that $\alpha = 0$ (just-in-time availability of cargoes at loading ports) and $\beta = PR/365$ with P the CIF value of the cargo and R the cargo owner's cost of capital, Table 9.8 exhibits the optimal speeds per leg for various values of the cargo, assuming $R = 3\ \%$. The zero value case corresponds to the case that cargo inventory costs are not factored in. The table also shows all cost components, total tonnes of CO_2 emitted and trip time in each case.

One can observe that, with the exception of the first leg, which is in ballast, all other legs are sailed at a higher speed for more expensive cargoes. In fact, even though there is an initial downward trend in speed along the route as the ship becomes more heavy with cargoes loaded on to it, above a certain value of cargo (about $15,000/tonne here) this trend is reversed and speed increases with payload, hitting its upper bound of 14 knots in the last two legs of the trip if the value of the cargo is $25,000/tonne. Further increases in the value of the cargo would set all leg speeds (except that of the first leg) to this upper bound.

One can also observe that expensive cargoes induce more CO_2, as they encourage higher speeds for the ship.

9.8.4 Sailing the Minimum Distance Route at Minimum Speed May Not Minimize Emissions

In the quest for environmentally optimal solutions, one might assume that if the minimum distance route is sailed at the minimum possible speed in all legs, this

Table 9.9 Interport distances (nautical miles)

i\j	0	1	2	3
0	–	255	175	10
1	255	–	200	250
2	175	200	–	170
3	10	250	170	–

Table 9.10 Cargo O/D matrix [d] (1,000 tonnes)

i\j	1	2	3
1	–	5	3
2	2	–	4
3	11	1	–

Table 9.11 Minimum trip time solution

Port stop	Pickup and delivery operations	Next leg	Payload at beginning of leg (000 tonnes)	Speed (knots)	Distance (nm)	Trip time (days)
0	–	0–3	0	14.00	10	0.03
3	P31	3–1	11	14.00	250	0.74
1	D31, P12, P13	1–3	8	14.00	250	0.74
3	D13, P32	3–2	6	14.00	170	0.51
2	D12, D32, P21, P23	2–1	6	14.00	200	0.60
1	D21	1–3	4	14.00	250	0.74
3	D23	3–0	0	14.00	10	0.03
0	–	–	–	–	–	
Total					1,140	3.39

Source: Psaraftis and Kontovas (2014)

would minimize emissions. After all, daily emissions are an increasing function of ship speed, and more days at sea would seem to imply more emissions. However, it turns out that this is not necessarily the case, as shown in the rudimentary example below, involving a pickup and delivery scenario.

Assume a 4-port problem (the home port 0 plus 3 other ports) with the distance matrix given by Table 9.9 as follows:

Also assume an asymmetric O/D table for six cargoes to be transported among ports 1–3 as given by Table 9.10:

We again assume the same feeder ship of the previous examples. The ship starts and ends at port 0, and has to visit the three ports as many times as necessary in order to carry all cargoes as shown in the O/D table. Note that one of the cargoes (from port 3 to port 1) is of size equal to the capacity of the ship. In this example we ignore cargo inventory costs, meaning that $\alpha = \beta = 0$.

If the objective is minimum trip time (this is achieved if we set $P_{FUEL} = 0$), all legs are sailed (as expected) at the maximum speed of 14 knots, and the ship makes a total of 6 port calls (once at port 2, twice at port 1 and three times at port 3) as follows (Table 9.11):

Table 9.12 Minimum emissions solution

Port stop	Pickup/delivery operations	Next leg	Payload at beginning of leg (000 tonnes)	Speed (knots)	Distance (nm)	Trip time (days)
0	–	0–3	0	8.00	10	0.05
3	P31	3–1	11	8.00	250	1.30
1	D31, P12	1–2	5	8.00	200	1.04
2	D12, P21	2–1	2	8.00	200	1.04
1	D21, P13	1–3	3	8.00	250	1.30
3	D13, P32	3–2	1	8.00	170	0.89
2	D32, P23	2–3	4	8.00	170	0.89
3	D23	3–0	0	8.00	10	0.05
0	–	–	–	–	–	
Total					1,260	6.56

Source: Psaraftis and Kontovas (2014)

Table 9.13 Solutions for non-zero fuel price and varying freight rates

Port stop	Pickup & delivery operations	Next leg	Payload at beginning of leg (000 tonnes)	Speed (knots) F = \$5,000 per day	Speed (knots) F = \$20,000 per day	Trip time (days) F = \$5,000 per day	Trip time (days) F = \$20,000 per day
0	–	0–3	0	9.39	14.00	0.04	0.03
3	P31	3–1	11	8.00	11.51	1.30	0.91
1	D31, P12, P13	1–3	8	8.00	12.05	1.30	0.86
3	D13, P32	3–2	6	8.00	12.51	0.89	0.57
2	D12, D32, P21, P23	2–1	6	8.00	12.51	1.04	0.67
1	D21	1–3	4	8.24	13.08	1.26	0.80
3	D23	3–0	0	9.39	14.00	0.04	0.03
0	–	–	–	–	–	–	–
Total						5.87	3.87

Source: Psaraftis and Kontovas (2014)

In Tables 9.11, 9.12 and 9.13, by "Pxy" we mean "at port x pick up cargo destined to port y," and by "Dxy" we mean "at port y deliver cargo originating from port x."

In this case total distance traveled is also minimized and equal to 1,140 nautical miles, and total CO_2 emitted is 260 tonnes. Total trip time is equal to 3.39 days. This solution is independent of F, so long as F is not zero.

At the other extreme of this example is if we examine the minimum emissions (or minimum fuel consumption) solution. We can do this by setting $F = 0$ and assuming any nonzero fuel price. If this is the case, the ship will make 7 port calls instead of 6 (twice at ports 1 and 2 and three times at port 3), and will sail all legs at the minimum speed of 8 knots. The solution will be as follows (Table 9.12):

Total distance traveled in this case will be 1,260 nautical miles and total trip time will be 6.56 days, both higher than before. But total CO_2 emitted will only be

80 tonnes, much lower. Obviously the lower emissions are mainly due to the lower speed. However, it is interesting to note that the amount of CO_2 emitted in this case is lower than the 84.90 tonnes of CO_2 that would be emitted if the ship had sailed the minimum distance route of Table 9.11 at the minimum speed of 8 knots (for a cubic fuel consumption function, total fuel consumed, and hence CO_2 produced, are proportional to the square of the speed, everything else, including payloads at each leg, being equal. Then $260(8/14)^2 = 84.90$).

The reason that sailing the minimum distance route at minimum speed is suboptimal with respect to emissions is that it involves more legs in which the ship is more laden as compared to the case it sails the alternate, longer route. A heavier load profile results in higher fuel consumption (and emissions) overall, even though the route is shorter. So in this case what would intuitively seem like an optimal policy is actually suboptimal.

Other solutions may be produced for different values of the input data. Table 9.13 shows two cases where $P_{FUEL} = \$600$/tonne (in both cases) and F is either \$5,000/day or \$20,000 day. Both cases produce the same optimal route as that of Table 9.11, but speeds along the legs of the route will vary for different values of F.

As expected, the ship goes faster when F is higher, with the lower speed bound active in 4 legs of the $F = \$5,000$/day case and the upper speed bound active in 2 legs of the $F = \$20,000$/day case.

In all of the above cases the combined speed-routing problem is solved by Dynamic Programming, as an extension of the approach of Psaraftis (2011). Details can be found in Psaraftis and Kontovas (2014).

9.9 Conclusions and Possible Extensions

This chapter has examined speed optimization in maritime transportation from various angles, including its interaction with route optimization. It has confirmed, among other things, that solutions for optimal environmental performance are not necessarily the same as those for optimal economic performance. Also policies that may seem at first glance optimal from an environmental viewpoint may actually be suboptimal. As a private operator would most certainly choose optimal economic performance as a criterion, if policy-makers want to influence the operator in his decision so as to achieve results that are good from a societal point of view, they could play with parameters that would internalize the external costs of CO_2 produced and move the solution closer to what is deemed more appropriate for the environment and for the benefit of society.

To that effect, a levy on bunker was seen as something that can be used to produce such a result, and in fact induce a lower speed and therefore reduced emissions. This confirms the assertion of Chap. 8 of this book, to the effect that the GHG Fund is the only among the MBM proposals submitted to the IMO that can have such impact on slow steaming in an 'automatic' fashion. This means with no additional information needed for the person responsible for the speed decision (other than the information on the fuel price including the levy).

A related policy issue is that mandating direct speed limits. If emissions can be reduced by reducing speed, can someone achieve this desirable outcome by imposing speed limits? This is an argument that is heard frequently these days. Among various lobbying groups, the Clean Shipping Coalition, a Non-Governmental Organization, advocated at IMO/MEPC 61 that "*speed reduction should be pursued as a regulatory option in its own right and not only as possible consequences of market-based instruments or the EEDI.*" However, that proposal was rejected by the IMO. In spite of this decision, lobbying for speed limits has continued by CSC and other groups. Recipients of this lobbying activity have included the IMO and the European Commission.

Our own position on this issue is not in favor of speed limits. It is clear that slow steaming and speed limits are two different things, as the first is a voluntary response and the second is a mandated measure. If the speed limit is above the optimal speed that is voluntarily chosen, then it is superfluous. If it is below, it will cause (perhaps massive) distortions in the market, particularly in boom periods, and costs that may exceed the benefits of speed reduction. Possible side-effects include, among others, building more ships to match demand, with possible increase of emissions during shipbuilding and recycling, increasing cargo inventory costs, producing more GHGs if low-powered ships are forced to speed up in boom periods, and having adverse implications on ship safety.

We have seen no comprehensive analysis of the possible market distortions of a speed limit. But in a recent paper, Cariou and Cheaitou (2012) investigate policy options contemplated by the European Commission and compare speed limits versus a bunker levy as two measures to abate GHGs, with a scenario from the container trades. They conclude that the former measure is counterproductive because it may ultimately generate more emissions and incur a cost per tonne of CO_2 which is more than society is willing to pay and because it is sub-optimal compared to results obtained if an international bunker-levy were to be implemented.

We have also seen in previous sections that if laden leg ship speeds are not constrained by charter party speed clauses, lower emissions are likely to occur. Conversely, a charter party agreement specifying a prescribed speed, explicitly or implicitly, might entail significant costs, both in terms of additional fuel (which is a private cost matter) and in terms of additional emissions (which is a cost to society). Our analysis strongly suggests that regulatory action to prevent such clauses in charter party agreements could very well be worth looking into as a policy alternative.

Psaraftis and Kontovas (2015), among other things, provide a discussion on the possible impact of slow steaming on port operations. If a port is congested, it would clearly make no sense to sail there at full speed, wasting money on fuel and producing emissions that can be avoided if ship speed were slower. A recent initiative is the so-called 'Virtual Arrival', which has been employed firstly by tankers in order to manage the vessels' arrival time based on the experience of congestion at some discharging ports. This initiative recognizes known inefficiencies in the supply chain, such as waiting to discharge because of port delays and reduces fuel consumption and, consequently, emissions by implementing a mutually-agreed reduction in a vessel's speed in order to achieve an agreed arrival time at a port. This scheme in order to work needs a mutual agreement by both the owner and charterer to agree a speed to meet the terminal booking that maximizes fuel efficiency and minimizes port waiting time. To ensure the accuracy and independence of the calculations and to avoid the risk of disputes, it is proposed to use a weather routing analysis company. After the agreement of both parties the ship slows to the economical speed based on the revised arrival time. Once the voyage is completed, demurrage is calculated based on the original plans and bunker savings are spilt between the parties.

'Virtual Arrival' seems profitable especially given the fact that there are indeed serious delays in discharging in some ports in the world. According to the Global Ports Congestion Index (GPCI) and its weekly newsletter publication that provides details on berthing delays at the major coal and ore ports worldwide, the average delay can be as high as 5 days. Obviously there is no point for vessels to steam at full speed when the have to wait a couple of days in order to discharge. Sailing at a slower speed and arriving on time entails benefits both for the owner and the charterer but also for the environment.

In another direction, Magirou, Psaraftis, and Bouritas (2015) have recently developed models that optimize speed in a dynamic and stochastic setting. It was found that for freight rates that depend on a state of the market Markovian random variable, economic speed depends on the market state as well, with increased speed corresponding to good states of the market. Also, the authors and their colleagues have extended the combined speed and routing approach of Psaraftis and Kontovas (2014) into a multiple ship scenario. The results of this analysis were still incomplete as this book was being finalized and will be reported in future publications.

Last but not least, and as already mentioned earlier, in the Appendix we update the taxonomy of Psaraftis and Kontovas (2013) and display an amended set of speed models, classified according to a set of criteria.[6]

Acknowledgments Work on this chapter has been supported in part by various sources, including the Lloyd's Register Foundation (LRF) in the context of the Centre of Excellence in Ship Total Energy-Emissions-Economy at the National Technical University of Athens (NTUA), the authors' former affiliation. LRF helps to protect life and property by supporting engineering-related education, public engagement and the application of research. This work has also been supported in part by an internal grant at the Technical University of Denmark (DTU).

[6] As this book was being finalized, an unprecedented decrease in oil prices was taking place. However, as charter rates fell too, a definitive statement on the effect of this development on average ship or fleet speeds was not possible.

Appendix: Taxonomy of Speed Papers, Amended from Psaraftis and Kontovas (2013)

Table has 7 parts, of 7 entries each. Two entries have two references each. Total references: 51

Taxonomy part I

Taxonomy parameter\ paper	Alderton (1981)	Alvarez (2009)	Andersson, Fagerholt, and Hobbesland (2014)	Bausch, Brown, and Ronen (1998)	Benford (1981)	Brown, Graves, and Ronen (1987)	Cariou (2011)
Optimization criterion	Profit	Cost	Cost	Cost	Cost	Cost	Cost
Shipping market	General	Liner	Ro/Ro	Tanker/barge	Coal	Tanker	Container
Decision maker	Owner	Owner	Owner	Owner	Owner	Owner	Owner
Fuel price an explicit input	Yes	Yes	No	Yes	No	No	Yes
Freight rate an input	Input	No	Implicit	No	No	No	No
Fuel consumption function	Cubic	Cubic	General	Unspecified	Cubic	Unspecified	Cubic
Optimal speeds in various legs	Yes	Yes	Yes	No	No	Only ballast	No
Optimal speeds as function of payload	Yes	Yes	Yes	No	No	No	No
Logistical context	Fixed route	Joint routing and fleet deployment	Fleet deployment	Routing and scheduling	Fleet deployment	Routing and scheduling	Fixed route
Size of fleet	Multiple ships	Multiple ships	Multiple ships	Multiple ships	Multiple ships	Multiple ships	Multiple ships
Add more ships an option	Yes	No	Yes	No	No	No	Yes
Inventory costs included	Yes	No	No	No	No	No	No
Emissions considered	No	No	No	No	No	No	Yes
Modal split considered	No	No	No	No	No	No	No
Ports included	Yes	Yes	No	Yes	No	No	No

Taxonomy part II

Taxonomy parameter\ paper	Cariou and Cheaitou (2012)	Chang and Wang (2014)	Corbett, Wang, and Winebrake (2010)	Devanney (2007)	Devanney (2010)	Doudnikoff and Lacoste (2014)	Du, Chen, Quan, Long, and Fung (2011) / Wang, Meng, and Liu (2013a, b)
Optimization criterion	Cost	Cost	Profit	Profit	Cost or profit	Cost	Fuel consumption
Shipping market	Container	General	Container	Tanker	Tanker (VLCC)	Liner	Container
Decision maker	Owner	Owner	Owner	Owner	Owner or charterer	Owner	Owner
Fuel price an explicit input	Yes	Yes	Yes	Yes	Yes	Yes	No
Freight rate an input	No	Yes	Input	Computed	Computed	No	No
Fuel consumption function	Cubic	Cubic	Cubic	Cubic	General	Cubic	Non-linear
Optimal speeds in various legs	No	No	No	Yes	Yes	Yes	Yes
Optimal speeds as function of payload	No	No	No	No	No	No	No
Logistical context	Fixed route	Fixed route	Fixed route	World oil network	Fixed route	Fixed route in SECAs	Berth allocation
Size of fleet	Multiple ships	One ship	Multiple ships	Multiple ships	One ship	Multiple ships	Multiple ships
Add more ships an option	Yes	No	Yes	Yes	Yes	No	No
Inventory costs included	Yes	No	No	Yes	Yes	No	No
Emissions considered	Yes	No	Yes	No	No	Yes	Yes
Modal split considered	No	No	No	No	No	No	No
Ports included	Yes	Yes	No	Yes	No	No	No

Taxonomy part III

Taxonomy parameter\ paper	Eefsen and Cerup-Simonsen (2010)	Faber, Freund, Köpke, and Nelissen (2010)	Fagerholt (2001)	Fagerholt, Laporte, and Norstad (2010)	Fagerholt, Gausel, Rakke, and Psaraftis (2015)	Fagerholt and Ronen (2013)	Gkonis and Psaraftis (2012)
Optimization criterion	Cost	N/A	Cost	Fuel consumption	Cost	Profit	Profit
Shipping market	Container	Various	General	Liner	Ro/Ro	Tramp	Tanker, LNG, LPG
Decision maker	Owner	N/A	Owner	Owner	Owner	Owner	Owner
Fuel price an explicit input	Yes	No	No	No	Yes	No	Yes
Freight rate an input	No	No	No	No	No	Implicit	Input
Fuel consumption function	Cubic	Cubic	Cubic	Cubic	Cubic	General	General
Optimal speeds in various legs	No	No	Yes	Yes	Yes	Yes	Yes
Optimal speeds as function of payload	No	No	No	No	No	No	No
Logistical context	Fixed route	Fixed route	Pickup and delivery	Fixed route	Route & speed selection in SECAs	Pickup and delivery	Fixed route
Size of fleet	Multiple ships	Multiple ships	Multiple ships	One ship	One ship	Multiple ships	Multiple ships
Add more ships an option	Yes	Yes	No	No	No	No	Yes
Inventory costs included	Yes	No	No	No	No	No	Yes
Emissions considered	Yes	Yes	No	Yes	Yes	No	Yes
Modal split considered	No	No	No	No	No	No	No
Ports included	Yes	No	No	No	No	No	Yes

Taxonomy part IV

Taxonomy parameter\paper	Hvattum et al. (2013)	Kapetanis et al. (2014)	Kontovas and Psaraftis (2011)	Lang and Veenstra (2010)	Lindstad, Asbjørnslett, and Strømman (2011)	Lo and McCord (1998)	Magirou et al. (2015)
Optimization criterion	Fuel consumption	Profit	Cost	Fuel costs	Pareto analysis	Fuel consumption	Profit
Shipping market	General	Drybulk	Container	Container	All major ship types	General	General
Decision maker	Owner	Owner	Charterer	owner	Owner	Ship's master	Owner
Fuel price an explicit input	No	Yes	Yes	No	Yes	No	Yes
Freight rate an input	No	Yes	Input	No	No	No	Yes
Fuel consumption function	Convex	General	Cubic	linearized	Cubic	Cubic	Cubic
Optimal speeds in various legs	Yes	Yes	Yes	No	No	N/A	Yes
Optimal speeds as function of payload	No	Yes	Yes	No	Yes	No	No
Logistical context	Fixed route	Fixed route	Fixed route	Vessel arrival planning	Fixed route	Weather routing	Fixed route
Size of fleet	One ship	Multiple ships	Multiple ships	Multiple ships	Multiple ships	One ship	One ship
Add more ships an option	No	Yes	Yes	No	Yes	No	No
Inventory costs included	No	Yes	Yes	No	Yes	No	No
Emissions considered	Yes	Yes	Yes	No	Yes	No	No
Modal split considered	No	No	No	No	No	No	No
Ports included	No	Yes	Yes	Yes	Yes	No	

Taxonomy part V

Taxonomy parameter\ paper	Meng and Wang (2011)	Norlund and Gribkovskaia (2013)	Norstad et al. (2011)	Notteboom and Vernimmen (2010)	Perakis and Papadakis (1989)	Perakis (1985)	Perakis and Jaramillo (1991)
Optimization criterion	Cost	Cost	Cost	Cost	Cost	Cost	Cost
Shipping market	Liner	Offshore supply vessels	Tramp	Container	Tramp	Tramp	Liner
Decision maker	Owner	Owner	Owner	Owner	Owner	Owner	Owner
Fuel price an explicit input	Yes	No	No	Yes	Yes	No	Yes
Freight rate an input	No	No	No	No	No	No	Yes
Fuel consumption function	Cubic	Cubic	Cubic	Unspecified	General	Cubic	Cubic
Optimal speeds in various legs	No	Yes	Yes	No	Yes	No	Yes
Optimal speeds as function of payload	No	No	No	No	No	No	No
Logistical context	Fleet deployment	Set covering	Pickup and delivery	Fixed route	Fleet deployment	Fleet deployment	Fleet deployment
Size of fleet	Multiple ships	Multiple ships	Multiple ships	Multiple ships	Multiple ships	Multiple ships	Multiple ships
Add more ships an option	No	No	No	Yes	No	Yes	Yes
Inventory costs included	No	No	No	No	Yes	No	No
Emissions considered	No	Yes	No	No	No	No	No
Modal split considered	No	No	No	No	No	No	No
Ports included		No	No	Yes	Yes	No	No

Taxonomy part VI

Taxonomy parameter\paper	Perakis and Papadakis (1987a, b)	Perakis and Papadakis (1989)	Psaraftis and Kontovas (2009b)	Psaraftis and Kontovas (2010)	Psaraftis and Kontovas (2014)	Qi and Song (2012)	Ronen (1982)
Optimization criterion	Cost	Time	Cost	Cost	Cost	Fuel consumption	Profit
Shipping market	Tramp	General	Tramp	General	General	liner	Tramp
Decision maker	Owner	Ship's master	Charterer	Charterer	Charterer	Owner	Owner
Fuel price an explicit input	Yes	No	Yes	No	Yes	No	Yes
Freight rate an input	No	No	Input	Input	Input	No	Input
Fuel consumption function	General	N/A	Cubic	General	General	cubic	Cubic
Optimal speeds in various legs	Yes	N/A	Yes	No	Yes	Yes	Yes
Optimal speeds as function of payload	No	No	Yes	No	Yes	No	No
Logistical context	Fleet deployment	Weather routing	Fixed route	Fixed route	Fixed or flexible route	Scheduling	Fixed route
Size of fleet	Multiple ships	One ship	Multiple ships	Multiple ships	One ship	Multiple ships	One ship
Add more ships an option	No	No	Yes	Yes	No	No	No
Inventory costs included	Yes	No	Yes	Yes	Yes	No	No
Emissions considered	No	No	Yes	No	Yes	Yes	No
Modal split considered	No	No	No	Yes	No	No	No
Ports included	Yes	Yes	No	No	No	Yes	Yes

Taxonomy part VII

Taxonomy parameter\paper	Ronen (2011)	Stopford (2009)	Wang and Meng (2012a)	Wang and Meng (2012b)	Wang and Meng (2012c)	Wang et al. (2014)	Yao, Ng, and Lee (2012)
Optimization criterion	Cost	Profit	Cost	Total cost and fuel cost	Cost	Cost	Fuel cost
Shipping market	Container	General	Container	Liner	Liner	Container	Container
Decision maker	Owner	Owner	Owner	Owner	Owner	Owner	Owner
Fuel price an explicit input	Yes	Yes	Yes	Yes	Yes	Yes	Yes
Freight rate an input	No	Input/ computed	No	No	No	No	No
Fuel consumption function	Cubic	Cubic	linearized	cubic	linearized	General	cubic
Optimal speeds in various legs	No	No	Yes	Yes	Yes	Yes	Yes
Optimal speeds as function of payload	No	No	No	No	No	No	No
Logistical context	Fixed route	Fixed route	Scheduling	Scheduling	Scheduling	Schedule design	Bunker fuel management
Size of fleet	Multiple ships	Multiple ships	Multiple ships	Multiple ships	Multiple ships	Multiple ships	Multiple ships
Add more ships an option	Yes	No	No	No	No	No	No
Inventory costs included	No	Yes	No	No	No	Yes	No
Emissions considered	No	No	No	No	No	No	No
Modal split considered	No	No	No	No	No	No	No
Ports included	Yes	Yes	Yes	Yes	Yes	Yes	Yes

References

Aas, B., Gribkovskaia, I., Halskau, Ø., Sr., & Shlopak, A. (2007). Routing of supply vessels to petroleum installations. *International Journal of Physical Distribution & Logistics Management, 37*(2), 164–179.

Agarwal, R., & Ergun, Ö. (2008). Ship scheduling and network design for cargo routing in liner shipping. *Transportation Science, 42*(2), 175–196.

Alderton, P. M. (1981). The optimum speed of ships. *Journal of Navigation, 34*, 341–355.

Alphaliner. (2013, October). Extra and super slow steaming help absorb 7.4 % of fleet. *Alphaliner Weekly Newsletter, 2013*(44).

Alvarez, J. (2009). Joint routing and deployment of a fleet of container vessels. *Maritime Economics & Logistics, 11*, 186–208.

Alvarez, J. F., Tsilingiris, P., Engebrethsen, E. S., & Kakalis, N. M. (2011). Robust fleet sizing and deployment for industrial and independent bulk ocean shipping companies. *INFOR: Information Systems and Operational Research, 49*(2), 93–107.

Andersson, H., Duesund, J. M., & Fagerholt, K. (2011). Ship routing and scheduling with cargo coupling and synchronization constraints. *Computers & Industrial Engineering, 61*(4), 1107–1116.

Andersson, H., Fagerholt, K., & Hobbesland, K. (2014). Integrated maritime fleet deployment and speed optimization: Case study from RoRo shipping. *Computers & Operations Research, 55*, 233–240.

Barrass, C. B. (2005). *Ship design and performance for masters and mates*. Oxford: Butterworth-Heinemann.

Bausch, D., Brown, G., & Ronen, D. (1998). Scheduling short-term marine transport of bulk products. *Maritime Policy and Management, 25*, 335–348.

Benford, H. (1981). A simple approach to fleet deployment. *Maritime Policy & Management, 8*, 223–228.

Brouer, B. D., Alvarez, J. F., Plum, C. E., Pisinger, D., & Sigurd, M. M. (2013). A base integer programming model and benchmark suite for liner-shipping network design. *Transportation Science, 48*(2), 281–312.

Brown, G., Graves, G., & Ronen, D. (1987). Scheduling ocean transportation of crude oil. *Management Science, 33*, 335–346.

Cariou, P. (2011). Is slow steaming a sustainable means of reducing CO_2 emissions from container shipping? *Transportation Research Part D, 16*(3), 260–264.

Cariou, P., & Cheaitou, A. (2012). The effectiveness of a European speed limit versus an international bunker-levy to reduce CO_2 emissions from container shipping. *Transportation Research Part D, 17*, 116–123.

CBO. (2006). *The economic costs of disruptions in container shipments*. Washington, DC: U.S. Congress, Congressional Budget Office.

Chang, C. C., & Wang, C. M. (2014). Evaluating the effects of speed reduce for shipping costs and CO_2 emission. *Transportation Research Part D: Transport and Environment, 31*, 110–115.

Christiansen, M., Fagerholt, K., Nygreen, B., & Ronen, D. (2007). Maritime transportation. In C. Barnhart & G. Laporte (Eds.), *Transportation, handbooks in operations research and management science* (Vol. 14, pp. 189–284). Amsterdam: Elsevier.

Christiansen, M., Fagerholt, K., Nygreen, B., Ronen, D. (2013). Ship routing and scheduling in the new millennium. *European Journal of Operational Research 228*, 467–483.

Christiansen, M., Fagerholt, K., & Ronen, D. (2004). Ship routing and scheduling: Status and perspectives (review). *Transportation Science, 38*(1), 1–18.

Corbett, J., Wang, H., & Winebrake, J. (2010). The effectiveness and costs of speed reductions on emissions from international shipping. *Transportation Research Part D: Transport and Environment, 14*, 593–598.

Cordeau, J.-F., Laporte, G., Legato, P., & Moccia, L. (2005). Models and tabu search heuristics for the berth-allocation problem. *Transportation Science, 39*(4), 526–538.

Devanney, J. W. (2007). *Solving elastic transportation networks*. Center for Tankship Excellence. Retrieved from www.c4tx.org

Devanney, J. W. (2010, October). The impact of bunker price on VLCC spot rates. *Proceedings of the 3rd International Symposium on Ship Operations, Management and Economics*. SNAME Greek Section, Athens, Greece.

Devanney, J. W. (2011). *The impact of charter party speeds on CO_2 emissions*. Center for Tankship Excellence. Retrieved from www.c4tx.org

Doudnikoff, M., & Lacoste, R. (2014). Effect of a speed reduction of containerships in response to higher energy costs in sulphur emission control areas. *Transportation Research Part D: Transport and Environment, 27*, 19–29.

Du, Y., Chen, Q., Quan, X., Long, L., & Fung, R. Y. K. (2011). Berth allocation considering fuel consumption and vessel emissions. *Transportation Research Part E, 47*, 1021–1037.

Eefsen, T., & Cerup-Simonsen, B. (2010, July). Speed, carbon emissions and supply chain in container shipping. *Proceedings of the Annual Conference of the International Association of Maritime Economists, IAME 2010*, Lisbon, Portugal.

Faber, J., Freund, M., Köpke, M., & Nelissen, D. (2010). *Going slow to reduce emissions: Can the current surplus of maritime transport capacity be turned into an opportunity to reduce GHG emissions?* Seas at Risk.

Fagerholt, K. (2001). Ship scheduling with soft time windows—An optimization based approach. *European Journal of Operational Research, 131*, 559–571.

Fagerholt, K., Gausel, N., Rakke, J., & Psaraftis, H. (2015). Maritime routing and speed optimization with emission control areas. *Transportation Research Part C: Emerging Technologies, 52*, 57–73.

Fagerholt, K., Laporte, G., & Norstad, I. (2010). Reducing fuel emissions by optimizing speed on shipping routes. *Journal of the Operational Research Society, 61*, 523–529.

Fagerholt, K., & Ronen, D. (2013). Bulk ship routing and scheduling: Solving practical problems may provide better results. *Maritime Policy & Management, 40*(1), 48–64.

Gkonis, K. G., & Psaraftis, H. N. (2012, October). Modelling tankers' optimal speed and emissions. *Proceedings SNAME 2012 Annual Meeting*, Providence, RI.

Goodchild, A. V., & Daganzo, C. F. (2007). Crane double cycling in container ports: Planning methods and evaluation. *Transportation Research Part B: Methodological, 41*(8), 875–891.

Grønhaug, R., Christiansen, M., Desaulniers, G., & Desrosiers, J. (2010). A branch-and-price method for a liquefied natural gas inventory routing problem. *Transportation Science, 44*(3), 400–415.

Halvorsen-Weare, E. E., & Fagerholt, K. (2011). Robust supply vessel planning. In J. Pahl, T. Reiners, & S. Voß (Eds.), *Network optimization* (pp. 559–573). Heidelberg: Springer.

Halvorsen-Weare, E. E., & Fagerholt, K. (2013). Routing and scheduling in a liquefied natural gas shipping problem with inventory and berth constraints. *Annals of Operations Research, 203*(1), 167–186.

Hsu, C. I., & Hsieh, Y. P. (2005). Direct versus terminal routing on a maritime hub-and-spoke container network. *Journal of Marine Science and Technology, 13*(3), 209–217.

Hvattum, L. M., Norstad, I., Fagerholt, K., & Laporte, G. (2013). Analysis of an exact algorithm for the vessel speed optimization problem. *Networks, 62*(2), 132–135.

Hwang, H.-S., Visoldilokpun, S., & Rosenberger, J. M. (2008). A branch-and-price-and-cut method for ship scheduling with limited risk. *Transportation Science, 42*(3), 336–351.

IMO. (2009). *Second IMO GHG study 2009—doc. MEPC59/INF.10*. London: International Maritime Organization (IMO).

IMO. (2014, June). *Third IMO GHG Study 2014*, Co-authored by Smith, T. W. P., Jalkanen, J. P., Anderson, B. A., Corbett, J. J., Faber, J., Hanayama, S., et al. London: International Maritime Organization (IMO).

Imai, A., Shintani, K., Papadimitriou, S. (2009). Multi-port vs. Hub-and-Spoke port calls by containerships. *Transportation Research Part E 45*, 740–757.

Jetlund, A. S., & Karimi, I. A. (2004). Improving the logistics of multi-compartment chemical tankers. *Computers & Chemical Engineering, 28*(8), 1267–1283.

Kapetanis, G. N., Gkonis, K., & Psaraftis, H. N. (2014, April). Estimating the operational effects of a bunker levy: The case of handymax bulk carriers. *TRA 2014 conference*, Paris, France.

Kontovas, C. A., & Psaraftis, H. N. (2011). Reduction of emissions along the maritime intermodal container chain. *Maritime Policy & Management, 38*(4), 455–473.

Lang, N., & Veenstra, A. (2010). A quantitative analysis of container vessel arrival planning strategies. *OR Spectrum, 32*, 477–499.

Lin, D.-Y., & Liu, H.-Y. (2011). Combined ship allocation, routing and freight assignment in tramp shipping. *Transportation Research Part E: Logistics and Transportation Review, 47*(4), 414–431.

Lindstad, H., Asbjørnslett, B. E., & Strømman, A. H. (2011). Reductions in greenhouse gas emissions and cost by shipping at lower speeds. *Energy Policy, 39*(6), 3456–3464.

Lloyds List. (2009, November 16). CKYH carriers agree to super-slow steaming. *Lloyds List.*

Lo, H. K., & McCord, M. R. (1998). Adaptive ship routing through stochastic ocean currents: General formulations and empirical results. *Transportation Research Part A: Policy and Practice, 32*(7), 547–561.

Maersk. (2013). *Building the world's biggest ship*. Retrieved from http://www.maersk.com/innovation/leadingthroughinnovation/pages/buildingtheworldsbiggestship.aspx

Magirou, E. F., Psaraftis, H. N., & Bouritas, T. (2015). The economic speed of an oceangoing vessel in a dynamic setting. *Transportation Research Part B, 76*, 48–67.

Meng, Q., & Wang, S. (2011). Optimal operating strategy for a long-haul liner service route. *European Journal of Operational Research, 215*, 105–114.

Meng, Q., Wang, S., Andersson, H., & Thun, K. (2013). Containership routing and scheduling in liner shipping: overview and future research directions. *Transportation Science, 48*(2), 265–280.

Mills, J., Donnison, A., & Brightwell, G. (2014). Factors affecting microbial spoilage and shelf-life of chilled vacuum-packed lamb transported to distant markets: A review. *Meat Science, 98* (1), 71–80.

Moccia, L., Cordeau, J. F., Gaudioso, M., & Laporte, G. (2006). A branch-and-cut algorithm for the quay crane scheduling problem in a container terminal. *Naval Research Logistics, 53*(1), 45–59.

Norlund, E. K., Gribkovskaia, I. (2013). Reducing emissions through speed optimization in supply vessel operations, *Transportation Research Part D 23*, 105–113

Norstad, I., Fagerholt, K., & Laporte, G. (2011). Tramp ship routing and scheduling with speed optimization. *Transportation Research Part C, 19*, 853–865.

Notteboom, T. E., & Vernimmen, B. (2010). The effect of high fuel costs on liner service configuration in container shipping. *Journal of Transport Geography, 17*, 325–337.

Perakis, A. N. (1985). A second look at fleet deployment. *Maritime Policy & Management, 12*, 209–214.

Perakis, A. N., & Jaramillo, D. I. (1991). Fleet deployment optimization for liner shipping. Part 1: Background, problem formulation and solution approaches. *Maritime Policy & Management, 18*(3), 183–200.

Perakis, A. N., & Papadakis, N. A. (1987a). Fleet deployment models, part 1. *Maritime Policy & Management, 14*, 127 144.

Perakis, A. N., & Papadakis, N. A. (1987b). Fleet deployment models part 2. *Maritime Policy & Management, 14*, 145–155.

Perakis, A. N., & Papadakis, N. A. (1989). Minimal time vessel routing in a time-dependent environment. *Transportation Science, 23*(4), 266–276.

Powell, B. J., & Perakis, A. N. (1997). Fleet deployment optimization for liner shipping: An integer programming model. *Maritime Policy & Management, 24*(2), 183–192.

Psaraftis, H. N. (2011). A multi-commodity, capacitated pickup and delivery problem: The single and two-vehicle cases, *European Journal of Operational Research 215*, 572–580.

Psaraftis, H. N. (2012). Market based measures for greenhouse gas emissions from ships. *WMU Journal of Maritime Affairs*. doi:10.1007/s13437-012-0030-5.

Psaraftis, H. N., & Kontovas, C. A. (2009a). CO_2 emissions statistics for the world commercial fleet. *WMU Journal of Maritime Affairs, 8*(1), 1–25.

Psaraftis, H. N., & Kontovas, C. A. (2009b, May 26–29). Ship emissions: Logistics and other tradeoffs. *Proceedings of 10th International Marine Design Conference*, Trondheim, Norway.

Psaraftis, H. N., & Kontovas, C. A. (2010). Balancing the economic and environmental performance of maritime transportation. *Transportation Research Part D, 15*(8), 458–462.

Psaraftis, H. N., & Kontovas, C. A. (2013). Speed models for energy-efficient maritime transportation: A taxonomy and survey. *Transportation Research Part C: Emerging Technologies, 26*, 331–351.

Psaraftis, H. N., & Kontovas, C. A. (2014). Ship speed optimization: Concepts, models and combined speed-routing scenarios. *Transportation Research Part C: Emerging Technologies, 44*, 52–69.

Psaraftis, H. N., & Kontovas, C. A. (2015). Slow steaming in maritime transportation: Fundamentals, trade-offs, and decision models. In C.-Y. Lee & Q. Meng (Eds.), *Handbook of ocean container transportation logistics: Making global supply chains effective*. Cham, Switzerland: Springer.

Qi, X., & Song, D.-P. (2012). Minimizing fuel emissions by optimizing vessel schedules in liner shipping with uncertain port times. *Transportation Research Part E: Logistics and Transportation Review, 48*(4), 863–880.

Rana, K., & Vickson, R. G. (1991). Routing container ships using Lagrangian relaxation and decomposition. *Transportation Science, 25*(3), 201–214.

Reinhardt, L. B., & Pisinger, D. (2014). A branch and cut algorithm for the container shipping network design problem. *Flexible Services and Manufacturing Journal, 24*(3), 349–374.

Ronen, D. (1982). The effect of oil price on the optimal speed of ships. *Journal of the Operational Research Society, 33*, 1035–1040.

Ronen, D. (2011). The effect of oil price on containership speed and fleet size. *Journal of the Operational Research Society, 62*(1), 211–216.

Song, D. P., & Xu, J. J. (2012). CO2 emission comparison between direct and feeder liner services: A case study of Asia-Europe services interfacing with the UK. *International Journal of Sustainable Transportation, 6*(4), 214–237.

Stahlbock, R., & Voß, S. (2008). Operations research at container terminals: A literature update. *OR Spectrum, 30*(1), 1–52.

Stopford, M. (2009). *Maritime economics*, Third Edition, Routledge, Taylor and Francis, London and New York.

Tirado, G., Hvattum, L. M., Fagerholt, K., & Cordeau, J.-F. (2013). Heuristics for dynamic and stochastic routing in industrial shipping. *Computers & Operations Research, 40*(1), 253–263.

TradeWinds. (2009, October 30). Maersk insists on slow speeds. *TradeWinds Magazine*.

TradeWinds. (2010, December 13). Slow spur for Maersk VLCCs. *TradeWinds Magazine*

Ulysses. (2012). *EU FP7 project Ulysses web site*. Retrieved from http://www.ultraslowships.com/index.html

Wang, S., & Meng, Q. (2012a). Sailing speed optimization for container ships in a liner shipping network. *Transportation Research Part E, 48*(3), 701–714.

Wang, S., & Meng, Q. (2012b). Liner ship route schedule design with sea contingency time and port time uncertainty. *Transportation Research Part B, 46*(5), 615–633.

Wang, S., & Meng, Q. (2012c). Robust schedule design for liner shipping services. *Transportation Research Part E, 48*, 1093–1106.

Wang, S., Meng, Q., & Liu, Z. (2013a). Bunker consumption optimization methods in shipping: A critical review and extensions. *Transportation Research Part E: Logistics and Transportation Review, 53*, 49–62.

Wang, S., Meng, Q., & Liu, Z. (2013b). A note on "Berth allocation considering fuel consumption and vessel emissions". *Transportation Research Part E, 49*, 48–54.

Wang, S., Alharbi, A., Davy, P. (2014). Liner ship route schedule design with port time windows. *Transportation Research Part C 41*, 1–17.

Yao, Z., Ng, S. H., & Lee, L. H. (2012). A study on bunker fuel management for the shipping liner services. *Computers & Operations Research, 39*(5), 1160–1172.

Zeng, Q., & Yang, Z. (2007). Model integrating fleet design and ship routing problems for coal shipping. In *Computational science–ICCS 2007* (pp. 1000–1003). Heidelberg: Springer.

Chapter 10
Being Green on Sulphur: Targets, Measures and Side-Effects

Christos A. Kontovas, George Panagakos, Harilaos N. Psaraftis, and Eirini Stamatopoulou

Abstract Green House Gas (GHG) emissions are not the only emissions of concern to the international transport community. SO_x emissions are non-GHG emissions that are caused by the presence of sulphur in the fuel. As the maximum percentage of sulphur in automotive and aviation fuels is strictly regulated in most countries around the world, much of the attention in recent years has focused on maritime transport. The attention mainly stems from the fact that in marine fuels the percentage of sulphur can be very high: it can be as high as 4.5 % in Heavy Fuel Oil (HFO), which is the fuel typically used in all deep-sea trades. Even though the amounts of SO_x produced by ships are substantially lower than CO_2, SO_x emissions are highly undesirable as they cause acid rain and undesirable health effects in humans and animals. To mitigate these adverse environmental effects, the international shipping community has taken substantial policy measures. With the introduction of new limits for the content of sulphur in marine fuels in Northern European and North American sea areas, short-sea companies operating in these areas will face substantial additional cost. As of 1/1/2015, international regulations stipulate, among other things, a 0.1 % limit in the sulphur content of marine fuels, or equivalent measures limiting the percent of SO_x emissions to the same amount. As low-sulphur fuel is substantially more expensive than HFO, there is little or no room within these companies current margins to absorb such additional cost, and thus significant price increases must be expected. Unlike its deep-sea counterpart, in short-sea shipping such a freight rate increase may induce shippers to use land-based alternatives (mainly road). A reverse shift of cargo would go against the EU policy to shift traffic from land to sea to reduce congestion, and might ultimately (under certain circumstances) increase the overall level of CO_2 emissions along the entire supply chain. The purpose of this chapter is to investigate the potential effect

C.A. Kontovas (✉) • G. Panagakos • H.N. Psaraftis
Department of Transport, Technical University of Denmark, Bygningstorvet 1 115 Room 108, Kgs. Lyngby 2800, Denmark
e-mail: kontova@transport.dtu.dk; geopan@transport.dtu.dk; hnpsar@transport.dtu.dk

E. Stamatopoulou
National Technical University of Athens, Athens, Greece
e-mail: eirini_stamatopoulou@hotmail.com

H.N. Psaraftis (ed.), *Green Transportation Logistics*, International Series in Operations Research & Management Science 226,
DOI 10.1007/978-3-319-17175-3_10

of sulphur regulations on the share of cargo transported by the waterborne mode vis-à-vis land-based alternatives.

Abbreviations

CO_2	Carbon dioxide
ECA	Emissions Control Area
ECSA	European Community Shipowners Association
EMSA	European Maritime Safety Agency
EU	European Union
FYROM	Former Yugoslav Republic of Macedonia
GHG	Green House Gas
GIS	Geographical Information System
HFO	Heavy Fuel Oil
IFEU	Institute for Energy and Environmental Research
ICS	International Chamber of Shipping
IMO	International Maritime Organization
KPI	Key Performance Indicator
LNG	Liquefied Natural Gas
MDO	Marine Diesel Oil
MGO	Marine Gas Oil
NO_x	Nitrogen oxide
O-D	Origin-Destination
PM	Particulate Matter
Ro-Ro	Roll on roll off vessel
Ro-Pax	Ro-Ro passenger vessel
SECA	Sulphur Emissions Control Area
SO_x	Sulphur oxides
SO_2	Sulphur dioxide
SO_3	Sulphur trioxide
TEN-T	Trans-European Transport Network
UK	United Kingdom

10.1 Introduction

As noted already in previous chapters of this book, Green House Gas (GHG) emissions are not the only emissions of concern to the international transportation community. It is reminded (see Chap. 3) that, at least in the green corridor context, SO_x emissions were selected along with CO_2 as the two most important environmental KPIs. SO_x is an abbreviation for sulphur oxide emissions (SO_2 and SO_3). These are non-GHG emissions that are caused by the presence of sulphur in the fuel. Much of the attention on sulphur in recent years has focused on maritime transportation. This mainly stems from the fact that in marine fuels the percentage of sulphur can be very high: it can be as high as 4.5 % in Heavy Fuel Oil (HFO), which is the fuel typically used in all deep-sea trades. The percentage is lower in distillates such as Marine Gas Oil (MGO) or Marine Diesel Oil (MDO), where it can go down to 0.1 %. It practically goes to zero in Liquefied Natural Gas (LNG).

Even though the amounts of SO_x produced by ships are substantially lower than CO_2, SO_x emissions are highly undesirable as they cause acid rain and undesirable health effects in humans and animals. To mitigate these adverse environmental effects, the international shipping community has taken substantial policy measures. To that effect, in 2008 the International Maritime Organization (IMO), under MARPOL's Annex VI, designated the Baltic Sea, the North Sea and the English Channel as 'Sulphur Emissions Control Areas' (SECAs), with the purpose of limiting SO_x emissions, and in 2010 it designated the entire North American and US-Caribbean coastal zone as an 'Emissions Control Area' (ECA), with ambitious goals to reduce SO_x, NO_x, and PM emissions. A SECA is an ECA in which only SO_x is regulated (situation only pertaining to European ECAs). For the purposes of this chapter we will use the more general term ECA, with the understanding that a SO_x ECA is called a SECA.

With the introduction of new limits for the content of sulphur in marine fuels within the European ECAs (see also Chap. 1, Sect. 1.6.2, for more details), short-sea companies operating in these ECAs will face substantial additional cost. As of 1/1/2015, IMO's MARPOL Annex VI and EU Directive 2012/33/EU (amending Council Directive 1999/32/EC) stipulate, among other things, a 0.1 % limit in the sulphur content of marine fuels, or equivalent measures limiting the percent of SO_x emissions to the same amount. As low-sulphur fuel (MGO or MDO) is substantially more expensive than HFO, there is little or no room within these companies current margins to absorb such additional cost, and thus significant price increases must be expected. Unlike its deep-sea counterpart, in short-sea shipping such a freight rate increase may induce shippers to use land-based alternatives (mainly road). A reverse shift of cargo would go against the EU policy to shift traffic from land to sea to reduce congestion, and might ultimately (under certain circumstances) increase the overall level of CO_2 emissions along the entire supply chain. If the shipping freight rate is no longer competitive with road transportation, this will likely have one or more of the following ramifications:

1. Shifts and congestion to road transport
2. Loss of cargo to the shipping company
3. Reduced profits or increased losses
4. (Potentially) more CO_2 in the overall supply chain
5. Increased cost of the produced goods, making these products uncompetitive as compared with sourcing from other areas, including areas outside the EU
6. The loss of business to the shipping lines as a consequence of 1, 2 and 3 above, makes the shipping routes non-viable and thus candidates for closure. A consequence is that all of the remaining cargoes on such routes will need to find alternative transportation routes, most likely road.

The purpose of this chapter is to investigate the potential effect of sulphur regulations on the share of cargo transported by the waterborne mode vis-à-vis land-based alternatives. It is largely based on the work of the authors and their colleagues on this topic, and specifically on Algaba (2014), Psaraftis and Kontovas (2009, 2010), and Panagakos, Stamatopoulou, and Psaraftis (2014). For additional information, the reader is also referred to a recent issue of *Transportation Research Part D* on ECAs and their impact on maritime transport (Cullinane & Bergqvist, 2014), where a number of related papers are presented.

In addition to shipping, other industries such as the manufacturing, mining and forest industries in the European ECA areas are likely to be impacted. The fear is that many of these industries may be forced to relocate because of the side-effects of such operational and regulatory changes. Such loss of business might force the marginally viable ship operators and ports out of business, channeling even more cargoes towards land-based modes. Among the various maritime sectors, the Ro-Ro sector in the Baltic and North Sea is particularly prone to be impacted by this situation. Ro-Ro operators such as DFDS and Stena have or are contemplating shutting down some routes as unprofitable.

The stricter standards on the sulphur content of marine fuels as introduced by MARPOL Annex VI in 2008, particularly the 0.1 % limit applicable to the ECAs as of 1 January 2015, cause serious concerns mainly within the shipping industry in Northern Europe, as they are expected to have a negative impact on the competitiveness of shipping operations, potentially leading to a shift to other less environmentally friendly modes of transportation. A number of studies were undertaken to examine the impact of these stricter requirements. Four of them were performed by countries within ECAs: Finland (Kalli, Karvonen, & Makkonen, 2009), Sweden (Ljungström, Leyendecker, & Lemieszewski, 2009), the UK (Stavrakaki et al., 2009) and Germany (Hader, Hübscher, Maatsch, & Tasto, 2010). Four more studies were commissioned by stakeholder organizations: one by the European Community Shipowners Association (ECSA) (Notteboom, Delhaye, & Vanherle, 2010) and three by the European Commission (Bosch et al., 2009; Delhaye et al., 2010; and Kehoe, Nikopoulou, Liddane, Ramstedt, & Koliousis, 2010). Two additional studies were commissioned to assess and compare the results of the previous studies: one by a group of northern shipowner associations, endorsed by ECSA and the International Chamber of Shipping (ICS) (Grebot et al., 2010), and one by the European Maritime Safety Agency (EMSA, 2010).

For as long as the new IMO requirements had not been transposed into European law, shipping and other industries in the ECAs developed aspirations of either relaxed sulphur content limits or prolonged enforcement dates. Several interest groups even attempted to reopen negotiations at the IMO level. Yet others hoped that this could be a chance at least to make the rules equal within the EU territory, as the increased cost of shipping in the ECAs was not borne by the southern EU operators. These intense lobbying efforts came to an end in November 2012 with the adoption of Directive 2012/33/EU. None of the aspirations of the industry materialized. The Directive would bring the 0.5 % limit into force on 1 January 2020 for all EU sea territory, even if on a global scale this limit gets postponed to 2025 (Malmqvist & Aldén, 2013).

Among other studies, Kronbak (2006) developed a GIS-based tool to assess the competitive position of the maritime mode vis a vis the road alternative. Jiang, Kronbak, and Christensen (2014) examined the costs and benefits of scrubbers versus marine gas oil and concluded that the determining factor is the price difference between marine gas oil and heavy fuel oil. Algaba (2014) investigated the impact of the new sulphur regulations on a North Sea short sea route and analysed possible alternatives to mitigate undesirable effects from the side of an operator. One of the conclusions was that reducing speed could be one of the tools to mitigate the effects of the modal shifts. Another conclusion was that, under certain conditions, the scrubber option is preferable to switching to low sulphur fuel.

A scrubber is a device that allows burning HFO as a fuel, by filtering out sulphur emissions at the exhaust stage. It is an option that maintains the lower HFO bunker prices but involves substantial capital investment. Danish Ro-Ro operator DFDS has heavily invested in scrubbers for its fleet. Its current investment program is USD 125 m for 21 ships. DFDS has recently received funding of USD 7.9 m for scrubber installation in five of its ships in the context of funding for the EU Trans-European Transport Network (TEN-T). Other operators are also likely to receive such funds. Funding from the TEN-T programme is limited to 20 % of relevant investment costs, so in that sense even though it is helpful it will not completely solve the problem.

Temporarily switching fuel from HFO to MDO and vice versa is a necessity for deep-sea vessels that cross in and out of ECAs, so these ships need to keep two fuel tanks, one for HFO and another for MDO. Technically it is easy to do (only caution is in the fuel pump and in the fuel switch phase, as HFO is preheated and MDO is cold). The corresponding investment costs are minor as compared to scrubbers, and scrubbers are not cost effective for deep-sea vessels as the portion of time they spend in ECAs is low.

Last but not least, switching to LNG fuel is also an option contemplated by many companies, and is being implemented by some. LNG would virtually eliminate SOx emissions, and it is promoted by the EU as an alternative clean fuel in the context of the TEN-T programme. However, the LNG option would require substantial costs to retrofit a ship and its engine(s), and would also require even more substantial infrastructure costs to be widely available in major ports.

To respond to pressure from industry, in 2014 the European Commission has also set up a 'subgroup on competitiveness' within ESSF, the European Sustainable Shipping Forum. The mandate of ESSF is to examine sustainability and competitiveness of maritime transportation in the EU. The mandate of the subgroup is to assist the ESSF to assess the critical success factors for a competitive EU maritime transportation sector and propose recommendations to increase its competitiveness. The specific issue of impact of the sulphur regulations on the short sea sector of Northern Europe has been the focus of the subgroup, the Ro-Ro sector being one of the critical industry sectors.[1]

[1] The Editor of this book has been a member of this subgroup.

The subgroup identified three main impacts as a result of increasing shipping costs: shift to other modes in the ECA, shifts from long to shorter sea routes, and cruise lines operating in north European areas that will shift to other, non-European routes. A three step approach was recommended by the sub-group:

1. A first step concerns the identification of routes and cargo segments that are expected to be the most sensitive. This includes liner traffic as well as spot traffic.
2. A second step involves the monitoring of the most sensitive routes/segments through surveys complemented by other means.
3. A third step, devoted to a comprehensive analysis and comprehensive methodology (including effects on employment, overall emissions, etc.) needs more time and could be carried out in 2015.

It was recognized that the impact on the competitiveness of shipping of higher fuel prices will differ greatly between the various market segments and affect certain types of goods/passengers and certain routes much more than others. In segments where shipping is dominating today, the chances of passing on the incremental cost to customers are likely to be greater compared to segments where intermodal competition is already fierce and shippers can choose between several existing options. Among major factors influencing the cargo owner's choice of mode or inter-modal chain are:

- The price of the transportation service, including (extra) handling costs at terminals;
- The availability, reliability and punctuality of the transportation service;
- The value of the commodity traded;
- The volume of the shipment;
- The total distance to the destination and the distance in ECAs;
- The required operation frequency;
- Access to (or lack of) idle capacity in competing modes, including terminals and rolling stock, and the time and cost involved in expanding capacity;
- The choice of transportation solution looked at from the shippers' point of view.

A survey has been prepared by ECSA and has been circulated to shipping lines to gather on a confidential basis information about the economic impact of the low sulphur limits effective 1/1/2015.

The rest of this chapter is structured as follows: Sect. 10.2 studies the potential use of speed reduction as a means to reduce SO_x emissions in an ECA. Sections 10.3 to 10.5 examine the effects of modal shifts by presenting a case study on the possible designation of the Mediterranean as an ECA. The model and its calibration are presented in Sect. 10.3. Section 10.4 is devoted to the model results and their sensitivity to changes in variables exhibiting a high degree of uncertainty. The environmental implications of a Mediterranean ECA are discussed in Sect. 10.5, while Sect. 10.6 presents various extensions and the conclusions of the chapter.

10.2 Speed Reduction as a Measure to Reduce SOx Emissions

Before we investigate the potential effect of higher fuel prices on modal split, one question that is relevant in an operational setting is the following: Can speed reduction at ECAs work, as a measure to reduce SO_x emissions? After all, and following the considerations of Chap. 9, a ship is likely to reduce speed within an ECA, due to the higher price of the fuel, and reducing speed would reduce all emissions, SO_x included.

To address this issue, it should be noted first that speed reduction, in and of itself, will not change the proportion of SO_x in a ship's exhaust. Therefore, in a strict sense, this measure will not help a ship comply with the new sulphur regulations. But speed reduction will change the total amount of SO_x produced within the ECA, much in the same way as this happens for CO_2 and other exhaust gases. It will also reduce the ship's fuel bill within the ECA. In that sense, speed reduction within an ECA may be worthy of note.

To find the amount of SO_x produced by a ship's exhaust, one has to multiply total bunker consumption by the percentage of sulphur present in the fuel (for instance, 4.5, 1.5, 0.1 %, or other) and subsequently by an appropriate sulphur coefficient. For pure SO_2 this coefficient is 0.02 and for pure SO_3 the coefficient is 0.025. In both cases the coefficient is exact, as it comes from the chemical reaction between sulphur and oxygen. The SO_x coefficient is between these two values as it depends on the SO_2/SO_3 mix, which in turn depends on the quality of the combustion and on other factors.

That said, let us assume a ship that goes from port A to port B, sailing a total distance of L. At the beginning or the end of the trip, there is an ECA Let d ($<L$) be the distance the ship will have to sail within the ECA. The same fuel is assumed to be used outside and within the ECA.

Assume there are two options: Option I is to sail the entire trip at a constant speed of V. Option II is to reduce speed to v ($<V$) within the ECA, but go at a slightly higher speed of V^* ($>V$) outside the ECA, so that *total transit time is the same*.

We assume that total transit time is kept the same so that we do not need more ships in the supply chain, and shippers do not lose money on in-transit inventory costs, as per the considerations of Chap. 9.

Let us now pose the question, with total transit time being the same, which option burns less fuel, I or II? The one that does so would also cost less in fuel costs, and would also produce less total emissions, not only in SO_x, but also CO_2 and all other pollutants.

The analysis is straightforward and goes as follows (see also Psaraftis and Kontovas (2009)):

Let the transit time in both scenarios be $T = L/V$ (in days). If within the ECA the speed is v ($<V$) for distance d, then

$$\frac{L}{V} = \frac{d}{v} + \frac{L-d}{V^*}$$

Therefore $V^* = \dfrac{L-d}{\frac{L}{V} - \frac{d}{v}}$ ($> V$)

The assumption here is that $L/V > d/v$, otherwise making up the time lost in the ECA would be impossible.

Without loss of generality we assume that fuel consumption per day obeys a cube law, that is, is equal to kV^3. Since we have to multiply by total days, the law becomes quadratic, as total fuel consumption is $T_{FC} = k\,V^3(L/V) = kLV^2$.

Option I: total fuel consumption $T_{FC}(V) = kLV^2$
Option II: total fuel consumption $T_{FC}\ (V^*, v) = k(L-d)V^{*2} + kdv^2$

Substituting, we get $T_{FC}(V^*, v) = \dfrac{k(L-d)^3}{\left(\frac{L}{V} - \frac{d}{v}\right)^2} + kdv^2$

Define the ratio $R = \dfrac{T_{FC}(V^*, v)}{T_{FC}(V)} = \dfrac{(L-d)^3}{L\left(L - d\frac{V}{v}\right)^2} + \dfrac{d}{L}\left(\dfrac{v}{V}\right)^2$

It can be shown that always $R > 1$ (assuming again that $L/V > d/v$). The proof of this is straightforward.

Let us illustrate this with an example:
Let $L = 2{,}000$ nautical miles
$d = 200$ nm within the ECA
$V = 20$ knots
$v = 18$ knots within the ECA
Then

$$V^* = \frac{1800}{\frac{2000}{20} - \frac{200}{18}} = 20.25 \ \textit{knots} \ \text{ outside SECA}$$

and

$$R = \frac{1800^3}{2000\left(2000 - 200\frac{20}{18}\right)^2} + \frac{200}{2000}\left(\frac{18}{20}\right)^2 = 0.9226 + 0.081 = 1.0036$$

Other speed and distance combinations will produce other ratios, but all will be >1.

The conclusion from this analysis is that speed reduction in ECAs will reduce emissions (of all gases, including SO_x) within the ECA, but result in more total emissions and more total fuel spent if speed is increased outside the ECA to make up for lost time. The reduced emissions within the ECA will be more than offset by higher emissions outside (for all gases). The total fuel bill will also be higher. Of course, whether or not society may mind polluting the areas outside ECAs more in order to make conditions in ECAs more friendly to the environment is a non-trivial issue that is outside the scope of this chapter.

We close this section by noting the work of Fagerholt et al. (2015), who investigate alternative speed-routing combinations for a ship that has to sail through an ECA and wants to minimize its fuel bill. The ship will have to switch to the more expensive fuel within the ECA. It is seen that the option to sail a circuitous path to avoid using the expensive fuel within the ECA as much as possible may produce lower cost solutions than sailing the shortest path route, but that these solutions may not necessarily decrease overall emissions including SO_x.

10.3 A Modal Shift Model and Case Study

10.3.1 Methodology

First we say a few words about modal split models. In general, these models determine the number of trips on different modes given the travel demand between an origin-destination (O-D) pair. They try to mathematically describe the mode choice mechanism, based on the assumption that the probability of choosing a particular mode is the probability that the perceived utility from that mode is greater than the perceived utility from each of the other available modes.

There are various forms of modal split models but by far the most common one is the logit model, which has been found to fit the mode choice behaviour quite well. The binomial form of the logit model, where there are only two alternative modes of transportation to choose from, is the simplest of these models. Assume that for a given O-D pair, mode 1 is a pure road mode that can be used as an alternative to mode 2, which is a route that includes the Ro-Ro mode. If x_i is the fraction of the cargo that will choose mode i ($i = 1, 2$), assuming there is available capacity to do so ($0 \leq x_i \leq 1$), the binomial logit model defines x_i as:

$$x_i = \exp(-\lambda C_i) \,/\, [\exp(-\lambda C_1) + \exp(-\lambda C_2)] \tag{10.1}$$

where C_i is the generalized cost associated with mode i and λ is a positive constant to be estimated at model calibration.

C_i can be expressed as a function of a number of variables. Depending on the particular application, the following variables have been proposed in the literature as the main determinants of C_i (see also Psaraftis and Kontovas (2010)):

- total monetary costs (freight rates and other direct or indirect costs);
- total transport time (in-vehicle, idling, border-crossing, etc.);
- reliability and regularity (in terms of on-time delivery);

- flexibility (ability to adapt to changes in annual demand/volume, size of consignment and time table);
- resilience (ability to cope with serious disruptions);
- safety and security; and
- environmental performance (e.g. emissions of GHG and air pollutants).

Many applications, including ours, consider only the first two variables. It that sense,

$$C_i = p_i + kt_i (i = 1, 2) \tag{10.2}$$

where

- p_i stands for the total transport cost associated with mode i (USD/tonne, €/lane meter, or other),
- t_i is the corresponding total transport time (days) and
- k is a positive constant, which is proportional to the value of the cargo (and in fact equal to $Pr/365$, where P is the value of the cargo and r is the investor's opportunity cost of capital) (USD or €/tonne/day).

In the ECA situation, it is clear that if p_2 for mode 2 (the Ro-Ro mode) should increase due to the fuel price increase, the share x_2 of mode 2 will decrease.

For the purposes of this chapter, the model is applied to estimate shares of cargo moved not just by alternative transportation modes but along alternative routes between a given O-D pair. Each available option, then, concerns a particular supply chain involving one or more transportation modes. It is important to keep in mind that the model results are being used to assess the emissions associated with each alternative route examined. It is, therefore, necessary all alternative routes to concern door-to-door services between the same O-D pair.

10.3.2 The Case Study

The specific case study draws from Panagakos et al. (2014) and examines the possible designation of the Mediterranean as a SECA, in terms of its possible impacts on modal shift. The Med is a sea with significant short-sea traffic, in both tramp and liner trades. In addition, there is deep-sea traffic between the Far East and Europe, much of which bypasses the Med as much of it calls at ports in Northern Europe, such as Rotterdam, Antwerp and Hamburg. The Med is not designated as an ECA, even though there have been considerations to that effect. In fact during the inter-institutional debates that preceded the adoption of the sulphur directive, the European Commission was asked by the legislators to consider extending the stricter ECA limits to all EU territorial waters. This renewed the discussion on possible designation of the Mediterranean Sea as an ECA (Bosch et al., 2009; Delhaye et al., 2010; Kehoe & Woxenius, 2010). The Committee on Transport and Tourism of the European Parliament commissioned an assessment of a possible extension of ECAs to the entire European coastline (Schinas & Bani, 2012), which confirms the wider societal benefits of any reduction in the environmental burden and concludes that the extension of an ECA around the EU would level the playing field for all stakeholders, enhance the technical compatibility of the short-sea fleet servicing European ports, provide a stimulus to technical research in the area of abatement and alternative fuels, and enable effective enforcement of the regulations under the existing Port State Control regime.

The hypothetical scenario examined in this chapter concerns consolidated cargoes transported by truck between Thessaloniki, Greece and industrial hubs of northern Germany. Clothing products, agricultural products and marble are the usual exports to Germany originating in the Thessaloniki region. Road is the exclusive transportation mode used for clothing products due to their high value and product nature (timely positioning in the marketplace is critical). The industry norm is that exports to Germany are scheduled for the last day of each calendar week (Friday). Products that are for some reason delayed are in most cases flown to destination at the expense of the party responsible for the delay.

There are two routes from Thessaloniki to northern Germany that trucks follow. The first one ('road-only') is Thessaloniki—Skopje—Belgrade—Budapest—Bratislava—Prague—Dresden—Berlin. Route 1 appears in Fig. 10.1. It is the most common one for destinations in northern Germany, as it avoids the restrictions of Austria and minimizes the high road tolls of Germany.

The second route, Thessaloniki—Igoumenitsa—Ancona—Bologna—Verona—Munich—Nurnberg—Berlin ('combined-transport') involves crossing the Ionian/Adriatic Seas by ferry boats. Route 2, depicted in Fig. 10.2, is faster than the first one but, as far as exports are concerned, it is selected only in cases of short delivery

Fig. 10.1 The road-only route (Route 1)

Fig. 10.2 The combined-transport route (Route 2)

times as ferry services result in higher total costs, while the security problems created by illegal immigrants trying to reach Italy onboard westbound trucks from Igoumenitsa are not negligible.

It is noted that Igoumenitsa is also connected via ferry services to the Italian ports of Bari and Brindisi. However, these connections are rarely used by trucks heading towards northern Germany. The combined-transport route examined here involves also a truck-on-train service between Brenner and Wörgl.

The information needed for model calibration was obtained through interviews with a small-size Greek truck operating company that specializes in services between Thessaloniki and destinations in Italy and Germany. It uses 40 tonne EURO III, IV and V trucks that are mainly owned by third parties and usually operated by their owners themselves. Data gathered cover year 2010, which was selected as the basis for the analysis due to the fact that the present financial and economic crisis in the country makes the more recent information rather atypical.

In 2010, the company arranged 250 round trips from Thessaloniki to Germany. Only 73 of them (29.2 %) concern full loads. The remaining are loads consisting of less-than-full cargoes consolidated in Thessaloniki. The transport chain examined here belongs to the latter type (177 trips in 2010). It is noted that before the present crisis, full cargoes were much more frequent, comprising about 70 % of the total.

The vehicle examined is a EURO III 40 tonne truck with a maximum payload of 24 tonnes and 85 cubic meters. Provided that consolidated freight consists of a mix of light and heavy cargoes, it is estimated that the average full payload is about 18 tonnes. Due to cargo consolidation, the trucks leave Thessaloniki almost full. Load factors are above 95 %.

In the typical case analysed here, the truck carries three cargoes on its outbound leg: The first batch is clothing products that have to be delivered directly to cargo owner's facilities in Berlin. A typical consignment is 350 cartons 60 cm × 40 cm × 30 cm, which in total amounts to 25.2 cubic meters and weighs about 3.5 tonnes. A typical value of such cargo can be in the area of 145,000 €. The transport cost for this type of cargo from Thessaloniki to Berlin is 37 €/cubic meter or 932.40 € in 2010 prices. An additional amount of about 50.00 € is charged for agency fees. The insurance cost is 0.15 % of the value of goods or 217.50 €. So the total cost for this first batch is about 1,200 € (although the case study examined here focuses on this first batch of cargo, data on all three cargoes are provided in order to be able to calculate the distances and load factors needed for allocating emissions along this multi-load multi-drop vehicle trip).

The second batch consists of 10 tonnes of olives destined to the warehouse of a freight forwarder in Hannover. They are packaged in metal canisters 24 cm × 24 cm × 35 cm weighing 20 kg each. Each pallet is 120 cm × 80 cm in dimensions and carries 45 canisters (in 3 layers of 15 canisters each) or 900 kg in total. The consignment consists of 11 pallets. Total transport cost is estimated at about 1,100 €.

The third batch of cargo is 3.6 tonnes of clothing products to be delivered to the facilities of the cargo owner in Bremen. A figure of 1,250 € is estimated as above for the transport cost of this batch. So, the total transport cost for the outbound leg is 3,550 €.

It is noted that when consignment is above 18 cubic meters in volume or 3 tonnes in weight, the cost estimate provided above includes the cost of cargo collection from the site of the exporter. In this case the same truck that will do the long haul goes to the exporter's site and picks up the cargo. When cargo volume is below 18 cubic meters or 3 tonnes, it is the exporter's obligation to bring the cargo to the facilities of the service provider, located 8 km away from the centre of Thessaloniki, where the consolidation process takes place. In these cases exporters use their own light trucks or can arrange for a pick up by a third party for about 60 €. The average distance that a truck needs to travel in the Thessaloniki area to pick up cargoes before it starts its main journey is about 80 km.

10.3.3 The Road-Only Route

A typical itinerary of the truck along the road-only route, which accounted for 121 of the 177 trips in year 2010 (68.36 %) is described in detail in Table 10.1. Although cargo is picked up on a Friday afternoon, the truck doesn't leave prior to Sunday 5:00 in order to avoid the weekend traffic restrictions. It reaches Berlin on Wednesday at 8:00 in the morning. The general pattern consists of driving for 4.5 h, pausing for 1 h, driving for another 4.5 h and resting for 11 h.

On the Thessaloniki—Berlin segment, the truck crosses three borders involving non-EU states, those of Greece—FYROM, FYROM—Serbia, and Serbia—Hungary. The average time spent for clearing these borders is 1.5, 3.0, and 3.5 h, respectively. Depending on the season of the year, the day of the week and the time of the day, border clearance times can be as long as twice the average figures

Table 10.1 Typical itinerary along the road-only route

No.	Activity	Origin	Destination	Dist. (km)	Tonnes	Tkm	Start		Finish		Duration (hours)	Comments
							Day	Time	Day	Time		
1	Driving from depot to Site 1 in Thessaloniki	Thessaloniki	Thessaloniki	32	0	0	Fr.	11:30	Fr.	12:30	1	0.5 h delay due to congestion
2	Pick up Consignment 1 of 3.5 tonnes to Berlin	–	–	0	3.5	0	Fr.	12:30	Fr.	14:00	1.5	
3	Driving from Site 1 to Site 2 in Thessaloniki	Thessaloniki	Thessaloniki	15	3.5	53	Fr.	14:00	Fr.	14:30	0.5	
4	Pick up Consignment 2 of 10 tonnes to Hannover	–	–	0	10	0	Fr.	14:30	Fr.	15:30	1	
5	Driving from Site 2 to Site 3 in Thessaloniki	Thessaloniki	Thessaloniki	18	13.5	243	Fr.	15:30	Fr.	16:30	1	0.5 h delay due to congestion
6	Pick up Consignment 3 of 3.6 tonnes to Bremen	–	–	0	3.6	0	Fr.	16:30	Fr.	18:30	2	
7	Driving from Site 3 to depot in Thessaloniki	Thessaloniki	Thessaloniki	15	17.1	257	Fr.	18:30	Fr.	19:00	0.5	
8	Off duty	–	–	0	17.1	0	Fr.	19:00	Su.	5:00	34	
9	Driving from Thessaloniki to Skopje	Thessaloniki	Skopje	236	17.1	4,036	Su.	5:00	Su.	8:30	4.5	1.5 h to clear the GR-FRM borders 1 h time difference
10	Driving from Skopje to Belgrade	Skopje	Belgrade	393	17.1	6,720	Su.	8:30	Mo.	4:00	19.5	3 h to clear the FRM-SRB borders 11 h idle

11	Driving from Belgrade to Budapest	Belgrade	Budapest	373	17.1	6,378	Mo.	4:00	Mo.	12:00	8	3.5 h to clear the SRB-HU borders
12	Driving form Budapest to Bratislava	Budapest	Bratislava	200	17.1	3,420	Mo.	12:00	Mo.	15:30	3.5	1 h idle
13	Driving from Bratislava to Prague	Bratislava	Prague	338	17.1	5,780	Mo.	15:30	Tu.	8:00	16.5	12 h idle
14	Driving from Prague to Dresden	Prague	Dresden	158	17.1	2,702	Tu.	8:00	Tu.	10:00	2	
15	Driving form Dresden to Berlin	Dresden	Berlin	196	17.1	3,352	Tu.	10:00	We.	8:00	22	18.5 h idle 1 h delay due to congestion
16	Unloading in Berlin	–	–	0	3.5	0	We.	8:00	We.	9:30	1.5	
17	Driving from Berlin to Hannover	Berlin	Hannover	291	13.6	3,958	We.	9:30	We.	13:30	4	
18	Unloading in Hannover	–	–	0	10	0	We.	13:30	We.	14:30	1	
19	Driving from Hanno-ver to Bremen	Hannover	Bremen	132	3.6	475	We.	14:30	We.	16:30	2	
20	Unloading in Bremen	–	–	0	3.6	0	We.	16:30	We.	18:00	1.5	
21	Off duty	–	–	0	0	0	We.	18:00	Th.	18:00	24	
	Total road-only route			**2,397**	**km**	**37,372**	**tkm**				**151.5**	**hours**

indicated. The truck also stops to rest three times; one outside Belgrade, one close to the Slovakian—Czech border, and one outside Berlin.

The first batch of cargo is unloaded in Berlin. Note that the type of consignee influences delivery time. If the consignee is the cargo owner himself, as is the case here, the delivery takes place at the consignee's facilities that most of the times are open only during working hours. If the consignee is the warehouse of a freight forwarding company, delivery can be made at any time of the day at advance notice. In our case, the truck has to wait outside Berlin for 7.5 h on top of the 11 h stipulated by the regulations, so as to reach the cargo owner's facilities at 8:00 in the morning.

Following unloading in Berlin, the truck drives to Hannover to unload the second cargo batch, and then to Bremen for the final consignment. The unloading operation is finalized on Wednesday at 18:00 h. Delivery by the end of working day Wednesday is the contractual obligation of the service provider. This is achieved in 99 % of the cases. In fact, the company's statistics show delivery 5 h ahead of schedule on the average. Note that the 7.5 h of idling in Berlin serves as a buffer for unexpected delays.

After all cargoes have been delivered, the truck goes off duty for 24 h according to the regulations. The road-only statistics are summarized below (see Panagakos et al., 2014 for more details):

Total distance:	2,397 km		
• Total time:	151.5 h, of which:	Driving:	34.5 h
		Loading/unloading:	8.5 h
		Border crossing:	8.0 h
		Onboard other means	0.0 h
		Idling	100.5 h
• Total tkm:	37,372 tkm, of which:	7,649 tkm are attributed to the first consignment of clothing products (3.5 tonnes) from Thessaloniki to Berlin (=37,372 tkm * 3.5 tonnes/17.1 t)	
• Nominal distance[a]:	2,186 km (=7,649 tkm/3.5 tonnes), of which:		
– First mile:	16 km (=80 km in Thessaloniki * 3.5 tonnes/17.1 tonnes)		
– Main journey:	1,894 km (Thessaloniki—Berlin)		
– Last mile:	276 km (=2,186—1,894—16)		
• Average nominal speed:	15.8 km/h (=2,397 km/151.5 h)		
• Average speed driving:	69.5 km/h (=2,397 km/34.5 h)		
• Actual cost (for the consignment of interest):	342.86 €/tonne (=1,200 €/3.5 tonnes)		
• Actual time (for the consignment of interest):	4.92 days (=118 h/24)		
• Unit value (for the consignment of interest):	41,430 €/tonne (~145,000 €/3.5 tonnes)		

[a]For the consignment of interest

10.3.4 The Combined-Transport Route

A typical itinerary of the truck along the combined-transport route, which in 2010 accounted for 56 of the 177 trips (31.64 %), is described in Table 10.2. There is a 10 % surcharge for this service, meaning that the cost for moving clothing products from Thessaloniki to Berlin is now 40.70 €/cu.m. (=37 €/cu.m * 1.10). The additional charges for agency fees and insurance remain unchanged, bringing the total cost for the first batch to 1,293.14 € (=40.70*25.2 + 50 + 217.50).

According to this schedule, the collection of cargoes has to be over by 15:00 on Friday afternoon, so that the truck can reach Igoumenitsa the same evening at least 2 h prior to the departure of the ferry boat. The ship arrives in Ancona at 16:00 of the following day. The truck reaches Brenner (through Bologna and Verona) at 01:30 on Sunday morning and stays in the ROLA terminal there until 00:30 Monday morning, when it gets on the train for the Brenner-Wörgl link. It is noted that this itinerary cannot be followed from June 1 to September 20, when the usual Sunday ban on heavy goods vehicles in Italy is extended to cover the entire weekend from Friday 18.00 to Sunday 24:00.

After driving off the train, the truck reaches its first destination in Berlin at about 14:00 of Monday afternoon, 42 h earlier than with the road-only option. Note that the time gain for the other two deliveries is lower, as the driver has to rest for 11 h right after the first stop in Berlin. Once again, the truck goes off duty for 24 h after all deliveries are made in order to ensure comparability between the two itineraries.

The statistics of the combined-transport route are as follows:

Total distance:	3,016 km		
• Total time:	124 h, of which:	Driving:	32.0 h
		Loading/unloading:	8.5 h
		Border crossing:	0.0 h
		Onboard other means:	20.5 h
		Idling:	63.0 h
• Total tkm:	47,957 tkm, of which:	9,816 tkm are attributed to the first consignment of clothing products (3.5 tonnes) from Thess. to Berlin (=47,957 tkm * 3.5 tonnes/17.1 tonnes)	
• Nominal distance:	2,805 km (=9,816 tkm/3.5 tonnes), of which		
– First mile:	16 km (=80 km in Thessaloniki * 3.5 tonnes/ 17.1 tonnes)		
– Main journey:	2,513 km (Thessaloniki—Berlin)		
– Last mile:	276 km (=2,805−2,513−16)		
• Average nominal speed:	24.3 km/h (=3,016 km/124 h)		
• Average speed driving:	68.3 km/h (=2,184 km/32 h)		

(continued)

• Actual cost (for the consignment of interest):	369.47 €/tonne (=1,293.14 €/3.5 tonnes)
• Actual time (for the consignment of interest):	3.31 days (=79.5 h/24)

10.3.5 Coefficient k

The coefficient k in the generalized cost function (10.2) is known in the literature as 'value of time.' It is usually expressed as:

$$k = Pr/365 \tag{10.3}$$

where:

P = the unit cargo value (USD or €/tonne), and
r = the opportunity cost of capital (%).

In most applications, r is taken as the annual yield of a risk-free investment. The interest rate of long-term government bonds is a usual indicator for r. While for years the yields of 10-year Greek government bonds fluctuated below 5 % in the pre-crisis era, starting from 2009 they exhibit a meteoric rise reaching 29.24 % in February 2012 just before the second 'haircut' of the Greek debt (Bank of Greece, 2012). A more meaningful indicator is thus needed. The interest rate on outstanding amounts of Euro-denominated deposits with agreed maturity of up to 2 years by non-financial corporations with domestic financial institutions, as reported by the Bank of Greece for year 2010 (2.97 %) has been selected for this purpose.

It should be mentioned, however, that the value of time for fashion items like clothing products can be much higher. Nordås, Pinali, and Geloso Grosso (2006) argue that labour-intensive products such as clothing are increasingly time-sensitive forcing suppliers to shorten lead time in order to stay competitive. Furthermore, in recent decades, the so-called "fast fashion" strategy, a concept developed in Europe to serve customers who desire trendy and relatively inexpensive clothing, is followed by many fashion retailers. Critical in fast fashion is the lead time, which has been reduced in just a few weeks (Sull & Turconi, 2008). Discounts in the area of 10 % for a 2-week delay in delivery are not unusual for manufacturing contracts in this sector, while for delays of 3 weeks and more the retailer has the right to cancel the contract altogether. Although such rates are inconceivable for deliveries within contractual margins, are nevertheless indicative of a value of time much higher than that implied by an opportunity cost of 2.97 %. The sensitivity of model results to different r values is examined in Sect. 10.4.5.

Table 10.2 Typical itinerary along the combined-transport route

No.	Activity	Origin	Destination	Dist. (km)	Tonnes	Tkm	Start		Finish		Duration (hours)	Comments
							Day	Time	Day	Time		
1	Driving from depot to Site 1 in Thessaloniki	Thessaloniki	Thessaloniki	32	0	0	Fr.	8:00	Fr.	9:00	1	0.5 h delay due to congestion
2	Pick up Consignment 1 of 3.5 tonnes to Berlin	–	–	0	3.5	0	Fr.	9:00	Fr.	10:30	1.5	
3	Driving from Site 1 to Site 2 in Thessaloniki	Thessaloniki	Thessaloniki	15	3.5	53	Fr.	10:30	Fr.	11:00	0.5	
4	Pick up Consignment 2 of 10 tonnes to Hannover	–	–	0	10	0	Fr.	11:00	Fr.	12:00	1	
5	Driving from Site 2 to Site 3 in Thessaloniki	Thessaloniki	Thessaloniki	18	13.5	243	Fr.	12:00	Fr.	13:00	1	0.5 h delay due to congestion
6	Pick up Consignment 3 of 3.6 tonnes to Bremen	–	–	0	3.6	0	Fr.	13:00	Fr.	15:00	2	
7	Driving from Site 3 to depot in Thessaloniki	Thessaloniki	Thessaloniki	15	17.1	257	Fr.	15:00	Fr.	15:30	0.5	Change drivers
8	Driving from Thessaloniki to Igoumenitsa	Thessaloniki	Igoumenitsa	350	17.1	5,985	Fr.	15:30	Fr.	21:30	6	1 h idle
9	Waiting to board ferry	–	–	0	17.1	0	Fr.	21:30	Fr.	23:30	2	
10	Crossing the Adriatic Sea	Igoumenitsa	Ancona	735	17.1	12,569	Fr.	23:30	Sa.	16:00	17.5	1 h time difference
11	Driving form Ancona to Bologna	Ancona	Bologna	227	17.1	3,882	Sa.	16:00	Sa.	19:00	3	
12	Driving from Bologna to Verona	Bologna	Verona	151	17.1	2,582	Sa.	19:00	Sa.	22:00	3	1 h idle
13	Driving from Verona to Brenner	Verona	Brenner	239	17.1	4,087	Sa.	22:00	Su.	1:30	3.5	
14	Waiting in ROLA terminal	–	–	0	17.1	0	Su.	1:30	Mo.	0:30	23	

(continued)

Table 10.2 (continued)

No.	Activity	Origin	Destination	Dist. (km)	Tonnes	Tkm	Start		Finish		Duration (hours)	Comments
							Day	Time	Day	Time		
15	Crossing Austria by train	Brenner	Wörgl	97	17.1	1,659	Mo.	0:30	Mo.	3:30	3	
16	Driving from Wörgl to Munich	Wörgl	Munich	109	17.1	1,864	Mo.	3:30	Mo.	5:00	1.5	
17	Driving from Munich to Nurnberg	Munich	Nurnberg	167	17.1	2,856	Mo.	5:00	Mo.	7:00	2	
18	Driving from Nurnberg to Berlin	Nurnberg	Berlin	438	17.1	7,490	Mo.	7:00	Mo.	14:00	7	1 h idle
19	Unloading in Berlin	–	–	0	3.5	0	Mo.	14:00	Mo.	15:30	1.5	
20	Driving from Berlin to Hannover	Berlin	Hannover	291	13.6	3,958	Mo.	15:30	Tu.	6:30	15	11 h idle
21	Unloading in Hannover	–	–	0	10	0	Tu.	6:30	Tu.	7:30	1	
22	Driving from Hannover to Bremen	Hannover	Bremen	132	3.6	475	Tu.	7:30	Tu.	9:30	2	
23	Unloading in Bremen	–	–	0	3.6	0	Tu.	9:30	Tu.	11:00	1.5	
24	Off duty	–	–	0	0	0	Tu.	11:00	We.	11:00	24	
	Total combined—transport route			**3,016**	**km**	**47,957**	**tkm**				**124**	**hours**

10.3.6 *Estimation of Parameter λ*

Equation (10.1) can be transformed into:
$x_1/x_2 = \exp[-\lambda(C_1 - C_2)]$, leading to:

$$\lambda = -\ln(x_1/x_2) / (C_1 - C_2) \tag{10.4}$$

If $k = 41{,}430 * 0.0297/365 = 3.37$ €/tonne/day, then (10.2) results in:
$C_1 = 342.86 + 3.37 * 4.92 = 359.45$ €/tonne, and
$C_2 = 369.47 + 3.37 * 3.31 = 380.63$ €/tonne,

Taking into consideration that $x_1 = 0.6836$ and $x_2 = 0.3164$, (10.4) produces a value of $\lambda = 0.036368$ €$^{-1}$.

It is worth mentioning that the value of k estimated above implies an in-transit inventory cost of 0.1404 €/tonne/h, almost identical to the 0.1350 €/tonne/h figure that Delhaye et al. (2010) borrow from the TRANS-TOOLS model for manufactured articles.

10.4 Modal Split and Sensitivity Analysis

The model as calibrated above is used to estimate the potential impact of designating the Mediterranean Sea as an ECA. As with other ECAs, the maritime industry has three alternative ways to react to such a development: (i) install an exhaust gas scrubber system and continue burning Heavy Fuel Oil (HFO), (ii) switch fuel from HFO to Marine Gas Oil (MGO) with sulphur content below 0.1 % or (iii) use Liquefied Natural Gas (LNG) as marine fuel.

Although LNG is the cleanest fossil fuel and reduces SO_x, PM and NO_x emissions drastically and even CO_2 emissions significantly, the scarcity of LNG refuelling stations in Europe and the necessary conversion of the propulsion system involving twice as big fuel tanks renders only the first two options feasible in the short run. On 24 January 2013, the Commission announced its Clean Power for Transport Package that includes an action plan for the development of LNG in shipping. According to this plan, LNG refuelling stations should be installed in all 139 maritime and inland ports on the trans-European core network by 2020 and 2025 respectively.

Furthermore, it appears that the shipping industry, up and until recently, considered scrubbers as a rather immature technology (Kehoe et al., 2010; Ljungström et al., 2009; Malmqvist and Aldén (2013); Notteboom et al., 2010). Even though companies such as DFDS have invested in this option, switching from HFO to MGO is the only real option in the immediate future for many companies and is the only scenario examined in our case study.

Increased transportation cost is a certain outcome of this fuel switch. However, the quantitative assessment of the cost rise is associated with a number of uncertainties. A 'basic scenario' is, thus, developed reflecting a set of assumptions, while the role of parameters exhibiting a high degree of ambiguity is examined through sensitivity analysis later on.

In terms of the time frame, all previous studies focus on 1 January 2015, when the 0.1 % limit will be enforced in the existing ECAs. There is no indication for the time the Mediterranean Sea will (if ever) become an ECA. The only certainty we have is that, according to Directive 2005/33/EC, the limits apply 12 months after the date of entry into force of the ECA designation. Provided that designating a new ECA involves rather cumbersome procedures, the scenario of applying the ECA limits in the Mediterranean Sea by 1 January 2015 is not very probable. However, solely for the sake of comparability, we hereby assume that the new limits will become effective in this part of the world together with the other ECAs.

10.4.1 Fuel Prices

The scientific community seems to agree on the following facts:

- There is a strong correlation between the prices of marine fuels and the price of crude oil.
- During the last 25 years the price of crude oil follows an upward trend due to increased demand (especially from Asia) and depletion of conventional oil fields.
- Forecasting fuel prices is not easy.

A number of studies project fuel prices to 2015 and beyond (Delhaye et al., 2010; Hader et al., 2010; Kehoe et al., 2010; Stavrakaki et al., 2009), whereas others have simply applied historic prices (Kalli et al., 2009). It appears that forecasting directly the price differential between the HFO (1 % sulphur) and MGO (0.1 % sulphur), which is actually what we need, is probably safer than independent price projections for the two fuel qualities, as the former approach takes advantage of the existing correlation between prices. Ljungström et al. (2009), Malmqvist and Aldén (2013) and Notteboom et al. (2010) follow this approach.

Based on Oct-Nov 2008 figures, Ljungström et al. (2009) report a differential of 297 USD/tonne between MGO and HFO in the port of Rotterdam. According to Malmqvist and Aldén (2013), this figure was dropped to 240 USD/tonne on 16 July 2012 but elevated to 330 USD/tonne 3 months later, on 16 October 2012. By the end of February 2013, the www.bunkerworld.com site was reporting a differential of 305 USD/tonne, while the price difference between these two fuel qualities in the port of Piraeus was 317 USD/tonne.

On the basis of this information and the expectation that the demand for MGO will be increased by 2020, when all EU countries would need to meet the stricter sulphur limits, Malmqvist and Aldén (2013) predict a 500 USD/tonne differential for that year. Thinking along the same lines, we accept the present differential of 330 USD/tonne as the default value for our basic scenario. The effects of higher price differences will be assessed in the sensitivity analysis part of the chapter. It is noted that the exchange rate of the end of February 2013 (1.34 $/€) is used for converting USD prices into Euro-denominated ones.

It needs to be added that a second order effect of designating the Mediterranean Sea as an ECA might be a price increase of the diesel oil used by road transportation, triggered by a potential inability of the oil refining industry to cope with the increased demand for distillates that another ECA might cause. However such effects are outside the scope of the present chapter and are not pursued further.

10.4.2 Fuel Consumption

The Ro-Pax vessel SUPERFAST XI is selected as the representative vessel employed on the Patras—Igoumenitsa—Ancona route. According to the company's website, she was built in Germany in 2002. The 199.9 m long ship carries on its 10 decks up to 1,639 passengers and 653 vehicles. Her four 12,000 KW Wärtsilä engines allow her to sail at a maximum speed of 29.3 knots.

The Face^{3}ts (2008) report provides an estimate of the fuel being consumed by SUPERFAST XI on the one-way sailing Patras—Igoumenitsa—Ancona. She is burning 167 tonnes of fuel oil and 350 L of lubricants. A quantity amounting to 30 % of the fuel oil is consumed while in port and, back in 2008, was of the 1.5 % sulphur quality. The sulphur content of the remaining quantity (70 %) was 2.7 %. The fuel qualities used today are HFO (1 % sulphur) at sea and MGO (0.1 % sulphur) in port. Therefore, the additional fuel cost associated with the Mediterranean ECA concern only the 70 % of the fuel oil consumed, since the remainder is of the 0.1 % quality anyway.

It needs to be clarified that the figures mentioned above include the segment Patras—Igoumenitsa that does not actually belong to the journey examined here. However, the company in its pricing policy treats both origins as a single one (the freight rates are identical regardless of the port of embarkation). The additional fuel costs that will eventually be allocated to our truck, then, basically concern a fictitious average truck originating somewhere between Patras and Igoumenitsa. This is not unreasonable, however, should one consider that a truck getting on board in Igoumenitsa has reserved space that remains unexploited during the Patras—Igoumenitsa segment (not allowed by Greek legislation).

10.4.3 Allocation of Additional Costs

The additional costs related to the switching of fuel from HFO to MGO need to be allocated to the vessel's payload. For Ro-Pax vessels carrying a mixture of trucks/trailers, passengers, cars, caravans etc., this is easier said than done. Different proportions and significance of passengers lead to substantially different cost structures.

After acknowledging that "it is hardly feasible to make a valid allocation of the costs—and particularly the fuel costs—to individual cargo units,... [nor is it] possible to make any accurate assignment on the basis of fares", Hader et al. (2010) provide for a number of representative Ro-Ro vessels indicative estimates of the share of total voyage costs that is being borne by the cargo. Moreover, they consider passenger volumes to be more elastic than cargo with respect to price increases, allocating to the latter an over proportional share of the additional costs and the associated price rises. The estimates of Hader et al. (2010) appear in Table 10.3.

Interestingly enough, the vessel SUPERFAST VII (renamed to STENA SUPERFAST VII as of Sept. 2011 and operated by Stena Line on the Belfast Cairnryan route) which is similar in dimensions, capacity and modus operandi with our SUPERFAST XI, has been selected by Hader et al. (2010) as the representative ship for the route Rostock-Helsinki. The share of costs estimated to be borne by the cargo for this ship is 35 %. Taking the elasticity into consideration, the proportion of additional costs estimated to be borne by the cargo becomes 45 %. This is the default value used in our basic scenario.

Another entry in Table 10.3 that deserves our attention is the LISCO GLORIA ship serving the Kiel-Klaipeda route. Although this vessel did not resemble SUPERFAST XI, it used to serve a route identical to the Igoumenitsa—Ancona one in terms of length (~400 nautical miles). To the extent that distance is an important factor in shaping voyage costs, the 95 % figure of Table 10.3 for the proportion of additional costs to be borne by LISCO GLORIA's cargo is taken as the maximum value for the sensitivity analysis performed later on.

The only piece of information still needed is the average number of trucks/ trailers on the SUPERFAST XI on her voyages across the Ionian and Adriatic Seas. Face[3]ts (2008) reports that in year 2007, the four SUPERFAST ships employed on the Greece-Italy routes executed 1,372 one-way voyages transporting 576,000 passengers and 150,000 trucks/trailers. The average figure per voyage was, thus, 420 passengers and 109 trucks.

10.4.4 *Modal Shift*

The model of Sect. 10.3.1 can be depicted schematically by the graph of Fig. 10.3. The X-axis in this graph is the difference (percent) in the transport cost along the combined-transport route (Route 2) resulting from the fuel switch necessitated by the Mediterranean ECA under study. The Y-axis is the corresponding share of the road-only route (Route 1). Note that for $X = 0$ (no difference in transport costs), $Y = 0.6836$ (the initial share of Route 1).

The new modal split resulting from the requirement to switch fuels is assessed as follows:

Table 10.3 Estimated proportion of total costs assigned to cargo (Hader et al., 2010)

Corridor	Routes	Ship	Est.no.of trailers/ FEUs per roundtrip	Costs attributable to/borne by trailers/trucks/ FEUs today (%)	Share of additional costs to be borne by the cargo (%)
German Baltic Sea ports-Western Sweden	Kiel-Gothenburg	STENA HOLLANDICA	340	40	60
German Baltic Sea ports-Norway	Kiel-Oslo	COLOR FANTASY	120	17	25
German Baltic Sea ports-Southern Sweden	Travem.-Trelleborg	ROBIN HOOD	160	80	100
	Travem.-Malmo	FINNEAGLE	200	95	100
	Rostock-Trelleborg	ROBIN HOOD	100	80	100
German Baltic Sea ports-Finland	Lübeck-Finland[a]	FINNSTAR	320	95	100
	Lübeck-Hanko	TIMCA	280	100	100
	Rostock-Helsinki	SUPERFAST VII	120	35	45
German Baltic Sea ports-Russia	Kiel-St. Petersburg	TRANSLUBECA	190	100	100
	Lübeck-Hamina-St. P.	PAULINE RUSS	140	100	100
	Lübeck-Sass.-St.P.	TRANSLUBECA	190	100	100
German Baltic Sea ports-Baltic States	Kiel-Klaipeda	LISCO GLORIA	180	80	95
	Rostock-Ventspils	URD	140	80	100
Belgium-Western Sweden	Gent-Gothenburg	TOR MAGNOLIA	250	100	100
	Zeebrügge-Gothenburg	SCHIEBORG	120	100	100

[a]Rauma/Turku/Hels./Kotka

- Total fuel consumption: 167 tonnes/voyage
- Of which, exceeding S limits: 116.9 tonnes/voyage (=167 tonnes* 70 %)
- Additional fuel cost per voyage: 38,577 $/voyage (=116.9 tonnes * 330 $/t)
- Or in Euro/voyage: 28,788.81 €/voy. (=38,577 $/voy. ÷ 1.34 $/€)

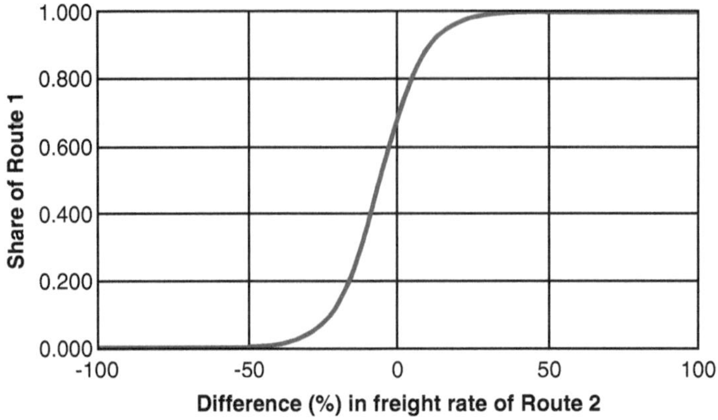

Fig. 10.3 The effect of changes in the transport cost along Route 2 on the share of Route 1

- Borne by the cargo: 12,954.96 €/voy. (=28,788.81 €/voy. * 45 %)
- Additional cost per truck: 118.85 €/truck (=12,954.96 (€/voy) ÷ 109 (trucks/voy))
- Additional cost per cargo tonne: 6.95 €/tonne (=118.85 €/truck ÷ 17.1 tonne/truck)

It is assumed that in the long run, actors operating in a competitive market will be forced to pass on this additional cost to their customers (the truck operators in our case). In turn, truck operators will pass it on to shippers, who will see the transport cost along the combined-transport route increasing to 376.42 €/tonne (=369.47 + 6.95).

This cost rise produces a new share for Route 1 equal to 0.7356, meaning that 5.2 % of the traffic will shift itineraries from Route 2 (combined-transport) to Route 1 (road-only).

10.4.5 Sensitivity Analysis

The robustness of model results with respect to the opportunity cost of capital, the price difference between HFO and MGO and, the share of the additional fuel cost that is being borne by the cargo of a Ro-Pax vessel is examined here.

The opportunity cost of capital, r, proves to be a very significant factor in forming modal shares. Figure 10.4 shows the S-curves of the modal split model for three different values of r: 2.97 % (default value of the basic scenario), 5 % and 10 %. The additional cost of 6.95 €/tonne calculated above leads to a modal shift of 5.2 % for $r = 2.97$ %; 6.2 % for $r = 5$ %; and 12.1 % for $r = 10$ %. In line with the discussion of Sect. 10.3.5, opportunity costs in the area of 10 % are closer to the realities of the fashion industry. In such case, shifts in the region of 12 % should be expected.

Figure 10.5 exhibits the effect of price difference between HFO and MGO on model results. In the basic scenario a differential of 330 USD/tonne has been selected leading to a modal shift of 5.2 % towards the road-only option. This shift

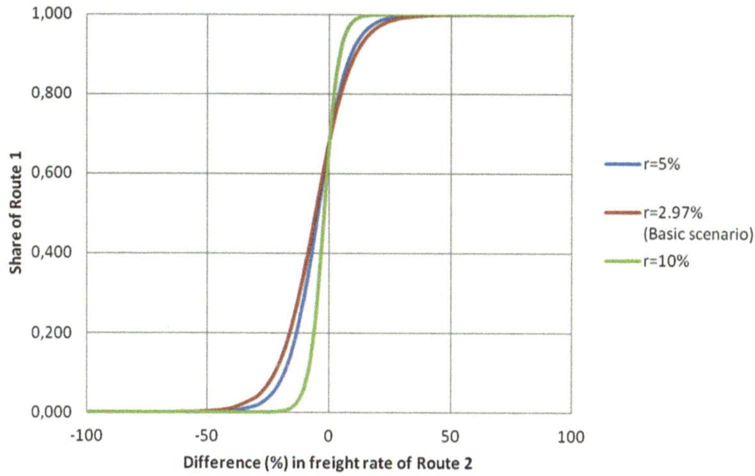

Fig. 10.4 The effect of the opportunity cost of capital on model results

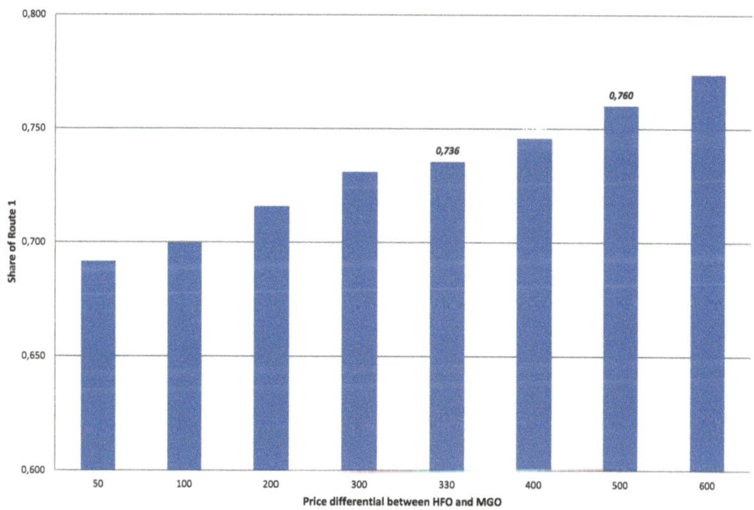

Fig. 10.5 The effect of fuel price differential on model results

escalates to 7.6 % should the price difference between the two fuel qualities become 500 USD/tonne, as assumed by Malmqvist and Aldén (2013).

The effect on modal shift of the share of additional costs that the cargo of a Ro-Pax vessel has to bear is presented in Fig. 10.6. This parameter depends on the significance of passenger traffic on the route and the price elasticities of both passengers and cargoes with respect to price increases, which in turn depend on the existence and price/quality relation of alternative transportation solutions.

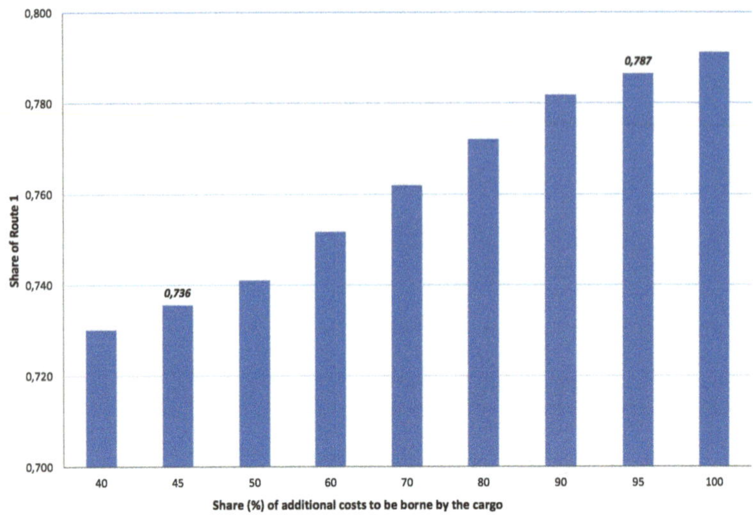

Fig. 10.6 The effect of the cost bearing capacity of cargo on model results

For the Greece-Italy ferry trades under study, all operators give priority to passengers and private cars over trucks/trailers. In fact, during the summer months, when the touristic traffic between the two countries is quite dense, trucks often encounter difficulties booking space on the ships. The 45 % estimate of the basic scenario corresponds to a modal shift of 5.2 % in favour of the road-only route, which gets doubled to 10.3 % in case 95 % of the additional fuel cost is allocated to the cargo.

It is noted that an average value of 70 % for this parameter, combined with an r value of 10 % leads to a shift in the area of 17.1 %, which is comparable to the results of Hader et al. (2010) for Germany.

10.5 Environmental Implications

The environmental consequences of the modal shift estimated above are discussed in this section. The web-based EcoTransIT World[2] tool has been used for the necessary calculations. Table 10.4 presents the emissions as they stand today prior to the designation of the Mediterranean Sea as an ECA. The emissions reported concern the CO_2-eq, PM_{10}, NO_x and SO_2 and correspond to the transport of 3.5 tonnes of clothing products from Thessaloniki to Berlin.

Due to the cargo collection and distribution operations performed by the same vehicle used for the long haul, average distances and load factors have been calculated for the first and last miles. However, this part of the operation is identical in the two alternative routes and does not produce any differences.

[2] http://www.ecotransit.org/index.en.html

Table 10.4 Emissions without a new ECA (for 3.5 tonnes of cargo)

Link	Distance (km)	Load factor (%)	CO_2-eq (tonnes)	PM_{10} (kg)	NO_x (kg)	SO_2 (kg)
Route 1						
First mile in Thessaloniki	16	38.54	0.0051	0.0011	0.0040	0.0059
Thessaloniki—Skopje	236	95.00	0.0389	0.0078	0.2968	0.0452
Skopje—Belgrade	393	95.00	0.0624	0.0127	0.4778	0.0731
Belgrade—Budapest	373	95.00	0.0588	0.0122	0.4561	0.0699
Budapest—Bratislava	200	95.00	0.0315	0.0064	0.2463	0.0374
Bratislava—Prague	338	95.00	0.0533	0.0102	0.4097	0.0625
Prague—Dresden	158	95.00	0.0253	0.0053	0.2001	0.0295
Dresden—Berlin	196	95.00	0.0309	0.0063	0.2373	0.0371
Last mile in Germany	276	58.22	0.0625	0.0134	0.4648	0.0741
Total route 1	**2,186**		**0.3687**	**0.0753**	**2.7930**	**0.4348**
Route 2						
First mile in Thessaloniki	16	38.54	0.0051	0.0011	0.0040	0.0059
Thessaloniki—Igoumenitsa	350	95.00	0.0562	0.0110	0.4312	0.0662
Igoumenitsa—Ancona	735	95.00	0.2780	0.1441	4.7051	1.0193
Ancona—Bologna	227	95.00	0.0357	0.0073	0.2732	0.0420
Bologna—Verona	151	95.00	0.0236	0.0048	0.1821	0.0278
Verona—Brenner	239	95.00	0.0381	0.0076	0.2884	0.0443
Brenner—Wörgl	97	95.00	0.0031	0.0006	0.0029	0.0029
Wörgl—Munich	109	95.00	0.0172	0.0035	0.1312	0.0202
Munich—Nurnberg	167	95.00	0.0270	0.0053	0.2000	0.0310
Nurnberg—Berlin	438	95.00	0.0697	0.0139	0.5375	0.0816
Last mile in Germany	276	58.22	0.0625	0.0134	0.4648	0.0741
Total route 2	**2,805**		**0.6161**	**0.2127**	**7.2207**	**1.4153**
Average (without ECA)			**0.4470**	**0.1188**	**4.1939**	**0.7450**

It is noted that, according to IFEU (2011), the EcoTransIT default values for the sulphur content of marine fuels outside ECAs are taken equal to 2.37 % for main engines burning HFO; 1.5 % for main engines burning MDO/MGO; 1.5 % for the auxiliary engines at sea; and 0.5 % for the auxiliary engines in port. This is not the case for the vessels operating on the Greece-Italy routes. To overcome this difficulty, the emissions along the Igoumenitsa-Ancona segment were calculated on the basis of the Kiel-Klaipeda link, which happens to be in an ECA and of an almost identical distance.

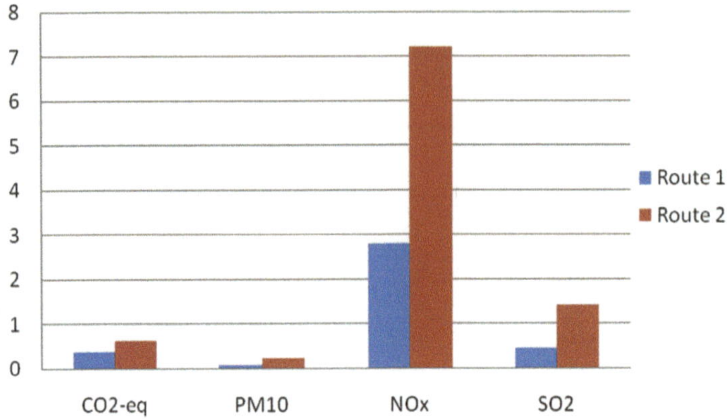

Fig. 10.7 Comparison of the two alternative routes without a new ECA

When compared with the figures that Delhaye et al. (2010) extracts from Notteboom et al. (2010) concerning the large Ro-Pax vessel ToR Petunia (employed on the Gent-Gothenburg route), the emissions of SUPERFAST XI (expressed in kg/tonne-km) appear much higher. The only exception relates to PM_{10} emissions that basically coincide. The service speed is certainly an explanatory parameter, since Notteboom et al. (2010) use 18.5 knots as the basis for their calculation, while SUPERFAST XI sails at almost 23 knots. The load factor of the vessels is another parameter of immense significance when it comes to relative (per tonne-km) figures. Nevertheless, it was decided to retain the EcoTransIT estimates as they are much closer to the actual fuel consumption figures (167 tonnes per voyage) provided by the Face^{3}ts (2008) report.

The emissions produced by the two alternative routes are compared in Fig. 10.7. Perhaps surprisingly, it is seen that the road-only option (Route 1) exhibits considerable environmental advantages over the combined-transport alternative. This is due to the fact that:

- Route 2 is longer than Route 1 by more than 28 %;
- the sea leg is characterized by impressively poor performance in terms of all GHG and air pollutants examined; and
- the train leg, which appears to be the most environmentally friendly mode, comprises a very small part of Route 2 unable to make a difference.

The last row of Table 10.4 calculates the average emissions produced by the 3.5-tonne consignment after accounting for the existing shares of the two alternative options.

The emissions pertaining to the basic scenario after the designation of the Mediterranean Sea as an ECA appear in Table 10.5. The only differentiation concerns the SO_2 and PM_{10} figures of Route 2. The new SO_2 amount is based on the old one after taking into consideration that 70 % of the total fuel consumption will need to be switched from the 1 % to the 0.1 % sulphur quality and that the latter

Table 10.5 Emissions with a new ECA (for 3.5 tonnes of cargo)

Link	Distance (km)	Load factor (%)	CO_2-eq (tonnes)	PM_{10} (kg)	NO_x (kg)	SO_2 (kg)
Route 1						
First mile in Thessaloniki	16	38.54	0.0051	0.0011	0.0040	0.0059
Thessaloniki—Skopje	236	95.00	0.0389	0.0078	0.2968	0.0452
Skopje—Belgrade	393	95.00	0.0624	0.0127	0.4778	0.0731
Belgrade—Budapest	373	95.00	0.0588	0.0122	0.4561	0.0699
Budapest—Bratislava	200	95.00	0.0315	0.0064	0.2463	0.0374
Bratislava—Prague	338	95.00	0.0533	0.0102	0.4097	0.0625
Prague—Dresden	158	95.00	0.0253	0.0053	0.2001	0.0295
Dresden—Berlin	196	95.00	0.0309	0.0063	0.2373	0.0371
Last mile in Germany	276	58.22	0.0625	0.0134	0.4648	0.0741
Total route 1	**2,186**		**0.3687**	**0.0753**	**2.7930**	**0.4348**
Route 2						
First mile in Thessaloniki	16	38.54	0.0051	0.0011	0.0040	0.0059
Thessaloniki—Igoumenitsa	350	95.00	0.0562	0.0110	0.4312	0.0662
Igoumenitsa—Ancona	735	95.00	0.2780	0.0728	4.7051	0.1396
Ancona—Bologna	227	95.00	0.0357	0.0073	0.2732	0.0420
Bologna—Verona	151	95.00	0.0236	0.0048	0.1821	0.0278
Verona—Brenner	239	95.00	0.0381	0.0076	0.2884	0.0443
Brenner—Wörgl	97	95.00	0.0031	0.0006	0.0029	0.0029
Wörgl—Munich	109	95.00	0.0172	0.0035	0.1312	0.0202
Munich—Nurnberg	167	95.00	0.0270	0.0053	0.2000	0.0310
Nurnberg—Berlin	438	95.00	0.0697	0.0139	0.5375	0.0816
Last mile in Germany	276	58.22	0.0625	0.0134	0.4648	0.0741
Total route 2	**2,805**		**0.6161**	**0.1414**	**7.2207**	**0.5357**
Average (ECA)			**0.4341**	**0.0928**	**3.9637**	**0.4615**

one produces 10 times less SO_2 emissions that the former. Similarly, the correction concerning PM emissions is based on the PM_{10} emission factors for S-content 1 % (0.72 g/kWh) and S-content 0.1 % (0.30 g/kWh) provided by IFEU (2011).

Despite the significant improvements in terms of SO_2 and PM_{10} emissions that the stricter regulations will trigger along the Igoumenitsa-Ancona segment (refer to Fig. 10.8), Route 1 continues being friendlier to the environment.

Figure 10.9 compares the average emissions for the 3.5-tonne consignment when the Igoumenitsa-Ancona connection lies outside and inside an ECA. The designation of the Mediterranean Sea as an ECA brings about significant improvements in

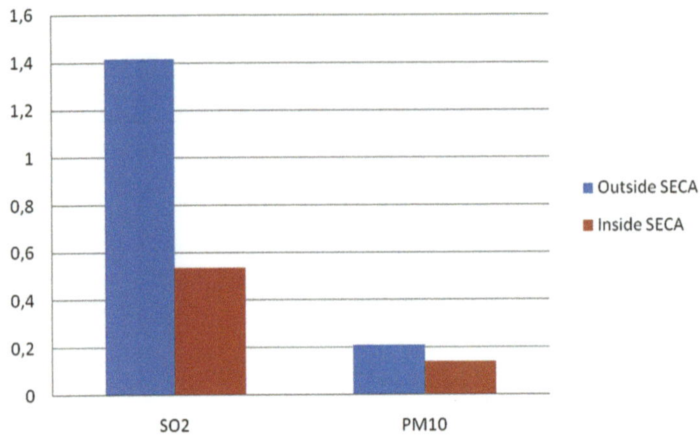

Fig. 10.8 Improvements in SO_2 and PM_{10} emissions along Route 2 due to fuel switch

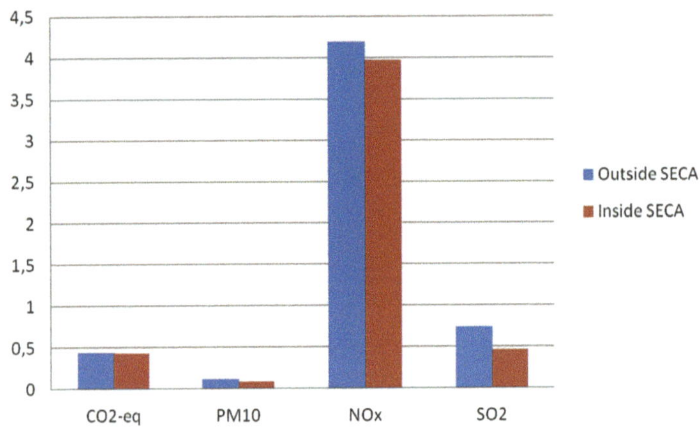

Fig. 10.9 Average emissions for a 3.5-tonne consignment

all fronts. Given that in most relevant studies the potential backshift from sea- to land-based routes is associated with a deterioration of the environmental performance, this is a rather unexpected result. It is explained by the fact that the switching to a cleaner and more expensive fuel:

- leads to a modal shift in favour of the road only option, which in this case exhibits better environmental characteristics; and
- the SO_2 and PM_{10} emissions of the remaining Route 2 traffic are substantially improved compared to the status quo.

Knowing that shipping is the friendliest transportation mode to the environment, isn't this result a paradox? The answer is no. Shipping is not a just a single service offered in the same way around the world. There are many types of vessels employed on many different operations meeting a broad range of transportation needs for people and freight. In our specific case, the ship had to sail at an average speed of 22.9 knots in order to reach Ancona in 17.5 h. The negative repercussions of speed to both ship economics and the environment are well known (see also Chap. 9). A question is whether the business concept that a Ro-Pax ship serves can make sense at lower speeds. This is certainly an issue worthy of investigation in a high fuel price regime. In any event, we believe that the stricter sulphur limits of the regulation examined in this chapter would provide the right incentives.

10.6 Conclusions and Possible Extensions

10.6.1 Conclusions

This chapter applied a modal split model on a case study that investigates the impact of a possible designation of the Mediterranean Sea as an ECA. The model is of a binomial logit type, taking into consideration transportation cost and time as explanatory variables of the choices made. The method demonstrates how the effects of the sulphur regulations on modal split can be calculated. The results of the scenario are scenario-specific and do not necessarily generalize to other scenarios.

The specific case study examined concerned the transportation of clothing products from Thessaloniki, Greece to northern Germany. The small size of the consignment, which is the norm under the present economic conditions, necessitates cargo consolidation/ distribution at both ends of the voyage. The two alternative routes examined involve a road-only solution along the path Thessaloniki-Skopje-Belgrade-Budapest-Bratislava-Prague-Dresden-Berlin and a combined-transport solution following the path Thessaloniki-Igoumenitsa-Ancona-Bologna-Verona-Munich-Nurnberg-Berlin. In addition to crossing the Ionian/Adriatic Seas with a Ro-Pax vessel, the latter solution involves also a truck-on-train operation along the segment Brenner-Wörgl.

Information was gathered through interviews with a small-size Greek truck operating company that specializes in services between Thessaloniki and destinations in Italy and Germany. It concerns actual trips made in year 2010. As such, the application follows the revealed preference approach.

Under certain assumptions comprising the 'basic scenario', the designation of Mediterranean as an ECA will cause an increase of transport costs by 6.95 €/tonne (equivalent to 1.9 %). According to our model, this rise will result in a modal shift of 5.2 % in favour of the road-only route.

The shift grows to 12.1 % for an opportunity cost of 10 %, which is much closer to the realities of the clothing industry than the 2.97 % value assumed in the basic scenario. Should this figure be combined with a 70 % cost bearing ratio by the cargo of a Ro-Pax ship (in contrast to the 45 % ratio of the basic scenario), the modal shift reaches the level of 17.1 %, which is comparable to previous results for the existing ECAs.

As for the environmental implications of this shift, it turns out that the stricter regulations bring about significant improvements in relation to all emissions examined (CO_2-eq, PM_{10}, NO_x and SO_2). This is attributed to the longer (by 28 %) distance of the combined-transport solution in comparison to the road-only route and the poor performance of the Ro-Pax vessels basically due to the need to maintain a relatively high speed (22.9 kn). The railway involved in the combined-transport case appears to be the most environmentally friendly mode but comprises a very small part of this route in order to make a difference.

It is of course necessary to note that designing the Mediterranean Sea as an ECA comprises a political decision that should adopt a broader scope and cannot be based simply on a case study like the present one. However, we think the results of this chapter are worthy of note and believe that the methodology used in the chapter could form the basis for such a broader analysis. Independent of geographical context, we also think that this model, or extensions of it, can be used as a basis to assess the possible side effects of sulphur regulations. The geographical area that seems to be under the most pressure at this point in time is the Northern European ECA.

10.6.2 Possible Extensions

The following considerations can be embedded in the model to extend it and make it more realistic.

1. Ro-Ro routes profitability can be an unstable variable as even small shifts of traffic away from the maritime mode can make a route unprofitable and subsequently shut it down. The modal split model described above does not capture this fact so it will have to be appropriately enhanced.
2. Another possible side-effect is if the Ro-Ro carrier reduces speed as a possible measure to mitigate the effects of modal shifts on route profitability. Reducing speed will decrease cost, but will also increase transit time, and as transit time enters the generalized cost calculation, some cargoes may be tempted to use the road mode as it travels faster.
3. The new freight rate the Ro-Ro carrier will charge as a result of the fuel price increase (price surcharge) is a decision variable that will have two counter-balancing effects: (a) increase of revenue for the cargo carried, and (b) decrease of cargo carried due to the surcharge. Whether the effect of (a) will be able to offset the effect of (b) will depend on the surcharge and on the scenario.
4. The scrubber option will increase capital costs but decrease fuel costs, vis-à-vis the non-scrubber, low sulphur fuel option. Both should be taken into account when this option is analyzed.
5. The value of the cargo will impact these calculations, as more expensive cargoes will be generally encouraged to use the faster mode.

6. Another measure the Ro-Ro carrier may adopt is reducing the number of vessels and/or the frequency of service on the route. Doing so could increase utilization of the fleet and hence profitability, but may have the side-effect that ship capacity may not be enough, in which case cargo carried (and hence revenue) is lost, and if a surcharge is applied it can lead to further modal shifts.
7. Another side-effect of frequency reduction is that cargoes will have to wait more in port to use the Ro-Ro mode, and if this waiting time is factored into the Ro-Ro mode's total transport time, the generalized cost of the Ro-Ro mode will increase and its share will decrease.

In Algaba (2014), analysis for a short-sea route in Northern Europe was performed and some (not all) of the above considerations were included. The best strategy that was found was reducing the number of yearly trips per vessel to 83 % of the initial yearly trips, installing a scrubber (operating with high sulphur content fuel) and keeping the original fleet without applying any surcharge. However, only one route was considered and not all side-effects from frequency reduction were considered. This work will be extended in a recent research award to the Technical University of Denmark funded by the Danish Maritime Fund, on the possible impacts of sulphur regulations on the Ro-Ro sector in Northern Europe. The results this project will be reported in future publications.

Acknowledgements Work on this chapter has been supported in part by The Lloyd's Register Foundation (LRF) in the context of the Centre of Excellence in Ship Total Energy-Emissions-Economy at the National Technical University of Athens (NTUA). LRF helps to protect life and property by supporting engineering-related education, public engagement and the application of research. This work has also been supported in part by a gift agreement from Det Norske Veritas to NTUA and by an internal grant at the Technical University of Denmark.

References

Algaba, O. B. (2014). *Impact study of the new sulphur regulations on a North Sea short-sea route* (Master's thesis). DTU, Department of Transport, Kongens Lyngby, Denmark.

Bank of Greece. (2012, November–December). *Bulletin of the Conjunctural Indicators*, Number 147.

Bosch, P., Coenen, P., Fridell, E., Åström, S., Palmer, T., & Holland, M. (2009). *Cost benefit analysis to support the impact assessment accompanying the revision of directive 1999/32/EC on the sulphur content of certain liquid fuels*. Didcot, England: AEA report to European Commission.

Cullinane, K., & Bergqvist, R. (2014). Emission control areas and their impact on maritime transport. *Transportation Research Part D: Transport and Environment, 28*, 1–5.

Delhaye, E., Breemersch, T., Vanherle, K., Kehoe, J., Liddane, M., & Riordan, K. (2010). *The competitiveness of European short-sea freight shipping compared with road and rail transport*. COMPASS final report.

EMSA. (2010). *The 0.1% sulphur in fuel requirement as from 1 January 2015 in ECAs—An assessment of available impact studies and alternative means of compliance*. Technical report of the European Maritime Safety Agency.

Face^{3}ts. (2008). *Environmental performance of the Greece-Italy ferry lines in relation to climate change and air pollution (in Greek)*. A Face^{3}ts (Further action on climate, environment, energy, economy, technology and sustainability) project report.

Fagerholt, K., Gausel, N., Rakke, J., & Psaraftis, H. N. (2015). Maritime routing and speed optimization with emission control areas. *Transportation Research Part C: Emerging Technologies, 52*, 57–73.

Grebot, B., Scarbrough, T., Ritchie, A., Mahoney, C., Noden, R., Sobey, M., et al. (2010). *Study to review assessments undertaken of the revised MARPOL Annex VI regulations*. London, England: Entec UK Limited.

Hader, A., Hübscher, A., Maatsch, S., & Tasto, M. (2010). *Reducing the sulphur content of shipping fuels further to 0.1 % in the North Sea and Baltic Sea in 2015: Consequences for shipping in this shipping area*. Bremen, Germany: Institute of Shipping Economics and Logistics.

Jiang, L., Kronbak, J., & Christensen, L. P. (2014). The costs and benefits of sulphur reduction measures: Sulphur scrubbers versus marine gas oil. *Transportation Research Part D, 28*, 19–27.

Kalli, J., Karvonen, T., & Makkonen, T. (2009). *Sulphur content in ships bunker fuel in 2015—A study on the impacts of the new IMO regulations on transportation costs*. Prepared by the Centre for Maritime Studies, University of Turku, for the Ministry of Transport and Communications of Finland, Helsinki, Finland.

Kehoe, J., Nikopoulou, Z., Liddane, M., Ramstedt, L., & Koliousis, I. (2010). *Impact study of the future requirements of annex VI of the MARPOL convention on short sea shipping. SKEMA study prepared for directorate-general for energy and transport*. Brussels, Belgium: European Commission.

Kehoe, J., & Woxenius, J. (2010). *Assessment of the impact of the application of new sulphur limits to the Mediterranean and the Atlantic European seas. SKEMA study prepared for directorate-general for mobility and transport*. Brussels, Belgium: European Commission.

Kronbak, J. (2006). *The use if GIS for modelling and visualization of cost-competition between land and sea transport*. Presented at the IAME 2006 conference, Melbourne, Australia.

Ljungström, T., Leyendecker, J., & Lemieszewski, S. (2009). *Consequences of the IMO's new marine fuel sulphur regulations*. Norrköping, Sweden: Swedish Maritime Administration.

Malmqvist, G., & Aldén, B. (2013). *Sulfur regulation in the Baltic Sea: Scenarios for the mid Nordic region—threats and opportunities. A North East Cargo Link II report*. Sundsvall, Sweden: Midnordic Green Transport Corridor.

Nordås, H. K., Pinali, E., & Geloso Grosso, M. (2006). *Logistics and time as a trade barrier* (OECD Trade Policy Working Paper No. 35). Paris, France: OECD.

Notteboom, T., Delhaye, E., & Vanherle, K. (2010). *Analysis of the consequences of low sulphur fuel requirements. Study commissioned by the European Community Shipowners' Associations (ECSA)*. Antwerp, Belgium: University of Antwerp.

Panagakos, G., Stamatopoulou, I. V., & Psaraftis, H. N. (2014). The possible designation of the Mediterranean as a SECA: A case study. *Transportation Research Part D, 28*, 74–90.

Psaraftis, H. N., & Kontovas, C. A. (2009, May 26–29). Ship emissions: Logistics and other tradeoffs. In *10th International Marine Design Conference (IMDC), Trondheim, Norway*.

Psaraftis, H., & Kontovas, C. (2010). Balancing the economic and environmental performance of maritime transportation. *Transportation Research Part D, 15*(2010), 458–462.

Schinas, O., & Bani, J. (2012). *The impact of a possible extension at EU level of ECAs to the entire European Coastline*. Note to the Committee on Transport and Tourism, Directorate-General for Internal Policies, European Parliament, Brussels, Belgium.

Stavrakaki, A., Ritchie, A., Pueyo, A., Lavric, L., Green, C., Scarbrough, T., et al. (2009). *Impact assessment for the revised Annex VI of MARPOL*. London, England: Entec UK Limited.

Sull, D., & Turconi, S. (2008). Fast fashion lessons. *Business Strategy Review, Summer*, 5–11.

Chapter 11
Critical Analysis of Air Emissions from Ships: Lifecycle Thinking and Results

Stefanos D. Chatzinikolaou and Nikolaos P. Ventikos

Abstract The authors have incorporated the life cycle thinking approach in their research activities during the past few years with the aim to conduct environmental assessments of ship air emissions from a life cycle perspective. This chapter presents some illustrative findings from this work. These include a presentation of a life cycle ship framework which considers the ship as a system that may be detailed into sub-systems for which: (a) inputs, (b) processes, and (c) outputs, are identified and elaborated. Important ship life cycle stages are built-in in this model; namely the shipbuilding stage, ship operation including major maintenance activities, and finally the stage of ship dismantling/recycling. This chapter also presents illustrative numerical results of a Life Cycle Assessment study conducted for important air emissions occurring throughout the life cycle of an ocean going ship. Finally, the chapter discusses the main difficulties observed during the period of experimenting with the Life Cycle Assessment method and some limitations of this method to sufficiently cover the case of shipping. The main issues discussed in this context are: the adaptation of the methodology to maritime transport scenarios, the system boundaries selection, the establishment of life cycle inventories and the availability of data and most importantly the impact assessment step of Life Cycle Assessment which (for the case of ships) has shown to have more difficulty and wider uncertainty than any other step of the method.

Abbreviations

CF	Characterization factor
CH_4	Methane
CO	Carbon monoxide
CO_2	Carbon dioxide
DALY	Disability adjusted life years
dwt	Dead weight of the ship

S.D. Chatzinikolaou (✉) • N.P. Ventikos
Laboratory for Maritime Transport, School of Naval Architecture and Marine Engineering, National Technical University of Athens, Athens, Greece
e-mail: stefanos@deslab.ntua.gr; niven@deslab.ntua.gr

H.N. Psaraftis (ed.), *Green Transportation Logistics*, International Series in Operations Research & Management Science 226,
DOI 10.1007/978-3-319-17175-3_11

ESA	Environmental system analysis
EU	European Union
GHG	Greenhouse gasses
HFO	Heavy fuel oil
IMO	International Maritime Organization
ISO	International Organization for Standardization
LCA	Life cycle assessment
LCI	Life cycle inventory
LCIA	Life cycle impact assessment
LNG	Liquefied natural gas
MARPOL	International convention for the prevention of pollution from ships
MGO	Marine gas oil
NMVOC	Non methane volatile organic compounds
NO_x	Nitrogen oxides
PM	Particulate matter
SETAC	Society for Environmental Toxicology and Chemistry
SO_2	Sulphur dioxide
SO_x	Sulphur oxides
UNEP	United Nations Environmental Program
VOC	Volatile organic compounds

11.1 Introduction

This chapter presents and discusses methods for applying the theoretical concept of life cycle thinking into maritime transport's air emissions assessments. The life cycle thinking concept has continuously gained acceptance in environmental assessments over the past few years since it represents a transition from traditional environmental protection schemes as a response to the growing awareness of modern societies about the long term impacts of human activities.

A well structured technique behind the life cycle thinking approach is the Life Cycle Assessment (LCA), a systematic method standardized under the International Organization for Standardization (ISO), which considers the full life cycle of the system it examines: from the extraction of resources and raw material production, through transportation, assembly, operational life and maintenance, up to the recycling and final disposal of wastes. When studying industrial systems in this respect, some important benefits emerge such as, the avoidance of the unwanted shifting of impacts from one stage of the life cycle to another, or altogether to another system. This way the LCA is particularly helpful in avoiding the externalization of environmental costs.

The rest of this chapter is structured as follows. In Sect. 11.2 we initially describe the fundamentals of LCA and in Sect. 11.3 go on with a literature survey of expressive applications of this method in the maritime transport sector. Next, in Sect. 11.4, a ship life cycle framework employing the basic elements of the LCA method is presented. This framework has the capability to deliver air emissions inventories for all important processes throughout a ship's life cycle. In addition, the framework goes one step further by providing an assessment of the impact of ship air emissions to the environment and human health. Major impacts of shipping emissions occur at the local level (i.e. damages to human health, air quality and land and marine ecosystems), as well as at the global level (i.e. climate change). Assessing such impacts poses certain challenges since the available methods and tools have not been developed for the specific case of shipping. The framework incorporates a combination of algorithms and user inputs for the emissions calculations and uses one widely applied damage model for the impact assessment step. A case study of application of the life cycle framework is being presented in Sect. 11.5, which has been conducted using data from a Panamax tanker (75,000 tonnes of dwt). Moreover, the framework has been tested against various alternatives scenarios of operation and results are presented and discussed accordingly. Section 11.6 discusses the main difficulties observed during the period of experimenting with the LCA method and points out a number of limitations of this method to thoroughly cover the case of maritime transport. The main issues discussed in this context are: the adaptation of the methodology to maritime transport scenarios, the system boundaries selection, the difficulties in the establishment of life cycle inventories and the availability of data and most importantly the impact assessment which has shown to be the most challenging step of LCA. Finally, in Sect. 11.7 the overall conclusions from this work are presented together with some ideas for making then LCA more robust and adaptable for the case of shipping in the future.

11.2 Life Cycle Assessment

The field of Environmental System Analysis (ESA) addresses the interaction between human-made systems and the environment. Different ESA tools are available which are generally divided into procedural and analytical tools. Procedural tools focus on improving the procedures leading to decision-making, while analytical tools provide information that may be utilized as means of communication, optimization of the studied system, comparisons of different alternatives for the system, etc. (Finnveden & Moberg, 2005).

The Life Cycle Assessment (LCA) method is an analytic ESA technique, which is used for assessing the possible environmental impacts of technologies and products. The LCA method was first developed for the environmental assessment of industrial products in the 1960s. The term 'product' can include not only product systems but also service systems, or processes. Since its beginning, the method has been improved considerably and numerous LCA studies have been conducted in different industries including transportation in general. ISO standards (ISO 14040–14044) are available for LCA, which provide the theoretical framework, terminologies and methodological choices for this method. An often-quoted definition of LCA is the one provided by the ISO 14040 standard: LCA is the "compilation and evaluation of the inputs, outputs and the potential environmental impacts of a product system throughout its life cycle" (ISO, 2006a).

The LCA method is continuously gaining acceptance as a suitable tool for analysing the impact that different solutions have on their external environment throughout the duration of their lifetime. One of the most important benefits of an LCA is that it allows studying an entire product system hence avoiding potential sub-optimization that could result if only a single process were the focus of the study. An effective LCA allows analysts to (EPA, 2006):

- Calculate a product's environmental impact
- Identify the positive or negative environmental impact of a process or product
- Find opportunities for process and product improvement
- Compare and analyse several processes based on their environmental impacts
- Quantitatively justify a change in a process or product

The growing interest for LCA in recent years is also demonstrated by the fact that there are important initiatives launched for this concept at European and global level. The European Platform of Life Cycle Assessment (http://lct.jrc.ec.europa.eu/index_jrc), run by the Joint Research Centre (JRC) is the official EU initiative created to facilitate communication on life-cycle data and commence a co-ordination scheme involving both ongoing data collection efforts in the EU and existing harmonization projects. Another major initiative was launched cooperatively by the United Nations Environment Programme (UNEP) and the Society for Environmental Toxicology and Chemistry (SETAC) namely the UNEP/SETAC Life Cycle Initiative (http://lcinitiative.unep.fr). The mission of this initiative is to bring together different science-based Life Cycle approaches worldwide and explore the possibilities to achieve a global consensus on how the method should be conducted.

11.2.1 LCA Steps

The LCA process usually consists of the following four main steps:

1. Goal Definition and Scoping: Definitions of the product, process or activity. Establishment of the context in which the assessment is to be made and identification of the boundaries and environmental effects to be reviewed for this assessment.
2. Inventory Analysis: Identification and quantification of energy and materials use and environmental releases (e.g., air emissions, solid waste disposal, waste water discharges).
3. Impact Assessment: Assessments of the potential human and environmental effects of energy and material usage from emissions and releases identified in the inventory analysis step.

4. Interpretation: Evaluation of the results of the inventory analysis and impact assessment in order to select the preferred product, process or service with a clear understanding of the uncertainty and the assumptions used to generate these results. Communication of the results to the interested parties.

11.2.2 Life Cycle Impact Assessment

Impact assessment is one of the most interesting features of LCA and the most challenging one as well. Once all the required emission and resource data is collected in an inventory list, a life cycle impact assessment (LCIA) can be performed to calculate the potential environmental impact of the inventory data. The outcomes of the assessment (the impact score) can be interpreted and further analysed to reduce uncertainties from vague inventory data, data gaps and important assumptions taken during the data collection and impact assessment (ISO, 2006b).

The LCIA step of LCA is a field of active development and impact methods have been continuously elaborated and improved over the past few years (Finnveden et al., 2009; Hauschild et al., 2013). The LCIA consists of mandatory elements (1, 2, and 3) as well as optional elements (4 and 5) as shown in the following figure (Fig. 11.1).

Selection of impact categories involves the identification of relevant categories of impact for the particular study's needs (i.e. climate change, eutrophication acidification etc.) Classification is the assignment of inventory results to impact categories. This should be done by assigning the inventory results that are not only exclusive to one impact category but also relate to more than one impact categories, including distinction between parallel mechanisms (e.g. SO_2 is apportioned between the impact categories of human health and acidification), or relation to serial mechanisms (e.g. NO_x can be classified to contribute to both ground-level ozone formation and acidification).

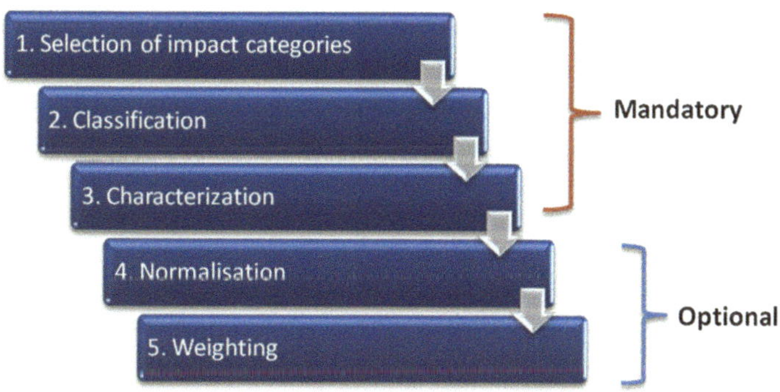

Fig. 11.1 LCIA steps according to ISO standards

Within LCA, a number of different characterization methods are formed together to address different environmental impacts (the impact categories) covered by the methodology. Each one of these characterization methods uses a cause-effect pathway and impact indicator, to produce the so-called characterization factors (CFs). Therefore, CFs are weighting factors that aggregate life cycle emissions reflecting the relative contribution of an LCI result to the impact category score. In general, characterization models can use two types of impact indicators, midpoint indicators and endpoint indicators. Midpoint indicators are normally expressed as equivalent values. Examples are kg CO_2-equivalents for climate change, SO_2-equivalents for acidification and MJ-equivalents for resource use. A well known example is the contribution of CH_4 to the global warming impact category which is 25 times higher than the CO_2 emissions of the same quantity. The characterization factor of CH_4 is therefore 25 (De Schryver, 2010).

The characterization step in LCIA makes use of damage assessment models. Damage assessment is a relatively new step in impact assessment. Its purpose is to combine a number of impact category indicators into a damage category (area of protection). The last decade has seen numerous new impact assessment methods covering many different impact categories and providing characterization factors that often deviate from each other for the same substance and impact (Hauschild et al., 2012).

11.3 Life Cycle Studies in Shipping

Within the maritime transport sector, LCA studies have been initially conducted during the 1990s. These studies have demonstrated that the LCA method may be employed for an environmental life cycle evaluation of a ship. An illustrative study in this respect is the Screening LCA (which is a simplification of the LCA method) applied on a RO-RO passenger vessel (Johnsen & Fet, 1998). One of the main findings of this study was that the LCA method is very time consuming and methodological simplification is needed. However, the authors stated that by following a detailing approach which essentially means breaking the ship-system into sub-systems (i.e. hull, machinery, equipment for cargo etc.) the assessment may become more straightforward and effective. It is highlighted though, that this approach might not always be helpful since other problems such as bad data quality and inconsistency in the system boundaries might lead to uncertain results.

A number of LCA software applications have been developed to assess the environmental impact of ships. The Norwegian University of Science and Technology developed a life cycle tool for fishing vessels (Ellingsen, Fet, & Aanondsen, 2002). The National Maritime Research Institute of Japan has developed a tool to investigate the environmental impact of different cargo vessels using impact assessment methods applicable specifically for the case of Japan (Kameyama, Hiraoka, Sakurai, Naruse, & Tauchi, 2004). A consortium of Swedish maritime organizations has launched a life cycle design tool for evaluating the energy efficiency of ships (Jiven et al., 2004). The above LCA tools are based on the SimaPro® software platform which is a widely used commercial tool for life cycle analysis applications. The software SSD after "Sustainable Ship Design" which is also based on the SimaPro® software platform has been developed to evaluate different "green" technologies in terms of environmental impacts in a life cycle perspective (Tincelin, Mermier, Pierson, Pelerin, & Jouanne, 2010).

LCA studies often analyse specific compartments and processes of the ship or assess the impact of specific emissions. Hou (2011) has studied the life cycle environmental impact of different superstructure materials (traditional steel and aluminium vs. new-type composite sandwich material). Chatzinikolaou and Ventikos (2013a, 2013b) have used elements of the LCA to develop a holistic framework capable of producing ship air emission inventories in a life cycle perspective.

Recent life cycle studies have focused on the area of marine fuels. Ryste (2012), has used the LCA framework to conduct a life cycle analysis of the bunkering process of LNG as marine fuel looking, in particular, to the climate change impacts of this type of fuel. The International Council on Clean Transportation, ICCT (2013) has recently published an analysis of the life cycle greenhouse gasses and the possible benefits of using LNG as an alternative marine fuel. A comparative LCA study has examined, in a life cycle perspective, the impact of LNG and HFO used as marine fuels (Laugen, 2013). Comparisons of different options of marine fuels (HFO, MGO, gas-to-liquid fuel, and LNG, combined with two exhaust abatement techniques) have been performed in another study, using the life cycle approach from extraction of raw material to transportation of 1 ton cargo in 1 km on a Ro-RO vessel (Bengtsson, Andersson, & Fridell, 2011).

Finally, there are studies dealing with environmental loads of maritime transport in a life cycle perspective but without making use of the standardized LCA method or any dedicated software. One of these studies investigated how the average annual cost of ship transport varies with the corrosion margins selected at the design stage. The results clearly indicated that ships built with sufficient corrosion allowances, truly adequate for the ship's design life, have a lower life cycle cost per annum despite the fact that such ships would carry a slightly smaller quantity of cargo (Gratsos, Psaraftis, & Zachariadis, 2010).

11.4 Ship: LCA Framework

The objective of the LCA—framework is to model and assess the important processes in terms of air emissions during the ship's life cycle. The approach followed for this challenge has made use of basic knowledge from Systems Theory and the LCA method. It is acknowledged that there is a clear distinction between the real system (ship) and the developed framework. Essentially, the framework that will be presented in the following has been developed explicitly for the specific purpose of air emissions analysis and inevitably is only a simplification of the real ship system.

11.4.1 Description of the Framework

Four major ship life cycle stages are considered and analysed i.e. the shipbuilding, the operation phase, the maintenance phase (only the off-duty repairs are taken into account) and the recycling phase. The usual life span of a cargo ship is 20–25 or more years from 'cradle to grave'. During this period emerging technology, policy solutions or other reasons may drastically modify the environmental footprint of the ship. Some of the anticipated future developments are known (e.g. the changes of sulphur limits in marine fuels) and hence can be integrated in the analysis.

One of the important anticipated changes in the ship's life cycle refers to the quality of maritime fuels. In order to cope with this challenging issue the EX-TREMIS/EUROSTAT database has been utilized. This is a web-based reference system and inventory of fleet data, transport activity data, energy consumption factors, emission factors and total emissions for rail, maritime and aviation for European countries (Chiffi, Schrooten, & De Vlieger, 2007). This database projects the developments in various emission factors for different ship types and sizes in the following years until 2030. The projection provided by the EX-TREMIS/EUROSTAT database is used as a basis for the calculation of emissions for the model Inventory. Other factors that may have an important impact on the life cycle of the ship (such as the prices of steel and fuel and freight market conditions) are more difficult to model and they are not taken into account.

The ship (the System) is viewed as a series of subsystems in a concept which is very often used as reference in the shipbuilding industry. A subsystem is defined as an individual step that is part of the defined total system. With respect to air emissions only two subsystems are qualified as important throughout the life cycle of ocean going ships; namely the hull subsystem, and the machinery subsystem. Each one of these two subsystems is further detailed into system elements as shown in Fig. 11.2. At the system element level a distinction is made between the different components of the subsystem that may be individually elaborated. In the process level all the important processes are identified per system element in the context of inputs (energy and raw materials) and outputs (air emissions). This identification is performed per life cycle stage since one system element may not have the same processes in different life cycle stages.

The hull subsystem is divided in the hull material and hull protection system elements. This partition has been used in previous LCA studies (Johnsen & Fet, 1998). For the hull subsystem no important processes (with respect to air emissions) are considered in the operational life phase.

For a cargo ship, steel is the main hull material with respect to air emissions production. In the shipbuilding life cycle stage important processes of the hull material system element are steel welding, cutting, and abrasive blasting. The boundaries of the shipbuilding include the production of steel and a transportation scenario of the steel material from the production site to the shipyard. In the life cycle stage of operation no important processes in terms of emissions production are considered for the hull material system element. The processes included in the

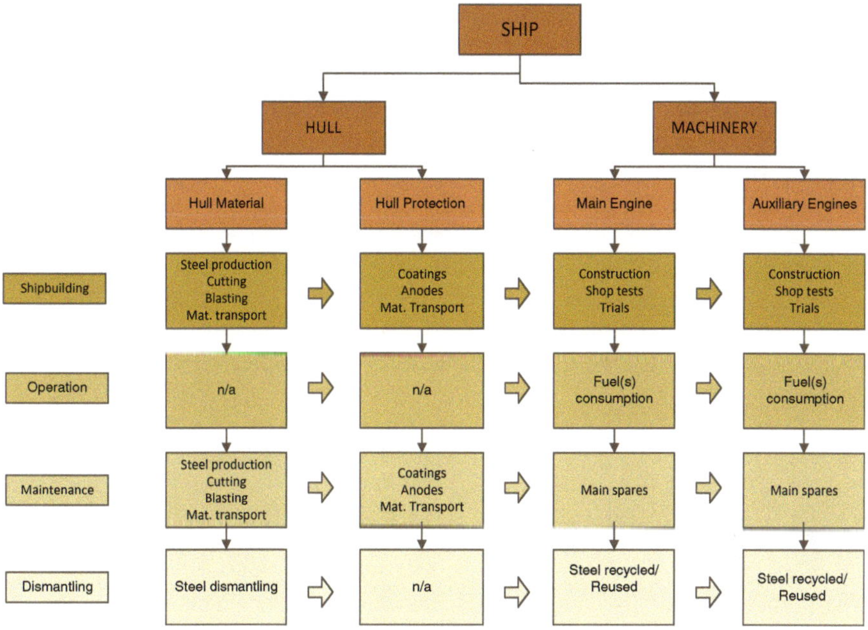

Fig. 11.2 Ship—LCA framework for the assessment of air emissions in a life cycle perspective

maintenance life cycle stage are identical to the shipbuilding stage although quantities of materials and resulting emissions are considerably less. The processes included in the recycling stage is steel recovery which takes into consideration the specific way that the steel is being recovered (re-rolling of steel) in the selected site (Alang, India) for which data was available.

Coating is a major process in the hull protection system element. This concerns mainly the life cycle stages of shipbuilding and ship maintenance. Some painting activities are performed also during the operational life of the ship (usually by the onboard personnel) but they are not considered important air emission contributors.

The process of anodes installation (on the hull, rudder and water ballast tanks) is also included in the hull protection system element in the stages of shipbuilding and ship maintenance. A specific scenario is also considered for transporting the relevant materials to the shipyard. For the ship recycling stage the fate of materials used for hull protection is not known. Therefore for this life cycle stage no process has been incorporated in the hull protection system element.

The machinery subsystem is divided in two system elements: the main engine(s) which provide the propulsion power to the ship, and the auxiliary engines which offer electrical power for accommodation, cargo and other ship needs. These two system elements have identical processes in the framework.

The great portion of ship environmental impact derives from the operational life when the consumption of fuels takes place. Detailed information on the initial stages of the life cycle of diesel engines before they are installed into the ship is not widely available. However, the study has collected data for the construction and testing processes of engines before they are transported to the shipyard for onboard installation. Therefore, the processes of engine construction and testing are included in the analysis. The operational phase includes the consumption process and the maintenance phase considers some major replacement of main engines parts for which data were available. The dismantling phase considers the specific practice adopted in the selected recycling site (Alang, India). The production of fuels to be used in ship engines is a matter of boundaries selection. The fuel production process is not included in the life cycle boundaries of this framework although the authors have tested elsewhere (Chatzinikolaou, Psaraftis, & Ventikos, 2011) the possible contribution of this process to the overall emissions impact. More on this issue and on the general issue of boundaries selection in LCA are discussed in Sect. 11.6.

11.4.2 Framework Capabilities

The framework initially comprises a series of algorithms which calculate the air emissions during the life cycle of the ship. The calculations lead to the development of the Life Cycle Inventory (LCI) of ship air emissions. The LCI is then utilized with the adaptation of an Impact Assessment technique to calculate the environmental impact of air emissions. The Framework's output main capabilities are the following:

- Inventory of air emissions from any identified process
- Emissions covered: CO_2, CO, SO_2, NO_x, PM (all), CH_4, VOCs
- Air emissions per life cycle stage
- Air emissions per process, system element, subsystem, and total
- Annual air emissions analysis
- Emission comparisons between different operational ship profiles
- Examination of different operational scenarios (initial scenario, slow steaming, speed limit, fleet distribution, etc.)

Some basic naval architecture calculations are initially performed in the framework for determining important ship details which are going to be useful for the air emissions calculations. These refer to the calculations of wetted surface, hull, deck and superstructure surfaces, cargo holds surfaces, water ballast tanks surfaces and steel weight. The study has made effort to avoid using generic or databases data and developed algorithms that model important processes in the ship life cycle. Unique features from this effort are the algorithms developed explicitly for the calculation of emissions during welding and coating operations in shipyards, the algorithms for assessing the added resistance effect due to marine growth on the ship's hull, and the algorithms for assessing air emissions in different scenarios of the operational life (Chatzinikolaou & Ventikos, 2013a, 2013b).

The framework also includes an algorithm for calculating the various time periods of ship operation and the trips accomplished per year of the life cycle. These calculations lead to the estimation of the transport work accomplished (throughput) in trip and year basis. The user should determine some basic variables such as the distance covered per trip, the relevant speeds (ballast, laden leg), the waiting times (outside port, manoeuvring) the number of ships used for the required transport work and the life cycle years. The round trips per year are calculated using formulas for the unavailability of tankers which as function of the age of the ship (Turan, Ölçer, Lazakis, Rigo, & Caprace, 2009).

11.4.3 Ship Emissions Impact Assessment

The motivation for studying the impact of ship air emissions derives from the fact that there is not much knowledge on the real impacts of shipping to the environment and human health although there is plenty of scientific effort put on the quantification of ship emissions.

There is strong evidence showing that the impact of shipping emissions may be underestimated so far. Since the 70 % of global shipping emissions occur within 400 km from land (Endresen et al., 2003) ships are potentially significant contributors to air pollution impacts. Moreover, studies (Corbett et al., 2007) claim that PM emissions of ships are responsible for approximately 60,000 premature deaths annually, most of them occurring near coastlines in East Asia, Europe and South Asia. A recent study, (EMEP, 2012) has estimated that emissions from shipping in the Mediterranean Sea can contribute to more than 10 % of sulphur deposition in Cyprus (14 %), Italy (15 %) and Malta (56 %) and to more than 10 % of nitrogen deposition in Cyprus (30 %), Greece (21 %), Italy (15 %), and Malta (51 %).

Assessing the impact of air emissions is a challenging task. Frequently this kind of assessment is omitted in maritime studies and the impacts are measured on the basis of emissions quantities. However, the impact of an emission is not directly proportional to the quantity emitted. The following four drivers are jointly determining the impact of an emission (Finnveden et al., 2009):

- The emission quantity;
- The properties of the substance emitted;
- The characteristics of the source of the emission; and
- The features of the receiving environment.

The majority of impact assessments available today take into account only the first two impacts drivers of the above list. This is logical when addressing impacts at global scale (e.g. climate change), since the impact is independent of where the emission occurs. However, for air pollution impacts (e.g. acidification, eutrophication, human health effects etc.) which have local or regional characteristics, the situation can be very different and all four drivers of the impact should be adequately considered in order to arrive at reliable results. Hence, the assessment of emissions real impacts from shipping would allow for cross evaluations with other modes of transport which is a practice very often used in decision making. Impact assessment of shipping emissions is also useful for measuring the effectiveness of alternative mitigation measures and technologies which are continuously launched onboard ships, especially with respect to energy efficiency and environmental protection.

The Eco-Indicator 99 is a damage oriented method which is widely used for the impact assessment step of LCA. The method identifies eleven exposure and effect analysis categories at the midpoint level. These 11 midpoint effects are then allocated into three areas of protection or damage categories (i.e. human health, ecosystem health and resource availability). Databases are available for the mid-point and endpoint factors which are used in the method.

The final three categories of damages are described below (Eco-Indicator Manual, 2013):

1. Damage to Human Health, expressed as the number of year life lost and the number of years lived disabled. These are combined as Disability Adjusted Life Years (DALYs), an index that is also used by the World Bank and the World Health Organization;
2. Damage to Ecosystem Quality, expressed as the loss of species over a certain area, during a certain time; and
3. Damage to Resources, expressed as the surplus energy needed for future extractions of minerals and fossil fuels.

The Eco-Indicator 99 method produces one final indicator (the Eco-indicator) as a result of the weighting of impacts of the three types of damages that have been described previously. The unit of the final indicator is the Eco-indicator point (Pt) or milli-point (mPt). One point (1 Pt), corresponds to the 1/1,000 of the yearly environmental load caused by the average European inhabitant (Eco-Indicator Manual, 2013).

11.5 Case Study

11.5.1 Case Study Ship

The LCA framework has used as a basis for its development a Panamax tanker (75,000 dwt) built in S. Korea in 2009 for which the required information has been available. This ship is assumed to have a round trip between two known ports with the same speed of 14 knots at both legs (laden and ballast leg). Within this round trip the waiting times at ports for loading and unloading operations as well as manoeuvring times are also considered. This is referred to as the initial scenario since life cycle emissions calculations have been first performed for this particular scenario. Details of the ship are shown in Table 11.1.

Table 11.1 Oil tanker 75,000 dwt, life cycle inventory of emissions—initial scenario

Ship of the initial scenario: Panamax tanker	
Year of built	2009
Country	S. Korea
General data	
Displacement (tonnes)	88,221
DWT (tonnes)	74,296
Lightship (tonnes)	13,925
Steel weight (tonnes)	12,022
Design speed (knots)	15.30
Main particulars	
L_{BP} (m)	219.00
Breadth (m)	32.24
Depth (m)	20.60
Draught (m)	14.17
Block coefficient C_B	0.85
Main engine	
STX-MAN B&W 6S60MC	2 stroke
Country	S. Korea
Power of ME (kw)	12,240
RPM	105
Weight of ME (tonnes)	368
Auxiliary engines (3)	
MAN B&W	4 stroke
Country	S. Korea
Power of ME (kw)	740
RPM	720
Weight of ME (tonnes)	19.7

11.5.2 Life Cycle Inventory

The Life Cycle Inventory (LCI) includes results of air emissions from any identified process of the system. The emissions covered in this LCI are: carbon dioxides (CO_2), carbon monoxides (CO), sulphur dioxides (SO_2), sulphur oxides (SO_x), nitrogen oxide (NO_x), particulate matter (PM), methane (CH_4), volatile organic compounds (VOC), and non methane volatile organic compounds (NMVOC). Air emissions results can be provided for any ship system element, subsystem, and life cycle stage. Results are also provided per year and total life cycle.

The following table shows the LCI results of the case study ship (Panamax Tanker 75,000 tonnes of dwt). CO_2 emissions are by far the largest emissions category produced during the ship's life cycle. For a life cycle of 25 years the overall CO_2 emissions are over 1 million tonnes. The results have been crosschecked with similar results from previous life cycle studies for validation purposes (Fet, 2002; Johnsen & Fet, 1998; Jiven et al., 2004; Gratsos et al., 2010) (Table 11.2).

The results justify the dominance of the operational life of the ship in the emissions production. However, the importance of the life cycle stage is subject to the emission category examined. For emissions directly connected to the combustion of fuels in engines (i.e. CO_2, SO_2, NO_x, PM) the operational emissions account for more than 90 % of the total. Emissions of CO however, are not negligible in other life cycle stages. Higher concentrations of CO emissions are observed in shipbuilding and dismantling which are attributed to steel processing (welding, cutting etc.) (Table 11.3).

In shipbuilding the hull subsystem produces larger amount of emissions compared to the construction of the machinery subsystem as shown in the Fig. 11.3.

Table 11.2 Oil tanker 75,000 dwt, life cycle inventory of emissions—initial scenario

Emissions		Operation	Shipbuilding	Maintenance	Dismantling	Total life cycle
CO_2	Tonnes	1.06E+06	2.29E+04	9.62E+03	8.51E+03	1.10E+06
CO	Tonnes	3.17E+03	4.53E+02	8.16E+01	7.72E+02	4.48E+03
CH_4	Tonnes	2.81E+01	4.06E+00	1.48E+00	2.13E+00	3.58E+01
NO_x	Tonnes	3.04E+04	1.28E+02	9.20E+01	1.07E+02	3.07E+04
PM (all)	Tonnes	2.45E+03	2.29E+01	8.69E+00	2.25E+01	2.51E+03
SO_2	Tonnes	1.57E+04	1.02E+02	7.39E+01	1.28E+02	1.60E+04
VOC	Tonnes	–	2.00E+01	5.78E+01	2.99E−01	7.81E+01

Table 11.3 Distribution of LCI emissions in the life cycle

	Shipbuilding (%)	Operation (%)	Maintenance (%)	Dismantling (%)
CO_2	2.08	96.28	0.87	0.77
CO	10.11	70.84	1.82	17.23
CH_4	78.58	11.34	4.13	5.95
NO_x	0.42	98.94	0.30	0.35
PM (all)	0.92	97.84	0.35	0.90
SO_2	0.64	98.10	0.46	0.80

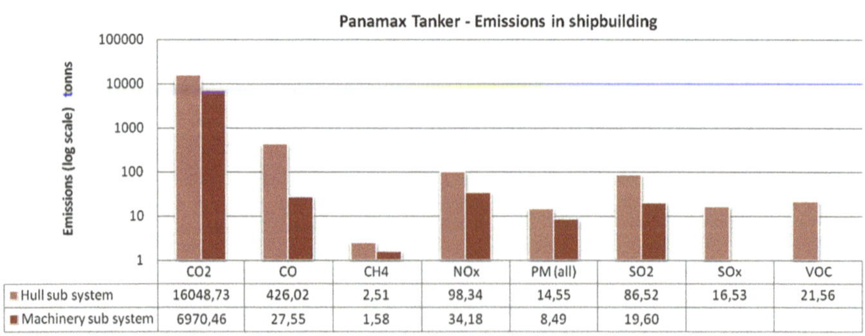

Fig. 11.3 Emissions of hull subsystem vs. machinery subsystem in shipbuilding

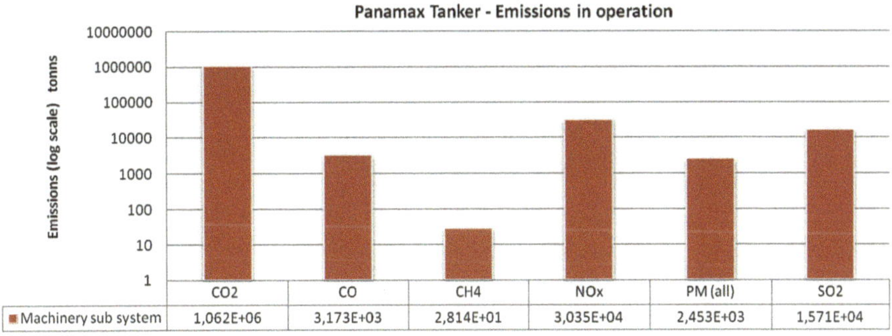

Fig. 11.4 Machinery subsystem—emissions during operational life of 25 years

In the hull subsystem, the hull material system element produces the larger amount of emissions. However, VOC emissions are almost totally attributed to the hull protection system element and more explicitly to the paint application processes. In the hull material system element the dominant process is the steel production process.

Considerable emissions in the phase of operation are produced only from the machinery subsystem due to the combustion of fuels in main and auxiliary engines. The LCI of the machinery subsystem in this phase is shown in Fig. 11.4.

Emissions in the phase of maintenance are mainly produced from the hull sub system. The same processes as in shipbuilding are considered but with different demands of energy and materials. Coating activities are important processes in this phase connected with VOC emissions. The framework uses information from specific repair sites and makes emission calculations of primer, antifouling and paints for different areas of the ship (i.e. wetted surface, deck, cargo holds, and ballast tanks). Information was also obtained from coating manufactures.

As has been demonstrated in previous life cycle studies, the overall environmental impact of a product could be drastically reduced if the recycling of materials associated with this product is taken into account. However, the literature reveals that assessing emissions (or any other environmental parameter) during the final phase of the ship's life cycle is a very challenging if not an impossible task (Tilwankar et al., 2008). Currently, the vast amount of ship dismantling tonnage ends up in S. Asian sites (nearly the 80 % of the world's scrapping volume), owing to certain natural, regulatory and cost advantages. In the majority of these sites the ship dismantling takes place following the principle of maximum separation of the ship's structure without making use of any technology similar to shipbuilding. The practices followed in the majority of these sites could be hardly qualified as industrial processes, thus the health and safety and environmental practices applied in these sites are often characterized as critical. South Asian countries utilize a technique of re-rolling scrap into producing construction steel without having to first cast scrap as billets and ingots. Information on emissions from the above process is generally not available. One source (Tilwankar et al, 2008) has indicated that the contribution to global warming of the virgin sheet metal steel obtained from

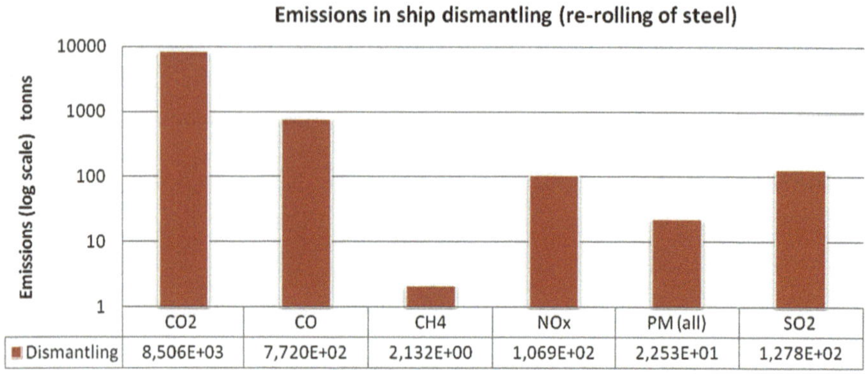

Fig. 11.5 Emissions in ship dismantling (re-rolling processing of steel)

iron ore mining is near about 2.7 times more compared to the sheet metal steel obtained from dismantled ships. The amount of steel of the ship treated and recovered with this specific method of re-rerolling varies subject to the type of ship. According to Mahindrakar, Das, Asolekar, and Kura (2008), for a tanker it ranges from 72 to 81 % of the recycled steel. Hence, the steel weight loss for tankers ranges from 8 to 10 % of the lightship weight which is attributed to the corrosion effect (Tilwankar et al., 2008).

A simple model has been incorporated in the framework using the above information and results are provided in Fig. 11.5. These results provide only an indication of the emissions related to the re-rolling process which is largely applied in S. Asian dismantling sites and by no means is capable to cover the true environmental impacts of the ship recycling process. It is also acknowledged that the current mechanisms established in the majority of these ship recycling sites pose severe threats to the environment and human health at the local level. In fact, the example of these practices could be used in the future to promote the clear benefits from studying the ship system with the life cycle thinking approach.

11.5.3 Scenarios of Operation

The Framework has the flexibility to cover various scenarios of operation. It includes a routine for calculating various time periods of ship operation and ship trips accomplished per year. These calculations lead to the estimation of the transport work accomplished (throughput) in trip and year basis. The user enters the values for some basic variables such as the distance covered per trip, the relevant ship speeds (in ballast and laden leg), the waiting times (i.e. outside port, maneuvering) the number of ships used for the required transport work and the life cycle years. An example is given in Table 11.4 where three scenarios with different operating characteristics (i.e. a slow steaming scenario, a cold ironing scenario and

Table 11.4 LCI emissions comparison of three alterative operational scenarios with the initial scenario

	Slow steaming	Cold ironing	Speed limit
	Initial mode: years 1–15	Availability of short side electricity in all port calls	Initial mode: years 1–5
	Slow steaming: years 16–25		Speed limit: years 6–25
	Speed (Laden): 11.5 knots		Speed (Laden): 12 knots
	Speed (Ballast): 13 knots		Speed (Ballast): 12 knots
CO_2 (%)	−7.65	−0.96	−10.57
CO (%)	−5.40	−0.88	−10.23
CH_4 (%)	+0.68	−0.72	−1.80
NO_x (%)	−8.95	−0.58	−12.65
PM (all) (%)	−11.67	−0.01	−20.12
SO_2 (%)	−15.09	−1.24	−15.55

a speed limit scenario) are compared to the initial one. Results show that there is a clear positive effect in emissions for the two scenarios with lower speeds (speed limit and slow steaming) compared to the initial scenario. The cold ironing scenario however, has resulted in minor benefits which support the rational that this solution is not very attractive for the particular case of tanker ships.

11.5.4 Assessment of Fleet Distribution

The developed framework has been used to examine the influence of fleet distribution in life cycle emissions. For demonstration purposes a simplified scenario has been formulated and comparisons have been made between two different fleet compositions. The first option is to employ two Panamax ships which are considered sister ships to the ship of the initial scenario shown previously (Panamax tanker of 75,000 tonnes dwt). The second option is to employ one Suezmax ship to carry the same throughput in a life cycle scenario of 25 years. Details of the trip, speeds and throughput are provided in Table 11.5.

The results obtained from this comparison reveal that employing one Suezmax ship (option 2) will produce less overall emissions in the life cycle of 25 years. Looking at the overall CO_2 emissions of the two options it is observed that the Suezmax ship produces 732,850 tonnes less CO_2 in 25 years of life than the two Panamax ships together. This can be also rephrased as follows: the Suezmax has an average rate of 6.94 tonnes CO_2 per tonne of dwt while the two Panamax ships have an average of 12.33 tonnes CO_2 per tonne of dwt. Hence, the Suezmax has lower emissions results for all for all emissions categories (Fig. 11.6).

Table 11.5 LCI emissions comparison of three alterative operational scenarios with the initial scenario

Scenario details	Units	Option 1	Option 2
Ship type		**2 Panamax ships**	**1 Suezmax ship**
Port of departure (A)		A	A
Port of arrival (B)		B	B
Distance covered (A – B)	n.m.	2,464.00	2,464.00
Speed laden (A – B)	Knots	14.00	15.00
Speed at ballast (B – A)	Knots	11.00	11.00
Days (A – B) Laden	Days	7.33	6.84
Days (B – A) Ballast	Days	9.33	9.33
Days at see	Days	16.67	16.18
Days at port (loading)	Days	1.50	2.50
Days at port (unloading)	Days	1.50	2.50
Total days at port	Days	3.00	5.00
Duration of 1 trip	Days	20.67	22.18
Days off/year	Days	15.00	15.00
Days outside port per trip	Days	1.00	1.00
Number of trips/year	Trips	16.89	15.84
Ship life cycle	Years	25.00	25.00
Total trips in life cycle	Trips	422.18	396.11
DWT	Tonnes	74,296.00	158,370.00
Throughput/year (1 ship)	Tonnes	1,254,643.74	2,509,287.48
Throughput in life cycle (1 ship)	Tonnes	31,366,093.55	62,732,187.10
Fleet (number of ships)		2.00	1.00
Total throughput (fleet)	Tonnes	62,732,187.10	62,732,187.10

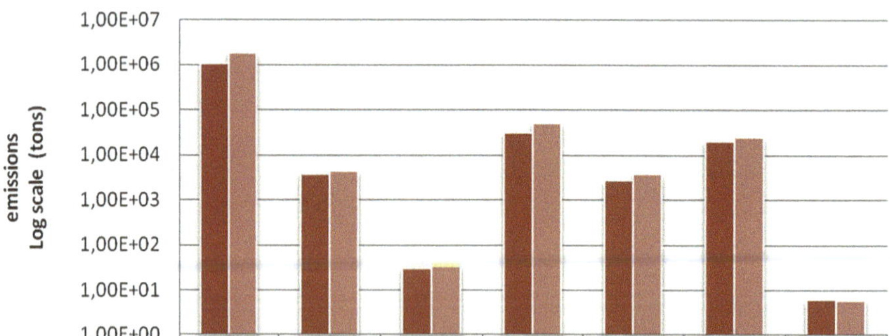

	CO2	CO	CH4	NOx	PM (all)	SO2	SOx
1 suezmax ship	1,10E+06	3,84E+03	3,00E+01	3,21E+04	2,71E+03	2,03E+04	6,17E+00
2 panamax ships	1,83E+06	4,42E+03	3,36E+01	5,05E+04	3,74E+03	2,50E+04	6,12E+00

Fig. 11.6 One Suezmax vs. two Panamax tankers. Comparisons of total life cycle emissions

11.5.5 Case Study of Ship Life Cycle Impact Assessment

Utilizing the developed LCI an impact assessment of the hull subsystem has been conducted with the application of the Eco-Indicator 99 damage assessment method. Illustrative results of these calculations are presented in this paragraph.

The inventory of this study includes information on the following air emissions: CO_2, CH_4, SO_x, NO_x, PM, VOCs, NMVOC, and CO). The allocation of emissions to midpoint impact categories follows the damage model of Eco-Indicator 99, which is a widely used impact assessment method in the context of LCA. According to this model, CH_4 is an important GHG and its impact pathway matches with the pathway of CO_2. NMVOCs are non-methane volatile compounds that are treated in the same manner as the VOCs in the Eco-Indicator 99. The land use and land conservation impacts have no relevant input. Although it is acknowledged that using non renewable resources (steel, fuels) during the ship's life cycle might have an important contribution to the environmental impact this is not considered in this particular impact assessment scenario.

The example of air emissions impact assessment considers only the emissions of the hull subsystem inventory. These emissions occur at shipbuilding, ship repair and recycling yards therefore they can be assumed similar to emissions of industrial land based sites for which the available LCIA techniques such as the Eco-Indicator 99 have been created. The impact of air emissions produced during the operational phase of the ship's life cannot be assessed by any LCIA technique in their present form. For example, the contribution of air pollutants of ships such as NO_x and SO_x to the acidification is not comparable to the impact from land based air pollutants. Currently, in the context of LCA there is no available damage model to cover the environmental impact of ship air pollutants (NO_x, SO_x, PM, VOC). This is in fact an area where future research should focus on. The development of LCA damage models explicitly for the case of maritime transportation would allow incorporating the emissions occurring away from land (in open sea) to the impact assessment procedure.

The emissions are first allocated to midpoint impact categories according to the damage model presented in Fig. 11.7, and then the characterization calculations are performed to arrive at the Impact Score. Table 11.6 summarizes the results of the impact assessment of the ship hull subsystem.

The optional steps of normalization and final weighting, between the three areas of protection (damage categories) have been also performed despite that they are not mandatory and involve the greatest uncertainty due to the limited knowledge of the contribution and relative importance of the impact categories. The single score however is being presented since it allows the aggregation of different effects making the results of the LCA study more understandable to decision makers and the public (EC-JRC, 2011).

The impact score in the category of land use reflects the low contribution of shipping to land occupation which may form a significant advantage compared to other modes of transport that make considerably larger use of land. This midpoint impact category should be further elaborated in order to make comparisons of

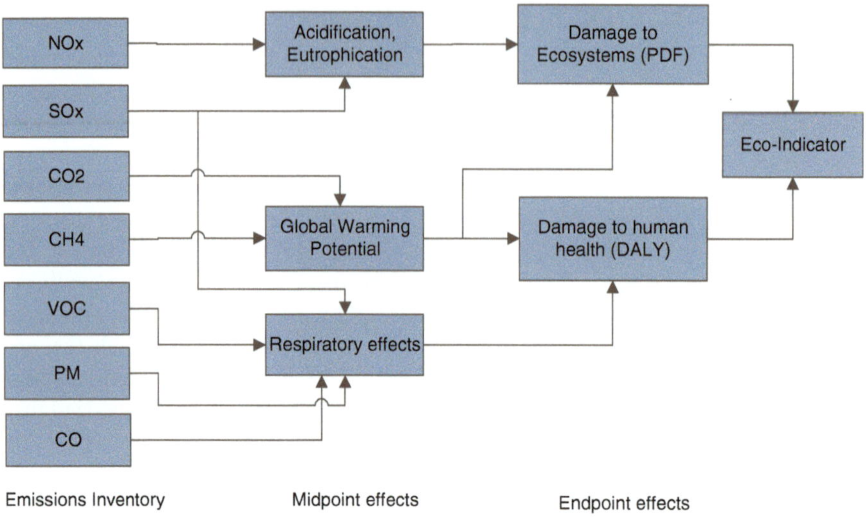

Fig. 11.7 Damage model for life cycle impact assessment of ship emissions

Table 11.6 Results of the impact assessment of the hull subsystem

Midpoint impact	Impact score	Units	Normalization	Weighting	Single score
Human health	5.41E+01	DALY	3.52E+03	400	
Ecosystem quality	1.73E+06	PDF × m^2/year	3.37E+02	400	**1.54E+06 Pt**
Land use	435,662.6	Mj surplus	5.18E+01	200	

maritime transport and other modes of transport more robust in the future. Interpreting the results of the impact assessment it could be commented that the climate change impact of the emissions produced by the hull subsystem during the life cycle of the ship is equivalent to 10.6 years of disability adjusted years (DALYs). The impact of the hull subsystem on human health (midpoint category) over the 25 years of ship life is equal to 54.1 years of human life lost. Finally, the final score of the impact assessment should be interpreted taking into account the value of the one Eco-indicator point. Considering that according to the Eco-Indicator 99 method, one point (1 Pt) corresponds to the 1/1,000 of the yearly environmental load caused by an average European inhabitant the result can be translated as follows: The impact of emissions produced by the ship hull over 25 years of life cycle is of equal magnitude to the impact produced by a small European town of 1,540 inhabitants in only 1 year.

Another interesting result is that the impact of the hull subsystem is quite fairly distributed between the life cycle phases. The shipbuilding phase is responsible for 40 % of the total environmental impact while maintenance and recycling phases are responsible for 35 % and 25 % respectively.

11.6 Discussion

This paragraph discusses issues concerning the applicability of the LCA method to the case of shipping and notes some topics which require more elaboration in the future. The main issues discussed in this context are: the adaptation of the methodology to maritime transport scenarios, the system boundaries selection, and most importantly the impact assessment which has shown to be the most challenging step of LCA.

In general, LCA is applicable for analysing ships in a life cycle perspective. The LCA has been very supportive in the development of the framework since it represents a well structured analytic tool for that purpose. However, the implementation of LCA in ship air emissions evaluation has revealed issues that need to be further addressed in order to enhance its applicability to the maritime transport case.

LCA is a product oriented methodology; thus it has been developed to cover adequately land based industrial products. Therefore, its application to the ship system which is in fact a very complex system of systems is a challenging task. Simplification of the system is often required which normally involves narrowing of its boundaries in order to include processes that can be better controlled. In practice, since there is no agreed reference system for the selection of boundaries the choice is left to the analyst to decide in accordance with the goals and needs of the study, the availability of data and resources and time constrains.

In the case of a ship LCA, one crucial issue of boundaries selection concerns the production stage of important materials used in the ship life cycle such as the steel and the marine fuels. A typical life cycle of ship's steel is illustrated in Fig. 11.8. The red line in Fig. 11.8 highlights the stages of the life cycle of ship's steel which are connected with the ship's life, beginning from stage 4, which is shipbuilding until stage 6, which is ship dismantling. Nevertheless, with regard to the steel used for the steel structure of a ship the environmental impacts (i.e. CO_2 emissions) in

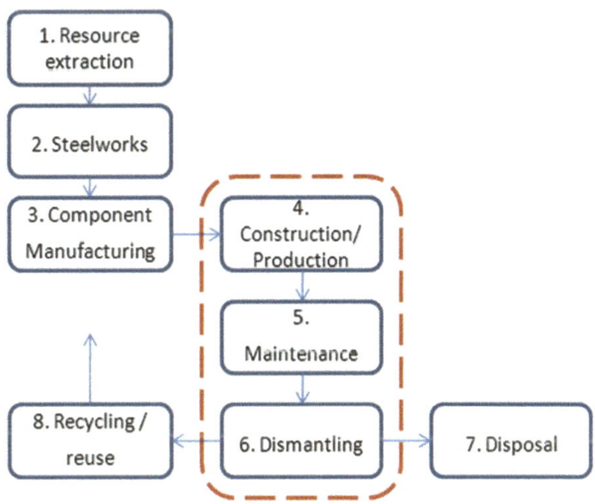

Fig. 11.8 Life cycle of ship's steel (adapted from Fet, 2002)

the first three stages of the steel life cycle are comparable with the impacts in other stages and hence they should not be overlooked (Fet, 2002).

Including the life cycle of marine fuels in a ship LCA is another matter in dispute. The production of fuels may have a non negligible effect on the emissions inventory results especially for specific air emissions. More explicitly, for the case study ship (Panamax tanker), the authors of this study have shown elsewhere (Chatzinikolaou et al., 2011) that counting the production stage of fuels which are going to be used in the ship's life cycle would actually increase by 98 % the CH_4 emissions and add 85 % to the VOC emissions. In addition, CO_2 would rise up another 10 %, owing to the production stage of marine fuels. However, if the environmental cost of oil production has been internalized (i.e. it has been already included in the product's price) in another system, perhaps it would be fair to exclude it from the ship life cycle boundaries. Since there is no reference system on how to define boundaries in an LCA, the criterion of internalization of environmental costs might be a reasonable suggestion in this respect.

For a complicated LCA system like the ship, the inventory results can be quite excessive, both with respect to the number of parameters identified and also with respect to how the results can be broken down into sub-systems and system elements. A good prioritization in entries and outputs is therefore required. For the case of ship air emissions this study has limited the LCI demands to include air emissions that are mostly subject to the MARPOL international regulating framework.

The most challenging task in a ship—LCA is the impact assessment step. As was already stated, there is a variety of available impact assessment techniques which have been developed for use in LCA but none of them has been developed to explicitly cover the needs of a maritime transport scenario. These impacts techniques are damage oriented methods which estimate the impacts on the environment, the human health and the use of resources. Existing damage methods suitable for use in the context of LCA usually include an environmental mechanism for each emission which is essentially a pathway from the inventory of emissions to midpoint and endpoint effects on the environment or/and the human health. None of these methods is capable of thoroughly covering the specific case of shipping damages. The impact assessment in LCA takes into account only two of the four drivers determining the impact of the substance emitted. The two remaining drivers (i.e. the characteristics of the source of the emission and the features of the receiving environment) are important for air emissions such as NO_x and SO_x, which have different impact when emitted in the open sea. However, it is noted that the impact of ship GHGs (CO_2, CH_4 in this study) is adequately covered by the available damage models of LCA.

Owing to the aforementioned reasons, this study has made assessments of impacts only for the hull subsystem which except for GHG emissions also causes local emissions of air pollutants that can be covered with available LCA damage models. The impact of the machinery subsystem has not been analysed because it occurs during the sailing of the vessel and the available models in LCA are not capable of covering this particular case of impacts coming from a moving emissions source. The impact assessment especially for the case of shipping is an area where further research should focus on.

11.7 Conclusions

A new framework for ship life cycle emissions quantification and impact assessment (on the environment and human health) was presented in this chapter. This framework has been developed with the employment of important features of the LCA method which is a widely applied ESA technique standardized under the ISO standards.

The framework has been tested in a ship case study (Panamax tanker), which provided estimations of air emissions for all important processes during the life cycle and an impact assessment of the hull subsystem of that ship in a life cycle perspective. The results justify the dominance of the operational life of the ship in emissions releases. However, the importance of the life cycle stage is subject to the emission type examined. For emissions which are directly connected to the combustion of fuels in engines (i.e. CO_2, SO_2, NO_x, PM) the share of operational emissions is well over 90 % of the total. However, there are no negligible emissions in other life cycle stages (e.g. VOC in shipbuilding and maintenance). The results of the impact assessment study reveal that the hull sub system of the case study ship has relatively low life cycle impact which is quite fairly distributed between its four life cycle phases.

One of the main goals of this study was to limit the use of average or generic data from databases and to increase the utilization of data explicitly referring to the specific case of shipping. Although some information for processes and material use was finally obtained from LCA databases and the literature, this study has managed to produce specific algorithms for some important processes (i.e. the calculations of welding length, coatings, anodes) and has modelled the operational profile taking into account important parameters such as the added resistance effect in order to better represent the real ship life.

"Cradle to grave" information of various substances and materials used during the life cycle is not always possible to identify or asses for the case of ships. Especially problematic is the data availability for shipbuilding and ship dismantling phases. For ship dismantling the analysis in this study cover only a small fraction of the environmental impact which is connected with a common technique for steel recovering applied in specific scrapping sites at S. Asia. Whereas the overall environmental impact of the ship could be reduced if the recycling of materials could be taken into consideration, the prevailing mechanisms in ship recycling constitute mostly negative effects to local environments and human health which have not been evaluated here. Only recycling processes hosted in shipyards could be candidates for making reliable impact estimations (i.e. perhaps if it is assumed that they are based on "reverse shipbuilding" operations).

The results of the ship—LCA have been crosschecked with other similar studies although comparisons of different LCA studies are not always possible due to different scope, boundaries selection, and functional units. A widely available LCA reference system for the aforementioned parameters is currently missing. With regard to the boundaries selection of LCA, a future reference system could consider the concept of internalization of environmental costs.

Areas of improvement in ship LCA studies concern the establishment of inventory databases specifically for maritime scenarios and the development of impact assessment methods to cover the specific characteristics of shipping. Uncertainty analysis or risk theory should be also considered as a useful incorporation perspective for LCA in the future.

In concluding, it is stated that studying systems with the life cycle approach offers some important benefits such as the identification of weak environmental processes within the life cycle and areas of improvement, the prevention of shifting of environmental impacts from one stage of the life cycle to another and most importantly the avoidance of creating externalities of environmental costs out of the system boundaries. Therefore, the life cycle thinking should be further integrated in maritime transport sector activities for continuously improving the environmental performance and increasing the benefits of this sector to the society.

Acknowledgments Part of this work has been carried out in the context of the research activities of the Centre of Excellence in Ship Total Energy-Emissions-Economy of the National Technical University of Athens, School of Naval Architecture and Marine Engineering. The financial support provided by the Lloyds Register Foundation (LRF) to the Centre is gratefully acknowledged. LRF helps to protect life and property by supporting engineering-related education, public engagement and the application of research.

References

Bengtsson, S., Andersson, K., & Fridell, E. (2011). *Life cycle assessment of marine fuels: A comparative study of four fossil fuels for marine propulsion*. Gothenburg: Chalmers University of Technology.

Chatzinikolaou, S., Psaraftis, H., & Ventikos, N. (2011, June). *Assessment of environmental impact in marine transportation and related activities—Deliverable 25* (Report No: ABS-NTUA-D25). Athens, Greece: National Technical University of Athens, Laboratory for Maritime Transport.

Chatzinikolaou, S. D., & Ventikos, N. P. (2013). Assessment of ship emissions in a life cycle perspective. In *Proceedings of the 3rd International Exergy, Life Cycle Assessment, and Sustainability Workshop & Symposium (ELCAS3), July 7–9, 2013, Nisyros, Greece*

Chatzinikolaou, S. D., & Ventikos, N. P. (2013, February). *Ship life cycle models, Chapter 4, LRF—CoE* (3rd Annual Report). Athens, Greece: National Technical University of Athens, Laboratory for Maritime Transport.

Chiffi, F., Schrooten, E., & De Vlieger, I. (2007). *EX-TREMIS—Exploring non road transport emissions in Europe—Final report*. Seville, Spain: IPTS—Institute for Prospective Technological Studies, DG-JRC.

Corbett, J. J., Winebrake, J. J., Green, E. H., Kasibhatla, P., Eyring, V., & Lauer, A. (2007). Mortality from ship emissions: A global assessment. *Environmental Science and Technology, 41*(24), 8512–8518.

De Schryver, A. M. (2010). *Value choices in life cycle impact assessments*. Ph.D. thesis, Radboud University, Nijmegen, Netherlands. ISBN: 978-94-91066-02-3.

EC-JRC. (2011). *Recommendations based on existing environmental impact assessment models and factors for life cycle assessment in European context*. ILCD Handbook—International Reference Life Cycle Data System. European Union EUR24571EN. ISBN 978-92-79-17451-3.

Eco-Indicator 99. (2013). *Manual for designers*. Retrieved January 17, 2013, from http://www.pre-sustainability.com/download/manuals/EI99_Manual.pdf

Ellingsen, H., Fet, A. M., & Aanondsen, S. (2002). *Tool for environmental efficient ship design*. Retrieved January 17, 2012, from http://www.iot.ntnu.no/users/fet/Publi-Forfatterskap/publikasjoner/Ensus-ellingsen-fet-2002.pdf

Endresen, Ø., Sørgård, E., Sundet, J. K., Dalsøren, S. B., Isaksen, I. S. A., Berglen, T. F., et al. (2003). Emission from international sea transportation and environmental impact. *Journal of Geophysical Research, 108*, 4560. doi:10.1029/2002JD002898.

Environmental Protection Agency (EPA). (2006). *Life cycle assessment: Principles and practices* (Contract No. 68-C02-067). McLean, VA: Scientific Applications International Corporation, SAIC.

European Monitoring and Evaluation Programme. (2012). *Transboundary acidification, eutrophication and ground level ozone in Europe 2010* (EMEP Status Report 1/2012). ISSN: 1504-6109 (print), ISSN: 1504-6192 (online).

Fet, A. M. (2002). Environmental reporting in marine transport based on LCA. In Proceedings of the Institute of Marine Engineering, Science and Technology. *Journal of Marine Design and Operations* B (B1), 1476–1556.

Finnveden, G., & Moberg, Å. (2005). Environmental systems analysis tools: An overview. *Journal of Cleaner Production, 13*, 1165–1173.

Finnveden, G., Hauschild M., Ekvall T., Jeroen Guinée J., Heijungs R., Hellweg S., Koehler A., Pennington D., and Suh S. (2009). Recent developments in life cycle assessment. *Journal of Environmental Management, 91*(2009), 1–21. doi:10.1016/j.jenvman.2009.06.018.

Gratsos, G., Psaraftis, H. N., & Zachariadis, P. (2010). Life-cycle CO_2 emissions of bulk carriers: A Comparative Study. Trans RINA, Vol 152, Part A3, *International Journal of Maritime Engineering*, July–September 2010, pp. A 119–A 134. ©2010: The Royal Institution of Naval Architects.

Hauschild M., Goedkoop M., Guinée J., Heijungs R.,Huijbregts M., Jolliet O., Margni M., & De Schryver A., Humbert S., Laurent A., Sala S., and Pant R. (2013) Identifying best existing practice for characterization modelling in life cycle impact assessment. *Life Cycle Impact Assessment*. doi:10.1007/s11367-012-0489-5

Hou, Q. (2011). *Life cycle assessment of cruising ship superstructure*. Master thesis in sustainable development, Uppsala University, Uppsala, Sweden.

International Council on Clean Transportation (ICCT). (2013). *Assessment of the fuel cycle impact of liquefied natural gas as used in international shipping*. Retrieved May 14, 2014, from http://www.theicct.org/sites/default/files/publications/ICCTwhitepaper_MarineLNG_130513.pdf

ISO. (2006a). ISO 14040 International Standard. In *Environmental management—Life cycle assessment—Principles and framework*. Geneva, Switzerland: International Organization for Standardization.

ISO. (2006b). ISO 14044 International Standard. In *Environmental management— Life cycle assessment—Requirements and guidelines*. Geneva, Switzerland: International Organization for Standardization.

Jiven, K., Sjöbris, A., Nilsson, M., Ellis, J., Trägårdh, P., & Nordström, M. (2004). *LCA-ship, design tool for energy efficient ships—A life cycle analysis program for ships* (Final report: 2004-08-27). Retrieved April 15, 2014, http://www.mariterm.se/download/Rapporter/Final%20report%20LCA-ship.pdf

Johnsen, T., & Fet, A. M. (1998). *Screening life cycle assessment of M/V color festival* (Report No. 10/B101/R-98/009/00). Retrieved May 15, 2013, www.iot.ntnu.no/users/fet/Forskningsrapporter/.../hia-dnv-screening.doc

Kameyama, M., Hiraoka, K., Sakurai, A., Naruse, T., & Tauchi, H. (2004). Development of LCA software for ships and LCI analysis based on actual shipbuilding and operation. In *Proceedings of the 6th International Conference on EcoBalance October 25–27, 2004, Tsukuba, Japan* (pp. 159–162).

Laugen, L. (2013). *An environmental life cycle assessment of LNG and HFO as marine fuels*. Master thesis final report of the M.Sc. degree at the Norwegian University of Science and Technology (NTNU), Department of Marine Technology. Retrieved May 14, 2014, from: http://www.diva-portal.org/smash/get/diva2:648752/FULLTEXT01.pdf

Mahindrakar, A. B, Das, S., Asolekar, S. R., & Kura, B. (2008). Environmental issues in the ship breaking industry in India. In *Proceedings of A&WMA's Annual Conference in Port-land, Oregon, June 24–27, 2008*

Ryste, J. M. (2012). *Screening LCA of GHG emissions related to LNG as ship fuel*. Master thesis final report of the M.Sc. degree at the Norwegian University of Science and Technology (NTNU), Department of Marine Technology. Retrieved from May 14, 2014, from http://www.diva-portal.org/smash/get/diva2:566151/FULLTEXT01.pdf

Tilwankar, A. K., Mahindrakar A. B., & Asolekar S. R. (2008). Steel recycling resulting from ship dismantling in India: Implications for green house gas emissions. In *Proceedings of the Second International Conference on Dismantling of Obsolete Vessels, September 15–16, 2008, Glasgow, Scotland*.

Tincelin, T., Mermier, L., Pierson, Y., Pelerin, E., & Jouanne, G. (2010). A life cycle approach to shipbuilding and ship operation. In *Proceedings of the International Conference on Ship Design and Operation for Environmental Sustainability, March 10–11, 2010, London*.

Turan, O., Ölçer, A., Lazakis, I., Rigo, P., & Caprace, J. D. (2009). Maintenance/repair and production-oriented life cycle cost/earning model for ship structural optimization during conceptual design stage. *Ships and Offshore Structures, 4*(2), 107–125. doi:10.1080/17445300802564220.

Chapter 12
Green Rail Transportation: Improving Rail Freight to Support Green Corridors

Paulus T. Aditjandra, Thomas H. Zunder, Dewan Md Zahurul Islam, and Roberto Palacin

Abstract The objective of this chapter is to illustrate the current state of European rail freight research and how this can improve rail freight to support green and sustainable transport, as promoted by the governments of the European Union (EU). Since much of EU policy is based on EU funded research, a review of such rail- and freight-related research, carried out over the past 20 years under the EU/EC framework programmes, will help to contextualize the challenges faced in promoting the governance of green corridors. International rail freight systems, improvement of rail-road modal transfer points, unexploited markets, vehicle design, train productivity, energy use, signalling and governance of multimodal corridors, are all discussed. We argue that green corridors, as a strategic freight transport policy, can only realistically be implemented alongside improvements to the long distance transportation sector, where rail becomes the backbone network of land surface transportation. Additionally, with the trend towards urbanization, where road would serve pretty much the last mile freight operation, a reliable, efficient, safe and environmentally sound inter-urban long distance rail freight system is needed, to support economic growth and to maintain the quality of life of every EU citizen.

Abbreviations

CEF	Connecting Europe Facility
CER	Community of European Railway and Infrastructure Companies
CO_2	Carbon dioxide
COMECON	Council for Mutual Economic Assistance (Russian: *Совет Экономической Взаимопомощи, Sovet Ekonomicheskoy Vzaimopomoshchi, СЭВ, SEV*)
CORDIS	Community Research and Development Information Service
EEC	European Economic Community

P.T. Aditjandra (✉) • T.H. Zunder • D.M.Z. Islam • R. Palacin
NewRail—Newcastle Centre for Railway Research, School of Mechanical and Systems Engineering, Newcastle University, Newcastle upon Tyne, UK
e-mail: paulus.aditjandra@ncl.ac.uk

H.N. Psaraftis (ed.), *Green Transportation Logistics*, International Series in Operations Research & Management Science 226,
DOI 10.1007/978-3-319-17175-3_12

EC	European Commission
EU	European Union
EU27	Was the number of member states in the EU in 2007 after the joining of Romania and Bulgaria
ERRAC	The European Rail Research Advisory Council
ERTMS	European Transport Management System
ETCS	European Train Control System
FP	Framework Programme—EU research and technological development funding mechanism
GHG	Greenhouse gas emissions
GSM-R	Global System for Mobile Communications-Railway
ICT	Information and communication technology
ISO	International Organization for Standardization
KPI	Key performance indicators
kWh/t/km	Kilowatt hour per tonne per kilometer
LDHV	Low density high value (goods)
NO_x	Mono-nitrogen oxides NO and NO_2
OICA	International Organization of Motor Vehicle Manufacturers (French: *Organization Internationale des Constructeurs d'Automobiles*)
PM	Particulate Matter PM10 are particles of 10 mm or less, PM25 of 2.5 mm or less in diameter
RFID	Radio-frequency identification
SO_x	Sulphur oxides, the two major ones being sulphur dioxide (SO_2) and sulphur trioxide (SO_3)
SME	Small Medium Enterprise
TEU	Twenty foot equivalent unit
TEN-T	Trans-European Transport Networks
TRIP	Transport Research & Innovation Portal
UK	United Kingdom
USA	United States of America
WWII	World War Two

12.1 Introduction

"Periculum Privatum Utilitas Publica" (at Private Risk for Public Service) was the motto designed to promote the first proven steam railway, opened in 1825, between Stockton and Darlington, in England. The history of rail transportation had begun—and so had a shifting relationship between the public and private domains. Transport performed by early British railways was primarily involved in the movement of coal, with passenger traffic considered a by-product of steam locomotion (Clapham, 1915). It was only as revenue from passengers increased, and trunk networks became established to connect the British rail network, that the 'golden era' of rail as a mass

transportation of freight and people emerged, across Europe (Lardner, 1850). The opening of the Stockton–Darlington rail line reduced the price of a tonne of coal by more than half, within 2 years—from 18 shillings to 8—and the movement of coal to ships, via ports, rapidly became a lucrative business that led to new line extensions (Tomlinson, 1914). The growth of the British rail network, from the late 1830s, through the 1870s up to the start of the twentieth century, was developed in parallel with population growth along the rail network corridors and the dispersion of city dwellers into the suburbs. The number of stations grew over time, from 24 stations in 1831 up to 5,500 stations, by 1911 (Gregory & Henneberg, 2010). Freight traffic became increasingly varied and innovative, with perishable goods such as milk, flowers and fish being carried into the Victorian metropolises, at speed, from distant locales. Britain's railways were built by private, profit driven companies, but always with parliamentary approval. They were often opposed by landowners, but supported by merchants and industrialists, who benefited from the new trade corridors and therefore bypassed those with no lobby power (Wolmar, 2007).

The privately financed, planned and designed railways were further popularized across Europe until the early-mid twentieth century, when some undertakings fell into bankruptcy and the great European wars introduced a requirement for military control. Although the companies still remained nominally private, state planning and control developed for various reasons, including military paranoia, parochial engineering approaches, safety standardization—and divergent approaches across the European rail network for rail gauges, loading gauges, voltages, signalling systems, safety requirements and even the positioning of lights on locomotives. It is worth noting however that this intervention represented standardization at a national level and was an advance from the earlier days, where variation existed at the regional or even rail line level.

Back in Britain, the rapid growth of the rail network development achieved its peak in 1914, before road transport became an increasingly effective competitor in the interwar period (Wolmar, 2007). Later on, the aeroplane exacerbated competition, at a pan-European and eventually domestic level (Smith, 2012). Similarly, in Germany, after the invention of the automobile by Carl Benz (patented in 1886, first long-distance test in 1888 by his wife Berta Benz) it took almost a decade for the 'road vehicle without horse' to gain a market niche, but post 1900 the development took off at high speed (Rothengatter, 2014). The number of motorized road vehicles (passenger cars and trucks) in Germany increased from 884 in 1901 to 16,939 in 1911 (OICA statistics). Travel patterns changed and, as the twentieth century dawned, the convenience, flexibility and aspirational value of cars became ever more dominant, with the geospatial shape of cities and nations adapting, to suit. Mobility increased and rail and public transport in general took a smaller proportional share of transport activity.

However, by the closing decades of the twentieth century the disbenefits of the 'car culture'—increased pollution and congestion, the relatively higher level of accidents resulting in fatalities and injuries, the health impact of the automobile (Banister, 2005)—had served to place train travel at the heart of the transport agenda (Hall, 2007; Hall & Banister, 1994), surrounded by the sustainability benefits of this transport mode (Givoni & Holvad, 2009). From the 1992 Rio Summit and the 1994 Kyoto Protocol agreements, a combination of oil depletion theories (Gilbert & Perl, 2008) and real scientific consensus—that climate change (Stern, 2006) is driven by hydrocarbon use and that transport is one of the key users of such energy—has pushed society, and hence research, into the search for transportation systems that are economically efficient, socially beneficial and environmentally sound. Rail transportation is now at the forefront of a new sustainable transport policy in Europe (European Commission, 2011).

In terms of freight, one of rail's biggest technical advantages is its ability to transport larger volumes, over long distances, more efficiently than almost any other mode except bulk shipping (Jackson, 2014). Rail's disadvantages are how to interlink this with smaller volumes, over shorter distances, although the extent to which this is a result of a series of engineering, financial and planning choices is open to debate.

Through the twentieth century, as competition with road freight transport evolved, the railways progressively lost market share and, with it, the profit from increasingly state run monopolies which had hitherto been cash cows (Rothengatter, 2006). This unwelcome situation was responded to with heavy-handed regulation for the trucking industry, followed, post WWII, by high subsidies to the rail companies—deemed necessary for accessibility but which arguably worsened the railway market situation (Rothengatter, 2006).

The railway companies in Germany, for example, had been merged in 1920 to establish the *Deutsche Reichsbahn* which, as a federal railway company, was a monopoly, paid monopoly fees and was charged with a number of social obligations (Monopolkommission, 2007). In the course of the diffusion of motorized road vehicles, this monopolist position came under attack, with resulting loss of market share and associated reduction in tariffs (Rothengatter, 2014). This lowering of profits undermined its ability to pay monopoly fees and to meet its social obligations. In 1925 and again in 1928 and 1931, the federal state in Germany reacted by introducing strict regulations for the market activities of trucks and buses, along with road service laws, car regulation and taxes (Zimmermann, 1990). Despite this apparently very restrictive regulation (quotas for truck registration, obligatory tariffs, prohibition of long distance freight trucking), the market share of road transport increased rapidly, while protecting and subsidizing the railways equally failed to halt rail's decline (Rothengatter, 2014). A particularly dramatic fall in rail freight traffic was seen in the EU27 between 1980 and 1995, with a 40 % drop in volume (in tkm—tonne-kilometres) (DG-TREN, 2008, p. 115). Decline was also evident in terms of modal share, plummeting from 20 % in 1970, to a mere 8 %, by 2000 (Preston, 2009).

In the 1980s, as the European economy began to lag behind the rest of the developed world, the Delors Commission took the initiative to attempt a re-launch of the common market, publishing, in 1985, a well-received White Paper identifying 300 measures to be addressed in order to complete a single market and which led to the adoption of the Single European Act—a treaty which reformed the decision making mechanisms of the European Economic Community (EEC) and set a deadline of 31 December 1992 for the completion of a Single European Market (eventually launched 1 January 1993). Transport policy in the EU had been largely quiescent until the 1980s, but the burgeoning Single European Market began to liberalize road transport, specifically unpicking the restrictions on truck distances and permitting *cabotage*—the transport of goods or passengers between two points in the same country by a vessel or an aircraft registered in another country. These measures further served to increase the market share gap between rail and road.

In the final three decades of the twentieth century, rail freight in the EU declined, due to a failure to adapt to modern logistics demands and the abandonment of many early rail successes, such as the transport of perishable goods. In the former COMECON countries and states of Central and Eastern Europe, collapse began following adaptation to liberal economies. Overall rail market share in the EU27 has been flat since 1995; localized trends can be seen only at a national level and are not always intuitively understood, being influenced by a variety of factors. From the mid-1990s some limited growth was evident in some countries (Božičnik, 2009; Wolff, 2006), the strongest reported coming from states that perform a transit role, such as the Netherlands, Austria and Switzerland (Preston, 2009), though other transit countries such as Poland and Hungary evidence falling shares, as the logistics networks have remapped from east facing COMECON systems, to western facing EU economies. These trends demonstrate differences in railway infrastructure, trade flows, economic planning and levels of rail deregulation, as discussed in recently completed research addressing the current pan-European rail freight market (Zunder, Islam, Mortimer, & Aditjandra, 2013). (Examples can be seen in Fig. 12.1.)

In this chapter we aim to give an outline of European rail freight, with an overview of research and innovation promoted by EU government to address the sustainability agenda. The shift in focus from road to rail in EU transport policy is driven by the problems now arising from the comprehensively developed road network, that has benefitted extensively from public policies over the past

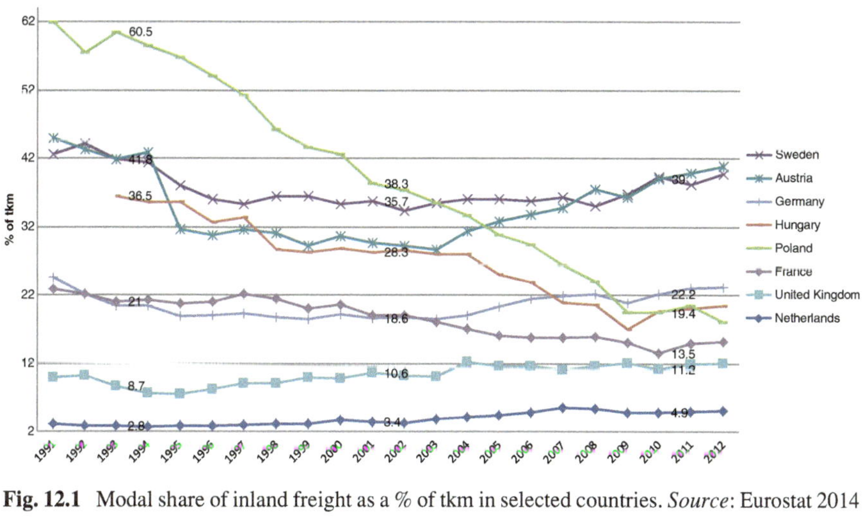

Fig. 12.1 Modal share of inland freight as a % of tkm in selected countries. *Source*: Eurostat 2014

100 years. Such problems take the form of spatial constraint (that has consequently led to urban road congestion), environmental pollution (from internal combustion engine emissions) and the extent to which certain routes should be prioritized—within acceptable environmental limits—for the benefit of demand and supply efficiency. Rail has once more turned out to be the answer to optimizing surface transportation over medium to long distance, between hubs (e.g. cities), and we report on the current state of knowledge. Following this introduction is a brief review of rail freight policy, in Sect. 12.2. Section 12.3 considers the green credentials of rail freight, while Sect. 12.4 reviews EU funded research projects with a rail and freight focus and Sect. 12.5 draws some conclusions.

12.2 European Rail Freight Policy

We turn now to the discussion of rail freight policy development in the EU, intended to drive the resurrection of a new rail freight era. In that sense, this section complements Sect. 1.5 of Chap. 1.

From the end of the 1980s, Sweden, the UK and Germany individually commenced railway reform, for a variety of ideological, financial and operational reasons. Other countries, especially the Baltic States (following their independence from the Soviet Union), adopted wholesale, monopoly-based privatizations of vertically integrated railways. Of the different approaches taken, the ones which dominated the Railway Directives promoted by the EU were based on liberal economic approaches, with separation of infrastructure from operations and, within that, between freight and passenger rail operations, with separate balance sheets and open access to the whole rail freight network on a non-discriminatory basis—very much the approach adopted by the UK for rail after some early privatizations that had created problematic privatized monopolies in telecoms and energy (Finger, Groenewegen, & Künneke, 2005). To overcome the structural obstacles, a complete separation of infrastructure and transport services operation was proposed (Stehmann & Zellhofer, 2004) and legal instruments were adopted by each of the EU Member States, in order to allow free flow of goods and services (Kirchner, 2006). However, these new developments were challenged by a number of structural issues (Di Pietrantonio & Pelkmans, 2004; Wolff, 2006):

- national and state owned monopolistic operations; economies of density; safety; asymmetry of information;
- the intense competition with road haulage, barges/inland waterways and coastal shipping;
- a structural shrinkage of rail-friendly primary industries, as production was offshored;
- the mixed usage of rail networks, where operational and strategic priority is clearly given to passenger trains; and

- an inflexible response to emerging and changing requirements and expectations from the logistics industry and large commercial customers.

The Railway Directive 91/440/EEC of 1991 was the key turning point for rail liberalization in Europe. The Directive required that the national railways be operated on a commercial basis, be market demand driven, and be managerially independent from the member states. Since then a series of railway reform packages have been introduced that have mainly centered on the commercialization of railway companies, market opportunities and technical harmonization, in order to achieve a single European railway market. The main features of railway reform policies have been vertical separation of infrastructure management and operational services, along with other unbundling of railway functions, infrastructure charging, specification of access rights for different market segments and the introduction of appropriate regulatory authorities, including appeals bodies (Holvad, 2009). This is in contrast to the USA, where deregulation led to the development of regional, vertically integrated monopolies, which have steadily grown in scale since the 1980s (Vassallo & Fagan, 2007).

In the meantime, following the Maastricht Treaty of 1992 (EU formulation of the Euro single currency), the Common Transport Policy of 1992 was established, to strengthen the deregulation and liberalization of the transport market in the EU, with the development of Trans-European Transport Networks (TEN-T)—strategic, cross border transport corridors, highlighted for mutual development across EU Member States. TEN-T is a key element in the process of European integration between the poorer peripheral states, as well as the congested central regions, of the existing (European) Community (Vickerman, 1995). Between the major metropolitan regions (that act as the major traffic generators and attractors) lie (infrastructure) corridors, the potential development of which was expected to lead to economic growth. A further TEN-T guideline was introduced, in 1996, to promote the interconnection and the interoperability of national networks, as well as access thereto. Interconnection aims to improve the efficiency of transport systems, as well as producing effective synergy across different EU nations and different networked modes. Interoperability, on the other hand, aims to improve the capability of a network for safe through-traffic, with respect to administrative, technical and operational preconditions, so that barriers to free access are kept to a minimum level (with 'zero' as the ultimate). These preconditions were essential—particularly with respect to the development of high-speed rail networks (Sichelschmidt, 1999)—and put passenger traffic as priority over freight.

Despite the progress made, through policy supports and measures, to achieving an open access rail freight system, a number of issues arose, including the delayed and varied implementation of required national legislation, or even failure to start the national implementation process of EU Directives (Holvad, 2009). In addition, without a complete separation of train 'ownership' operations from infrastructure, some state owned freight operators have prospered and developed dominant, Europe-wide businesses on the basis of state funding, echoing the experience of the telecommunications and electricity industries (Stehmann & Zenger, 2011).

However, evidence of the emergence of new freight operators across Europe (including Russia) has reported that new entrants benefit from the 'vertical separation' (Pittman, Diaconu, Sip, Tomova, & Wronka, 2007). Furthermore, a benchmarking study to understand the impact of reform initiatives on railway operations in 23 European countries during 1995–2001 demonstrated that vertical separation generally improves the technical efficiency of railway systems, especially in regard to both material and staffing costs (Asmild, Holvad, Hougaard, & Kronborg, 2009).

A comparison of the introduction of railways competition in Sweden, the UK and Germany demonstrated that, in spite of complete successful separation of infrastructure and operations, Sweden and the UK have seen a major increase in costs and government support, while in Germany, where infrastructure and operations have remained part of the same holding company, growth in public financial support for the railways has been the slowest, as has growth in fares (Nash, Nilsson, & Link, 2013). It should be noted that this conclusion can be biased towards passenger rail and also that the period required for rail freight growth to emerge can be longer. In the UK, government support for rail freight (e.g. intermodal grant, first introduced on the sale of GB Freightliner) has been gradually reduced, while recent rail freight traffic has been increasing (Rail Freight Group, 2014).

12.2.1 Rail and the Transport White Paper

Rail transport has been explicitly supported in EU transport policy, since the 2001 Transport White Paper promoting a pro modal shift (from road to rail/inland waterways transport). This proposed an interventionist approach to achieving demand changes and the active promotion of intermodality (Zunder, Aditjandra, & Islam, 2012). In 2006 the EU published a mid-term review of the 2001 Transport White Paper that was an attempt to recognize that all transport needs to be sustainable and that policy should optimize each mode, as well as integrating the modes seamlessly, and then look for modal shift, for long-distances (for example, over 300 km), for urban areas and for congested corridors. The 2011 White Paper set targets for rail to play a dominant role in long-haul freight above 300 km, with 30 % of road freight over 300 km shifting to other modes (rail or inland waterways) by 2030, increasing to more than 50 %, by 2050. This is to be achieved by better modal choice, greater integration of the modes' networks and hubs and the development of pan-European green freight corridors, as illustrated in Fig. 12.2.

12.2.2 Rail and Green Corridors Policy

The concept of green freight transport corridors was introduced in 2007, following the launch of the Freight Transport Logistics Action Plan (European Commission,

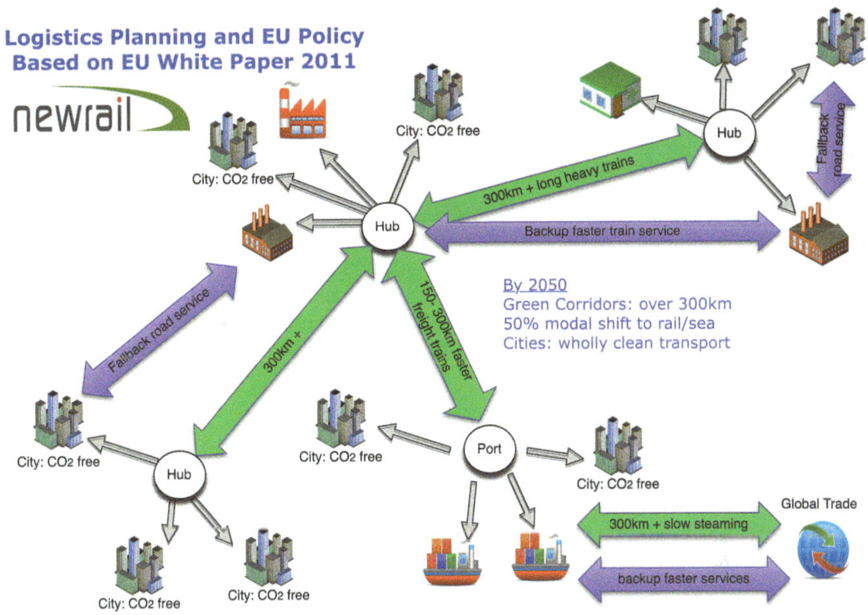

Fig. 12.2 Logistics planning and EU policy based on EU Transport White Paper 2011 (*Source*: Authors)

2007), to act as catalyst to concentrate freight traffic between major hubs over relatively long distance. The root of the examination is the development of the Trans-European Transport Network (TEN-T) (Islam, Meier, Aditjandra, Zunder, & Pace, 2013, p. 11), which began with a strong passenger focus and only latterly addressed the benefits of pan-European freight networks. The TEN-T programme has now undergone three phases and the majority of the funded projects are on rail priorities, but the only TEN-T funded freight corridor was the development of the *Betuweroute*, between Rotterdam and the Ruhr. As of 2014, investigations into the feasibility of green freight corridors, as strategic planning and governance tools, are only just starting to be discussed (Panagakos & Psaraftis, 2014; Psaraftis, Minsaas, Panagakos, Palsson, & Salanne, 2013; Psaraftis & Panagakos, 2012) and the coherence at national level has been assessed through research (see for example Aditjandra, Zunder, Islam, & Vanaale, 2012). Despite the novelty of green corridor governance, the theoretical grounding is associated with green supply chain management, in the framework of network and stakeholder governance (Hunke & Prause, 2013).

Furthermore, the benefits of green corridor policy were suggested, for small and medium enterprises (SMEs) in the logistics sector (Prause & Hunke, 2014). In 2014 the EU launched the Connecting Europe Facility (CEF) and announced green corridor managers, as a tool to invigorate the concept at a planning, financing and operational level. It is yet to be seen how effective this initiative will be, given that it has to mould pan-European corridors from deeply parochial infrastructure network managers.

Not all policy that affects rail freight is explicitly for rail, but is in other areas of transport policy. Many EU policies and interventions are based on EU funded research, which is in turn further developed in line with policy so, to better understand rail freight policy, it is helpful to review the EU research projects. However, let us first consider rail's 'green' credentials.

12.3 How Green Is Rail Freight?

Railway revitalization, in European transport policy, aims to promote the role of rail by tackling the impact of transport on the environment and on community. Rail is usually considered to be a 'green' mode of transport—greener at least than cars and planes, in terms of relative negative impacts on climate change. But how green is green? 'Green' is most usually reversely associated with greenhouse house gas emissions (GHG) and thus climate change. CO_2 is one of the main contributors to GHG, which is particularly relevant for transport. A study comparing rail passenger operations with air and automotive demonstrated that, on some routes and under certain operating conditions, old existing rolling stock can lead to a similar level of CO_2 emissions as aircraft (Givoni, Brand, & Watkiss, 2009). This is particularly relevant, as rolling stock has a life cycle in excess of 30 years. Electrification of the rail network would certainly reduce CO_2 emissions from rail, but from a carbon cost abatement perspective (see for example: McKinsey & Company, 2009) in the UK, the economic justification is not met, although it is suggested that the inclusion of local and regional air and noise pollution effects may change this conclusion (Givoni et al., 2009). In addition, electrification of the railways only partly solves the green paradigm of railways, as the 'green' question moves from the transport sector to the energy sector. Indeed, the energy mix of a given country would determine how 'green' is the electricity used by the railways.

From the perspective of freight operations, an overview of aspects and issues of green logistics demonstrated that transportation, alongside facilities and inventory, is defined as a physical driver to greening the supply chain (Dekker, Bloemhof, & Mallidis, 2012). In terms of transportation mode choice (plane vs. ship vs. truck vs. rail vs. barge vs. pipeline), freight operations have a limited choice, as the transport mode is often determined by the type of products (e.g. liquid, bulk or package) and the distance to be travelled. For continental transport chains, the choice is mainly between truck, airplane, train or short sea shipping (and in some cases inland waterways). Time sensitive goods are often supplied by air (and in some cases truck), while large volumes of commodities (i.e. coal, iron ore) are transported by rail, inland barge or pipeline (i.e. gas or oils). While generally shippers' preferences are determined by cost, quality and speed, the CO_2 impact from each of the transport modes chosen depends heavily on the way they are

Table 12.1 Energy use and emissions for typical transport units of different modes

Energy use/emissions (g/t/km)	PS-type container vessel (11,000 TEU)	S-type container vessel (6,600 TEU)	Rail-electric	Rail-diesel	Heavy truck	Boeing 747-400
kW h/t/km	0.014	0.018	0.043	0.067	0.18	2.00
CO_2	7.48	8.36	18	17	50	552
SO_X	0.19	0.21	0.44	0.35	0.31	5,369
NO_x	0.12	0.162	0.10	0.00005	0.00006	0.17
PM	0.008	0.009	n/a	0.008	0.005	n/a

Source: Dekker et al. (2012)
Note: TEU is the standard for containers; one TEU is equivalent to a 20 ft container; PM is particulate matters, also called fine dust

calculated and the assumptions made (McKinnon, 2007, 2010). Table 12.1 illustrates a comparison of emissions between several transport modes.

12.4 Review of EU Funded Rail Freight Focussed Research Projects

In this section we review EU funded projects with a rail freight focus. The European transport research portals CORDIS (http://cordis.europa.eu) and TRIP (http://www.transport-research.info) were used to retrieve the reports of past projects. Using the keyword 'freight' to search the projects, 288 projects were identified using CORDIS and 535 in TRIP (TRIP includes national based funded projects in its database).[1] When using the keyword 'rail' to search the projects, 484 projects were identified in CORDIS and 983 in TRIP. Given such huge numbers, only selected, completed projects were reviewed to describe the efforts made to promote green corridor policy. The review centers on the characteristics of the corridors under assessment, the identification of technology innovations and on lessons learned. Appendix 1 shows a number of completed European funded projects that focus on promoting rail freight competitiveness, within the context of green corridor policy.

The origin of a freight dedicated network can be traced back to the FP4 EUFRANET (1997–1999) project that modelled a hypothetical dedicated rail freight network, across the EU, that demonstrated the differences in freight volume across the rail network and, in turn, defined the key axes of a core freight transport network. There are other FP4 projects that address rail freight transport, but their focus was on technological innovation, rather than corridors, e.g. HISPEEDMIX (1997–2000)—testing the feasibility of high speed freight trains; INTELFRET (1997–1999)—defining the functional specification of freight trains with advanced

[1] Accessed online as of July 2014.

technology e.g. automated train internal communication, brake test, traction control, etc.; FIRE (1998–1999)—an information service platform (e.g. train schedules, wagon monitoring and commercial offers for rail-based international freight transport; and OPTIRAILS (1999–2000)—the safety and technology oriented aspects of signalling and telecommunication, based on the European Train Control System (ETCS)/European Rail Traffic Management System (ERTMS).

There is no evidence of any projects in FP5 that specifically address a corridor approach. Some FP5 projects focus on technological innovations, e.g. EDIP (2002–2005)—an on-board radio based control of multiple-locomotive freight trains for trans-European operation; F-MAN (2001–2004)—tracking systems, data processing and asset management; and PROMAIN (2000–2003)—an intelligent module in rail freight, or an infrastructural management programme for rail transport. The funded projects to address rail-road modal transfer are CARGOSPEED (2001–2004)—developing a prototype of a rail wagon and a transfer mechanism with pop-up lifting and turning to support smooth flow of cargo at intermodal terminals, including ports; and IN.HO.TRA (2003–2006)—focussing on the development of specific (horizontal transfer) technologies for the transfer of loading units between different modes, with lower costs, and faster and higher reliability. Despite the lack of freight corridor based study, innovative urban freight projects were funded to look at the issue of last mile freight, to anticipate the connectivity between rail and road at inter urban points e.g. BESTUFS (2000–2004); and CITYFREIGHT (2002–2004).

During the period of FP6, a number of projects were funded that focussed specifically on addressing cross-border, pan-European rail freight, towards rail competitiveness across the transport market. A number of these projects, as listed in Appendix 1, address liberalization of the rail freight market through the evaluation of pilot corridors, e.g. BRAVO (2004–2007) and TREND (2005–2006). REORIENT (2005–2007) was looking at corridor governance; RETRACK (2006–2011) was addressing reorganization; and NEWOPERA (2005–2008) and CREAM (2007–2012) were testing technical and operational innovations. In parallel, a number of technological innovation funded projects focussed on improving the bottleneck at terminals, using an intelligent transport system, e.g. FastRCargo (2006–2009)—supporting a fast transhipment process with continuous system health monitoring of rail freight; and CHINOS (2006–2010)—using RFID—an electronic radio-frequency-identification transponder—to track and identify containers in a multimodal transport system. Other technological innovation projects exist, e.g. TRIMOTRANS (2005–2008)—aiming to design and develop large ISO containers; ISTU (2003–2006)—aiming to design and integrate propulsion components to pull containers at terminals; and INTERGAUGE (2006–2008)—focussing on addressing gauge changes at cross-border railways.

Apart from the corridor based analyses and technological innovations, there were also funded projects on the strategic level of multimodal freight governance, e.g. CAESAR (2005–2007)—aiming to establish an intermodal research advisory council, so supporting research to define their strategic research agenda. Similarly, PROMIT (2006–2009) was funded to establish an active information and coordination platform for intermodal logistics, with specific reference to east–west corridors. A hypothetical transport model for scenario testing, similar to FP4's EUFRANET, was developed in TRANS-TOOLS (2004–2006). The model output of TRANS-TOOLS demonstrated the benefit of TEN-T investments in relieving bottlenecks at a number of hubs across the European transportation system, including the rail network. Strategic governance in rail is only being funded in FP7, e.g. ERRAC ROAD MAP (2009–2012) addressing the coordination of all types of passenger and freight rail services, with other modes.

With the launch of FP7, the focus of funded projects linked to rail freight corridors can be classified as inter-urban freight studies, rail-road modal transfer points, international rail development, and innovation beyond the traditional rail freight market. As at July 2014, many of these projects were not yet logged on the CORDIS or TRIP databases. References to all projects named in this chapter can be found in Appendix 2.

12.4.1 *Efforts to Develop an International Dimension in Rail Transport Systems*

Siim Kallas, Vice President of the European Commission, delivered a speech on 01 April 2014 in Brussels entitled 'Europe's railways at a junction: the future for freight', wherein he noted that, currently, around 50 % of rail freight tonne-kilometres are international. For all other transport services, European citizens move and work in a borderless community, but customers and shippers in the rail freight market have yet to practically experience borderless European transport services and logistics, in reality.

Despite numerous efforts (e.g. implementation of projects) and steps (e.g. Directives and Railway Reform Packages), one important feature of today's European rail transport is fragmented railways, originating from a number of national railways that were aimed to serve national markets and interests, with different infrastructure and systems. Even within a single country, there are multiple systems of, for example, signalling and gauge clearance. In order to develop a sustainable and competitive rail (freight and passenger) transport system, it is vital that all nationally focused railways be integrated into one European railway. This has become even more crucial with the expansion of the European Union from merely a trading bloc of 6 member countries to a political and administrative Union of 10, 15, 25, 27 and currently 28 countries. In such a situation, an efficient railway system can be achieved through, among others, harmonization and standardization of the technical and operating rules and regulations, resulting in uniformity in the entire EU Rail Freight transport network.

One example of the European Commission's effort in this regard is the FP6 REORIENT (2005–2007) project that assessed the progress of transforming European national railways into an internationally integrated and standardized system, through implementation by EU Member (and Accession) Countries, of EU interoperability legislation. To do so, it worked on three aspects: the identification of the political and administrative bodies responsible for interoperability implementation; capturing progress in interoperability between country blocs, to define their ability to remove interstate inter-rail discrepancies; and defining the tolerance margins of national politicians, for channelling scarce economic resources into the rail sector, in competition with other social needs. Building on these aspects, the effort of developing pan-European operations continued, for example in FP6's RETRACK (2006–2011), and CREAM (2007–2012) projects. The pilot RETRACK service was run by SME rail freight operators and is still being commercially run, having been expanded to other corridors beyond Cologne (Germany)-Gyor (Hungary). The RETRACK operators reveal that flexible and customer oriented operational ability are important aspects of achieving success for this venture. As a sister project, CREAM designed, developed, operated and validate advanced customer-driven business models for railway undertakings, intermodal operators and logistics service providers. The CREAM service included intermodal rail-road and rail-short sea-road service options, on the Trans-European mega-corridor, between the Benelux countries and Turkey. The summaries of all three projects promote green corridor policy and can be found in Appendix 1.

A more recent effort is the FP7 TIGER (2009–2012) project, aimed at exhibiting the role of a hub, in the form of a dry port/freight village, to promote rail freight connectivity between seaports and hinterland distribution hubs (see Appendix 1 for further detail). FP7's MARATHON (2011-date) project has implemented the delivery of an operational and visible solution via longer and heavier trains, at pan-European level, and further internationalization, that is badly needed in a market whose traditional players have been nationally focused, for a very long period of time.

12.4.2 Solutions to Enhance Rail Freight in Unexploited Markets

The EU Transport White Paper (European Commission, 2011, p. 9) suggests the optimization of the performance of multimodal (e.g. rail-road or road-waterways) logistics chains by making greater use of more energy-efficient transport modes (such as rail and waterways). Thus it aims to achieve a modal shift of 30 % of road freight over 300 km by 2030, and of more than 50 % by 2050, facilitated by efficient and green freight corridors. Meeting this goal will also require appropriate infrastructure to be developed. A recent market study suggests that around 37 % of total low density high value (LDHV) goods are transported by road, over distances of 300 km or more, in the EU-27 and Switzerland (Jackson, Islam, Zunder, Schoemaker, & Dasburg, 2013). In line with the White Paper 2011 policy, it is essential that rail freight operators capture some of these cargoes, by meeting customer requirements—for example in transit time, reliability and, of course, price.

Rail freight transport performs well for low value, high volume cargo, such as coal, steel and other raw materials used as inputs for manufacturing. In recent years, however, manufacturing industries have moved from West to East, in pursuit of cheaper input costs, such as labour. FP7's LOGMAN (2009–2011) project aimed to gain insights into the new logistics and manufacturing trends with regard to sustainability, using transport demand macro-economic modelling tools that hypothetically forecast transport growth, as developed in FP6's TRANS-TOOLS (2004–2006) project. LOGMAN concluded that a number of logistics and manufacturing trends (i.e. 'intermodal transport', 'transport consolidation' and 'local recycling') were found to have the highest impact in CO_2 reductions. Other scenarios with lower impacts are 'off/on/nearshoring' and 'decentralization'. Europe is a consumer society, with its market increasingly demanding transports of consumable cargoes, produced elsewhere. An important necessity for this type of cargo is a door to door service offering, rather than terminal-to-terminal, but rail freight services have not yet been able to respond to the new trend in logistics and supply chain concepts and practices. Currently, European freight trains—sharing track with passenger trains—are operated at around 40 miles per hour on average, whereas passenger trains—even local services—are run at much higher speed. Responding to political and social demand or pressure, passenger trains are given priority over freight train operations. Thus freight train path allocation is a difficult task for infrastructure managers—a problem that could be solved by making the operational characteristics of freight trains—such as acceleration and braking systems—equivalent to those of passenger trains. This type of cargo, typically LDHV goods, requires a faster and more reliable transit time. In Europe currently these cargoes are transported by road, generally in containers. FP7's SPECTRUM (2011-date) project developed a detailed design concept for a high performance freight train, that is efficiently lightweight, has driving performance characteristics that facilitate mixed running with passenger services, and is capable of accommodating the required types of freight container unit.

12.4.3 Efforts to Improve Rail-Road Modal Transfer Points

Modal transfer points are generally considered a bottleneck, in developing and using multimodal transport systems and services. The EU Transport White Paper (European Commission, 2011, p. 14) suggests developing and/or upgrading efficient, multimodal terminals, at different modal transfer points, including ports. The CARGOSPEED (2001–2004) and IN.HO.TRA (2003–2006) projects, discussed above and funded under FP5, were the predecessors of the more recent, FP7 funded project SPECTRUM (2011-date), that aims to develop solutions and processes to enhance the competitiveness of transport by rail in unexploited markets (i.e. low density, high value cargoes, that are non-rail). The SPECTRUM project has identified the following transhipment technologies as noteworthy:

- Traditional lift-on lift-off intermodal systems—the reference system;
- Light-combi system, with lift-on lift-off technology, probably based on fork lift trucks;
- Horizontal sideways handling equipment, mounted on road vehicles;
- Small terminals—for class C[2] load carriers;
- Horizontal sideways handling, as above—for class A[3] and C load carriers;
- RoRo wagons, for a system based on semi-trailers only, with roll-on roll-off shuttle trains;
- Side loaders;
- Diagonal transhipment system, for swap bodies.

12.4.4 Demand for New Rail Freight Vehicle(s)

The structural design of rail vehicles has been moulded by the development of new materials and manufacturing processes. The typology of rail vehicle design is driven by four design factors: market demand (e.g. type/value of goods and associated costs); technology (e.g. vehicle performance in acceleration, deceleration, vibrations, aerodynamics, structure and materials); government policy (e.g. environmental measures in CO_2, oil spillage and energy, noise, and standards measures—such as structural integrity and crashworthiness); and operational requirements (e.g. logistics, security, safety, infrastructure—in the form of track/loading gauge, catenary voltage—and timetabling) (Matsika, Ricci, Mortimer, Georgiev, & O'Neill, 2013, p. 44). FP7's SUSTRAIL (2011-date) project aims to improve the rail freight system, to regain market position, via innovations on rolling stock, vehicles (with a targeted increased in speed and axle load) and track components (for higher reliability and reduced maintenance). In parallel, FP7's D-RAIL (2011-date) project aims to address the root causes of derailment of freight vehicles, which have a wider range of operating parameters (as a result of the huge range in loads, speeds and maintenance quality).

A market and logistics requirement-based vehicle design is currently being investigated and identified, in the form of a multipurpose flat wagon design, incorporating a horizontal transhipment technology, for the transport of containers and swap bodies that enables loading/unloading without the need for a dedicated rail freight terminal (Zunder, Jackson, & Matsika, 2014).

[2] Road-rail container swap body with length of 7.15, 7.45 or 7.82 m (standard EN 284) (UIRR, n.d.).

[3] Swap body with lengths of 12.50 or 13.60 m (standard EN 452) (UIRR, n.d.).

12.4.5 Longer and Heavier Trains

The EU Transport White Paper suggested that there must be new patterns and solutions, for example higher volumes of freight transportation (European Commission, 2011, p. 15). It is particularly paramount in the sense that the policy document also suggests developing a transport system that is sustainable, as well as competitive. Along these lines, rail freight transport has to produce maximum productivity from the existing European rail infrastructure (as well as other assets) for producing efficiency and reducing operating costs, as well as attracting new traffic to rail. Achieving and implementing economies of scale, for example longer or double stack trains, is an important tool for rail freight transport. The operation of longer trains—e.g. 2,000 m long—is a normal feature in the USA, but is restricted to 700 m in Europe (even shorter in some countries, e.g. 630 m in Denmark), due to the shorter lengths of the bypass tracks (Clausen & Voll, 2013, p. 131). In this regard FP7's MARATHON (2011-date) project tested a longer freight train (1,500 m) in January 2014, marking the start of the demonstration phase of the project. Three regular intermodal (container) trains, running between Germany and Spain, were converted into a pair of 750 m long trains, which were in turn coupled together as a single, 70 wagon train set. Building on earlier work by the FP6 NEWOPERA (2005–2008) project, that aims to programme dedicated rail freight (please see Appendix 1 for further details of this project), MARATHON aimed to reduce operating costs by 30 % and increase the capacity of key freight corridors. To feed the longer trains for block or shuttle services, single wagon load business is vital. The rail network can be integrated with inland waterways to promote eco-innovation in freight transport services (European Commission, 2011, p. 8).

12.4.6 Axle Load, High Cube and Train Productivity

The axle load of a rail wagon is the total weight felt by the railway for all wheels connected to a given axle. It is the part of total wagon weight (empty wagon weight + load on the wagon) resting on the axle. The use of higher axle load means that an operator will require a lower number of wagons for a certain volume of cargo, resulting in higher productivity per train - good from the operator's point of view. The axle load of current European railways greatly vary, ranging 17–22.5 tonnes (in few cases 25 tones) and the TEN-T regulations stipulate to achieve at least 22.5 tonnes on the core network by 31 December 2030, whereas railways in the USA allow axle loads of 32–35 tonnes. Thus, the operators argue for a higher axle load. In contrast, rail infrastructure managers prefer a lower axle load, as higher axle load will damage the tracks and will thus generate maintenance and re-investment cost. A recent study suggested that 'the negative impacts of increased axle loads occur primarily in the areas of track and bridge maintenance and renewal, and freight car maintenance' (Kalay, Lopresti, & Davis, 2011). To achieve an efficient rail freight system, Europe wide, axle load remains an important engineering design factor. A further restriction to achieving higher productivity is the carriage of single stack containers, due to restrictions in gauge clearance and bridge/tunnel heights. The use of high cube (9′ 6″ high containers) is increasing, but many European railways are unsuited to transporting this container type. As an example, in response to increasing high cube movement demand, Britain's Network Rail developed W12 gauge clearance in 2011—connecting, for example, the Port of Southampton to the Midlands—to accommodate container dimensions of 2.9 m (9′ 6″) high × 2.6 m (8′ 6″) wide.

12.4.7 Energy Use for Rail

Rolling stock is one of the main energy users in railways. Generally, for all forms of transport, increasing speed uses more energy. The typology of rail energy use can be classified as propulsion system based, traction energy management based, and energy storage system based (Reis, Fabian Meier, Pace, & Palacin, 2013, p. 28–29). For urban passenger rail, the regenerative braking energy system is seen as the way forward—or at least one of the key strategies to optimize energy consumption (see for example González-Gil, Palacin, & Batty, 2013; González-Gil, Palacin, Batty, and Powell 2014). For freight however, rail electrification and capacity management are envisaged as the way ahead, given that one unit of CO_2 is produced per tonne-kilometre for an electric train, up to 8 units for a diesel train, 21 units for a large lorry (>32 t) and 140 units for a small lorry (3.4–7.5 t) (Smith, 2012). FP7's CLEANER-D (2009–2014) project highlighted the potential contribution of a sensible approach to policy and regulation that could lead to sustainability of railways. The project explored how European railways can address the requirements of tighter GHG emissions legislation (Stage IIIB and beyond),[4] from the technological and policy perspectives. Given that rail freight in Europe uses a significant proportion of the overall diesel locomotive rolling stock, this is highly relevant to achieving a better energy usage for rail. Alternative fuels for rail could have a role to play, and have been discussed within the industry, but only as a substitute if diesel fuel production ceased (Lustig, 2013). A complementary role could be considered, as an intermediate step, but it is not foreseen to realistically have an impact, given the trade-off between using these fuels and higher NO_x production.

Finally, we must also highlight the importance of aerodynamics and drag reduction in reducing energy consumption. This is exacerbated in the case of rail freight, as it uses train configurations (e.g. boxy containers mixed with empty wagons) that significantly increase drag forces and therefore energy consumption. Any attempts to improve the energy performance of rail freight must take this aspect into account.

[4] The emission standards regulation for non-road diesel engine adopted in EU (DieselNet, n.d.).

12.4.8 Signalling Systems for Rail

The importance of traffic management in rail, for safety and efficiency, was initiated by the development of the European Transport Management System (ERTMS), consisting of the European Train Control System (ETCS) and the Global System for Mobile Communications-Railway (GSM-R). ERTMS is expected to be the only European-wide accepted signalling and control system (Vinck, 2006). A business case for a Primary European Rail Freight Network with six designated rail freight corridors was developed and a cost benefit analysis drawn up, to demonstrate the productivity-induced volume growth to be brought by the investment (CER, 2007). There are three levels of ERTMS implementation, with levels 2 and 3 having the greatest impact on capacity, as noted in the FP6 NEWOPERA (2005–2008) project. ETCS level 1 potentially increases only 1 % of rail capacity, while levels 2 and 3 increase capacity by 16 % and 50 % respectively. Despite clear benefit from the implementation of ERTMS, progress has been reportedly slow and there are a number of technological, operational, human and costs factors to be addressed, before the full benefit of the implementation can be realized (Smith, Majumdar, & Ochieng, 2012). In addition, the challenge for successful innovation is to support the migration conditions from the old national networks, to the new system (Laroche & Guihéry, 2013). For a discussion of ERTMS vis-à-vis green corridors, please see Chap. 6 of this book.

12.4.9 Governance of Multi-Modal Corridors

As earlier described in Sect. 12.3, freight transport mode choices are limited and rail share in 2008 represented only about 11 % of goods moved within EU-27 (European Commission, 2013). The green corridor concept is the movement of multi-modal freight traffic, between major hubs, over relatively long distances, where cooperation between transport modes (via stakeholders) is at play to improve the quality and environmental performance (EWTC II, 2011). The SuperGreen project used the TEN-T structure to investigate priority freight corridors through benchmarking key performance indicators (KPI) among shippers, along multi-modal corridors. Transport efficiency (via cost and time), service quality (via reliability, frequency, security and safety), environmental sustainability (GHG), infrastructural sufficiency (congestion and bottlenecks) and social issues were used to measure a number of freight operators' performance, along the nine designated corridors. SuperGreen concluded a governance framework to support TEN-T guidelines in promoting sustainable freight transport in Europe and beyond (see also Chap. 4 of this book). Increasing transparency is envisaged, between freight operators along the corridors, to improve load factor in the system, to address green corridor policy (Blinge, 2014).

12.4.9.1 Review of Green Corridors Benchmarking

As already stated previously in this book, the overall objective of the SuperGreen project was to support the 2007 EC Directive on 'Freight Transport Logistics Action Plans' in addressing green corridor issues. At the time of the project launch the term 'green corridors' was perceived rather antipathetically by many stakeholders, as greening corridors can force existing supply chain and logistics networks into a regulated regime, against the free market spirit. An element of the project therefore aimed to benchmark the selected priority corridors against various transport and logistics performance measures (KPI—Key Performance Indicators) to examine the current practice within transport operators using the corridors.

Performance benchmarking of transport chains with multiple vendors, manufacturers, distributors and retailers is challenging, as it is difficult to attribute performance to one particular entity within the supply chain (Hervani, Helms, & Sarkis, 2005). FP7's BE-LOGIC (2009–2011) project, aimed at benchmarking European freight transport chains, demonstrated that cost, time flexibility, reliability, quality and sustainability were identified as the main aspects considered by logistics service users and providers (Islam, Zunder, & Jorna, 2013). In SuperGreen, these KPIs were used to characterize some of the selected corridors, through interviews with some of the transport operators using them. While this was far from an ideal benchmarking exercise, an illustration of the current practice in the selected corridors for rail was able to be determined, as can be seen in Table 12.2 below.

Comparing Table 12.2 with Table 12.1, it can be concluded that rail freight operation within some of the selected SuperGreen corridors is within the typical estimated emissions, except for corridors beyond Europe (Silk Way—which links to China).

12.4.9.2 Review of SuperGreen Conclusions, Results and Lessons, vis-a-vis Rail

Apart from the selected priority corridors benchmarking exercise, SuperGreen was also evaluating the bottlenecks along the corridors (SUPERGREEN, 2011b), technological improvement determination (Clausen, Geiger, & Behmer, 2012; Fozza & Recagno, 2012), and supply (and or transport) chain strategies and management considerations. The main drawback of rail within the supply chain is it is highly unlikely that rail will serve the first and last mile of the chain and therefore will benefit from improvement of the interchange points that facilitate multi-modality. The use of ICT and Technology in rail is envisaged to contribute to greener transport (see also Chaps. 5 and 6 of this book), but the critical question is how to optimize the route for minimum delay, with minimum energy consumption.

In the assessment of SuperGreen rail corridors it was concluded that cross-border interoperability systems, via the deployment of ERTMS (as discussed in the previous section), would improve the existing system, although this approach

Table 12.2 SuperGreen corridors benchmark (*source*: SUPERGREEN (2011a)

Corridor name	Mode of transport	CO_2 (g/tkm)	SO_x (g/tkm)	Cost (€/tkm)	Average speed (km/h)	Reliability (%)	Frequency × times/year
Brenner	*Intermodal*	10.62–42.11	0.020–0.140	0.03–0.09	9–41	95–99	26–624
	Road	46.51–71.86	0.050–0.080	0.05–0.06	19–40	25–99	52–2,600
	Rail	9.49–17.61	0.040–0.090	0.05–0.80	44–98	60–95	208–572
	SSS	16.99	0.050–0.120	0.04–0.05	23	100	52–520
Cloverleaf	*Road*	68.81	0.091	0.06	40–60	80–90	4,680
	Rail	13.14–18.46	0.014–0.021	0.05–0.09	45–65	90–98	156–364
Nureyev	*Intermodal*	13.43–33.36	0.030–0.150	0.10–0.18	13–42	80–90	156–360
	SSS	5.65–15.60	0.070–0.140	0.05–0.06	15–28	90–99	52–360
Strauss	*IWT*	9.86–22.80	0.013–0.031	0.02–0.44	–	–	–
Mare Nostrum	*SSS*	6.44–27.26	0.092–0.400	0.003–0.200	17	90–95	52–416
	DSS	15.22	0.22	–	–	–	–
Silk way	*Rail*	41.00	–	0.05	26	–	–
	DSS	12.50	–	0.004	20–23	–	–

would not necessarily be applicable to the entire rail network, due to various issues also discussed in the previous sections (see also Chap. 6 of this book on ERTMS vis-à-vis green corridors). Additionally, the reorganization of rail freight services e.g. a track access charge, for example in the Channel tunnel, would increase the competitiveness of rail against other modes. Green supply chain, as promoted by SuperGreen, has the objective of improving sustainability; it is the movement of goods—in the form of a load or unit—with the minimum environmental impact. Full loads make this target easier to achieve, whereas part loads form a more complex business, due to the higher number of stakeholders involved. For this reason the governance of priority corridors, as promoted by SuperGreen, would ease progress towards the achievement of a sustainable freight system.

12.5 Conclusions

The current share of rail freight in the EU is about 11 % and this is an increase from 8 % in 2000. Regaining the share level of the 1970s (i.e. 20 %), or greater, is one of the current EU policies in decarbonizing the transport system to meet its environmental objectives. The optimistic scenario pointed to a maximum potential share of rail freight in the range of 31–36 % (Den Boer, van Essen, Brouwer, Pastori, & Moizo, 2011) but this can only be realized if modal shift policy is strictly enforced across the EU countries.

The 2011 White Paper has set up the logistics roadmap for long distance freight (>300 km) to shift to rail, or inland waterway, and paved the way for rail to take the opportunity to increase its market share. This roadmap is well supported by the TEN-T structure, which focuses on the improvement of the core network corridors that allow the green corridor policy to be effective. FP7's FREIGHTVISION (2008–2010) project hypothetically suggested that investment in rail infrastructure is needed, for the future of rail transport to prosper, together with political support for the prioritization of freight, ERTMS, electrification, and innovation in longer and heavier trains (FREIGHTVISION, 2011). Similarly, SuperGreen, through benchmarking priority 'green' corridors and its associated governance policy, opens up opportunities for improving a synchronized, multimodal transport system, where rail should play bigger role in addressing the global sustainability agenda.

Green corridor policy, as a strategic and governance tool, can only realistically be implemented with the improvement of the long distance transportation sector, where rail becomes the backbone network of land surface transportation. With the trend of urbanization, where road would serve pretty much the last mile freight operation, reliable, efficient, safe and environmentally sound inter-urban long distance rail freight is needed to support economic growth and to maintain the quality of life of all EU citizens.

Despite the central governance of the EU in promoting the rail modal-shift policy, it may be the private sector that can help most in making this happen, in much the same way that the rail industry prospered in the past. The spirit of *Periculum Privatum Utilitas Publica* is once again needed, to resurrect our railways.

Appendix 1: Selected Completed Projects Under the European Commission Framework Programmes That Promote Green Corridor Policy

Project acronym	Objective	Corridor assessed	Management/technology innovations	Lessons learned and policy implications
FP4 **EUFRANET**—improving the competitiveness of rail freight services (1997–1999)	To identify and evaluate strategic options for the development of a trans-European rail network mainly dedicated to freight transportation	This is a simulation approach so, effectively, corridors examined were across the European rail network	– Shippers' survey with stated preference	*Source*: (EUFRANET, 2001)
			– Demand model	– Definition of different layer of freight priority network
			– Supply model	– Definition of a core network of freight, centred around BENELUX countries, with strands to Stockholm (SE), Bologna (IT), Marseille (FR) and Northern France, connecting to UK
			– A dynamic equilibrium process of supply–demand for forecasting transport demand	– Definition of intermediate network, mainly dedicated to freight, but also carrying passenger trains
				– A mixed network on which passenger trains would normally have priority
FP6 **BRAVO**—Brenner rail freight action strategy aimed at achieving a sustainable increase in intermodal transport volume (2004–2007)	To develop a coherent corridor management scheme enabling open access and act as a blueprint for an interoperable rail traction scheme	Munich (DE)—Verona (IT)	– Open corridor management system	*Source*: (BRAVO, 2007)
			– Train path availability and allocation process	– Cross-border operation of multisystem-locomotives and loco

	involving multi-current locomotives			drivers for efficient and reliable rail with more than 2,100 trains
			– Interoperable rail traction scheme	– Radio-remote control of pushing engines to haul heavy trains on the steep Brenner north ramp
			– Advanced customer information system	– Online train monitoring including estimated time of availability information accessible for all rail stakeholders
			– Development schedule for unaccompanied combined transport (containers) services	– Feasibility of mega semitrailer pocket wagon, for higher capacity, to support automotive industry
FP6 **TREND**—implementation of change in the European rail freight area (2005–2006)	To accelerate the development of predominantly nationally aligned systems into a single, integrated Pan-European system approach towards rail freight service competitiveness	Corridor A: IT (Adriatic coast)—SI–HU	– Evaluation and assessment of Trans European rail freight services	Source: (TREND, 2006)
		Corridor B-West: NL (seaports)—DE–CH–IT	– Web-based geo-referenced information system (GIS)	– An evaluation scheme for integration and interoperability of priority rail freight corridors
		Corridor B-East: Scandinavia (German border)–DE–AT–IT	– Terms of reference for Trans European rail freight services	– Different countries characterised by different rail freight issues
		Corridor C: DE (Ruhr area)—CZ/AT–SL–HU–RS/RO–BL–TR	– Experts workshops to identify potential pan	– Market forces need time to create competitive environment, despite

(continued)

Project acronym	Objective	Corridor assessed	Management/technology innovations	Lessons learned and policy implications
			European rail freight corridors	increasing compliance of EU countries in liberalising the rail freight market
		Corridor D: NL (seaports)—DE–PL–LT–LV–EE		– Rail freight is not relevant for every kind of traffic; focus on relevant market segments
		Corridor E: FR (seaports UK via tunnel)—CH–IT		– Rail freight flow was unbalanced in terms of contribution to the different corridor destinations (the range of rail mode share is between 8 and 24 % across the six corridors)
		Corridor F: DE (Ruhr area and possible branch to PL)—FR–ES		– Border crossing bottlenecks with infrastructural issues were identified

FP6 **REORIENT**—Implementing change in the European Railway System (2005–2007)	To assess progress in transferring European railways to an internationally integrated and standardised system	Nordic capitals [Oslo (NO)/Stockholm (SE)/Helsinki (FI)] to the South Eastern European Capitals [Bucharest (RO)/Sofia (BL)/Thessaloniki (GR)] via PL, AT, SL and HU	– Examining external driving forces for changes in rail liberalization	Source: (Ludvigsen, 2009; REORIENT, 2007)
			– North–south corridor cross border socio-political dialogue assessment	– Nordic countries exhibit high levels of compliance with rail liberalization but without much intra-rail market competition;
				– Rail in Nordic countries is not much of a rival to truckers
				– New Member States lagged behind on legal compliance despite increasing number of permits granted to new entrants that compete fiercely with national incumbents
				– Rail quality service divide between the Nordic and the South Eastern European countries
				– Potential unitised goods commodity within the corridor

(continued)

Project acronym	Objective	Corridor assessed	Management/technology innovations	Lessons learned and policy implications
FP6 **NEWOPERA**—the rail freight dedicated lines concept (2005–2008)	To provide grounds for the establishment of new and existing lines, predominantly dedicated to freight	Conceptually across European rail network that refers to the global trade growth market analysis; in practice there were four corridors conceptually examined and one corridor [Berlin (DE)—Madrid (ES)] was taken as the pilot reference case	– Global and European trade growth patterns and rail freight market analysis	Source: (NEWOPERA, 2008)
			– Longer and heavier train concept	– The international intermodal unaccompanied traffic is growing and a greater percentage of intermodal traffic is concentrated on well-defined axes
			– Train operations and management software and training	– Increase rail freight market share means an increase in capacity, via either infrastructure investment or longer and heavier trains
			– ERTMS (European Railway Traffic Management System) concept	– The need for standardization and integration of a system of cross border corridors including: gauge, axel-load, train length, maintenance strategy, capacity and priority management, emergency management, tolls and pricing,

				and ERTMS level 2 and 3 by 2015–2020
			– Rail network model assignment	– There are benefits gained from dedicated rail freight corridors, especially medium to long term returns (estimated at over 5 % IRR) but an increase in infrastructure cost is inevitable
FP6 **RETRACK**—reorganization of transport network by advanced rail freight concepts (2006–2011)	To design, develop and implement a new and innovative trans-European rail freight service concept (market-oriented rail freight operators)	Rotterdam (NL) to Constanza (RO) and to the Black Sea area and Turkey; effectively the successful pilot corridor in operation is between Cologne (DE) and Gyor (HU)	– Terminals and cargo handling	Source: (RETRACK, 2012; Zunder et al., 2013)
			– ICT training using railway management system	– Organizational separation between the train operating companies and the infrastructure management hinders private rail undertakings to complete efficiently in rail markets
			– Common code of language between cross border rail freight staff	– A single standardised optimal terminal design can't be suggested as there are many effective options to transfer cargo
				– Successful rail freight operations by new entrant
				– The importance of lead partner with experience in the freight transport

(continued)

Project acronym	Objective	Corridor assessed	Management/technology innovations	Lessons learned and policy implications
				and forwarding sector and with good operational links
FP6 **CREAM**—technical and operational innovations implemented on a European rail freight corridor (2007–2012)	To design and validate advanced customer-driven business models for railway undertakings and intermodal operators through analysis of the operational and logistic service providers	Rotterdam (NL)/Antwerp (BE) to Turkey (TR) via either the north bound corridor through Hungary (HU) or the south bound corridors through the Balkans; effectively the pilot successful corridor in operation is via the project partners established business (e.g. copper anode train between Bulgaria and Belgium)	– Quality Management System Manual	Source: (CREAM, 2012; Vleugel & Bal, 2012)
			– Tracking of trains and waggons by GPS	– 28 % reduction of rail transport time between Ljubljana (SI)—Halkali (TR) through Quality Manual introduction
			– Train monitoring system	– Putting interoperability into practice was often hampered by long border station stop times, long and inefficient turn-around times of locomotives due to low frequency of transports, insufficient availability of interoperable locomotives, long-lasting homologation procedures for locomotives, and inappropriate market conditions in some countries
			– Transhipment for conventional semi-trailers	– Reach stackers transhipment innovation is

				less efficient than gantry crane operations
			– Dual power locomotive	– Specifically designed semi-trailers for glass loads (floatrailer) can fill the rail freight market gap
FP7 **TIGER**—the co-modal role in industrialising the maritime traffic hinterland distribution (2009–2012)	To achieve a greater degree of effectiveness, efficiency and competitiveness on the rail freight network, from sea port to the hinterland (dry port)	Genoa Sea Port—RTE Rivalta (Dry Port)	– Testing rail freight capacity market scenario to anticipate huge influx of international logistics with increasing size of sea vessels (up to 18,000 TEU)	Source: (TIGER, 2012)
		Port of Gioia Tauro/Port of Taranto—Bari—Interporto Bologna	– ICT/software package, e-management, e-customs programming and control system to improve Dry Ports	– The space in Europe is limited resource; sea terminals are the most expensive places to keep containers, so the connectivity between ports and the hinterland can improve cost effectiveness of freight movement
		Port of Hamburg—Port of Bremerhaven		– The importance of Dry Ports connected by rail to the sea ports to anticipate a huge influx of

(continued)

Project acronym	Objective	Corridor assessed	Management/technology innovations	Lessons learned and policy implications
				containers from sea vessels
		Mega hub spider web based in Hamburg and Munich		– Lengthening marginally the trains, by one or two wagons, saves transport costs and additional capacity is generated at very marginal cost
				– Intermodal operators, who are competitors in the market place, joined forces in order to participate in train loading in the interests of the optimization of train costs
				– Rail freight constitutes only a part of the entire customer's supply chain and a regular industrialised shuttle service between sea ports and dry ports/mega hub/freight village would improve rail competitiveness

FP7 **SUPERGREEN**—supporting EU's Freight Transport Logistics Action Plan on Green Corridors Issues (2010–2013)	To promote the development of European freight logistics in an environmentally friendly manner through performance benchmarking and evaluations on multi-modal freight corridors	[1] Brenner: Rostock/Sassnitz (DE)—Palermo (IT)/Athens (GR)	– Freight stakeholders workshops to identify priority corridors	Source: (SUPERGREEN, 2013a, 2013b, 2013c, 2013d)
		[2] Finis Terrae: Lisbon (PT)/Gijon (ES)—Paris/St Nazaire (FR)	– KPI survey on freight operators across priority corridors	– Establishing the benefit of priority 'green' corridors to address sustainable freight transportation
		[3] Cloverleaf: Dublin (IE)/Glasgow (GB)—Duisburg (DE)	– Freight stakeholders interviews in response to technological challenge in greening freight transportation	– Identification of bottlenecks and priority technological deployments across the priority corridors
		[4] Edelweiss: St Petersburg (RU)/Helsinki (FI)/Oslo (NO)/Stockholm (SE)—Genoa (IT)		– Green corridors handbooks to support freight stakeholders (notably logistics business community) and policy makers in addressing sustainable freight transport through governance
		[5] Nureyev: St Petersburg—Moscow (RU) and Klaipeda (LT)—Minsk (BY)		– Research and Development (R&D) recommendations, for all

(continued)

Project acronym	Objective	Corridor assessed	Management/technology innovations	Lessons learned and policy implications
				modes of transport, to improve connectivity and interoperability of priority corridors
		[6] Strauss: Bratislava (SK)—Rotterdam (NL)/Paris (FR)		– Policy recommendations to prioritise investments on the state-of-the-art transportation technologies for all modes of transport across the priority corridors
		[7] Two seas: Hamburg (DE)—Athens (GR)		– Bringing the economic, social, and environmental benefit of priority corridors to the implementation of the Core Network Policy, promoted by the TEN-T function
		[8] Mare Nostrum: Algeciras (ES)—Lyon (FR)		
		[9] Silk way: Shanghai (CN)—Le Havre (FR)/Rotterdam (NL)/Hamburg (DE)/Gdansk (PL)		

Note: for country code definition please see: http//countrycode.org

Appendix 2: Selected Reviewed Framework Programme Projects Linked to Rail and Logistics

No	FP	Project acronym	Project name	Electronic reference where available
1	4	EUFRANET	Improving the Competitiveness of Rail Freight Services	http://www.transport-research.info/Upload/Documents/200310/eufranet.pdf
2	4	HISPEEDMIX	High Speed Freight on the European High Speed Railway Network	http://www.transport-research.info/Upload/Documents/200310/hispeedmix.pdf
3	4	INTELFRET	Intelligent Freight Train	http://www.transport-research.info/Upload/Documents/200310/intelfret.pdf
4	4	FIRE	Freight Information in the Railway Environment	http://www.transport-research.info/Upload/Documents/200310/fire.pdf
5	4	OPTIRAILS	Optimization of traffic through the European Rail Traffic Management Systems	http://www.transport-research.info/Upload/Documents/200310/optirails.pdf
6	5	EDIP	On-board radio-based control of multiple locomotive freight trains for trans-European operation	n/a
7	5	PROMAIN	Progress in Maintenance and Management of Railway Infrastructure	n/a
8	5	CARGOSPEED	Cargo rail road interchange at speed	http://www.transport-research.info/Upload/Documents/200607/20060727_143123_02411_CARGOSPEED_Final_Report.pdf
9	5	IN.HO.TRA	Interoperable intermodal horizontal transhipment	http://www.transport-research.info/Upload/Documents/200607/20060727_150345_76487_INHOTRA_Final_Report.pdf
10	5	BESTUFS	Best urban freight solutions	http://www.transport-research.info/Upload/Documents/200608/20060821_161755_24060_BESTUFS%20Best%20Practice.pdf
11	5	CITYFREIGHT	Inter- and intra- urban freight distribution networks	n/a

(continued)

No	FP	Project acronym	Project name	Electronic reference where available
12	6	BRAVO	Brenner rail freight action strategy aimed at achieving a sustainable increase of inter-modal transport volume	http://www.transport-research.info/Upload/Documents/201010/20101007_161556_82086_BRAVO%20-%20Final%20Report.pdf
13	6	TREND	Implementation of change in the European rail freight area	http://www.transport-research.info/Upload/Documents/201010/20101004_214444_36830_Trend_B7-brochure_Results_and_Recommendations.pdf
14	6	REORIENT	Implementing change in the European Railway System	http://www.transport-research.info/Upload/Documents/200904/20090420_122526_74196_Reorient%20Publishable%20Final%20Activity%20Report.pdf
15	6	NEWOPERA	The rail freight dedicated lines concept	http://www.transport-research.info/Upload/Documents/201301/20130116_143319_35717_web_newopera2.pdf
16	6	RETRACK	Reorganization of transport networks by advanced rail freight concept	http://www.transport-research.info/Upload/Documents/201208/20120801_133828_14885_Retrack_D7.3.pdf
17	6	CREAM	Technical and operational innovations implemented on a European rail freight corridor	http://www.transport-research.info/Upload/Documents/201207/20120710_095633_84456_CREAM(FP6-038634)_Publishable-Final-Activity-Report_07-2012.pdf
18	6	FastRCargo	Fast transhipment equipment and novel methods for Rail Cargo in Europe	n/a
19	6	CHINOS	Container handling in inter-modal nodes—optimal and secure	http://www.transport-research.info/Upload/Documents/201203/20120313_153302_17791_CHINOS%20Final%20Report_final.D0.4.pdf
20	6	TRIMOTRANS	Development of new inter-modal loading units and dedicated adaptors for the trimodal transport of bulk materials in Europe	n/a

(continued)

No	FP	Project acronym	Project name	Electronic reference where available
21	6	ISTU	Integrated Standard Transport Unit for self-guided freight container transportation systems on rail	n/a
22	6	INTERGAUGE	Interoperability, security and safety of goods movement with 1,435 and 1,520 (1,524) mm track gauge railways: new technology in freight transport including hazardous products	n/a
23	6	CAESAR	Coordination action for the European strategic agenda of research on intermodalism and logistics	n/a
24	6	PROMIT	Promote innovative inter-modal freight transport	n/a
25	6	TRANS-TOOLS	Tools for transport forecasting and scenario testing	http://www.transport-research.info/Upload/Documents/201003/20100304_172116_94411_TRANSTOOLS%20-%20Final%20Report.pdf
26	7	BE-LOGIC	Benchmarking logistics and co-modality to improve the efficiency within and across different modes of transport and to support the development of a quality logistics system	http://www.transport-research.info/Upload/Documents/201203/20120330_180131_96883_BE%20LOGIC_final_report_en_v04_publishable.pdf
27	7	TIGER	The co-modal role in industrialising the maritime traffic hinterland distribution	n/a
28	7	ERRAC Roadmap	Coordinating rail research activity	http://www.transport-research.info/Upload/Documents/201204/20120404_120654_619_errac_freight_roadmap_2011_final_draft_version.pdf
29	7	SPECTRUM	Solutions and processes to enhance the competitiveness of transport by rail in unexploited markets	n/a http://www.spectrumrail.info/
30	7	MARATHON	Make rail the hope for protecting nature	n/a http://www.marathon-project.eu/
31	7	CLEANER-D	Clean European Rail Diesel	http://www.cleaner-d.eu/deliverables.aspx

(continued)

No	FP	Project acronym	Project name	Electronic reference where available
32	7	SUPERGREEN	Supporting EU's Freight Transport Logistics Action Plan on Green Corridors Issues	http://www.supergreenproject.eu/
33	7	SUSTRAIL	The sustainable freight railway: designing freight vehicle track system for higher delivered tonnage with improved availability at reduced cost	n/a http://www.sustrail.eu/
34	7	D-RAIL	Development of the future rail freight system to reduce the occurrences and impact of derailment	n/a http://d-rail-project.eu/
35	7	LOGMAN	Logistics and Manufacturing trends and sustainable transport	http://www.transport-research.info/Upload/Documents/201204/20120404_170607_2733_LOGMAN_D6_final.pdf
36	7	FREIGHTVISION	Sustainable European Freight Transport 2050	http://www.transport-research.info/web/projects/project_details.cfm?id=36661

References

Aditjandra, P. T., Zunder, T. H., Islam, D. M. Z., & Vanaale, E. (2012), Investigating freight corridors towards low carbon economy: Evidence from the UK. In *4th Conference Transport Research Arena* (Vol. 48, pp. 1865–1876). Athens, Greece: Elsevier Procedia—Social and Behavioural Sciences. http://dx.doi.org/10.1016/j.sbspro.2012.06.1161

Asmild, M., Holvad, T., Hougaard, J. L., & Kronborg, D. (2009). Railway reforms: Do they influence operating efficiency? *Transportation, 36*(5), 617–638.

Banister, D. (2005). *Unsustainable transport city transport in the new century*. Abingdon, Oxford: Routledge Taylor & Francis.

Blinge, M. (2014). Policy measures to realise green corridors—A stakeholder perspective. *Research in Transportation Business & Management* (in press) Vol. 12, pp:55–62. doi:10.1016/j.rtbm.2014.06.007.

Božičnik, S. (2009). Opening the market in the rail freight sector. *Built Environment, 35*(1), 87–106.

BRAVO. (2007) *BRAVO—Brenner Rail Freight Action strategy aimed at achieving a sustainable increase of intermodal transport volume by enhancing quality, efficiency, and systems technologies. Final Report for Publication*. Retrieved from http://www.transport-research.info/Upload/Documents/201010/20101007_161556_82086_BRAVO%20-%20Final%20Report.pdf

CER. (2007). *Business cases for a primary European rail freight network*. Brussels, Belgium.

Clapham, J. H. (1915). Book reviews: the North-Eastern Railway: its rise and development by W. W. Tomlinson. *The Economic Journal, 25*(100), 572–574.

Clausen, U., Geiger, C., & Behmer, C. (2012). Green corridors by means of ICT applications. In *4th Conference Transport Research Arena* (Vol. 48, pp. 1877–1886). Athens, Greece: Elsevier Procedia—Social and Behavioural Sciences.

Clausen, U., & Voll, R. (2013). A comparison of North American and European railway systems. *European Transport Research Review, 5*(3), 129–133.

CREAM. (2012). *The CREAM project final report*. In V. Behrends, L. Deiterding, N. Galonske, M. Mascolo, K. Petri, & K.-U. Sonderman (Eds.), Hannover, Germany: HaCon Ingenieurgesellschaft mbH.

Dekker, R., Bloemhof, J., & Mallidis, I. (2012). Operations research for green logistics: An overview of aspects, issues, contributions and challenges. *European Journal of Operational Research, 219*(3), 671–679. Elsevier B.V.

Den Boer, E., van Essen, H., Brouwer, F., Pastori, E., & Moizo, A. (2011). *Potential of modal shift to rail transport Study on the projected effects on GHG emissions and transport*. Delft, Netherlands: Community of European Railway and Infrastructure Companies (CER).

DG-TREN. (2008). *EU energy and transport in figures: Statistical pocketbook 2007/2008*. Luxembourg: Office for Official Publications of the European Communities.

Di Pietrantonio, L., & Pelkmans, J. (2004, September). The economics of EU railway reform. *Bruges European Economic Policy Briefings, 8*.

DieselNet. (n.d.). *Emission Standards EU nonroad diesel engines*. Retrieved August 15, 2014, from https://www.dieselnet.com/standards/eu/nonroad.php

EUFRANET. (2001). *Final report for publication*. Retrieved from http://www.transport-research.info/Upload/Documents/200310/eufranet.pdf

European Commission. (2007). *COM (2007) 607 final—Freight and logistics action plan*.

European Commission. (2011). *2011 WHITE PAPER roadmap to a single European transport area*.

European Commission. (2013). *EU transport in figures: Statistical pocketbook 2013*. Luxembourg: Publication Office of the European Union. doi:10.2832/19314.

EWTC II. (2011). *Green Corridor Manual (draft)—Purpose, definition and vision for Green Transport Corridors*. Copenhagen, Denmark: Danish Transport Authority.

Finger, M., Groenewegen, J., & Künneke, R. (2005). The quest for coherence between institutions and technologies in infrastructures. *Journal of Network Industries, 6*(4), 227–259.

Fozza, S., & Recagno, V. (2012). Sustainable technologies and innovation for green corridors: Survey and application. In *4th Conference Transport Research Arena* (Vol. 48, pp. 1753–1763). Athens, Greece: Elsevier Procedia—Social and Behavioural Sciences.

FREIGHTVISION. (2011). *FREIGHTVISION—Sustainable European Freight Transport 2050*. In S. Helmreich & H. Keller (Eds.), Heidelberg: Springer.

Gilbert, R., & Perl, A. (2008). *Transport revolutions: Making the movement of people and freight work for the 21st century*. London: Earthscan.

Givoni, M., Brand, C., & Watkiss, P. (2009). Are railways climate friendly? *Built Environment, 35* (1), 70–86.

Givoni, M., & Holvad, T. (2009). The prospects for European railways is the second railway age still here or yet to begin? *Built Environment, 35*(1), 5–10.

González-Gil, A., Palacin, R., & Batty, P. (2013). Sustainable urban rail systems: Strategies and technologies for optimal management of regenerative braking energy. *Energy Conversion and Management, 75*, 374–388.

González-Gil, A., Palacin, R., Batty, P., & Powell, J. P. (2014). A systems approach to reduce urban rail energy consumption. *Energy Conversion and Management, 80*, 509–524.

Gregory, I. N., & Henneberg, J. M. (2010). The railways, urbanization, and local demography in England and Wales, 1825–1911. *Social Science History, 34*(2), 199–228.

Hall, P. (2007). Beyond the automobile? *Access, 30*, 10–15.

Hall, P., & Banister, D. (1994). The second railway age. *Built Environment, 19*(3/4), 157–162.

Hervani, A. A., Helms, M. M., & Sarkis, J. (2005). Performance measurement for green supply chain management. *Benchmarking: An International Journal, 12*(4), 330–353.

Holvad, T. (2009). Review of railway policy reforms in Europe. *Built Environment, 35*(1), 24–42.

Hunke, K., & Prause, G. (2013). Management of green corridor performance. *Transport and Telecommunication, 14*(4), 292–299.

Islam, D. M. Z., Meier, J. F., Aditjandra, P. T., Zunder, T. H., & Pace, G. (2013). Logistics and supply chain management. *Research in Transportation Economics, 41*(1), 3–16. Elsevier Ltd.

Islam, D. M. Z., Zunder, T. H., & Jorna, R. A. (2013). Performance evaluation of an online benchmarking tool for European freight transport chains. *Benchmarking: An International Journal, 20*(2), 233–250.

Jackson, C. (2014, April). Making the most of freight capacity. *Railway Gazette International*, p. 3.

Jackson, R., Islam, D. M. Z., Zunder, T. H., Schoemaker, J., & Dasburg, N. (2013). The potential of the low density high value rail freight market in Europe. In *13th World Conference on Transport Research, Rio de Janeiro* (pp. 1–23).

Kalay, S., Lopresti, J., & Davis, D. (2011). Development of enabling technologies for heavy axle load operations in North America. In *9th World Congress on Railway Research, Lille, France* (pp. 1–12).

Kirchner, C. (2006). Legal instruments for liberalising European rail freight markets from 2007. *Intereconomics, 41*(6), 295–298.

Lardner, D. (1850). *Railway economy*. New York: Harper & Brothers.

Laroche, F., & Guihéry, L. (2013). European Rail Traffic Management System (ERTMS): Supporting competition on the European rail network? *Research in Transportation Business & Management, 6*, 81–87.

Ludvigsen, J. (2009). Liberalisation of rail freight markets in the old and new EU-member states. *European Journal of Transport and Infrastructure Research, 9*(1), 31–45.

Lustig, D. (2013, June). In search of the tipping point. *Railway Gazette International*, pp. 30–34.

Matsika, E., Ricci, S., Mortimer, P., Georgiev, N., & O'Neill, C. (2013). Rail vehicles, environment, safety and security. *Research in Transportation Economics, 41*(1), 43–58. Elsevier Ltd.

McKinnon, A. (2007). *CO_2 emissions from freight transport in the UK*. Report prepared for the Climate Change Working Group of the Commission for Integrated Transport, Edinburgh.

McKinnon, A. (2010). Environmental sustainability: A new priority for logistics managers. In A. McKinnon, S. Cullinane, M. Browne, & A. Whiteing (Eds.), *Green logistics: Improving the environmental sustainability of logistics* (1st ed.). London: Kogan Page.

McKinsey & Company. (2009). *Pathways to a low-carbon economy version 2 of the global greenhouse gas abatement cost curve*. Retrieved from http://www.mckinsey.com/~/media/McKinsey/dotcom/client_service/Sustainability/cost curve PDFs/ImpactFinancialCrisisCarbonEconomics GHGcostcurveV21.ashx

Monopolkommission. (2007). *Wettbewerbs- und Regulierungsversuche im Eisenbahnverkehr*, Berlin, Germany.

Nash, C., Nilsson, J., & Link, H. (2013). Comparing three models for introduction of competition into railways. *Journal of Transport Economics and Policy, 47*(Part 2), 191–206.

NEWOPERA. (2008). *NEWOPERA—The rail freight dedicated lines concept final report*. In F. Castagnetti (Ed.). Brussels, Belgium: The European Freight and Logistics Leaders Forum (F&L), doi:10.1007/BF02472679

Panagakos, G. P., & Psaraftis, H. N. (2014). How green are the TEN-T core network corridors? In *5th Conference Transport Research Arena*. Paris: IFSTTAR. Retrieved from http://tra2014.traconference.eu/papers/pdfs/TRA2014_Fpaper_18260.pdf

Pittman, R., Diaconu, O., Sip, E., Tomova, A., & Wronka, J. (2007). Competition in freight railways: 'Above-the-rail' operators in Central Europe and Russia. *Journal of Competition Law and Economics, 3*(4), 673–687.

Prause, G., & Hunke, K. (2014). Sustainable entrepreneurship along green corridors. *Entrepreneurship and Sustainability Issues, 1*(3), 124–133.

Preston, J. (2009). Trends in European railways over the last two decades. *Built Environment, 35* (1), 11–23.

Psaraftis, H. N., Minsaas, A., Panagakos, G., Palsson, C., & Salanne, I. (2013). Green corridors in European surface freight logistics. In J. H. Bookbinder (Ed.), *Handbook of global logistics* (pp. 193–218). New York: Springer.

Psaraftis, H. N., & Panagakos, G. (2012). Green corridors in European surface freight logistics and the SuperGreen project. In *4th Conference Transport Research Arena* (Vol. 48, pp. 1723–1732). Athens, Greece: Elsevier Procedia—Social and Behavioural Sciences. http://dx.doi.org/10.1016/j.sbspro.2012.06.1147

Rail Freight Group. (2014). Personal communication Paulus T. Aditjandra with Maggie Simpson, June 8, 2014.

Reis, V., Fabian Meier, J., Pace, G., & Palacin, R. (2013). Rail and multi-modal transport. *Research in Transportation Economics, 41*(1), 17–30.

REORIENT. (2007). *Implementing change in the European Railway System selected findings from REORIENT*. Retrieved from http://www.transport-research.info/Upload/Documents/200904/20090420_122526_74196_Reorient Publishable Final Activity Report.pdf

RETRACK. (2012). *RETRACK final report*. In T. van Rooijen, P. van de Lande, M. Zhang, P. Mortimer, R. Jackson, D. Islam, et al. (Eds.) Rotterdam: TNO.

Rothengatter, W. (2006, November/December). Issues of interoperability in the European Railway System. *Intereconomics*, pp. 306–311.

Rothengatter, W. (2014). Personal communication with Paulus T Aditjandra, June 4, 2014.

Sichelschmidt, H. (1999). The EU programme 'trans-European networks'—A critical assessment. *Transport Policy, 6*(3), 169–181.

Smith, R. A. (2012). Energy for railways. In O. Inderwildi & S. D. King (Eds.), *Energy, transport & the environment* (pp. 561–575). London: Springer.

Smith, P., Majumdar, A., & Ochieng, W. Y. (2012). An overview of lessons learnt from ERTMS implementation in European railways. *Journal of Rail Transport Planning & Management, 2* (4), 79–87. Elsevier Ltd.

Stehmann, O., & Zellhofer, G. (2004). Dominant rail undertakings under European competition policy. *European Law Journal, 10*(3), 327–353.

Stehmann, O., & Zenger, H. (2011). The competitive effects of rail freight mergers in the context of European liberalization. *Journal of Competition Law and Economics, 7*(2), 455–479.

Stern, N. (2006). *Stern review: The economics of climate change*. London: HM Treasury, HMSO.

SUPERGREEN. (2011a). *Deliverable D2.4 Benchmarking of green corridors version 2*. Seventh Framework Programme. Retrieved from http://www.supergreenproject.eu/docs/public/D2.4-2 REVISED v.3.1 FINAL PUBLIC.pdf

SUPERGREEN. (2011b). *Deliverable D2.5—Definition of areas for improvement*. Seventh Framework Programme. Retrieved from http://www.supergreenproject.eu/docs/public/D2 5_ WP 2_Super Green_FINAL PUBLIC.pdf

SUPERGREEN. (2013a). *Deliverable D5.2—Define and submit R&D recommendations, Final version*. Seventh Framework Programme. Retrieved from http://www.supergreenproject.eu/docs/public/D5.2-2.pdf

SUPERGREEN. (2013b). *Deliverable D6.2—Policy recommendations version 2*. Seventh Framework Programme. Retrieved from http://www.supergreenproject.eu/docs/public/D6.2-2.pdf

SUPERGREEN. (2013c). *Green corridors handbook Vol. I*. Seventh Framework Programme. Retrieved from http://www.supergreenproject.eu/docs/public/Handbook-I.pdf

SUPERGREEN. (2013d). *Green corridors handbook: Vol. II*. Seventh Framework Programme. Retrieved from http://www.supergreenproject.eu/docs/public/Handbook-II.pdf

TIGER. (2012). *TIGER final report*. In F. Castagnetti (Ed.) Brussels, Belgium: NEWOPERA Aisbl.

Tomlinson, W. W. (1914). *The North Eastern Railway: Its rise and development*. Newcastle upon Tyne, England: Andrew Reid & Company.

TREND. (2006). *TREND, Deliverable Work Package B7? Conclusions and recommendations*. Hannover and Frankfurt. Retrieved from http://www.transport-research.info/Upload/

Documents/201010/20101004_214444_36830_Trend_B7-brochure_Results_and_Recommendations.pdf

UIRR. (n.d.). Loading units. *International Union For Road-Rail Combined Transport*. Retrieved August 15, 2014, from http://www.uirr.com/en/road-rail-ct/framework-conditions/loading-units.html

Vassallo, J. M., & Fagan, M. (2007). Nature or nurture: Why do railroads carry greater freight share in the United States than in Europe? *Transportation, 34*(2), 177–193.

Vickerman, R. W. (1995). The regional impacts of trans-European networks. *The Annals of Regional Science, 29*, 237–254.

Vinck, K. (2006). Developing European freight corridors with ERTMS—New European political initiatives. In *Competition in Europe's rail freight market* (pp. 79–95). Brussels, Belgium: CER.

Vleugel, J., & Bal, F. (2012). Some approaches to reduce transport time of intermodal services: Smart rail investments. *European Transport\Trasporti Europei, 52*(3), 1–15.

Wolff, C. (2006). Overview of rail freight markets—Europe and beyond. In *Competition in Europe's rail freight market* (1st ed., pp. 15–34). Brussels, Belgium: Community of European Railway and Infrastructure Companies (CER).

Wolmar, C. (2007). *Fire & steam: A new history of the railways in Britain*. London: Atlantic Books.

Zimmermann, W. (1990). *Auswirkungen staatlicher Regulierungseingriffe im Strassengüterverkehr und in der Binnenschiffahrt auf mittelständische Anbieter und Verlader*. Berlin, Germany: Duncker & Humblot.

Zunder, T. H., Aditjandra, P. T., & Islam, D. M. Z. (2012). Europe's freight transport policy White Paper 2001 to White Paper 2011. In *Proceedings of the 91st Annual Meeting of the Transportation Research Board*. Washington, DC: Transportation Research Board. doi:10.1007/1-4020-0613-6_21178.

Zunder, T. H., Islam, D. M. Z., Mortimer, P. N., & Aditjandra, P. T. (2013). How far has open access enabled the growth of cross border pan European rail freight? A case study. *Research in Transportation Business & Management, 6*, 71–80. Elsevier Ltd.

Zunder, T. H., Jackson, R., & Matsika, E. (2014). Charting the development of an innovative rail freight concept for low density, high value goods—from inception to design. In *5th Conference Transport Research Arena*. Paris: IFSTTAR. http://tra2014.traconference.eu/papers/pdfs/TRA2014_Fpaper_18021.pdf

Chapter 13
Emissions and Aviation: Towards Greener Air Transport

Antony Evans

Abstract Air travel revolutionized intercity transport, and today it contributes significantly to economic growth worldwide. However, aviation growth has a number of negative consequences, including environmental impacts on air quality, noise and the global climate. A number of new technologies and aircraft operating procedures are under development to mitigate these impacts. These, including revolutionary new aircraft and engine technology, biofuels, improvements in air traffic management, and new airline operating procedures, are discussed in detail in this chapter. A further topic of discussion is policy intervention. Because of the rapid growth in demand for air travel, policy intervention is likely to be required to drive much of the technological developments described and to speed up their uptake into the global fleet. With a combination of technological developments and policy intervention, however, significant progress towards greener air transport is possible.

Abbreviations

4-D	Four-dimensional
ADS-B	Automatic dependent surveillance-broadcast
BTL	Biomass-to-liquid
BWB	Blended wing-body
CCS	Carbon capture and storage
CDA	Continuous descent approach
CO_2	Carbon dioxide
CRC	Conceptual Research Corporation
CTL	Coal-to-liquid
DDA	Delayed deceleration approach
EPA	US Environmental Protection Agency
ERA	NASA's Environmentally Responsible Aviation project

A. Evans, Ph.D. (✉)
UCL Energy Institute, London, UK
e-mail: antony.evans@ucl.ac.uk

H.N. Psaraftis (ed.), *Green Transportation Logistics*, International Series in Operations Research & Management Science 226,
DOI 10.1007/978-3-319-17175-3_13

ETS	Emissions trading scheme
FAA	US Federal Aviation Administration
F-T	Fischer–Tropsch
GDP	Gross domestic product
GPS	Global positioning system
GTL	Gas-to-liquid
HEFA	Hydro-processed esters and fatty acids
HRJ	Hydro-processed renewable jet
ICAO	International Civil Aviation Organization
ICR	Intercooled recuperative cycle
IPCC	Intergovernmental Panel on Climate Change
LEAP	Leading Edge Aviation Propulsion
LUC	Land use change
MIT	Massachusetts Institute of Technology
NACRE	New Aircraft Concepts Research project
NASA	US National Aeronautics and Space Administration
NextGen	Next Generation Air Transportation System
NO_x	Nitrous oxide
RNAV	Area navigation
RNP	Required navigation performance
SPK	Synthetic paraffinic kerosene
US	United States

13.1 Introduction

The development of the powered aircraft in 1903 was to revolutionize intercity transport. For the first time, the possibility arose for passengers and freight to be transported by air instead of over land or sea, making previously impossible trips not only possible, but practical and even attractive. However, it was not until after the Second World War, more than 40 years later, that long distance intercity travel would cease to be dominated by rail. It took technological innovations during and after the Second World War, most notably the development of the jet engine, for aircraft to become sufficiently cost effective for aviation to compete with rail. By the 1950s, advancements in aircraft technology along with government investment in aviation infrastructure (particularly airports) led to aviation taking over from rail to become the most important mode of transport for long distance commercial intercity passenger travel in the United States (Schäfer, Heywood, Jacoby, & Waitz, 2009). The growth of aviation in other industrialized countries and regions followed similar trends. Between 1960 and 2011, worldwide scheduled passenger air travel grew from 109 billion passenger-kilometres travelled to 3.7 trillion—an average growth rate of over 7 % per year (ICAO, 2006; UN, 2013). Because of the growing demand to travel as income levels rise and people can afford to travel more, and because of the continuing trend to shift from slower to faster modes of transport across all distances (Schäfer et al., 2009), forecasts for future growth in passenger air transport are also high. The Airbus Global Market Forecast (Airbus, 2013) and the Boeing Current Market Outlook (Boeing, 2013) both predict growth rates of around 5 % per year from 2013 to 2032. By 2050 conservative estimates predict a 30–110 % growth in passenger kilometres travelled over 2005 levels (Berghof et al., 2005), while more aggressive estimates predict an increase of an order of magnitude (Schäfer, 2006).

Aviation makes a significant contribution to economic growth. ATAG (2012) estimate that the air transport industry, in 2010, contributed around 56.6 million jobs and US$1.4 trillion of Gross Domestic Product (GDP) to the world economy, including direct, indirect, induced and tourism catalytic effects. This is equivalent to 3.5 % of world GDP. By 2030, ATAG (2012) estimate that the air transport industry will directly contribute around 12.1 million jobs and US$1.4 trillion of GDP (at 2010 prices) to the world economy, with indirect and induced contributions around a further 32 million jobs and US$3.7 trillion in GDP. Furthermore, if tourism is added, the total contribution of the air transport industry will amount to over 82 million jobs and $6.9 trillion in GDP in 2030.

Despite these benefits, growth in aviation is also expected to have a number of negative consequences, including particularly significant environmental impact, including air quality and noise impacts, and global climate change, as reported by the Intergovernmental Panel on Climate Change (IPCC) (Cairns, Newson, Boardman, & Anable, 2006; Penner, 1999; Reynolds et al., 2007). Nitrogen oxides (NO_x) and particulate emissions in the vicinity of airports have significant impacts on local air quality, resulting in negative health effects and premature mortalities (Graham et al., 2009). Aircraft noise in the vicinity of airports also has negative health effects (Cohen, Evans, Krantz, & Stokols, 1980), while also reducing property values near airports and affecting children's abilities to learn (Haines, Stansfeld, Job, Berglund, & Head, 2001). The climate effects of aviation are more complex and widespread. Carbon dioxide (CO_2), the emissions of which are directly proportional to fuel burn, is a greenhouse gas that impacts the atmosphere for hundreds of years, increasing radiative forcing[1] and ultimately causing the global average temperature to rise. Non-CO_2 effects are also significant. Nitrous oxide (NO_x) emissions at aircraft cruise altitudes lead to the production of tropospheric ozone, which has a warming effect, and accelerates the removal of methane from the atmosphere, which has a cooling effect (Penner, 1999). Ozone and methane have different life times, so the warming effect is regional, while the cooling effect is global, although the warming effect is thought to dominate. Contrails from aircraft engines can also increase high altitude cloud cover, which tends to produce a net warming effect in the region where the aircraft was flown, although there is scientific uncertainty in its overall effect.

Aviation was estimated to be responsible for the demand for 214 million tonnes of fuel (dominated by Jet A-1) in 2012, which equates to approximately 2 % of total

[1] A change in average net radiation at the top of the troposphere (Penner, 1999).

anthropogenic CO_2 emissions (or about 12 % of CO_2 emissions from all transportation sources) (ATAG, 2012). This is consistent with estimates from the IPCC (Penner, 1999). By 2050, the contribution of aviation to anthropogenic CO_2 emissions is expected to increase to around 3 %. However, because of the significant non-CO_2 effects of aviation, aviation is estimated to have accounted for around 3.5 % of global anthropogenic radiative forcing in 2005, increasing to around 4.9 % if recent estimates for the uncertain effect from aviation induced cirrus clouds are considered (Lee et al., 2010).

The environmental impacts of aviation can be mitigated by two primary mechanisms: (1) inducing the introduction of new technology or operations into the air transport system, including through policy intervention, that reduces the environmental impact of aviation per passenger transported; and (2) introducing economic instruments that reduce the growth in demand for passenger air transport. The first is attractive because it enables economic growth, allowing increased demand for air transport as well as supporting the aircraft manufacturing industry and encouraging innovation. The second is significantly less attractive as it may limit economic growth. In this chapter, we focus on different aircraft and fuel technology innovations that have the potential to contribute to mitigating the environmental impact of aviation in Sect. 13.2. This is followed by discussions of operational innovations and policy intervention targeting the same goal in Sects. 13.3 and 13.4, respectively. In Sect. 13.5 we discuss system effects, followed by conclusions in Sect. 13.6.

13.2 Technological Developments

Historical improvements in fuel efficiency have been significant, driven by the fact that airline fuel costs, particularly for jet aircraft, are a large percentage of flight operating costs. Figure 13.1 shows historical fuel-burn since the 1950s, showing a reduction of 40 % by 2000. The reduction in fuel-burn per passenger transported is greater (70 %), with aircraft having also increased in size.

NO_x emissions and noise have also reduced over time. As described by Graham, Hall, and Vera Morales (2014), NO_x is produced due to oxidation of nitrogen at the high temperatures in the engine combustor. The amount of NO_x created is therefore not only a function of how much fuel is burnt, but also on the design of the engine combustor. Generally, the higher the temperatures and pressures at which combustion takes place, the higher the NO_x emissions. The current regulatory standard, CAEP/6, represents a reduction of about 40 % over the original limits set in the 1980s. Most modern engines meet this standard (ICAO, 2012). Aircraft noise, produced by both the engines and the airframe, has also reduced. Thrust-corrected noise has dropped by around 20 EPNdB since the 1950s (Hall, 2009). This is significant, and corresponds to a 100-fold reduction in sound power, and a four-fold reduction in perceived noise level.

It is clear that aircraft fuel-burn, NO_x generation and noise have been reduced significantly since the 1950s. Further improvements, however, are likely to be more

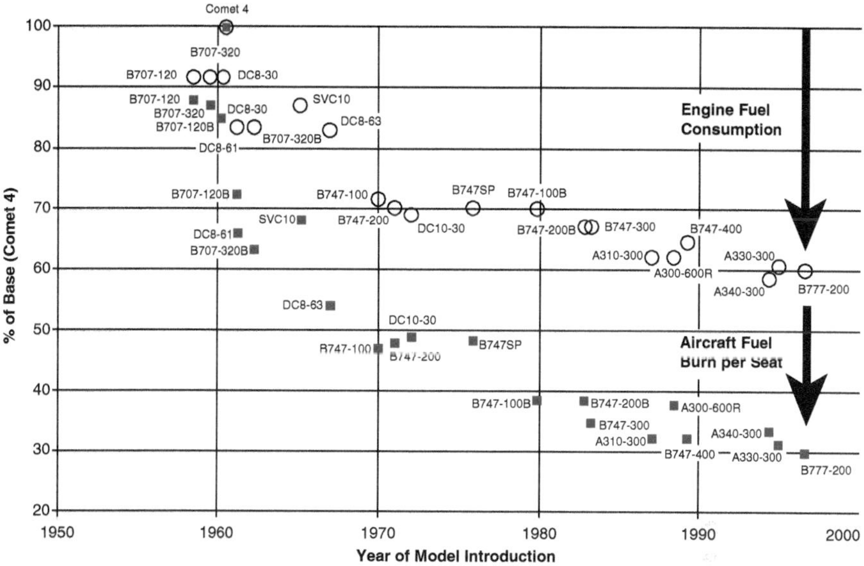

Fig. 13.1 Historical data on aircraft fuel-burn, *Source*: Henderson and Wickrama (1999)

difficult because of diminishing returns, meaning that 'business as usual' may not be sufficient to make significant future gains. This has driven a number of future aircraft concept studies over the past decade, described below. First, however, we examine the existing state-of-the-art.

13.2.1 Existing State-of-the-Art Aircraft Technology

The Boeing 787 Dreamliner, released into service in 2011, represents an example of the current state of the art in commercial passenger aircraft technology. While it still represents the traditional tube-and-wing design, it incorporates a number of new technologies compared to older aircraft. The most significant of these is its use of composite materials in its construction, which reduce weight and therefore fuel burn. The aircraft is 80 % composite by volume (IndustryWeek.com, 2007). While composite materials have been used extensively in airliner wing and tail construction in the past, the Dreamliner represents the first major airliner to use composite materials as the primary material in the construction of its fuselage as well as its wings and tail.

The other significant change from traditional aircraft design introduced by the Dreamliner is the replacement of bleed air (from the engines) and hydraulic power with electrically powered systems. According to Boeing, these electrical systems use 35 % less power than traditional systems (Sinnet, 2007). The nickel cadmium (NiCd) batteries used on traditional aircraft have also been replaced on the Dreamliner by lithium ion batteries, which are lighter, smaller and more powerful.

This modification initially lead to a number of electrical system problems, which in two cases resulted in on-board fires due to thermal runaway, ultimately leading to the grounding of the type by the US Federal Aviation Administration (FAA) in 2012. This has since been resolved, and the type is operational again, but this does illustrate the challenges that can accompany the introduction of new technology into modern aircraft.

The resulting improvement in fuel burn over the Boeing 767—the technology the Dreamliner is designed to replace—is 20 % for long-haul flights (Norris, 2012). This is significant. However, it is not only fuel burn that the Dreamliner is designed to reduce. Boeing has also adopted several noise-reducing technologies in the engines. The most significant of these includes incorporation of sound-absorbing materials in the engine inlet, and redesigned exhaust duct covers to enable quieter mixing of exhaust gases with the outside air. This is done through the use of chevrons—a toothed pattern on the rim of the exhaust duct covers (Zaman, Bridges, & Huff, 2010).

Airbus is soon to release a competitor to the Dreamliner—the Airbus A350 XWD—possibly by as early as the fourth quarter of 2014. This aircraft will incorporate similar technology, including the composite fuselage. Airbus claims that the A350 will be as much as 6 % more fuel-efficient than the Dreamliner (Business Traveller, 2014).

Possibly more widely anticipated than the new wide-body aircraft described above, is the development of replacements to the workhorses of the global aviation fleet—The Boeing 737 and Airbus A320. Both manufacturers have new models under development, the Boeing 737 MAX and the Airbus A320neo (NEO stands for New Engine Option), shown in Fig. 13.2. These designs include a number of new technologies, including new engines, weight savings, and aerodynamic improvements. The latter includes large curved 'sharklets' on the A320neo and a hybrid between a blended winglet, wingtip fence, and raked wingtip on the 737 MAX (as can be seen in Fig. 13.2). However, these designs do not incorporate many of the more revolutionary technologies of the 787 and A350, such as the composite fuselage and increased use of electrical systems. They are instead shorter-term improvements to the existing designs. The reason for this less ambitious approach from the manufacturers is the fact that more revolutionary technology that could significantly reduce the fuel burn of these aircraft, such as open rotor engines, is not yet mature. It is

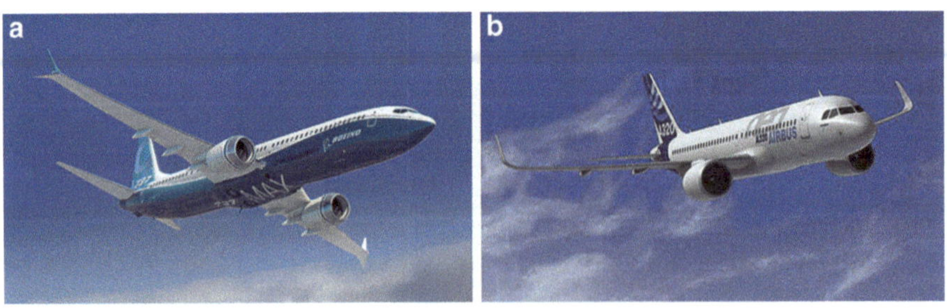

Fig. 13.2 (**a**) Boeing 737 Max, *Source*: voices.suntimes.com and (**b**) Airbus A320 NEO, *Source*: australianaviation.com.au

therefore expedient to delay investment in a completely new design until these technologies are available. This has, however, allowed other players into the market, including particularly Bombardier, with the CS100, and the Chinese COMAC C919, both of which are designed from the ground up. While the COMAC is unlikely to pose a significant threat to Boeing and Airbus outside China, the CS100 may.

Some of the most important developments of the 737 MAX and A320neo are in the engines. The 737 MAX will make use of the CFM International LEAP engine (LEAP stands for Leading Edge Aviation Propulsion). This includes a turbine with flexible blades which are designed to untwist as the turbine's rotational speed increases, greater use of composite materials, and a 'blisk' fan—a rotor disk and blades made as a single component. These developments allow for more efficient turbine operation and a lighter engine, reducing fuel burn. The A320neo will be available with both the LEAP engine, and the Pratt and Whitney Purepower PW1000G, a geared turbofan. This engine introduces a gearbox between the low-pressure turbine and the fan, allowing a larger fan to be used efficiently, increasing the bypass ratio and therefore reducing fuel burn. The gearbox also allows for a reduction in the size and weight of the turbine required to drive it, decreasing total engine weight. CFM International and Pratt and Whitney claim fuel burn improvements of up to 15 % over existing engines (CFM International, 2014; Pratt & Whitney, 2014).

13.2.2 Future Aircraft Technology

The blended wing-body (BWB) is one of the most common aircraft concepts studied to provide significant reduction in fuel-burn, NO_x emissions and noise. BWB aircraft have a flattened and aerofoil shaped fuselage, which produces lift along with the wings. Different BWB concepts have recently been studied by the Royal Aeronautical Society's Greener-by-Design initiative (Green, 2002), the MIT-Cambridge University Silent Aircraft Initiative (Dowling, 2007), the European NACRE (New Aircraft Concepts REsearch) project (Frota et al., 2011), the Boeing SUGAR program (Bradley & Droney, 2011), MIT (Greitzer et al., 2011), and a further Boeing study commissioned by NASA's 'Environmentally Responsible Aviation' (ERA) project (Bonet, 2012) (Fig. 13.3).

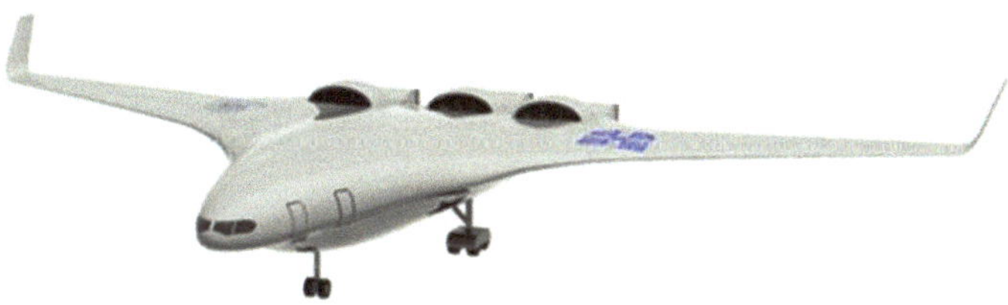

Fig. 13.3 MIT Cambridge University Silent Aircraft Initiative, *Source*: silentaircaft.org

Because of the impact of the flattened fuselage on the passenger cabin, these aircraft are generally studied as replacements for the larger twin-aisle conventional tube-and-wing aircraft. The BWB design results in an improved lift-to-drag ratio relative to these conventional aircraft, leading to significant fuel burn improvements. However, its shape is also favourable for noise reduction, hence its selection for the Silent Aircraft Initiative. The design provides space to embed the engines in the airframe, which allows for the use of high bypass ratios with smaller weight penalties than on conventional aircraft. Long inlet ducts can also be used to absorb sound, and, if the ducts are on the upper surface, ground observers can be shielded from general engine noise. Because the design has good low-speed lift characteristics, airframe noise can be significantly reduced on landing. The engine location also has fuel burn benefits, with engine airflow coming from the slow-moving region near the airframe surface. However, the engines must manage significant flow distortion as the fan passes through the ingested boundary layer, which is technologically challenging. The Boeing study commissioned by NASA's ERA project (Bonet, 2012) makes use of an open-rotor engine (described below) instead of the traditional turbofan, and replace the conventional, slotted, trailing-edge flap with a plain version, which also has noise benefits.

A number of other future aircraft concepts have also been proposed. These are typically conventional tube-and-wing designs, but incorporate significant changes to existing types. One of the most common is tail-mounted engines for noise shielding, such as designs by NACRE (Frota et al., 2011), the Conceptual Research Corporation (CRC) (Raymer et al., 2011), and TOSCA (Vera Morales et al., 2011). One of the most commonly proposed engines for this configuration is the 'open rotor' (or 'unducted fan'). This consists of two counter-rotating, advanced-design propellers, and no nacelle. It is more fuel-efficient than a turbofan, primarily because it has a very high effective bypass ratio, but with no nacelle drag. However, the removal of the nacelle has consequences for noise, hence its application only with acoustic shielding from the aircraft fuselage.

Another novel engine design proposed is the introduction of an 'intercooled recuperative' (ICR) cycle in a turbofan engine (Green, 2002). This would improve the fundamental thermodynamic efficiency of the design by cooling its airflow between compression stages. Intercooling is widely used in land–based gas turbines used for power generation. However, in an aircraft application, the heat exchangers required would add significantly to the weight of the engine, thereby compromising the overall benefits. This concept is therefore further from realization than those described above.

Modified tube-and-wing designs also typically incorporate forward-swept or unswept wings with greater wingspan (but the same weight) relative to conventional designs. This increases the aircraft's lift-to-drag ratio and allows for the possibility of 'natural laminar flow'. Laminar flow reduces surface friction drag significantly, but typically requires control of the flow next to the wing surface using suction. In contrast, on unswept or forward-swept wings it is possible to achieve laminar flow without suction. A consequence, however, is reduced cruise speeds of Mach 0.7–0.75, considerably lower than the typical values operated today

(Mach 0.8–0.85). There are economic costs due to the increased travel time, but Boeing's Current Market Outlook (Boeing, 2013) suggests that such cruise speeds would still be economically viable for a design range of up to 6,500 km (Graham et al., 2014).

CRC's design is tailless, instead using artificially stabilizing controllers. It also employs an active aeroelastic wing, which is deliberately constructed with greater flexibility than is conventional, and its shape is changed according to the flight condition by an automatic control system. These modifications lead to significant weight reductions.

Other technology advances proposed for this type of aircraft include the replacement of slats with leading-edge droop or a 'Krueger flap'; the use of 'riblet' surfaces on the fuselage which further reduce surface friction drag; advanced lightweight structures; and the use of a 'low-noise' undercarriage (Graham et al., 2014).

A concept that combines some of the benefits of both a BWB and the modified tube-and-wing designs described above, is the MIT 'double-bubble' concept (Greitzer et al., 2011). A 'double-bubble' fuselage generates lift, reducing the required size of the wing and horizontal-tail, and allowing for the elimination of leading edge high-lift devices, with corresponding weight and aerodynamic drag reductions. An unswept wing is used, and ultra-high-bypass ratio engines are mounted at the rear, providing acoustic shielding and similar fuel burn benefits associated with engine integration on the BWB (Fig. 13.4).

Other concepts, such as the Northrop Grumman 'SELECT' design (Bruner et al., 2010), instead use a conventional configuration, but apply a number of the novel technologies described above, such as ultra-high-bypass ratio engines, highly advanced materials, natural laminar flow on a significant proportion of its wing, active aeroelastic control, etc.

More revolutionary concepts have also been proposed. The Boeing SUGAR High design (Bradley & Droney, 2011) makes use of a braced-wing intended to achieve the greatest possible lift-to-drag ratio via induced-drag reduction. Lockheed Martin's Preferred System Concept (Martin, 2012), commissioned for the NASA ERA project, makes use of a box-wing configuration, on which a second

Fig. 13.4 MIT 'double-bubble' concept, *Source*: NASA (2014)

pair of wings links the tips of the conventional wings to the vertical tail surface. A number of the novel technologies described above are also applied.

Predicted reductions in fuel burn, NO_x emissions and noise for these aircraft concepts differ significantly. Based on the thorough summary compiled by Graham et al. (2014), the modified tube-and-wing concepts are predicted to have fuel burn improvements of up to 51 % compared to existing aircraft, while the more revolutionary concepts, including the BWB, tailless CRC tube-and-wing concept, MIT double-bubble, and Boeing braced-wing concepts, are predicted to have fuel burn improvements of up to 71 %. Predicted reductions in NO_x emissions are as high as 91 %, while predicted reductions in noise are as high as 75 %. It is noted that there is a wide range in predicted benefits across all environmental impacts, because there are trade-offs between reducing fuel-burn, NO_x, and noise. Hence, if a design is tailored to any one, its reductions in the other can suffer accordingly.

13.2.3 Fuel

A significant area of current technological development is in aviation fuel, and particularly the development of biofuels. Biofuels reduce life-cycle CO_2 emissions because the plant material from which they are made takes CO_2 from the air when it grows. This offsets the CO_2 released into the atmosphere when the biofuel is burned in the aircraft engine, forming a closed cycle. In contrast, the CO_2 released by burning traditional jet fuel made from oil comes from fossil fuels underground, so are not offset, and do not form a closed cycle. Hence biofuels can reduce life-cycle CO_2 emissions significantly. However, because of the energy use associated with land use changes, farming, biofuel harvest, transport of the raw materials for processing, biofuel processing itself, and transport to the end user, life-cycle CO_2 emissions are not eliminated.

Because of the extensive existing infrastructure for jet-fuel transport and combustion (i.e., in aircraft engines), "drop-in" biofuels are considered to be the most preferable for aviation. A drop-in fuel is essentially a like-for-like replacement of Jet A-1, and therefore must have similar characteristics to Jet A-1. The most important of these characteristics are energy content, freeze point, thermal stability, viscosity, general combustion characteristics, lubricity, material compatibility and flash point (Bauen, Howes, Bertuccioli, & Chudziak, 2009). An ideal drop-in biofuel has sufficiently similar characteristics to Jet A-1 that they can be mixed, making its introduction into commercial use easier. It should also ideally make use of existing airport infrastructure, and require minimal modifications to the engine in order to reduce costs. Finally, although the most difficult, drop-in biofuels should be sustainable, scalable and cost effective. This means that their production should not replace feedstock land; they should not require large quantities of water; and they should have the potential for large scale production, at a reasonable cost—ideally close to that of oil-based Jet A-1. This latter requirement is the biggest challenge for biofuels, with current costs far exceeding those of oil-based jet fuel.

The most widely studied aviation biofuel is Synthetic Paraffinic Kerosene (SPK). From its name it is clear that it is a drop-in fuel (Jet A-1 is a type of kerosene). SPKs are functionally similar to Jet A-1, except that they contain few aromatic compounds and no sulphur. They can be produced from various feedstocks, including biomass, renewable waste oil, vegetable oil, coal and natural gas. While SPK production from coal and natural gas (referred to as CTL—coal-to-Liquid and GTL—gas-to-liquid, respectively) does not strictly produce a biofuel, since the feedstock is a fossil fuel, its production using the Fischer–Tropsch process is similar to that used for SPK production from biomass (referred to as BTL—biomass-to-liquid), which is a true biofuel. SPK production from renewable waste oils and vegetable oils does not use this process, instead using hydrogen treatment. This produces an SPK referred to as Hydro-processed Renewable Jet (HRJ) or Hydro-processed Esters and Fatty Acids (HEFA), which is also a true biofuel.

HRJs and HEFAs are produced from oil-bearing biomass, such as jatropha, camelina, salicornia, soy, palm, rapeseed, and algae. These are converted into HRJs and HEFAs by pressing, refining and hydro-treatment (Rosillo-Calle, Thrän, Seiffert, & Teelucksingh, 2012). Each of the different oil-producing plants has pros and cons. For example, jatropha can grow on marginal land, while salicornia can be integrated with fish farming and mangroves, but both have relatively low yields (Garnham, 2011). Algae on the other hand can have very high yields, particularly if recycled CO_2 from a traditional power plant is added, but there is still high economic uncertainty about its viability (Garnham, 2011).

BTLs are produced from cellulosic biomass, which includes woody biomass, such as logging residues, urban wood wastes or short rotation coppice, and herbaceous biomass, such as miscanthus or straw. These are converted into BTL through a thermo-chemical process including pyrolysis or gasification, and Fischer–Tropsch synthesis (Rosillo-Calle et al., 2012). Cellulosic biomass is comparatively abundant and many feedstocks do not compete with food for land use. In fact, based on supply curves from the U.S. Department of Energy's "Billion-Ton Update" study (Department of Energy, 2011) and using a 50% conversion efficiency of biomass to jet fuel, waste feedstocks in the U.S. in 2012 were theoretically enough to power domestic operations of the entire U.S. commercial airline fleet. However, growth in demand for aviation is likely to outstrip the growth in waste feedstock generation, so this may not be a solution in the future. Similarly, aviation is not unique in its interest in biomass, so there are many other competing demands for it. Aviation may, however, have the highest willingness to pay, because so few alternative technologies exist.

Life-cycle CO_2 emissions, estimated by the PARTNER consortium for different biofuel feedstocks (Stratton, Wong, & Hileman, 2010), under different assumptions about land use change, are shown in Fig. 13.2. Land use change has a large impact on life-cycle CO_2 emissions. If the biofuel is grown on marginal land or water, which are not a significant sink for CO_2 emissions, the contribution of land use change to increasing life-cycle CO_2 emissions can be small. But if vegetation must be destroyed, and particularly if this is a large sink for CO_2 emissions such a rainforest, the contribution of land use change to increasing life-cycle CO_2

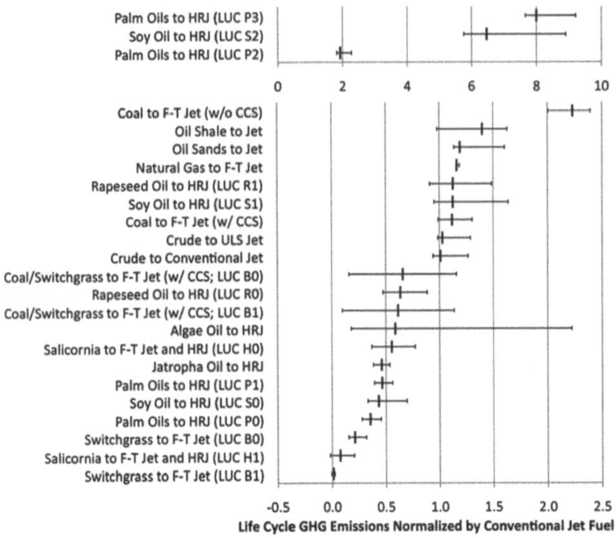

Soy HRJ Pathway Scenarios	
LUC-S0	No land use change
LUC-S1	Grassland conversion to soybean field
LUC-S2	World wide conversion of non-cropland
LUC-S3	Tropical rainforest conversion to soybean field
Palm HRJ Pathway Scenarios	
LUC-P0	No land use change
LUC-P1	Logged over forest conversion to palm plantation
LUC-P2	Tropical rainforest conversion to palm plantation
LUC-P3	Peatland rainforest conversion to palm plantation
Salicornia SPK Pathway Scenarios	
LUC-H0	No land use change
LUC-H1	Desert converted to salicornia cultivation
F-T Biomass Type (BTL and CBTL)	
LUC-B0	No land use change
LUC-B1	Carbon depleted converted to switchgrass cultivation

Fig. 13.5 Life cycle GHG emissions for the alternative jet fuel pathways under consideration. Uncertainty bars represent the low emissions, baseline, and high emissions scenarios. Please note the different scales for the *top* and *bottom* portions of the figure. *Note*: CCS denotes Carbon Capture and Storage, LUC denotes Land Use Change, and F-T denotes Fischer–Tropsch. *Source*: Stratton et al. (2010)

emissions can be large. Further variability, shown by the error bars in Fig. 13.2, is introduced by feedstock variation, the conversion technology used, the efficiency of the process used, the type of plant cultivation and harvesting, the efficiency of any carbon capture, and the emissions allocation methodology when co-products are created (Fig. 13.5).

Land use requirements for SPK biofuels are a particular concern. The biofuel yield determines the amount of land needed to supply a given amount of jet fuel. Figure 13.3, from Stratton et al. (2010), shows the amount of land, superimposed on a map of the United States, for different feedstock yields, to replace 100 and 50 % of conventional jet fuel in the United States. These can be compared to the biofuel yields for camalina (751 L/ha/year), salicornia (1,200 L/ha/year), palm (3,311 L/ha/year) and algae (16,969 L/ha/year). The amounts of land are clearly vast, even at the very high yields of algae (Fig. 13.6).

SPK-based biofuels, from a range of different feedstocks, have now been flown on commercial aircraft by a number of airlines, including on scheduled service. Certification of SPKs is therefore underway or complete, and is not expected to be an obstacle. However, the key outstanding issues, such as crop suitability and supply security, sustainability and life-cycle emissions, supply chain logistics, and of course cost, can only really be tackled once production on a larger scale is underway.

Other biofuels also exist. These include alcohols (ethanol, butanol, etc.), biodiesel and biokerosene. The chemical compounds making up these fuels contain

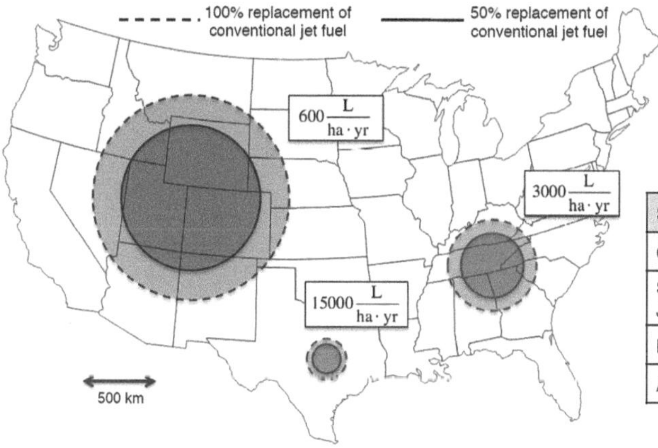

Synthetic Fuel	L / ha / yr
Camelina Jet	751
Salicornia Jet / Diesel	1,200 / 1,721
Palm Jet	3,311
Algae Jet	16,969

Fig. 13.6 Land area requirements to replace conventional jet fuel use within the US with 100 % SPK and 50/50 blend of SPK with conventional jet fuel. Average US conventional jet fuel consumption in 2009 is 1.4 million bbl/day. *Source*: Stratton et al. (2010)

oxygen, and are not therefore strict hydrocarbons. Most significantly, their gravimetric energy density (energy content per unit mass) and volumetric energy density (energy content per unit volume) are both lower than traditional jet fuel (and SPKs), so they are considered more appropriate for ground transport than aviation. There are also other incompatibilities regarding their use in aviation, including thermal stability concerns for biodiesel and biokerosene.

Cryogenic fuels (hydrogen) have also been proposed as an alternative to traditional jet fuel. While hydrogen powered aircraft have flow, such as a modified Tupolev Tu-154B more than 25 years ago, the use of cryogenic fuels operationally is very challenging because hydrogen is much less convenient to store than liquid hydrocarbons. The high volume (for a given energy content) is also a particular problem, leading to significant reductions in aircraft energy efficiency. Furthermore, hydrogen production is currently associated with high CO_2 emissions (using steam-reforming of natural gas), offsetting any climate benefit. This technology is thus only likely to become realistic if zero-CO_2 electricity becomes freely available.

13.3 Operational Developments

Aircraft technology is a key mechanism by which the environmental impacts of aviation can be mitigated. However, because the average life of commercial aircraft is in the order of 30 years, entry of this technology into the fleet can be slow, with the envisioned reductions in environmental impact only being realized well into the future. This is not the case, however, for operational improvements, which can generally be implemented on much shorter time scales. This is particularly true for operational measures implemented by airlines, and while operational measures in air traffic control and air traffic management take longer, benefits are still likely to be realized faster than from most of the technologies described above.

13.3.1 Air Traffic Management

Airspace congestion is currently only a significant issue in the world regions with the highest traffic levels, i.e., North America and Western Europe. However, because of the high traffic levels in these regions, the resulting impacts on fuel usage, and therefore emissions, are significant. Hence there are significant efforts to improve the efficiency of these airspace systems, to allow aircraft to fly closer to their optimal flight paths than is currently the case. This could result in significant reductions in fuel burn. In Europe alone, fuel burn is estimated to be as much as 23 % higher than the theoretical minimum, were each flight able to fly on a fuel optimal routing, based on an analysis of nearly 1,800 Airbus A320 flights in Europe in 2008 (Reynolds, 2009). While all deviations from the fuel-optimal routing will never be eliminated, significant efforts are underway in both Europe and the United States to minimize them by shifting towards a more efficient, flexible, and informed air traffic management.

The Single European Sky program, launched in 1999, is an initiative to organize the European airspace into functional blocks, according to traffic flows rather than to national borders, while also modernizing and optimizing the future European Air Traffic Management network through advanced technologies and procedures (Eurocontrol, 2007). Similarly, in the United States, the Next Generation Air Transportation System (NextGen), which is intended to make air travel more convenient, predictable and environmentally friendly, will be enabled by a shift to satellite-based and digital technologies and new procedures (FAA, 2008).

A key enabling technological development is the specification of four-dimensional (4-D) trajectories (three spatial dimensions plus time). This is achieved through a shift from ground-based (radar) position information, to a satellite-based system based on Global Positioning System (GPS) data using technology such as Automatic Dependent Surveillance-Broadcast (ADS-B), Area Navigation (RNAV) and Required Navigation Performance (RNP). These technologies provide more accurate and precise positioning data and enable data-sharing, making air traffic management more flexible and dynamic.

ADS-B is based on GPS technology that calculates the precise position of an aircraft and broadcasts this information to other aircraft as well as to the ground. This enables a reduction of vertical and lateral separation between aircraft, as well as the planning of more direct and fuel-efficient flight paths. The current en-route flight environment consists of a network of airways that simplify the management of air traffic flows for controllers, but which can take aircraft well away from the shortest path between two airports. This 'lateral inefficiency' could be reduced significantly by allowing more direct routing, enabled by ADS-B, leading to estimated cruise fuel burn and emissions savings of between 1 and 5 %. Similarly, by reducing inefficiencies in vertical flight profiles and speed profiles, cruise fuel burn and emissions could be reduced by between 1 and 1.5 % and 1.6 % and 2.4 %, respectively. Combined, this results in a savings of between 2.8 and 8.0 % in cruise fuel, equivalent to between 2.2 and 6.4 % in block fuel.[2]

RNAV and RNP enable flight paths in and out of airports to be optimized. Take-off and climb, particularly, require high engine thrust levels, and therefore, while the duration of the phases is short, the relative fuel burn is high. RNAV and RNP enable aircraft to fly more precise and flexible departure trajectories. Optimized for fuel burn, they can provide departure fuel burn benefits of between 10 and 30 %, equivalent to between 0.8 and 2.4 % block fuel (Muller, 2011). Flight descent and approach are less fuel intensive, but meaningful fuel burn reductions can still be achieved using RNAV and RNP. Continuous Descent Approaches (CDAs) keep aircraft at high altitude (and therefore lower thrust) for longer than traditional approaches, reducing approach fuel burn emissions by between 10 and 20 % (equivalent to 0.4–0.8 % block fuel), as well as reducing airport noise (Reynolds et al., 2007). Delayed Deceleration Approaches (DDAs) optimize the aircraft speed profile so that it remains in clean/low drag aerodynamic configurations for longer than traditional approaches, reducing approach fuel burn and emissions by between 20 and 40 % (equivalent to 0.8–1.6 % block fuel) (Dumont et al., 2011).

Aircraft also burn fuel, and therefore generate emissions, on the ground, particularly while taxiing from the gate to runway, and *vice versa*. Procedures are being developed to reduce these emissions. Surface congestion management techniques aim to hold aircraft at the gate or in a parking area, with engines off, when the airport is operating at capacity, instead of waiting, with engines running, in a departure queue. Savings of between 10 and 20 % of taxi fuel burn have been observed in trials (equivalent to 0.6–1.7 % block fuel) (Nakahara, 2013). Single-engine taxi is another approach to reduce taxi fuel burn, by reducing the number of engines running while taxiing. According to a recent survey of pilots (Clewlow et al., 2010), single engine taxiing would result in a 37% taxi fuel burn reduction, on average when it is used. There are, however, challenges associated with issues such as reduced manoeuvrability, problems starting the second engine, distractions and workload, meaning that not all flights currently use the technique, and it is used to a greater extent on arrival than on departure.

[2] Block fuel for a given flight refers to the total fuel used from the moment the aircraft pushes back from the departure gate, before takeoff, until the moment the aircraft arrives at the arrival gate, following its landing.

Table 13.1 Airline operational procedures for mitigating the environmental impact of aviation

Intervention	Percent of total annual sector emissions reduced
Better use of capacity	2.2
Increased turboprop use	3.0
Maintenance-engine wash	0.5
Fuel reserves	0.3
Reduce fuel tankering	2.9
Light-weighting	0.5

13.3.2 Airline Operations

Airlines can also reduce fuel burn and emissions by adjusting their flight operating procedures. A number of these are listed in Table 13.1, with estimates of the potential percentage emissions reduction possible, as estimated by Morris, Rowbotham, Angus, Mann, and Poll (2009).

Better use of capacity refers to increased seating densities and average load factors, combined with incentives to encourage passengers to carry less baggage. This is particularly popular with low cost carriers. Typically flights are not flown, on average, at full capacity to limit “spill”. Spill refers to the loss of customers who would have purchased a ticket had their been seats available, but were not able to because the flight was full. Because of daily variability in demand for air travel, the higher the average load factor, the more total passengers spilled. This represents lost revenue for the airline. However, modern revenue management techniques have significantly reduced the importance of this problem, because, with good revenue management, only those passengers with lowest value to the airline (i.e., with the lowest willingness to pay) are “spilled”. Hence average load factors can be increased with little concern for the potential revenue lost. Some costs do, however, remain, such as an increased cost of passenger re-accommodation in the case of a delay or cancellation. With higher average load factors, there are fewer seats available on later flights for passengers who had their flight cancelled or missed a connection because of a delay.

Turboprop aircraft generally have lower fuel burn per passenger than jet aircraft (and particularly regional jet aircraft), but at the cost of flying slower and with reduced cabin comfort. But the reduction in emissions possible, of nearly 3 %, is significant when compared to the other operational measures listed in Table 13.1. Hence there is an increased use of turboprop aircraft on ‘thin’ short haul routes, especially to replace regional jet aircraft.

A number of different maintenance procedures can be used to reduce fuel burn, such as more frequent engine maintenance, aerodynamic maintenance and engine washing. It is now possible to do the latter overnight while the aircraft is at the gate, reducing aircraft downtime.

When aircraft fuel is loaded onto an aircraft, fuel is added for a number of purposes – to complete the planned trip, for emergency holding and diversion (called reserve fuel), and as a buffer at the discretion of the pilot (called buffer or contingency fuel). Because reserve fuel can only be used in an emergency, pilots add contingency fuel in case of weather delays, unexpected headwinds, holding requirements, etc. However, any additional fuel added to an aircraft increases weight, and therefore fuel use and emissions. Many airlines are therefore encouraging their pilots to reduce contingency fuel to a minimum. This does, however, increase the risk of having to declare an emergency or of making an expensive diversion, so it has not been well received by all pilots for safety reasons.

Reduced tankering does not suffer from these problems. Tankering refers to carrying more fuel than required for a single flight leg, with the objective of not refuelling at the destination airport. This is generally because of fuel price differentials between airports, or to facilitate a quick turnaround at the destination airport. However, the weight of the increased fuel increases fuel burn and emissions. Hence, Morris et al. (2009) estimate that emissions can be reduced by nearly 3 % by eliminating fuel tankering, although this is highly dependent on the assumptions about the differences in fuel prices at different airports.

Finally, light weighting can also reduce fuel burn and emissions. This refers primarily to removal or replacement of components in the aircraft cabin, such as seats, galley trolleys, carpets, magazines and other cabin fittings. The most significant weight savings can generally be achieved by replacing older seats with modern lightweight seats.

13.4 Policy Intervention

As described at the start of this chapter, the environmental impacts of aviation can be mitigated by two primary mechanisms: the introduction of new technology or operations; and a reduction in the growth in demand for passenger air transport. The introduction of new technology may be induced by increasing oil prices or by policy intervention. The latter includes new regulations for aircraft noise or emissions levels, and economic instruments, such as the inclusion of aviation in an emissions trading scheme (e.g., European Commission, 2006). Reduced demand growth can be induced by economic instruments that increase costs to passengers, such as an air passenger duty (HM Revenue and Customs, 2008), and result in a reduction in passenger demand growth as passengers choose to travel using alternative modes of transport, or choose not to travel at all. Most policy interventions target the former mechanism, driving the introduction of new technology and operations. However, by increasing costs to air passengers, the latter mechanism is also active in many cases, reducing demand growth. In this section, we describe the most important policy interventions.

The local air quality impacts of aviation have traditionally been controlled by limits on NO_x concentrations in the vicinity of airports, but aircraft NO_x standards that have now been adopted by the International Civil Aviation Organization (ICAO) and the US Environmental Protection Agency (EPA) amongst others (greenaironline.com, 2012) will further reduce local air quality impacts. The NO_x standards, however, will also contribute to reducing the non-CO_2 climate impacts of aviation.

Aircraft noise impacts have historically been reduced through noise standards for aircraft, such as the FAA Stage 2 (1969), Stage 3 (1977) and Stage 4 (2006) noise regulations (corresponding to the ICAO Chaps. 2–4 noise requirements), which limit the amount of noise that can be generated by an aircraft. Airport noise impacts are also controlled through curfews, which limit when aircraft can land. For example, John Wayne Airport in Orange County bans take offs from 22h00 to 07h00, and landings from 23h00 to 07h00. Partial curfews are more common than complete curfews, with night-time operations restricted to certain types of aircraft. Alternatively, quota count systems, like that at London Heathrow airport, limit the total amount of noise permitted, but allow operators to choose to operate fewer noisy aircraft or a greater number of quieter aircraft.

Proposed policies to mitigate the climate impacts of aviation include carbon taxes, in which a fixed tax is levied for every tonne of CO_2 emitted, and an emissions trading scheme (ETS) or cap-and-trade scheme, a market-based approach that provides economic incentives to reduce emissions. In an ETS, a central authority (usually a governmental body) sets a limit or cap on the amount of emission (typically CO_2) that may be emitted. The limit or cap is allocated to firms in the form of emissions permits or credits. A firm that cannot reduce its emissions to the level of its allocated credits must buy credits from other firms that can reduce their emissions below the level of their allocated credits. The result is a market for trading emissions credits, with the price determined by the market. An ETS is a very effective way to reduce emissions in the most cost effective way possible. However, the initial limit or cap must be set low enough, or lowered over time.

The largest existing ETS is the European Emissions Trading Scheme, which was launched in 2005 and covers multiple sectors, including power generation, oil refineries, the production and processing of ferrous metals, the production of cement, the production of pulp, paper and board, etc. In 2012 the European ETS was to include aviation (European Commission, 2006), notably with the entire flight charged, and not just that portion of the flight in European airspace. However, there was significant opposition to this internationally, with a number of countries, including the United States and China, banning their airlines from paying into the scheme. The primary stated reasons for this were that it violated the Chicago Convention, that prohibits the taxation of jet fuel, and that such a scheme should be international in nature, and organized instead under the auspices of ICAO. The inclusion of aviation in the European ETS was therefore suspended for a year, subject to an international scheme being developed. Recently, there has been an agreement that such a scheme should be implemented by 2020, with a detailed plan to be drawn up by ICAO's next general assembly in 2016 (Alcock, 2013; theguardian.com, 2013). The European ETS has consequently been adapted to charge only for flights to and from European countries, and, until 2016, only for that portion of the flight in European airspace.

Another further policy that is under consideration for mitigating the climate impacts of aviation is an aircraft CO_2 standard. This would provide an upper limit in CO_2 emissions that future aircraft must adhere to, much like the NO_x and noise standards mentioned above. A key achievement so far is the development of a suitable metric by which the CO_2 emissions of existing and future aircraft can be consistently measured. There is still, however, significant debate about whether or not the limit should apply to both existing aircraft types in production, as well as future designs, or only to future designs, and at what level the actual limit should be set. There is hope that the CO_2 limit will be agreed upon by the end of 2016 (IATA, 2014).

Although not a policy intervention, a further contributor to mitigating the environmental impact of aviation is the growth of voluntary carbon offsetting schemes, now offered by many airlines. These schemes give the passenger the option to pay extra for their flight, with this money being used to compensate for the CO_2 emissions generated by paying for equivalent emissions savings or reductions elsewhere. While there are concerns about the principle of offsetting, because it could slow the development of low carbon technology, and issues associated with the design and performance of schemes, they do provide real funding to reduce CO_2 emissions, and provide an opportunity to engage directly with the passenger on environmental issues (Hooper, Daley, Preston, & Thomas, 2008). There is also demand. Brouwer, Brander, and Van Beukering (2008) estimate that passenger's stated willingness to pay for carbon offsetting is similar to the estimated marginal damage cost associated with CO_2, although significant difference exist between passengers in Europe, North America, Asia and the rest of the world.

13.5 System View

Estimating the potential for mitigating the environmental impacts of aviation, given the technologies, procedures and policies described above, is not trivial. The air transport system is complex and relies on decision making from a number of stakeholders, with different priorities, including passengers, airlines, aircraft manufacturers and policymakers. Even the fuel savings potential of any given technology is not only dependent on the fuel efficiency of the new technology itself, but also on the uptake of that technology into the fleet. The oil price, affecting the price of jet fuel, makes it more or less economical for airlines to purchase new, more fuel-efficient technology, as opposed to other technologies that, while less fuel efficient, may have lower purchase prices. Similarly, policy measures such as the carbon tax or ETS described above effectively increase the price of jet fuel, making technology with reduced fuel consumption more economical.

Demand effects are also important. Both increasing oil prices and policy measures such as taxes and emissions trading affect airline costs directly, which may be passed on to passengers in the form of higher fares. Passenger demand growth would slow accordingly, reducing the demand for new equipment. Alternatively, however, the reduced costs of operating new equipment could also be passed on to passengers, increasing passenger demand, with a resulting increase in the number of flights operated, and therefore emissions, relative to a case in which no new technology was added. This "rebound effect" could be as much as 20 % in the aviation sector (Evans & Schäfer, 2013).

Fleet composition is also relevant. An older fleet would include a higher number of aircraft ready for or close to retirement, allowing the new technology to not only grow the fleet, but also replace some of the existing fleet. A young fleet, however, would have less scope for this in the near term. With average aircraft life in the order of 30 years, and fleet retirement rates remarkably consistent over time (Morrell & Draym, 2009), complete fleet entry of new technology is generally slow. Hence the environmental benefit of new technology is generally only realized to its fullest extent in the long term. As described above, operational innovations suffer less from this problem.

A number of studies have been done to evaluate the potential for mitigating the environmental impacts of aviation, or to evaluate the potential to meet emissions targets set by different governments and organizations. Carbon neutral growth before 2050 is generally highly reliant on new technology, the implementation of a global policy such as emissions trading or a carbon tax, and the wide scale availability of biofuel, none of which will be trivial to accomplish. Even then, because of the non-CO_2 impacts of aviation on the global climate, climate neutral growth is unlikely (Krammer, Dray, & Köhler, 2013).

13.6 Conclusions

A number of promising technologies and operational improvements are under development that could see significant progress towards mitigating the environmental impacts of aviation. These include new and revolutionary engine and aircraft technologies, biofuels, enhancements in air traffic management, and new airline operating procedures, as described above. However, policy intervention is likely to be required to drive much of these developments and their uptake into the global fleet in sufficient timeliness to keep pace with the rapidly growing demand for air transport. With these policy interventions come increased costs and associated impacts on the aviation community, including the flying public. Significant challenges therefore exist for the aviation community to mitigate the significant environmental impacts of aviation. However, awareness has increased significantly, and, as described above, there are meaningful international efforts to tackle the problem, and significant progress towards greener air transport is possible.

References

Airbus. (2013). *Global market forecast 2013–2032*. Retrieved March 3, 2014 http://www.airbus.com/company/market/forecast/

Alcock. (2013). http://www.ainonline.com/aviation-news/ain-air-transport-perspective/2013-10-07/icao-agrees-global-emissions-cap-and-trade-scheme-2020

ATAG (Air Transport Action Group). (2012). Aviation: Benefits beyond borders report. Prepared by Oxford Economics for ATAG, Geneva, Switzerland. Retrieved February 14, 2014, from http://aviationbenefitsbeyondborders.org/download-abbb-report

Bauen, A., Howes, J., Bertuccioli, L., & Chudziak, C. (2009). *Review of the potential for biofuels in aviation*. London: E4tech.

Berghof, R., Schmitt, A., Eyers, C., Haag, K., Middel, J., Hepting, M., et al. (2005). CONSAVE 2050—executive summary, competitive and sustainable growth. In *Competitive and Sustainable Growth, G4MA-CT-2002-04013*. Retrieved June 27, 2008, from http://www.dlr.de/consave/CONSAVE 202050 20Executive 20Summary.pdf

Boeing. (2013). *Current market outlook 2013*. Retrieved March 3, 2014, from http://www.boeing.com/boeing/commercial/cmo/

Bonet, J. T. (2012). *Boeing ERA N+2 advanced vehicle concept results*. Oral presentation at the 50th AIAA Aerospace Sciences Meeting. Retrieved from https://www.aiaa.org/uploadedFiles/About-AIAA/Press-Room/Key_Speeches-Reports-and-Presentations/2012/Bonet-John-Boeing-AVC-AIAA-GEPC2.pdf

Bradley, M. K., & Droney, C. K. (2011). *Subsonic ultra green aircraft research: Phase I final report* (NASA/CR-2011-216847).

Brouwer, R., Brander, L., & Van Beukering, P. (2008). "A convenient truth": air travel passengers' willingness to pay to offset their CO_2 emissions. *Climatic Change, 90*(3), 299–313.

Bruner, S., Baber, S., Harris, C., Caldwell, N., Keding, P., Rahrig, K., et al. (2010). *NASA N+3 subsonic fixed wing silent efficient low-emissions commercial transport (SELECT) vehicle study* (NASA/CR-2010-216798).

Business Traveller. (2014, March 31). Will airbus deliver? Asia Pacific: Business Traveller. Retrieved April 1, 2014, from http://www.businesstraveller.asia/asia-pacific/archive/2014/april-2014/special-reports/can-airbus-deliver

Cairns, S., Newson, C., Boardman, B., & Anable, J. (2006). *Predict and decide—Aviation, climate change and UK Policy* (Final report). Oxford, England: Environmental Change Institute, University of Oxford.

CFM International. (2014). *CFM innovation: Fuel*. CFM International. Retrieved April 1, 2014, from http://www.cfmaeroengines.com

Clewlow, R., Balakrishnan, H., Reynolds, T. G., & Hansman, R. J. (2010). A survey of airline pilots regarding fuel conservation procedures for taxi operations. *International Airport Review, 3*, 2010.

Cohen, S., Evans, G. W., Krantz, D. S., & Stokols, D. (1980). Physiological, motivational, and cognitive effects of aircraft noise on children. *American Psychologist, 35*(3), 231–243.

Department of Energy. (2011). US billion-ton update: Biomass supply for a bioenergy and bioproducts industry. In R. D. Perlack & B. J. Stokes (Leads), *ORNL/TM-2011/224*. Oak Ridge, TN: Oak Ridge National Laboratory. p. 227.

Dowling, A. P. (2007). *The silent aircraft initiative—Overview* (AIAA-2007-0452). Presented at 45th AIAA Aerospace Sciences Meeting and Exhibit.

Dumont, J.-M., Reynolds, T. G., & Hansman, R. J. (2011, September 21–22). Fuel burn and emissions reduction potential of low power/low drag approaches. In *11th AIAA Aviation Technology, Integration, and Operations (ATIO) Conference, Virginia Beach, VA*.

Eurocontrol. (2007). *Trends in air traffic: Vol. 2. A matter of time: air traffic delay in Europe*. EATM CODA, Eurocontrol. Retrieved June 27, 2008, from https://extranet.eurocontrol.int/http://prisme-web.hq.corp.eurocontrol.int/ecoda/coda/public/standard_page/public_application.html

European Commission. (2006). *Proposal for a directive of the European Parliament and the Council Amending Directive 2003/87/EC as to include aviation activities in the scheme for greenhouse gas emission allowance trading within the community*. European Commission, Brussels, COM (2006) 818 final.

Evans, A. D., & Schäfer, A. (2013). The rebound effect in the aviation sector. *Energy Economics, 36*, 158–165. doi:10.1016/j.eneco.2012.12.005.

FAA (United States Federal Aviation Administration). (2008). *Operations and performance data, Aviation System Performance Metrics (ASPM)*. Retrieved June 27, 2008, from http://www.apo.data.faa.gov/aspm/ASPMframe.asp

Frota, J., Nicholls, K., Whurr, J., Müller, M., Gall, P.-E., Loerke, J., et al. (2011). *NACRE final activity report 2005–2010*. NACRE project report. Retrieved from http://www.transportresearch.info/Upload/Documents/201210/20121030_161035_66969_Final_Activity_Report_R1.0.pdf

Garnham, E. (2011). *Biofuels and the aviation industry*. M.Sc. thesis, Sustainable Energy Futures, Imperial College, London.

Graham, W. R., Hall, C., & Vera Morales, M. (2014). *The potential of future aircraft technology for noise and pollutant emissions reduction*. Transport Policy Special Issue on Aviation and the Environment. doi:10.1016/j.tranpol.2014.02.017.

Graham, M., Ohsfeldt, M., Ratliff, G., Sequeira, C., Thompson, T., Thrasher, T., et al. (2009, October). *Aircraft impacts on local and regional air quality in the United States*. Partnership for AiR Transportation Noise and Emissions Reduction (PARTNER) Project 15 Final Report, Report No. PARTNER-COE-2009-002. Retrieved November 29, 2009, from http://web.mit.edu/aeroastro/partner/reports/proj15/proj15finalreport.pdf

greenaironline.com. (2012, July 9). *US EPA aligns with ICAO on new NOx standards for aircraft engines as FAA pledges action on GA leaded fuel*. Retrieved April 8, 2014, from http://www.greenaironline.com/news.php?viewStory=1566

Green, J. E. (2002). Greener by design—The technology challenge. *Aeronautical Journal, 106*, 57–113.

Greitzer, E. M., Bonnefoy, P. A., De la Rosa Blanco, E., Dorbian, C. S., Drela, M., Hall, D. K., et al. (2011). *N+3 aircraft concept designs and trade studies, final report* (NASA/CR-2011-216794).

Haines, M. M., Stansfeld, S. A., Job, R. F. S., Berglund, B., & Head, J. (2001). Chronic aircraft noise exposure, stress responses, mental health and cognitive performance in school children. *Psychological Medicine, 31*(2), 265–277. Cambridge University Press.

Hall, C. A. (2009). Low noise engine design for the Silent Aircraft Initiative. *Aeronautical Journal, 113*, 599–607.

Henderson, S. C., & Wickrama, U. K. (1999). Figure 9-3 of 'Aircraft emissions: Current inventories and future scenarios'. In J. E. Penner, D. H. Lister, D. J. Griggs, D. J. Dokken, & M. McFarland (Eds.), *Aviation and the global atmosphere: A special report of IPCC working groups I and III*. Cambridge, England: Cambridge University Press.

HM Revenue and Customs. (2008, November 24). *2008 pre-budget report, air passenger duty*. HM revenue and Customs PBRN20. Retrieved December 21, 2009, from http://www.hmrc.gov.uk/pbr2008/pbrn20.pdf

Hooper, P., Daley, B., Preston, H., & Thomas, C. (2008). *An assessment of the potential of carbon offset schemes to mitigate the climate change implications of future growth of UK aviation*. Manchester, England: Centre for Air Transport and the Environment.

IATA (International Air Transport Association). (2014). *CO_2 standard*. Retrieved April 7, 2014, from http://www.iata.org/publications/airlines-international/august-2013/Pages/co2-standard.aspx

ICAO. (2012). *Emissions databank*. Retrieved from http://easa.europa.eu/environment/edb/aircraft-engine-emissions.php

ICAO (International Civil Aviation Organization). (2006). *World total traffic 1995–2006*. Retrieved from www.icaodata.com

IndustryWeek.com. (2007, December 1). *Boeing 787: A matter of materials—Special report: Anatomy of a supply chain.* IndustryWeek.com. Retrieved from http://www.industryweek.com/articles/boeing_787_a_matter_of_materials_--_special_report_anatomy_of_a_supply_chain_15339.aspx

Krammer, P., Dray, L., & Köhler, M. O. (2013). Climate-neutrality versus carbon-neutrality for aviation biofuel policy. *Transportation Research Part D: Transport and Environment, 23*, 64–72.

Lee, D. S., Pitari, G., Grewe, V., Gierens, K., Penner, J. E., Petzold, A., et al. (2010). Transport impacts on atmosphere and climate: Aviation. *Atmospheric Environment, 44*(2010), 4678–4734. doi:10.1016/j.atmosenv.2009.06.005.

Martin, K. C. (2012). *NASA environmentally responsible aviation.* Oral presentation at the 50th AIAA Aerospace Sciences Meeting. Retrieved from https://www.aiaa.org/uploadedFiles/About-AIAA/Press-Room/Key_Speeches-Reportsand-Presentations/2012/Martin-Lockheed-AVC-AIAA-GEPC2.pdf

Morrell, P., & Draym, L. M. (2009). *Environmental aspects of fleet turnover, retirement and life cycle* (Final Report for the Omega Consortium).

Morris, J., Rowbotham, A., Angus, A., Mann, M., & Poll, I. (2009). *A framework for estimating the marginal costs of environmental abatement for the aviation sector* (Final Report for the Omega Consortium).

Muller, D., Uday, P., & Marais, K. B. (2011, September 21–22). Evaluation of trends and potential environmental benefits of RNAV/RNP arrival and departure procedures. In *11th AIAA Aviation Technology, Integration, and Operations (ATIO) Conference, Virginia Beach, VA.*

Nakahara, A., & Reynolds T. G. (2013). Estimating current & future system-wide benefits of airport surface congestion management. In *10th USA/Europe Air Traffic Management Research and Development Seminar, Chicago.*

NASA. (2014). *The double bubble D8.* Retrieved August 21, 2014, from http://www.nasa.gov/content/the-double-bubble-d8-0/#.U_YbA0urHf0

Norris, G. (2012). *Operators reporting positive 787 fuel-burn results.* Retrieved August 18, 2014, fromhttp://aviationweek.com/awin/operators-reporting-positive-787-fuel-burn-results

Penner, J. E. (Ed.). (1999). *Aviation and the global atmosphere: Special Report of the IPCC Working Groups I and III in Collaboration with the Scientific Assessment Panel to the Montreal Protocol on Substances that Deplete the Ozone Layer.* Cambridge, England: Cambridge University Press.

Pratt & Whitney. (2014). *Burns less fuel: an insatiable appetite for savings.* Pratt & Whitney. Retrieved April 1, 2014, from http://www.purepowerengine.com/fuel_burn.html

Raymer, D. P., Wilson, J., Perkins, H. D., Rizzi, A., Zhang, M., & Ramirez Puentes, A. (2011). *Advanced technology subsonic transport study* (NASA/TM-2011-217130).

Reynolds, T. G. (2009). *Development of flight inefficiency metrics for environmental performance assessment of ATM.* Eighth USA/Europe Air Traffic Management Research and Development Seminar.

Reynolds, T., Barrett, S., Dray, L., Evans, A., Köhler, M., Vera-Morales, M., et al. (2007). Modelling environmental & economic impacts of aviation: Introducing the aviation integrated modelling tool. In *7th AIAA Aviation Technology, Integration and Operations Conference, Belfast, September 18–20, 2007.*

Rosillo-Calle, F., Thrän, D., Seiffert, M., & Teelucksingh, S. (2012). The Potential Role of Biofuels in Commercial Air Transport (BIOJETFUEL). In *IEA bioenergy task* (Vol. 40).

Schäfer, A. (2006)."Long-term trends in global passenger mobility. *The Bridge, 36*, 24–32. Washington, DC: The National Academies Press.

Schäfer, A., Heywood, J. B., Jacoby, H. D., & Waitz, I. A. (2009). *Transportation in a climate-constrained world.* Cambridge, MA: MIT Press.

Sinnet, M. (2007). *Saving fuel and enhancing operational efficiencies.* Boeing. Retrieved March 31, 2014, from http://www.boeing.com/commercial/aeromagazine/articles/qtr_4_07/AERO_Q407_article2.pdf

Stratton, R. W., Wong, H. M., & Hileman J. I. (2010). *Life cycle greenhouse gas emissions from alternative jet fuels*. Partnership for AiR Transportation Noise and Emissions Reduction. Cambridge, MA: Massachusetts Institute of Technology.

theguardian.com. (2013, September) *UN aviation agency 'dragging feet' on efforts to tackle emissions*. BusinessGreen, part of the Guardian Environment Network, theguardian.com.

UN (United Nations). (2013). *Statistical yearbook 2011, fifty-sixth issue*. New York: Department of Economic and Social Affairs, Statistics Division, United Nations. ISBN 978-92-1-061333-0.

Vera Morales, M., Graham, W. R., Hall, C. A., & Schäfer, A. (2011). *Techno-economic analysis of aircraft*. TOSCA project work-package 2 final report. Retrieved from http://www.toscaproject.org/FinalReports/TOSCA_WP2_Aircraft.pdf

Zaman, K. B. M. Q., Bridges, J. E., Huff, D. L. (2010). Evolution from 'Tabs' to 'Chevron Technology'—A review. In *Proceedings of the 13th Asian Congress of Fluid Mechanics 17–21 December 2010, Dhaka, Bangladesh* (pp. 47–63), 1.34 Mb. Cleveland, OH: NASA Glann Research Center.

Chapter 14
Emissions and Inland Navigation

Gernot Pauli

Abstract Inland navigation may seem to be an environmentally friendly while somewhat quaint mode of transport. However, it does in fact play an important role by supporting industrial development. While inland navigation does indeed have all the ingredients of a green mode of transport, its emissions to air must nonetheless be considered and reduced. This chapter examines the possibilities for reduction of greenhouse gas (GHG) and pollutant emissions while emphasizing win–win situations. This will be accomplished by first establishing the (European) policy context of inland navigation and explaining the basic facts and issues regarding emissions. The main part of this chapter is a presentation and analysis of possible measures to reduce emissions. The chapter concludes with recommendations to further the greening of inland navigation. The recommendations focus on the reduction of GHG and pollutant emissions from inland navigation, as these emissions stand in the way of inland navigation being a truly green mode of transport.

Abbreviations

ADEME	Agence de l'Environnement et de la Maîtrise de l'Énergie
AIS	Automatic Identification System
BTS	Barge Traffic System
CCNR	Central Commission for the Navigation of the Rhine
CH_4	Methane
CNG	Compressed natural gas
CO	Carbon monoxide
CO_2	Carbon dioxide
EBU	European Barge Union
EEDI	Energy Efficiency Design Index
EEO	Energy Efficiency Operational Indicator
EGR	Exhaust gas recirculation
EN	European Norm

G. Pauli (✉)
Central Commission for the Navigation of the Rhine (CCNR), Strasbourg, France
e-mail: G.Pauli@ccr-zkr.org

H.N. Psaraftis (ed.), *Green Transportation Logistics*, International Series in Operations Research & Management Science 226,
DOI 10.1007/978-3-319-17175-3_14

EPA	United States Environmental Protection Agency
ESO	European Skippers Organization
EU	European Union
Euromot	European Association of Internal Combustion Engine Manufacturers
GHG	Greenhouse gas
HC	Hydrocarbons
IFEU	Institute for Energy and Environmental Research
IMO	International Maritime Organization
INE	Inland Navigation Europe
IT	Information Technology
ITF	International Transport Forum
LNG	Liquefied natural gas
NO_X	Nitrogen oxides
OECD	Organization for Economic Co-operation and Development
PM	Particulate matter
PMF	Particulate matter filter
RIS	River Information Services
SCR	Selective catalytic reduction
SEEMP	Ship Energy Efficiency Management Plan
TEN-T	Trans-European Transport Networks
UIC	International Union of Railways
USA	United States of America

Sitting on the green shores of a river, watching barges slowly passing by, one may get the impression, that inland navigation is a mode of transport that is friendly to the environment, if rather stuck in the past. As pleasant as this picture is, it is nonetheless an illusion. Inland navigation is in principal a natural mode of transport for green corridors supporting the development of modern industries, but its emissions to air have to be of major concern. This chapter examines the possibilities for reducing greenhouse gas (GHG) and pollutant emissions from inland navigation, including the use of alternative fuels, with a particular view to win–win situations. In doing so, it explains the policy context particular to inland navigation mainly in Europe as well as basic facts and issues regarding emissions from inland navigation. Possible measures for reducing emissions are presented including the use of alternative fuels. This discussion constitutes the main part of the chapter. Finally, recommendations for policy makers as well as decision makers in the inland navigation sector are made.

14.1 Inland Navigation: A Natural Mode of Transport for Green Corridors

When compared to rail and road, inland navigation is the least important mode of transport. In the USA and in Europe, less than 10 % of all freight transport is performed on inland waterways. However, in China, the modal share of inland navigation has reached almost 25 %, as Fig. 14.1 shows, and is growing faster than rail.

In Europe too, there are regions where inland navigation plays an important role, in particular in the Netherlands, where one third of all freight is carried on inland waterways (OECD/ITF, 2013). One of the main advantages of Rotterdam and Antwerp, Europe's main sea ports, is that they are connected to the hinterland by a network of efficient inland waterways, of which the Rhine forms the backbone, as can be seen in Fig. 14.2.

Inland navigation is clearly suited to carry large amounts of dry or liquid bulk goods over long distances. In the USA, pushed convoys of some 15 barges, which carry some 20,000 tons of cargo, are common. Similar convoys are sailing on the Rhine and the Danube, in particular supplying steel works with raw materials. In China and the EU, but not in the USA, inland navigation has in recent years become a major carrier for containers, mostly connecting the large container terminals in the seaports with those in the hinterland (Bonnerjee et al., 2009; WorldBank, 2009). Similarly, in Rotterdam and Antwerp, the modal share of inland navigation in the transport of containers stands at roughly a third with the intent for this share to rise even further. Thus, inland navigation is also a viable solution for the transport of manufactured goods and has an important role to play in the development of modern industrialized countries.

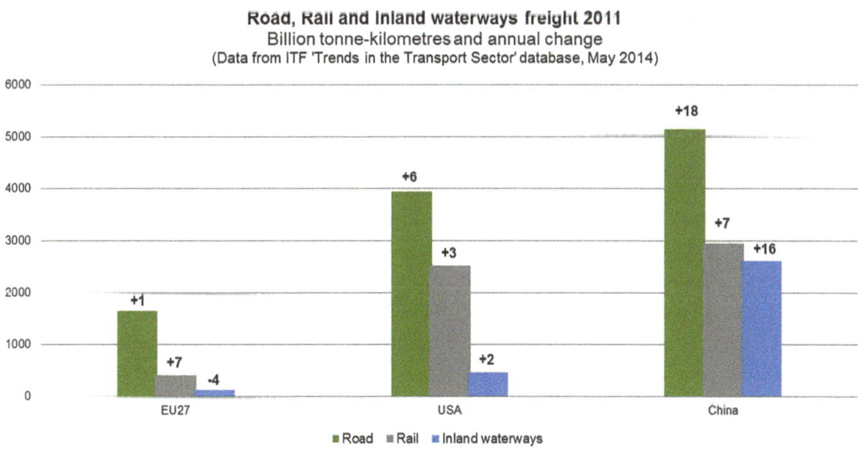

Fig. 14.1 Comparison of road, rail and inland navigation freight in China, EU and USA, in billion tonne-kilometres. *Source*: OECD/ITF (2013)

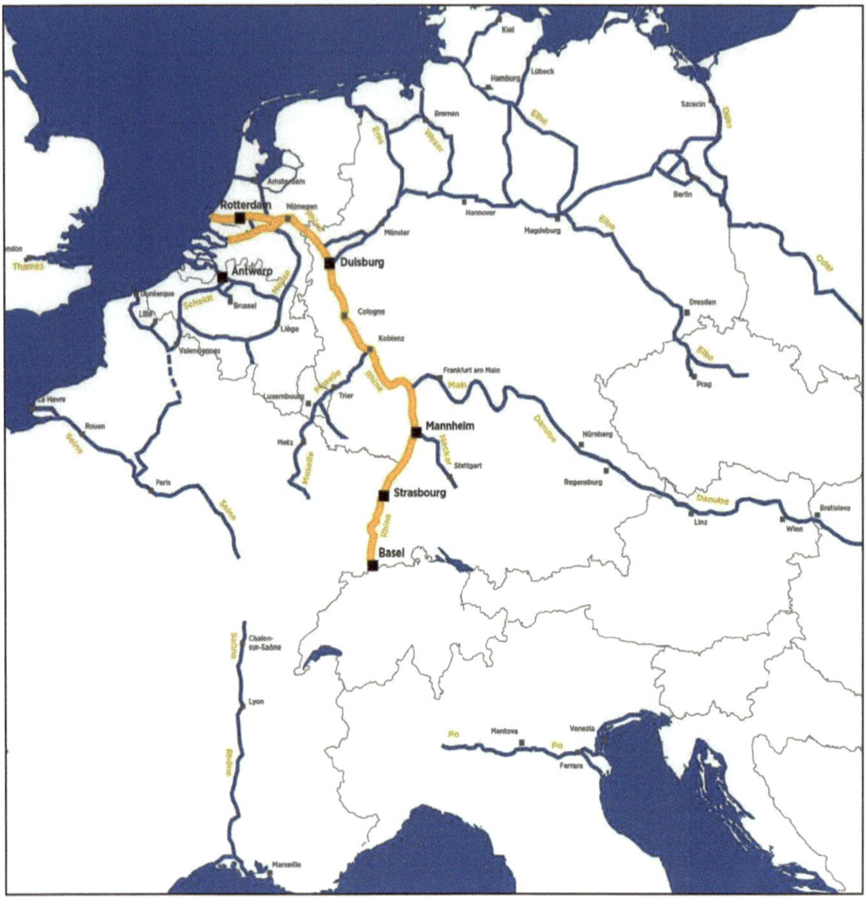

Fig. 14.2 Main inland waterways in Europe

Great rivers, such as Yangtze and Pearl River in China, Mississippi and Missouri in the USA, Rhine and Danube in Europe, are the backbone of the inland navigation network. These rivers form historical and natural transport corridors. The Rhine is a particularly good example. For over 2000 years the Rhine has provided vital transport and has evolved over time into a major transport artery for industrial areas and transport hubs in Switzerland, France, Germany and the Netherlands, accounting today for two thirds of all inland water transport in the EU. With the founding of the Central Commission for the Navigation of the Rhine (CCNR), 200 years ago, a unique governance system was set up. Its main tasks remain to guarantee freedom of navigation on the Rhine and to further its prosperity. In other words, the basis for a single transport market was laid with unified rules set by an international body, almost 150 years before the EU and its predecessors came into being. The idea for a modern transport corridor was born almost 200 years before the EU developed its concept of Trans-European Transport Networks (TEN-T), which calls for the creation of cross border transport corridors. The development of the physical infrastructure and the removal of bottlenecks, key criteria for the TEN-T projects, have been major objectives of the CCNR. The continuous growth of the average size of the vessels sailing on the Rhine is testimony to this. Also, as demanded today for transport corridors, sufficient intermodal points have been a characteristic for the Rhine. Today navigation on the Rhine is connected with other modes of transport by an average of three container terminals per 100 km of waterway. Therefore it is not surprising, that in the Rhine-Alpine Core Network Corridor inland navigation's share in international freight transport is more than half (Riebe, 2014).

Are inland waterways in general and the Rhine in particular also green corridors, as defined in Chap. 3 of this book? In Europe, the green credentials of inland navigation have been extensively researched, including calculation of external costs to allow for a comparison of different ecological and other impacts. Following the idea of transport corridors typical transport tasks, such as carrying bulk goods from Rotterdam to Duisburg, were examined (den Boer, Otten, & van Essen, 2011; PLANCO, 2007). Other studies developed average values, which subsequently are used for the calculation of the external costs (CE_Delft, Infras, & Fraunhofer ISI, 2011; IFEU, 2011). Additional insights have been gained by analysing the sustainability of inland navigation, taking ecological, but also economic and social criteria into account (Pauli, 2010). The green credentials of the different transport modes are also at the heart of decisions on shifting cargo from one mode to another, mostly from road to rail or inland waterways. This is a concern for policy makers as well as for shippers of cargo. Decisions are based not only on transport costs, but also increasingly on the environmental performance of the different transport modes. Therefore, the aforementioned studies and many more try to provide data for these decisions. The results of the studies often differ significantly. However, the conclusion seems justified that inland navigation is indeed a natural mode of transport for green corridors with few, albeit serious shortcomings:

- Inland navigation relies almost solely on mineral oil as fuel. Broadening its fuel base and at the same time decarbonization of the fuel is needed.
- By far inland navigation's largest externalities are caused by its pollutant emissions. Reducing its pollutant emissions must be its main objective, if inland navigation wants to be a truly green mode of transport.

The remainder of the chapter is organized as follows. Section 14.2 explains the policy context of inland navigation's "greening", in particular the actions and objectives of international organizations and regulation of airborne emissions. Section 14.3 presents basic facts and issues of greenhouse gas and pollutant emissions related to inland navigation. Section 14.4 discusses possible measures for the reduction of emissions to air from inland navigation, including measures regarding technical aspects of the vessels and their operation, measures relating to design and equipment of vessel engines, the use of alternative energy sources as well as infrastructure and transport management measures. Finally, Sect. 14.5 gives recommendations for policy makers as well as decision makers in the inland navigation sector.

14.2 The Policy Context for the Greening of Inland Navigation

The general policy context for green transportation logistics is described in Chap. 1 of this book. Therefore, in the following only the policy context, which is particular to inland navigation, will be explained.

14.2.1 Actions and Objectives of Europe Commission and CCNR

Probably the most important policy instrument and the EU level regarding inland navigation is the so-called NAIADES action programme. The European Commission demands in the second edition of this programme presented in September 2013 (EU, 2013),

- new and ambitious emission limits for new engines on inland navigation vessels,
- exploring further emission limits for existing engines,
- preparing the inland waterway infrastructure for liquefied natural gas (LNG) use, including adoption of technical standards for inland waterway LNG bunkering and use of LNG as a fuel.

The CCNR is aside from the EU, the most important governing body for inland navigation in Europe. With its "Vision 2018", it is setting itself a number of ambitious objectives, which will contribute to the sustainable development of inland navigation in ecological, social and economic terms (CCNR, 2013b). Of those objectives, the following address in particular the greening of inland navigation:

- an even more significant contribution of inland navigation to combating global warming by reducing both its fuel consumption and its emissions of greenhouse gases,
- an even more significant contribution of inland navigation to preserving air quality by further reducing the emissions of pollutants.

In general, the inland navigation industry seems to support the reduction of emissions to air from inland navigation. For example, the European shipping industry associations have set themselves a target of reducing pollutant emissions from inland navigation by 95 % by 2020 and CO_2 emissions by 50–70 % by 2050, albeit without specifying the point of reference (INE, EBU, et al., 2011).

14.2.2 Regulation of Inland Navigation Emissions to Air

The main driver for the reduction of transport's emissions to air are legal requirements, as can be observed in particular in the road sector. In inland navigation, the requirements are far less pronounced. Cargo vessels on inland waterways are typically equipped with internal combustion engines, burning diesel or gasoil. In contrast to maritime navigation, the legally admissible sulphur content of inland navigation fuels in the EU and the USA is so low, that these fuels can be seen as quasi sulphur free. Therefore, nitrogen oxides (NO_X), particulate matter (PM), hydrocarbons (HC) and carbon monoxide (CO) constitute the main pollutant emissions from engines for inland navigation (Pauli & Schweighofer, 2008). By far the most pertinent GHG is carbon dioxide (CO_2). Methane (CH_4), even though it is a relatively much more damaging GHG than CO_2, currently does not contribute to GHG emissions from inland navigation. As LNG and compressed natural gas (CNG) are used very rarely as fuel in inland navigation, there is no significant methane slip, the primary cause for CH_4 emissions in transport. However, this may change, in particular with the advent of LNG in inland navigation.

Well aware of the damaging effects of pollutant emissions on the health of people and the environment, the CCNR, the EU and the USA have regulated the pollutant emissions from inland navigation. The regulation has happened in stages, with the CCNR taking the lead in Europe with Stage I coming into force 2003, and Stage II coming into force in 2007. In the same year, the EU incorporated inland navigation into the already existing European emissions regulation for non-mobile machinery. The emission stage is called IIIa according to the original nomenclature of this regulation. In the USA, the emission stages are called tiers. Tier 1 for inland navigation came into force in 2004. Current and future emission limits given in Table 14.1 show that regulators in Europe have not yet decided on ambitious emission limits for pollutants as they have in the USA, and that GHG emissions remain unregulated, both in the EU and in the USA.

In contrast, road transport in the EU has seen tremendous progress in reducing its pollutant emissions. Successive emission regulations in recent years have enabled

Table 14.1 Current limits for pollutant emissions from engines installed on inland navigation vessels

Regulator		CCNR	EU	USA EPA	
Emission level		Stage II	Stage IIIa	Tier 3	Tier 4
Year of application		2007	2007–2009	2009–2014	2014–2017
CO	g/kWh	3.5–5.5	5.0	5.0–8.0	5.0–8.0
HC		1.0–1.5	7.2–11.0	4.7–11.0	0.19
NO_X		6.0–11.0			1.8
PM		0.2–0.8	0.2–0.5	0.11–0.40	0.04–0.12

Source: (CCNR, 2003; EPA, 2012; EU, 2004)
Note: The bandwidth of emission limits is due to different emission limits for engines of different swept volume or power range

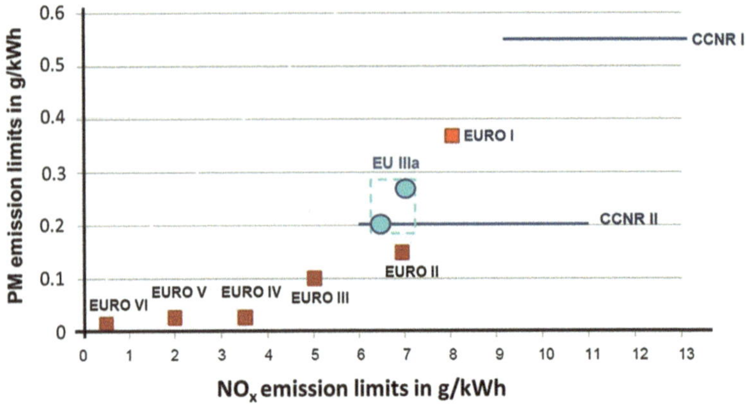

Fig. 14.3 Comparison of selected emission limits from emission regulations in Europe (EU heavy duty road transport—EURO I to EURO VI, EU inland navigation—EU IIIa, CCNR inland navigation—CCNR I and II)

road transport to "clean up". Inland water transport has clearly been left behind. It remains, for the time being, in its initial stages of emission regulations, as shown in Fig. 14.3.

14.3 Basic Facts and Issues Regarding Emissions from Inland Navigation

The basic facts and issues regarding emissions from transportation in general are explained in Chap. 2 of this book. Therefore, in the following only particularities for inland navigation will be dealt with.

14.3.1 Operating Conditions Particular to Inland Navigation that influence Energy Consumption

Inland navigation vessels can neither be put on a test bench, nor are there standards for measuring their energy consumption. Inland navigation vessels vary tremendously in size; in Europe the smallest vessels have a carrying capacity of 300 tons and the large convoys 15,000 tons and more. Such examples illustrate the difficulties when determining the energy consumption and carbon footprint of inland navigation vessels.

Particular for inland navigation are the rather complex, and therefore often omitted in comparative studies and applications, effects that result from inland navigation vessels operating in water as opposed to on land like lorries and trains do and in shallow waters as opposed to deep water in maritime navigation. The resulting effects can be described with the following examples, which were developed by (CCNR, 2012) and are based on (PLANCO, 2007; Renner & Bialonski, 2004):

- If the water depth decreases from 4.5 to 4 m, the power requirements of a large motor vessel of the kind typical for the Rhine increase by about one third (vessel speed: 16 km/h; loaded draught 2.5 m).
- If the water depth increases from 3 to 5 m, the speed of a large motor vessel with a loaded draught of 2.5 m and a constant power output of 200 kW, the vessel's speed more than doubles from about 6 km/h to about 13 km/h.
- If the same large motor vessel sailing with a speed of 17 km/h in a water depth of 5 m reduces its speed by only about 15 %, it requires only half as much power.

But there are even more surprising effects. For all modes of transport it can be observed, that the greater the transport vehicle's carrying capacity, the lower its specific power requirements are, which can be expressed in kW/tkm. However, for inland navigation and where the water depth is very high, the power requirements of a large vessel transporting large amounts of cargo can actually be smaller in absolute terms than those for a smaller vessel. For example, on a waterway with a depth of 5 m a large motor vessel requires only 230 kW to transport 1,900 tons of cargo at a speed of 13 km/h, whereas a smaller vessel of the type "Johann Welker" would need 420 kW to transporting 1,250 tons at the same speed (Zöllner, 2009).

In summary, vessel size and waterway parameters have a crucial effect on the energy consumption of inland navigation vessels, which is often not well understood or indeed neglected altogether.

14.3.2 Energy Consumption, Carbon Footprint and CO_2 Emissions from Inland Navigation

As shown above, it is a very challenging undertaking to calculate the energy consumption of inland navigation vessels. Consequently, different approaches have been employed. The top down approach tries to determine the amount of fuel used for the entire fleet in a certain region or country and then putting this into relation with the transport activities in the particular region. The outcome is typically a value such as litre per tkm or MJ per tkm. This method can be applied when the fuel consumption and the transport activities can be determined in a sufficiently exact way. The CCNR has undertaken an initial attempt to make such a calculation for inland navigation in western Europe (Knörr, Heidt, Schmiedt, & Notter, 2013). The top down method has also been applied for freight transport on the Mississippi/Missouri (Kruse, Protopapas, Olson, & Bierling, 2009). This approach is simple and reliable, if there is a clear perimeter for the area under examination. In Europe that is not the case, as inland navigation is essentially a cross border mode of transport and inland and coastal navigation overlap. As fuel for inland navigation is also tax free, a significant amount may therefore be syphoned off for other uses.

The bottom up approach determines the fuel consumption of individual vessels or individual journeys and integrates them into an aggregated value for a certain region. This method was used by the Second IMO GHG Study to determine the fuel consumption of maritime transport (Buhaug et al., 2009). Recently, it was also applied in an exemplary study to determine the fuel consumption of inland navigation in Germany. This study showed that an earlier study using the top down approach was erroneous. Subsequently the values for the fuel consumption had to be revised downwards (Knörr et al., 2013). The bottom up approach is highly complex, as it has to cope with a myriad of deviations regarding vessel types and sizes, load factors, infrastructure and other parameters that determine fuel consumption of inland navigation vessels in practice.

Some studies (den Boer et al., 2011; PLANCO, 2007) have tried to avoid the pitfalls of both the top down and the bottom up method by concentrating on specific transport tasks, meaning transport of one product with one vessel type on a particular corridor. These studies basically apply a bottom up approach for a specific transport task. Therefore, they are much less complex and their results more reliable. These studies can provide sufficiently meaningful answers to questions related directly to this transport task and perhaps to this corridor in general. However, if questions relate to other transport tasks or to general transport policies, their value is limited.

Very few actual values for fuel consumption of inland navigation vessels are known. Expenses for fuel represent a major part of the overall transport costs. Therefore, ship-owners may not want to make them known to their customers or competitors.

Figure 14.4 gives a comparison of the specific energy consumption from different inland navigation vessels. (Using the specific energy content of the fuel, the fuel

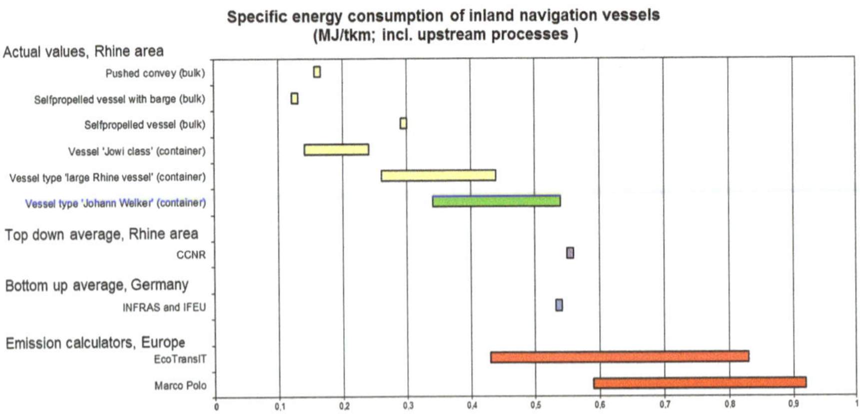

Fig. 14.4 Comparison of data on the specific energy consumption of inland navigation vessels. *Source*: Company data and (IFEU, Öko-Institut, IVE/RMCON, 2011; Knörr et al., 2013; Van Essen & den Boer, 2012)

consumption is converted into energy consumption.) The figure shows the significant differences in the actual values of energy consumption for inland navigation vessels, mostly due to large differences in vessel sizes. Figure 14.4 also highlights the large differences between the actual values on the one hand and the values used by emission calculators on the other.

Emission calculators—IT based systems to calculate different emissions from transport activities—play an important role in the comparison of emissions from different modes of transport. They are used as tools for policy development and decision making in companies. The best known emission calculators in Europe are the so-called Marco Polo calculator, created by the European Commission, and EcoTransIT, created by large companies operating mainly in the rail freight sector. Figure 14.4 shows the values for the specific energy consumption of inland navigation vessels used by these calculators as well as the value as determined by the Institute for Energy and Environmental Research (IFEU) in Germany and INFRAS, Swiss consulting company (Knörr et al., 2013). The differences are remarkable and may indicate, that the values used by these calculators are indeed far too high and therefore misleading, disadvantaging inland navigation. These values result from older studies, which were based on insufficient and outdated data and assumptions, as was shown for the Marco Polo Calculator (Van Essen & den Boer, 2012). A similar conclusion can be reached, when comparing the data used for EcoTransIT (IFEU, 2011) with the most recent study from IFEU and INFRAS (Knörr et al., 2013).

However, there is also surprising congruence, namely the average values calculated by the CCNR using a top down approach and the one calculated by IFEU and INFRAS using a bottom up approach. This may be an indication, that these values are more accurate and more reliable than those of other studies.

Mitigation of climate change places the focus on the carbon footprint of human activities. For transport, the carbon footprint can be described as the GHG emitted while performing a certain transport task. This is called the specific CO_2 emissions or CO_2 intensity. Calculation of specific CO_2 emissions makes it possible to compare GHG emissions from different modes of transport. However, great care is needed when undertaking such a comparison, in particular with regards to possible upstream effects, notably emissions from producing and transporting the fuel, and from GHG emissions other than CO_2, e.g. methane. These other emissions are ideally converted into CO_2 equivalent and added to the initial CO_2 emissions. For inland navigation, the carbon footprint and its specific CO_2 emissions (CO_2 intensity) are easy to determine, once the fuel consumption is known, as inland navigation uses quasi only one type of fuel. For rail transport this is often more complicated, because rail transport uses diesel as well as electric traction.

Figure 14.5 shows that the specific CO_2 emissions of inland navigation are on the same level or lower than those of rail transport and much lower than those of road transport.

A different method for comparing the GHG emissions of different modes of transport is the calculation of their external costs. In the study "External Costs of Transport in Europe—Update Study for 2008", commissioned by the International

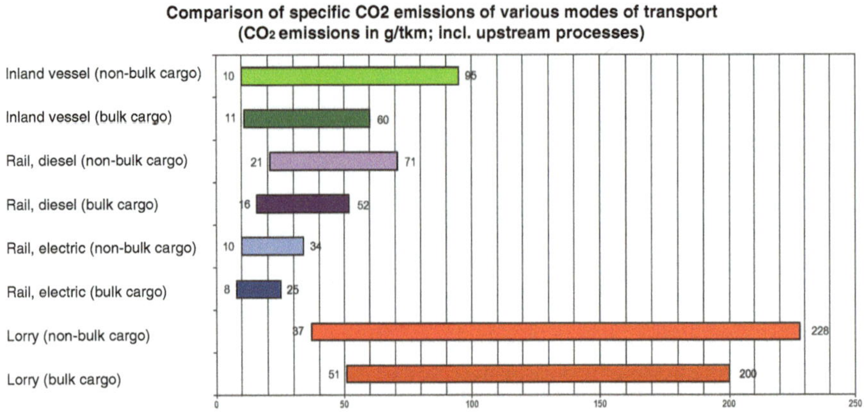

Fig. 14.5 Comparison of specific CO_2 emissions from different modes of transport (including upstream processes). *Source*: (ADEME, 2006; den Boer et al., 2011; PLANCO, 2007)

Union of Railways (UIC), the authors calculated for inland navigation slightly lower external costs with respect to climate change than for rail, when the upstream and downstream effects are taken into account (CE_Delft et al., 2011). The external costs for GHG emissions from road transport were calculated as more than two times higher than those from inland navigation. The latter coincides with the results of a study for the European Commission "Contribution to Impact Assessment of Measures for Reducing Emissions of Inland Navigation" (Panteia, 2013).

The European Standard EN 16258:2012 "Methodology for calculation and declaration of energy consumption and GHG emissions of transport services (freight and passengers)" establishes a common methodology for the calculation and declaration of energy consumption and greenhouse gas (GHG) emissions related to any transport service (of freight, passengers or both). As it is a European Standard, it is not associated with certain private companies, as for example is EcoTransIT. For the calculation, the company performing a transport service must provide actual data on fuel consumption. If that is not possible, average values for the specific fuel consumption of the transport in question can be used. However, these values are not provided for in the standard. Since the 1st of October 2013, the calculation of the quantity of CO_2 emitted during a transport service is mandatory in France and must be mentioned on invoices. For this reason, ADEME, the French Environment and Energy Management Agency, with the active support of transport professionals, has established a methodological guide, which also includes average values (ADEME, 2012).

14.3.3 Pollutant Emissions from Inland Navigation

Taking pollutant emissions into account adds substantial complexity to the analysis of the green credentials of inland navigation, mainly because there are several

Table 14.2 Aggregated emission factors from different studies for engines of inland navigation vessels

Source		Analysis for	CO	HC	NO_X	PM
			g/kWh			
PLANCO (2007)		2006	1.6	0.79	9.6	0.22
CBS (2011)		2009	1.9[a]	0.42[a]	8.9[a]	0.3[a]
TNO (2010)		2010	2	0.4	9.4	0.4
Knörr et al. (2013)		2010	1.9[a]	0.51[a]	10.3[a]	0.26[a]
Panteia (2013)	Vessel 1,500 tons	2012	n.a.	n.a.	10	0.47
	Vessel 2,750 tons		n.a.	n.a.	9	0.35

Source: Knörr et al. (2013)
[a]Calculated with a fuel consumption of 200 g diesel fuel per kWh

Table 14.3 Comparison of external cost factors for pollutant emissions from engines of inland navigation vessels

Source	Analysis for	Road (%)	Rail (%)	IWT (%)
PLANCO (2007), bulk goods	2006	267	42	100
	2025	83	42	25
CE_Delft et al. (2011)	2008	124	20	100
Panteia (2013)	2011	67	n.a.	100
	2018	38	n.a.	95
	2025	17	n.a.	38

pollutants or types of emissions, but also because the level of pollutant emissions depends on the fuel consumption as well as other factors, such as the type, age and condition of the vessels' engines. Thus, studies on pollutant emissions generally have to rely on a number of assumptions and simplifications. Table 14.2 shows aggregated emission factors of inland navigation from different studies. For CO and NO_X, the maximum deviations are some 20 %, which may be seen as acceptable bearing the complexity of the analysis in mind. For HC and PM, the values differ up to a factor of 2. As PM is after NO_X the most important pollutant, these deviances put the reliability of the studies into doubt and more importantly, any decisions based on them.

How does the performance of inland navigation compare to other modes of transport? This is an important question, as the answer may influence public perception, determine environmental regulation or guide companies' decision making. Alas, neither is the question easily answered, nor are the answers given by different studies clear. Answering the question is not easy, as contrary to transport GHG emissions, where there is in principle only one type of emission with one effect, there are several pollutant emissions, which in addition have a range of environmental and health effects. Therefore, external cost calculations are seen as a suitable method for comparing the effects of pollutant emissions.

Table 14.3 shows the results of external cost for pollutant emissions from three important European studies. To make the results comparable, the values are given in Euros per tkm. (In order to compensate for different calculation methods, the author has set the initial values for inland navigation as 1 for each study.) The comparison of Table 14.3 offers important information, but also highlights the difficulties of such a comparison:

- The analysis relating to the past decade shows a clear cost advantage for inland navigation, when compared to road, and a clear disadvantage, when compared to rail.
- The analysis relating to the future shows contradictory results. Whereas PLANCO concludes, that inland navigation will keep its clear advantage, expressed by threefold lower external costs, Panteia continues at least until the next decade the lower external costs with road transport.

How can this be? First of all, different emission standards for different modes of transport come into effect at different points in time. PLANCO starts for road with the EURO IV emission standard and assumes EURO V in the future. Panteia already starts with EURO V and assumes Euro VI coming into effect very soon. Both studies use for inland navigation CCNR II emission standard as a starting point, but assume different emission regimes in the future. The one chosen by Panteia is more stringent. These developments make adjustments of the PLANCO results necessary insofar as the external costs for road transport in the future should be significantly lower than stated in the table. But even when halving the value for road transport, it would still be higher than that for inland water transport in the calculations of PLANCO. Thus, the basic contradiction between the different studies remain. On the one hand those with inland navigation causing lower external costs than road transport and on the other hand the Panteia study, with road transport always in the lead. As the latter has been used for developing important policies for inland navigation, a sensitivity study may be called for, which starts from the assumption, that in 2011 the external costs for pollutant emissions from road and inland water transport where on an equal level.

14.3.4 Emissions Other Than Those from the Operation of the Vessels

The emissions as discussed above result solely from the operation of the transport vehicles. Taking a life cycle approach, the emissions from the construction, the maintenance and the decommissioning of the vehicles would be needed to be taken into account. Going one step further, taking a systems approach, emissions from infrastructure construction and operation would have to be investigated. However, taking a life cycle and a systems approach for inland navigation would go well beyond the scope of this chapter, not least, as there are enormous knowledge gaps (Hill et al., 2012). An analysis for Germany shows, that the GHG emissions from the construction and maintenance of inland navigation vessels are almost insignificant compared to those of the vessel operation. In contrast, the GHG emissions from the construction and maintenance of the inland waterway infrastructure are a quarter of those from the vessel operation (Mottschall & Bergmann, 2013).

14.4 Possible Measures for the Reduction of Emissions to Air from Inland Navigation

There is a wide range of measures available for the reduction of emissions to air from inland navigation. They are presented here, using a systems approach. This means that the vessels and their operation including their fuels will be taken into account, as well as the waterway infrastructure and its operation and finally the management of inland water transport.

All measures fall broadly into two categories, namely those that reduce the fuel consumption and thereby the GHG and the pollutant emissions, and those that only target the pollutant emissions. In general, the latter are much more effective in cutting pollutant emissions, but are at the same time very costly. The measures in the first category often enable cost reduction and thereby create win–win situations for shipowners and society. Because of this, these measures will be looked at in greater detail.

14.4.1 Measures for the Reduction of Energy Consumption and Emissions Regarding Technical Aspects of the Vessels

Technical measures are those relating to the vessel design and equipment as well as the propulsion system.

All measures listed in Table 14.4 aim at the reduction of fuel or energy consumption, thereby also reducing emissions to air. These measures always require additional cost upfront, either for additional planning, design and tests, or for additional investment in the vessel or its equipment. Some of the measures are well established, in particular optimization of vessel design using pilot projects and computer simulation; others are new relatively, such as the adjustable tunnel apron.

14.4.2 Measures for the Reduction of Energy Consumption and Emissions Regarding Vessel Operation

Measures related to the vessel operation also contribute indirectly to emission reduction, as their primary objective is the reduction of fuel and energy consumption. If operational measures cause any additional costs these are comparatively low

Table 14.4 Overview of measures for the reduction of energy consumption and emissions to air regarding technical aspects of inland navigation vessels

Area of influence	Principal measures	Examples
Vessel design	Optimization of vessel design using pilot projects and computer simulation	Hydrodynamic properties (optimization of the main dimensions, vessel hull form, speed, propulsion organs)
	Reduction of resistance	Air lubrication, vessel hull form optimization, exhaust flow plate, adjustable tunnel apron, coupling point optimization
	Weight reduction	Lightweight construction, medium and high speed engines
Propulsion system	Optimization of conventional propulsion systems	Energy efficient design, prevention of oversized engines, father-and-son engine configurations
	Diesel-electric propulsion	Combination of a diesel engine operating in the optimum speed range with an electric generator and an electric engine for driving the vessel
	Hybrid propulsion	Buffering of the propulsion energy as electrical energy, possibly in combination with a diesel-electric propulsion system
	More efficient or alternative propulsion organs	E.g. "whale-tail"
	Energy recovery	Heating, air conditioning, additional propulsion power
Vessel equipment	Energy efficient equipment	Auxiliary drives, loads

Source: CCNR (2012)

in relation to overall cost, they do however always contribute to cost savings as they reduce fuel consumption.

Of the operational measures listed in Table 14.5, reducing the speed of a vessel or adjusting the speed of a vessel to the navigation channel dimensions and water depth are possibly the most effective measures for reducing energy consumption and thereby cost as well as emissions to air.

14.4.3 Estimated Potential for Reducing Energy Consumption and Emissions to Air from Inland Navigation with Technical and Operational Measures

Reduced energy consumption translates to emission reductions of the same proportion. But how much energy can be saved with the measures presented in Tables 14.4 and 14.5? CCNR has estimated potential savings in relation to an "average" vessel

Table 14.5 Overview of measures for the reduction of energy consumption and emissions to air relating to inland navigation vessel operation

Area of influence	Measures	Explanations
Operation of vessel	General speed reduction	Possibly the most effective single measure in conjunction with appropriate speed
	Adjustment of speed to the navigation channel dimensions/water depth (smart steaming)	In principle, the larger the navigation channel dimensions, the lower the resistance of the vessel
	On-board information systems for fuel efficiency	Econometer, journey planning
	Optimized journey planning	Selection of most suitable routes, consideration of limitations
	Automatic track guidance systems	Prevents unnecessary movements of the rudder
	Optimized maintenance	Skin, propeller, engine
	Avoiding engine idling	E.g. before or in locks
	Optimizing the trim	Load, ballast
Maintenance of vessel	Optimally tuned and maintained engines	Maintenance according to manufacturer's instructions
	Undamaged propulsion organs	Damage can reduce efficiency
	Clean, undamaged underwater bodies	Fouling and serious distortion can increase resistance

Source: CCNR (2012)

of the current fleet on the Rhine (CCNR, 2012). These estimates are presented in Table 14.6. The figures are in general based on the educated guesses of experts. For a few some of the measures, the figures are based on trial results. Therefore, the figures contain significant uncertainties. The figure 0 indicates that the particular measure is already applied to a significant number of vessels or that there are types of vessels technically unsuitable for that particular measure. The figures given combinations of measures indicate, that individual savings do not add up. With each measure taken and fuel reduction realized, the remaining potential for reduction becomes smaller.

14.4.4 Measures Relating to Design and Equipment of Vessel Engines

Inland navigation vessels engines are most often regular diesel engines, similar to or identical with those of other applications. Consequently measures listed in Table 14.7 are not restricted to inland navigation vessels, nor were they developed with this application in mind.

Table 14.6 Estimated potential for reducing energy consumption and emissions to air from inland navigation with technical and operational measure

Measures		Reduction of energy consumption (%)	Combined (%)	Combined (%)
Ship technology	Increase in engine efficiency	2–5	10–25	10–50
	Diesel-electric propulsion	0–20		
	Hybrid propulsion	0–20		
	Waste heat recovery	0–5		
	More efficient propulsion organs	5–20	0–25	
	Alternative propulsion organs	0–25		
	Lightweight construction	0–5	5–25	
	Air lubrication	0–15		
	Ship hull form optimization	0–10		
	Exhaust flow plate	0–10		
	Adjustable tunnel apron	0–10		
	Coupling point optimization	0–15		
Operation	Smart steaming, just in time	0–30	5–30	10–40
	Speed optimization using decision support systems	0–15		
	Journey planning optimization	0–20		
	Automatic channel guidance optimization	0–10		
	Motor maintenance optimization	0–5	0–10	
	Optimization and maintenance of the propeller	0–5		
	Optimization and maintenance of the hull plating	0–5		
	Optimization of the vessel's trim	0–5	5–15	
	Optimization of locks/bridge passages	0–15		
	Optimization of vessel operation in ports	0–5		
	Shore-side electricity	0–5		

Source: CCNR (2012)

Table 14.7 Overview of possible measures regarding design and equipment of vessel engines

Area of influence	Principal measures	Examples
Design and equipment of engine	Engine internal measures	EGR (exhaust gas recirculation), advanced injection systems, inlet air humidification, in-cylinder water injection, homogeneous charge compression, H_2 injection
	Exhaust after treatment	SCR (selective catalytic reduction), PMF (particulate matter filter), diesel oxidation catalytic converter

The measures related to design and equipment of engines have in general as an objective the reduction of pollutant emissions. These measures are very effective, but also very costly, as shown in the Panteia study (Panteia, 2013). High performance SCR and PMF can reduce pollutant emissions by up to 95 % compared to modern engines without exhaust aftertreatment; but for a single engine of some 1,000 kW, which is a common engine size for a self-propelled inland navigation vessel in Europe, the price for the exhaust aftertreatment system alone are estimated at 130,000 €, which may be almost as much as the price of the engine itself.

14.4.5 Use of Alternative Energy Sources (Fuels) to Reduce Emissions

Broadening the fuel base for inland navigation may have multiple benefits:

- safety of supply vis-a-vis diminishing mineral oil reserves,
- reduction of fuel cost,
- reduction of GHG and pollutant emissions (Table 14.8).

In general, alternative transport fuels are neither specifically developed for inland navigation, nor do they provide inland navigation a technological lead. Some of the alternative transport fuels, such as liquid biofuels and synthetic liquid fuels, can be distributed with the existing supply infrastructure and be used with the current engine technology. Inland navigation can easily switch to these fuels, once they are available and affordable.

LNG is for the time being the most promising alternative fuel for inland navigation. The European Commission, CCNR and national governments support its deployment and are trying to make it commercially viable. A study on LNG as an alternative fuel for road transport and shipping, commissioned by the German Federal Ministry of Transport and Digital Infrastructure (Wurster & Heidt, 2014), came to the following conclusions:

- LNG increases fuel supply security through diversification of fuel base.
- LNG is a suitable alternative for today's diesel applications.
- LNG technology is readily available.
- Use of LNG for new and large vessels is technically possible and cost-efficient.
- LNG infrastructure for inland navigation has to be built from scratch; but sufficient infrastructure coverage is achievable by just a small number of bunkering stations along main waterways, in particular the river Rhine.
- Significant reduction of pollutant emissions is possible with LNG, but clean technologies for engines using gasoil are also available.
- Only limited reduction of GHG emissions can be achieved with fossil LNG and today's technology.
- Significant emission reductions will only be possible with synthetic methane produced with renewable energy; but competition with other in energy needs exists.

Table 14.8 Overview of alternative energy sources for inland navigation

Principal energy source	Example	Explanations
Methane	Liquefied natural gas (LNG)	Inland navigation as lead application for land based transport modes; suitable for large vessels and long distances; first freight vessels sailing in Europe and China
	Compressed natural gas (CNG)	Inland navigation as technology follower; suitable for small vessels and short distances; small number of passenger vessels sailing in Europe already for some time
	Biomethane	Possible replacement for natural gas
	Synthetic methane, produced from CO_2 using renewable energy, such as wind power (e-gas, power-to-gas)	Possible replacement for natural gas
Biofuels	First generation	Inland navigation as technology follower
	Second generation	
Synthetic fuels	Biomass to liquid (BTL)	Inland navigation as technology follower
	Coal to liquid (CTL)	
	Gas to liquid (GTL)	
Hydrogen	Fuel cells	Inland navigation as technology follower; experimental applications in Europe
	Combustion engines	Inland navigation as technology follower
Electricity	Rechargeable batteries	In combination with photovoltaic systems or as hybrid propulsion; experimental applications in Europe, mostly for passenger vessels
	Shore power supply	Standard practice in Europe for inland navigation vessels at berth

Source: (Fuels, 2011; CCNR, 2012)

LNG must become economically viable, if it is to replace gasoil as fuel in inland water transport. The price of LNG for small scale applications such as inland navigation depends on several factors, of which the commodity price is the most important one. The cost for supplying the LNG to the consumer needs also to be taken into account. This cost varies significantly, as the supply chain delivering the LNG from the large import terminals in the seaports to the inland navigation vessels can be complex at times. In Western Europe and in particular for the Rhine basin, the supply chain looks rather simple and short. At the import terminal the LNG is loaded on a truck and delivered directly to the consumer, which is appropriate for limited quantities of LNG delivered over short distances. For larger quantities and/or longer distances, supply chains involving transport on vessels come into play. A (inland navigation) bunker vessel may pick up the LNG at the import terminal and deliver it directly to the consumer in the same port or another port nearby. If bunkering of the inland navigation takes place farther away, an inland navigation tank vessel will bring the LNG to a terminal in the hinterland, where inland navigation vessels can refuel. For the Rhine, which is a rather short waterway and where inland navigation vessels frequently visit the large seaports of Rotterdam and Antwerp, both hosting LNG import terminals, the cost for supplying LNG can be expected to be low, when compared to the Danube or waterways in the USA and in China, which are much longer.

LNG technology for shipping applications is readily available, because for many years LNG has been used as fuel on maritime vessels. Nevertheless, bringing LNG on board inland navigation vessels requires substantial investment. For a vessel with a single engine of some 1,000 kW, the additional investment cost are estimated at some 500,000 €, compared to a vessel using gasoil and fulfilling the same stringent emission requirements. Therefore, LNG must be sufficiently cheaper than gasoil to compensate for the additional investment cost. Assuming a 20 % price difference between LNG and gasoil (in relationship to the energy content) and a medium oil price, the net present value of the total operational costs for a typical new vessel requiring an engine of the same size would be almost 100,000 €. For a smaller vessel, however, the net present value would become negative (Panteia, 2013). This example illustrates, that in general large inland navigation vessels with extensive operation times may have lower overall costs when using LNG rather than gasoil.

But will the additional investment cost be recouped within the timeframe prescribed by bank loans or business plans for investment projects in inland navigation? For the time being, this does not seem to be the case. Current LNG projects in European inland navigation rely on public subsidies, in particular on those provided by the "LNG Masterplan Rhine-Main-Danube", which is a large research project that has received more than 40 million Euros in EU funding. The project is based on the vision, that LNG will be transported by inland navigation vessels from LNG terminals in the seaports to LNG hubs in the hinterland. These hubs will serve as bunkering stations for LNG fuelled vessels and also as filling stations for trucks and railway engines. Some of these hubs may also be combined with degasification plants, where the gas is pumped into the national gas grid. When this vision becomes reality, a proper market for small scale LNG applications will be established and substantial synergies achieved, making LNG viable as a fuel for inland navigation.

The "LNG Masterplan Rhine-Main-Danube" is part of the European Commission's attempt to support the use of LNG as a fuel for inland navigation. In January 2013, the European Commission presented its clean fuel strategy, an ambitious package of measures to ensure the build-up of alternative fuel stations across Europe with common standards for their design and use. Among other measures, the Commission proposed that LNG refuelling stations be installed in all 139 maritime and inland ports on the Trans-European Core Network by 2020 and 2025 respectively (European Commission, 2013). However, EU Member States and

Parliament agreed that this proposal was unnecessarily ambitious and therefore decided that Member States shall ensure, through their national policy frameworks, that an appropriate number of refuelling points for LNG are put in place at maritime ports and inland ports by 2025 and 2030 respectively. This decision, reducing the number of refuelling points and extending the time frame, was probably a wise one, as

- less than ten refuelling stations along the Rhine and its major tributaries would be sufficient to cover the geographical area, where two thirds of all inland navigation activities in the EU take place;
- the uptake of LNG in European inland navigation is very slow, not least because of its very difficult economic situation, which has caused an almost complete stop to the building of new inland navigation cargo vessels in Europe.

LNG is often promoted on its perceived green credentials, namely low GHG and pollutant emissions. But do these hold for inland navigation? The answer seems to be no—which comes as a surprise, as methane has a much lower content of carbon and "dirty" components compared to gasoil. However, the lower carbon content translates into lower GHG emissions only if the combustion process of methane is close to perfect. Otherwise, methane slips occur, meaning methane escapes into the air, easily offsetting its lower carbon content, as the global warming potential of methane for a 100 year time horizon is approximately 28 to 34 times higher than that of CO_2 (Myhre et al., 2013), the GHG emitted from engines using gasoil. Indeed, Euromot, a European association of engine manufacturers representing the major manufacturers of inland navigation engines, warns not to underestimate the occurrence of methane slip. Current engine technology and lack of suitable catalytic converters for inland navigation applications makes it necessary for emission regulations to foresee higher HC emissions for LNG fuelled engines than for those using gasoil (Euromot, 2013). A study commissioned by the German Federal Ministry of Transport and Digital Infrastructure (Wurster & Heidt, 2014) examines the GHG emissions from well to propeller. The study shows, that at present no significant reduction of GHG emissions can be achieved by using LNG. This may change, if in the future improved engine technology minimizes methane slip. Assuming, that methane slip will be on the same level as for road transport, some 15 % decrease in GHG emission can be realized compared to current levels. GHG emissions would be dramatically lower, namely up to 90 %, if the methane is produced from CO_2 using renewable energy. This process is often called power-to-gas and the end product e-gas. This synthetic methane is currently being pioneered in the automobile sector by Audi.

The theoretical advantages of LNG regarding pollutant emissions also do not translate into real advantages, if the pollutant emissions from inland navigation vessels are already subject to stringent emission limits. This is already the case in the USA and can be expected for Europe. In both cases, LNG and gasoil, these stringent emission limits can only be met with exhaust aftertreatment, levelling out the initial advantage of LNG. Only in countries such as China, where there is no such regulation, the use of LNG actually reduces pollutant emissions.

Given no significant reduction of emissions to air and no compelling business case, when using LNG as fuel in inland navigation—why then is LNG seen as the most promising alternative fuel for inland navigation? First of all, LNG is just starting to be used in inland navigation; only a handful of LNG powered vessels are sailing in Europe and not many more in China. Secondly, GHG emissions will go down with evolving technology, reducing harmful methane slip to a minimum. Thirdly, investment cost will go down with more vessels using LNG and the price for small scale LNG supply will decrease with the establishment of efficient supply chains and reaping of synergies with other LNG applications. Finally, there does not seem to be another fuel easily available, which is suited for inland navigation and which could substitute gasoil, once the mineral oil supply dries up.

Nevertheless, much work needs to be done before LNG is really established as the alternative to gasoil. This is reflected in numerous projects, which are underway in Europe. For the information of all interested parties and as a basis for coordination, the CCNR has set up a data base (www.inland-navigation.org/observatory/innovation-technologies/lng/lng-database/) containing basic information on these projects.

14.4.6 Infrastructure Measures for the Reduction of Fuel Consumption and Emissions

Quality and dimension of waterway infrastructure influences strongly fuel consumption and emissions of inland navigation vessels. As explained in the section about basic facts and issues regarding emissions from inland navigation, power requirements and fuel consumption depend largely on available water depths. Therefore navigation channels should be designed and built to provide sufficient water depth, or more generally, optimal fairway parameters, for the vessels deemed most suitable for the particular waterway. However, the capacity of natural waterways depends basically on the water supply. In addition, building or enlarging waterway infrastructure is very costly. This often dictates incremental infrastructure development. Where fairway dimensions are insufficient, good maintenance becomes even more important. Good maintenance and further development of navigation channels are important infrastructure related measures that can reduce fuel consumption and emissions to air, but there is a range of other measures as Table 14.9 shows.

In the context of transport corridors and integration of inland navigation in logistical chains the height of bridges over the waterways is of particular importance, because it determines how many containers can be stacked on a vessel travelling on a particular waterway. The CCNR ensures that bridges over the

Table 14.9 Overview of important measures for the reduction of energy consumption and emissions to air relating to waterway infrastructure

Area of influence	Measures	Explanations
Waterway – Structures – Navigation channel	Consistent design and construction of fairway and navigation structures for most suitable vessels	Dimensions of vessels determine their specific energy consumption; fairway parameters determine optimum (most suitable) vessel dimensions for a particular waterway
	Ensuring sufficient height of bridges over waterways	High bridges increase carrying capacity, in particular of container vessels
	Minimization of required manoeuvres	Manoeuvres, such as sharp turns in tide bents, considerably increase fuel consumption
Waterway information	Provision of static or semi static information, in particular waterway parameters, conditions of riverbed and water level	Complete and accurate information supports journey planning by boat master for the most efficient use of waterway and vessel
	Timely provision of dynamic information, in particular traffic conditions	Complete and accurate information supports journey execution by boat master, in particular optimal speed
Waterway operation	Traffic management	Managing vessels in particular traffic situations for optimal vessel speed and best use of fairway
	Operation of hydraulic structures (lock management)	Preventing waiting times and vessel engines running idle

Source: CCNR (2012)

Rhine from Strasbourg to Rotterdam have a minimum height of 9 m, thus allowing for four or more layers of containers. This enables the largest vessels on the Rhine to carry up to 500 TEU, making container transport very efficient. Figure 14.6 shows that container transport on the Rhine, including transport of the containers to and from the port by truck, has a much lower transport cost than competing modes of transport. On other waterways, the cost advantage of inland water transport is less pronounced or may diminish altogether.

The cost for industry is lower when containers are carried on the Rhine and it is even more so for society. Figure 14.7 shows that container transport on the Rhine also has much lower external cost than carrying containers by road or rail. However, external cost for emissions to air may account for 90 % of total external cost of inland water transport, whereas for road and rail transport the share is only 25 % and 17 % respectively (PLANCO, 2007). Because of stringent emissions regulations, road transport's pollutant emissions may have decreased considerably since this data was compiled.

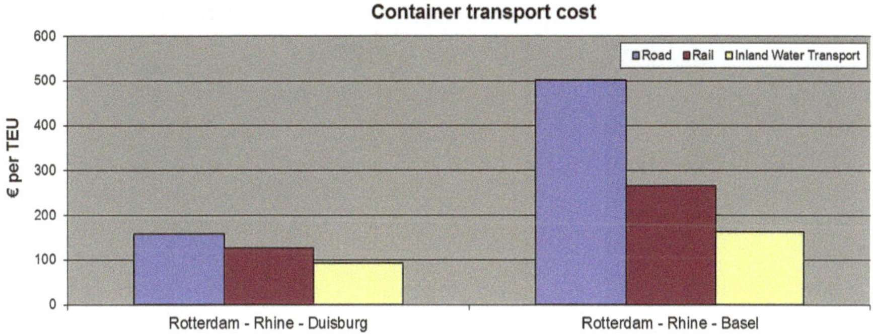

Fig. 14.6 Comparison of container transport cost for specific transport routes. *Source*: PLANCO (2007)

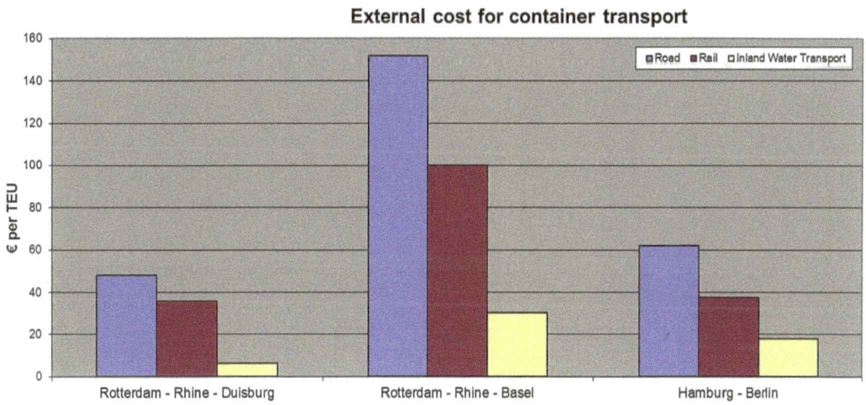

Fig. 14.7 Comparison of external cost of container transport for specific transport routes. *Source*: PLANCO (2007)

As Table 14.9 shows, provision and exchange of information is the basis for journey planning with the objective of most efficient use of waterways and vessels. In Europe and in particular on the Rhine, many information services for traffic and transport management have been implemented under the umbrella of the River Information Services (RIS). "RIS means the harmonized information services to support traffic and transport management in inland navigation, including interfaces with other transport modes. RIS aims at contributing to a safe and efficient transport process and utilizing the inland waterways to its fullest extent" (CCNR, 2011). The objectives of RIS include explicitly the reduction of fuel consumption and emissions to air. The availability of RIS can thus be seen as a defining characteristic for a green corridor, which includes inland navigation.

14.4.7 Transport Management Measures for the Reduction of Fuel Consumption and Emissions

Measures related to transport management can contribute considerably to the reduction of fuel consumption and emissions in inland water transport. Even though these measures are often taken with the objective of reducing cost or streamlining logistics operation, they generally result in lower fuel consumption and reduced emissions to air.

The large seaports of Antwerp and Rotterdam want to increase the modal share of inland navigation in their container hinterland transport to more than 40 %. In order to reach this goal, inland navigation vessels visiting the ports must seamlessly fit into the operation of the terminals, which in turn is largely dependent on the arrival and departure of the large seagoing vessels. To ensure this seamless fit, the Port of Antwerp has developed the so-called Barge Traffic System (BTS), which allows the operators of the inland navigation vessels to coordinate with the port authority and the terminal operators the visits to the port and calls on the different terminals. The vessels are equipped with Automatic Identification System (AIS) transponders, allowing for verification of the vessels' position and adjustments of their visits' schedule in real time. BTS's objectives are the smoothing of overall operation of the port, reduction of waiting times for inland navigation vessels and making hinterland services by inland water transport more attractive (Antwerp, 2012). NEXTLOGIC, the inland navigation vessel management system for container transport at the Port of Rotterdam, seems to have similar objectives, in particular improving operational efficiency, such as better deployment of port infrastructure and inland navigation vessels, improving reliability and predictability, in particular better alignment of supply and demand of inland navigation transport capacity, and enabling and stimulating growth of the inland navigation sector (NEXTLOGIC, 2014). These two examples illustrate, how transport management, by improving the operation at the interface of inland and maritime navigation, can contribute to the reduction inland navigation's energy consumption and emissions to air. Other important measures available for transport management are given in Table 14.10.

Larger vessels with greater carrying capacity have generally lower energy consumption per tkm, as already indicated in Fig. 14.4. Lower energy consumption leads to proportionally lower emissions to air, if all other aspects, which determine the emissions, remain the same. This imperative is demonstrated by Table 14.11, which shows the specific GHG emissions for European standard vessels under equal conditions. Thus, pooling loads and using one fairly large vessel instead of two or more smaller ones, can easily half the emissions. Of course, pooling of cargo is sometimes impossible, not least due to the fact that there is just not enough cargo to fill one vessel on a specific route at a given moment in time. However, where pooling can be achieved, reduction of emissions is rather dramatic.

Based on the fact, that larger vessels have in general lower specific emissions, some analysts suggest that the greatest potential for reducing fuel consumption and

Table 14.10 Overview of important measures for the reduction of energy consumption and emissions to air relating to transport management

Measures	Explanation
Improving operation at interface with other transport modes	Avoiding waiting times for vessels or unnecessary calls on terminals, in particular in sea ports; employment of ICT for complex real time planning processes
Reduction of vessel voyages without cargo	For specialized vessels or certain cargoes, in particular liquid cargo, empty return voyages are unavoidable; increased market transparency may support finding suitable cargo for otherwise empty vessel
Pooling cargo in order to fully use carrying capacity of vessels	High load factor reduces specific fuel consumption and emissions to air; carrying capacity (vessel draught) limited by waterway conditions, in particular water levels
Employing larger vessels	Larger vessels in principle lower specific fuel consumption and emissions to air, but physical and other limits for vessel size may be an issue

Table 14.11 Comparison of carrying capacity and specific CO_2 emissions of inland navigation vessels under equal conditions (diesel propulsion; maximum loaded draught: 2.5 m; water depth: 5 m; vessel speed: 12 km/h)

Vessel type	Carrying capacity in tons at draught of 2.5 m	CO_2 emissions (g/tkm)
Peniche	366	47.1
Gustav Koenigs	935	31.3
Johann Welker	1,272	17.6
Large Rhine vessel	1,900	6.4
Jowi class container vessel	3,335	7.7

Source: Zöllner (2009)

CO_2 emissions from inland navigation may lie in an increase of the average load carrying capacity (size) of the vessels (CCNR, 2012; Hazeldine, Pridmore, & Nelissen, 2009; Schweighofer, 2011). Data compiled by the Secretariat of the CCNR and presented in Fig. 14.8 shows, that between 1985 and 2010, the average load carrying capacity of inland vessels in Western Europe grew by approximately 20 tons each year for dry cargo vessels and that of tankers by approx. 25 tons, an average increase of some 1.5 % per year. This increase in average vessel loading capacity with the accompanying decrease in specific fuel consumption and emissions to air may—at least in principle—compensate for past and future increase in transport volume.

Infrastructure dimensions however, put a physical limit on maximum vessel size. Other constraints may come from wear and tear on the infrastructure or negative environmental impacts from large vessels, such as increased pressure on the aquatic environment. Growth in average vessel size is therefore only to be expected on

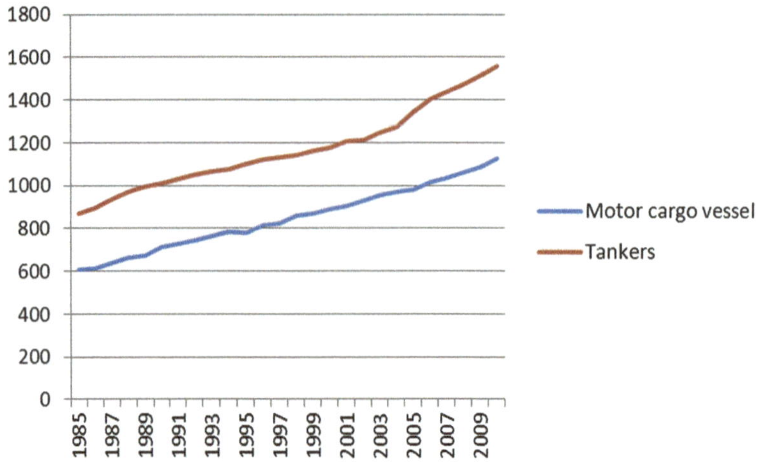

Fig. 14.8 Average load carrying capacity (in tons) of motor vessels of the West European fleet. *Source*: CCNR (2012)

those waterways, where many vessels are significantly smaller than those seen as optimal for that waterway. This is certainly the case for the Rhine: the average vessel carrying capacity of some 1,200 tons, taken from Fig. 14.8, equals roughly half the maximum carrying capacity of some 2,500 tons of the large Rhine vessel mentioned Table 14.11, which may be the optimum vessel size for the Rhine and its major tributaries.

IFEU and INFRAS, in their study on emissions to air from inland navigation in Germany, provide input data for an emission scenario for the year 2030 (Knörr et al., 2013). This input data does not take into account the potential for emission reductions from growth in average vessel size. Instead, this reduction potential is questioned on the grounds that vessels with a carrying capacity of more than 1,500 tons have a lower load factor than smaller vessels, which prevents a decrease of the specific energy consumption. IFEU and INFRAS support their reasoning with data on vessel load factors in the years 2010, 2011 and 2012. This raises the question, whether the analysts, who see in the increase of average vessel size perhaps the greatest potential for emission reductions, are wrong. Obviously, load factors are important; the specific emissions will only decrease with larger vessels, if these vessels carry a bigger load than the smaller vessels would have. This will only be the case, if there is enough transport demand to sufficiently fill the larger vessels. Incidentally, the years 2010–2012 coincided in Western Europe, including Germany, with a situation, where transport demand or cargo volume was roughly 10 % lower than just 5 years earlier, but available fleet capacity almost 10 % higher (CCNR, 2013a). This was due to the economic crisis on the one hand and a structural overcapacity of the fleet, which had been built up over the previous 10 years on the other. If IFEU and INFRAS had performed their analysis for the years leading up to 2008, the year the economic crisis took hold in Europe, their conclusions might have been different and they would acknowledged—at least to a certain extent—the emission reduction potential of an increase in the average vessel size. Incidentally, PLANCO did so in 2007, when estimating fuel consumption and emissions for inland navigation for the year 2025 (PLANCO, 2007).

14.5 Recommendations for Policy Makers and Decision Makers in the Inland Navigation Sector

This section provides recommendations and proposals for policy and decision makers, who wish to support the further greening of inland navigation. The recommendations focus on the reduction of GHG and pollutant emissions from inland navigation, as these emissions stand in the way of inland navigation being a truly green mode of transport. An intensive discussion of recommendations and possible actions aiming to reduce GHG emissions from inland navigation has already taken place within the CCNR (CCNR, 2012). Of those, the ones that seem especially pertinent in the context of this book, are included in this section.

There are significant knowledge and data gaps regarding inland navigation, which hinders policy development as well as decision making and often makes impact assessments meaningless. Therefore, a number of recommendations aim at closing these gaps. Measures or policies, developed for other modes of transport or for inland navigation in certain countries may also be useful for inland navigation on a European level. A number of recommendations are thus derived from other modes of transport or from certain countries. Table 14.12 contains these recommendations, together with a short description of the expected benefits. As all of the recommendations derive from a certain context, information is given on previous achievements upon which the proposed measures can build.

None of the recommendations given in Table 14.12 are politically sensitive or prohibitively costly to implement. Why then have they not been taken up yet? Neither industry nor politics alone can take up these recommendations and because inland navigation in Europe is in principal an international undertaking, the recommendations require cross border cooperation, strong leadership and the willingness of politicians, administration and industry in particular to work together. So far, this has not happened often, explaining the lack of results.

As shown in Sect. 14.2.2 of this chapter, legal requirements are the main driver for the reduction of transport's emissions to air. Therefore, the final recommendation concerns the emission regulations for engines used for inland navigation. In September 2014, the European Commission presented its proposal for the future emission regulations for non-road mobile machinery, which also includes inland navigation vessels (European Commission, 2014). These regulations concern many aspects of emission control techniques and type testing, but probably most important and certainly most discussed are the proposed emission limits. The limit values proposed for propulsion engines installed on inland navigation vessels are reproduced in Table 14.13.

Table 14.12 Recommendations for policy makers as well as decision makers in the inland navigation sector with the objective of reducing the sector's energy consumption and emissions to air

Recommendation	Possible benefits	Achievements so far
Measures to reduce knowledge gaps		
Develop a decision support system for European inland navigation policy decision making	Combining data on economy, ecology and technology with a comprehensive model of inland navigation to analyse the impact of policy measures, legal acts, economic developments etc.	Data and models for certain aspects and with limited geographical scope are available
Determine fuel consumption as well as pollutant and GHG emissions from inland navigation, both in absolute terms and relative to traffic and transport volume (tkm), using top down and bottom up approaches	Gain knowledge of fuel consumption as a basis for – emissions factors, emission calculators – voluntary or compulsory information by the shipping industry on greenhouse gas emissions – obligatory emission reporting of countries, such as specified in the Kyoto Protocol – formulation of political objectives – verification of inland navigation's "green" image	Numerous studies are available, often with different (geographical) scopes, deviating results/conclusions; limited statistics on fuel consumption are available yet often include unrealistic or implausible data/assumptions
Determine average values for energy consumption of inland navigation vessels	Supplement EN 16258:2012 "Methodology for calculation and declaration of energy consumption and GHG emissions of transport services (freight and passengers)" with standard values for energy consumption in order to ease the use of the methodology and to support the implementation of calculation and declaration of energy consumption and GHG emissions of inland navigation	Average values published in France by ADEME for French inland navigation
Prepare scenarios for the future developments of GHG and pollutant emissions from inland navigation	Build an effective tool for the development of climate protection and environmental objectives and of strategies, e.g. for the fuels to be used by inland shipping in future, or of programmes to promote climate-friendly inland shipping	Very basic scenarios used in limited number of studies already exist

(continued)

Table 14.12 (continued)

Recommendation	Possible benefits	Achievements so far
Examine CO_2 reduction potential through the use of LNG and other alternative energy sources (fuels) in inland navigation	Enable a focus on energy sources (fuels) that could substantially contribute towards reducing CO_2 emissions from inland navigation Avoid research in ineffective/counterproductive areas	A limited number of studies are available, albeit based on insufficient data or knowledge and never specific to inland navigation applications
Determine reduction in fuel consumption and GHG emissions of inland navigation vessels as a result of increased average carrying capacity of vessels, both in absolute terms and in relation to traffic and transport volume (tkm)	Increase in average capacity (size) is one of the most important factors in reducing fuel consumption and emissions to air; thus such a study would allow an estimation of the costs of improvements and the necessity of other avenues of reduction	Very few studies available, often with different (geographical) scopes, deviating results/conclusions
Measures to reduce fuel consumption and emissions to air		
Examine in a general manner the mandatory introduction of the Energy Efficiency Design Index (EEDI) for new inland navigation vessels or another energy classification	Develop a mandatory basis for determining whether the design of a new vessel is energetically favourable; provides ship-owners with a simple and transparent means of benchmarking	EEDI developed and introduced for maritime navigation; no significant work undertaken yet for inland navigation
Examine in a general manner the mandatory introduction of the Ship Energy Efficiency Management Plan (SEEMP) for all inland navigation vessels, possibly using the Energy Efficiency Operational Indicator (EEOI)	Develop a reliable basis for determining whether a vessel is operated in accordance with energy efficiency standards; provides ship-owners with a simple and transparent means of benchmarking	SEEMP developed and introduced for maritime navigation; no significant work undertaken yet for inland navigation
User-friendly provision of comprehensive relevant information on the main aspects of GHG and pollutant emissions from inland navigation and the reduction of these emissions, for example as apps for smart phones or computers	Overcome one of the main barriers to the implementation of measures by the shipping industry and other bodies by providing relevant information in a user-friendly manner	Individual attempts by governmental and non-governmental organizations; mostly as brochures
Europe-wide introduction of a common environmental label for inland navigation, either identical or similar to the Dutch "Green Award"	Environmental labels would support the adoption of measures to reduce emissions and protect the environment by the inland navigation industry	Different labels, which are not compatible, introduced in different countries, with a growing number of participating companies and vessels, in particular from Dutch and Belgian owners

(continued)

Table 14.12 (continued)

Recommendation	Possible benefits	Achievements so far
Europe-wide introduction of a programme to promote energy-efficient operation of inland vessels, similar to the Dutch "Smart Steaming" programme	Promotion of energy-efficient operation of inland vessels as a key element for the reduction of greenhouse gas emissions and protection of the environment	Programme successful in the Netherlands but not implemented elsewhere
Develop measures, not involving the building or equipping or operation of vessels, aimed at reducing greenhouse gas emissions from inland navigation	Reduces greenhouse gas emissions from the overall system	No significant work has yet been undertaken on GHG emissions from construction and operation of inland navigation infrastructure
Develop and implement a medium to long term Europe-wide programme, that is largely independent of public subsidies, to financially support those GHG and pollutant emissions reduction measures that either increase cost for vessel owners or require a substantial upfront investment	Stable support mechanism for financing of emission reduction measures, not dependant on political decisions	Very limited financial support by a small number of national programmes; programmes financed by contributions known from maritime navigation, such as the Norwegian NO_X fund; similar programme in preparation by Port of Rotterdam for inland navigation
Develop quantitative objectives for reducing GHG and pollutant emissions from inland navigation	Aligns political, economic, technical and other processes; creates a common foundation for a large number of compatible activities, hence minimizing uncertainties; contributes towards maintaining inland navigation's "green" image	Many objectives set so far on the EU and national levels yet not sufficiently specific to serve as a basis for concrete actions

Table 14.13 Future limits for pollutant emissions from propulsion engines installed on inland navigation vessels

Engine category	37–75 kW	75–130 kW	130–300 kW	300–1,000 kW	>1,000 kW
Emission stage[a]	IIIb	IIIb	IIIb	IVb	V
Year of application[b]	2019	2019	2019	2020	2020
CO (g/kWh)	5.00	5.00	3.50	3.50	3.50
HC (g/kWh)	4.70	5.40	1.00	0.19	0.19
NO_X (g/kWh)			2.10	1.20	0.40
PM (g/kWh)	0.30	0.14	0.11	0.02	0.01
PN	–	–	–	1×10^{12}	1×10^{12}
A	6.00				

Source: European Commission (2014)

[a]The proposed emission limits for the different engine categories vary considerably and can be attributed to different emission stages. Each stage stands for a certain technology mix or a certain level of environmental ambition

[b]These dates assume a coming into force of the proposed regulation in 2016

The European Commission's proposal contains two novelties for inland navigation emission regulations, namely a limit value for particulate number (PN) and a factor (A) for calculating hydro-carbon limits for fully and partially gas fuelled engines.[1] The particulate number can be seen as a tool to limit ultrafine particulates from diesel engines, which have very damaging effects on human health. The factor A is used to describe the maximum methane emissions of engines that use natural gas as a fuel. The value of 6 for this factor basically means that the methane slip of an engine running on methane may be up to 6 g/kWh.

Also remarkable are the large differences between the limit values for the different engine categories. They span three emission stages (IIIb, IVb, V) instead of one or two, as is the case in other emission regimes. The European Commission argues that the much more ambitious limit values for the large engines are justified, because those engines are installed in large vessels, which create significantly higher external costs than smaller vessels. The reason for that being in particular the different operating regimes for the vessels, with larger vessels often sailing round the clock, every day of the year. In contrast, the benefit-cost-ratio for very ambitious limit values for engines on small vessels would be so low that they could hardly be justified.

Even though the European Commission's proposal is largely based on a thorough impact assessment study, it was nevertheless strongly criticized by both the engine manufacturers and the ship-owners (EBU & ESO, 2014; Euromot, 2013; Panteia, 2013). They cited methodological shortcomings of the study as well as the extraordinarily high cost for the needed exhaust aftertreatment equipment. Therefore, they recommend the adoption of limit values in line with those already adopted by IMO and the EPA of the USA instead. Doing so, would allow the use of engines that are already under development for markets, which are much larger than the European one and would in particular avoid additional R&D cost.

In order to assess these arguments, the additional investment cost, which is the cost difference between the cost of an engine of a higher emission stage and an engine of today's emission stage, is compared in the following. According to the aforementioned impact assessment study, the additional investment cost for engines fulfilling the stage IVb emission limits is some 1.6 times higher and the cost for fulfilling stage V emission limits even some 3.5 times higher than those for stage IIIb. When R&D cost are included, the differences become even more dramatic. For certain large engines fulfilling emission limits according to stage V, inclusion of R&D cost would increase the additional cost almost fivefold. On the other hand, R&D cost for engines fulfilling stage IIIb emission limits are insignificant as those engines are already under development for maritime vessels and USA inland

[1] With the factor A the HC limit for fully and partially gaseous fuelled engines is calculated according to the formula $HC = 0.19 + (1.5 \times A \times GER)$, where GER is the average gas energy ratio over the appropriate cycle. If the calculated limit for HC exceeds the value of $0.19 + A$ the limit for HC shall be set to $0.19 + A$.

navigation vessels. For stage IVb engines this is not the case and some R&D will be necessary before they can become available.

How will the ship-owners cope with the additional cost? The European Commission estimates that its proposal would cause additional cost for ship-owners of some 280 million Euros over 20 years, compared to a scenario where the emission regulations remain unchanged. However, European inland navigation has for a number of years been in a very difficult economic situation, which shows no sign of significant improvement (CCNR, European Commission, & Panteia, 2014). Therefore, it is doubtful that the European inland navigation sector can bear these cost. Instead, ship-owners may be forced to adopt strategies to avoid the additional cost for installing cleaner engines on their vessels. These strategies may consist of

- advancing investment, i.e. finish the planned building of new vessels or replace engines before the deadline;
- the postponement of investments, i.e. operate existing vessels and engines as long as possible and have them repeatedly repaired to avoid having to buy new engines;
- shifting to smaller engines that are subject to less stringent emission regulations and therefore require less expensive exhaust aftertreatment systems. They might do this by opting for propulsion system with two or three engines instead of a single-engine.

From the point of view of the regulator, these avoidance strategies are undesirable as they make a greening of inland navigation impossible and also prevent engine manufacturers from developing the required engines. In short, the proposed regulation may not achieve its objectives and could even lead to opposite outcomes, because its associated costs force the ship-owners to avoid investing in cleaner engines.

Would emission limits as set by the EPA lead to a significant reduction of pollutant emissions from inland navigation and make the emission limits proposed with the new EU regulation unnecessary? A comparison of the emission limits proposed by the European Commission (EU new) with those of current regulations in Europe (EU IIIa) and the EPA regulations in the USA is given in Table 14.14.

Table 14.14 Comparison of limits for pollutant emissions from different emission regulations for inland navigation propulsion engines

Regulation	EU (today)	EPA (2017)	EU (proposed)	
Engine category	V 2:1	600–1,400 kW	300–1,000 kW	>1,000 kW
NO_X	6.8 g/kWh[a] (100 %)	1.8 g/kWh (26 %)	1.20 g/kWh (18 %)	0.40 g/kWh (6 %)
PM	0.27 g/kWh (100 %)	0.04 g/kWh (15 %)	0.02 g/kWh (7 %)	0.01 g/kWh (4 %)

[a]The emission limit for HC and NO_X combined is 7.8 g/kWh. To make the emission limits comparable, an emission limit of 1.0 g/kWh is assumed for HC

It shows that the emission limits proposed by the European Commission for this category of engines are only between 8 and 22 % lower than those of the future EPA regulation, when compared to the current EU regulation. Therefore, engine manufacturers and ship-owners seem to be right, when they argue that following EPA emission limits will reduce emissions considerably at much lower cost than the proposed EU regulation.

The proposal of the European Commission will be discussed in the European Parliament and Council. Not least because of the strong opposition by concerned industry and the member states' recognition of the difficult economic situation of the inland navigation sector, an alternative way forward may be called for. This alternative must ensure that significant progress in the greening of the inland navigation fleet will be achieved, while also accounting for the limited financial potential of the sector. The alternative may comprise the following elements:

- Introducing EU emissions regulations based on EPA/IMO limit values, as the required engines will soon be developed and on the market at much lower cost; therefore more engines will be sold to the European inland navigation sector early on and environmental benefits will kick in early;
- Introducing emission requirements for existing engines beyond their normal operating life to avoid extended operation of older engines and also achieve environmental benefits also from existing vessels;
- Introducing a review clause in legislation demanding the European Commission to look into more stringent emission requirements, including methane slip, over the next 5 years.

These actions as well as those described in Table 14.12 may be supported by a permanent forum encompassing representatives and experts from all concerned stakeholder groups as well as the national authorities and international organizations governing inland navigation in Europe. This forum, which could be called the European Sustainable Inland Navigation Forum, would not only coordinate actions, but also ensure the commitment of the participants to tact.

References

ADEME. (2006). *Etude sur le niveau de consommation de carburant des unites fluviales francaises*. ADEME, VNF, TL&A.

ADEME. (2012). *CO_2 information for transport services—Application of Article L. 1431-3 of the French transport code—Methodological guide*. SD & Energy, Ministry of Ecology. Paris: Ministère de l'Écologie, du Développement durable et de l'Énergie.

Antwerp, P. O. (2012). *Annual report 2012*. Antwerp, Belgium.

Bonnerjee, S., Cann, A., Koethe, H., Lammie, D., Lieven, G., Muskatirovic, J., et al. (2009). *Inland waterborne transport: Connecting countries (Insights)*. Paris: United Nations Educational, Scientific and Cultural Organization.

Buhaug, Ø., Corbett, J. J., Endresen, Ø., Eyring, V., Faber, J., Hanayama, S., et al. (2009). *Second IMO GHG Study 2009*. London: International Maritime Organization (IMO).

CBS. (2011). *Air pollution, actual emissions by mobile sources*. Retrieved December 14, 2011, from http://statline.cbs.nl/StatWeb/publication/?DM=SLEN&PA=80448eng&D1=a&D2=8,13,17-20&D3=l&LA=EN&HDR=T&STB=G1,G2&VW=T

CCNR. (2003). *Resolution 2003-II-27*. Strasbourg, France: Central Commission for the Navigation of the Rhine.

CCNR. (2011). *Guidelines and recommendations for River Information Services, Edition 3.0*. Strasbourg, France: CCNR.

CCNR. (2012). *Possibilities for reducing fuel consumption and greenhouse gas emissions from inland navigation*. Strasbourg, France: Central Commission for the Navigation of the Rhine.

CCNR. (2013a). *The inland navigation market in 2012*. Market Observation No. 17.

CCNR. (2013b). *Vision 2018*. Retrieved May 30, 2014, from http://www.vision-2018.org/pdf/VisionEN.pdf

CCNR, European Commission, & Panteia. (2014). *The inland navigation market in 2013 and perspective for 2014/2015*. Market Observation. Strasbourg, France.

CE_Delft, Infras, & Fraunhofer ISI. (2011). *External costs of transport in Europe: Update study for 2008*. Delft, Netherlands: CE Delft.

den Boer, E., Otten, M., & van Essen, H. (2011). *STREAM International Freight 2011: Comparison of various transport modes on a EU scale with the STREAM database*. Delft, Netherlands: CE Delft.

EPA. (2012). *Federal marine compression-ignition (CI) engines: Exhaust emission standards*. Emission Standards Reference Guide, 2014. Retrieved December 18, 2013, from http://www.epa.gov/otaq/standards/nonroad/marineci.htm

EU. (2004). Directive 2004/ 26/EC of the European Parliament and of the Council of 21 April 2004 amending Directive 97/68/EC on the approximation of the laws of the Member States relating to measures against the emission of gaseous and particulate pollutants from internal combustion engines to be installed in non-road mobile machinery.

EU. (2013). *Communication from the Commission to the European Parliament, the Council, the European Economic and Social Committee and the Committee of the regions: Towards quality inland waterway transport—NAIADES II*. Brussels, Belgium: European Commission.

Euromot. (2013). *Euromot comments on the Panteia Report of February 28th 2013*.

European Commission. (2013). *EU launches clean fuel strategy*. Brussels, Belgium.

European Commission. (2014). *Regulation of the European Parliament and of the Council on requirements relating to emission limits and type-approval for internal combustion engines for non-road mobile machinery*. Brussels, Belgium: European Commission.

Fuels. (2011). *Future transport fuels: European expert group on future transport fuels*.

Hazeldine, T., Pridmore, A., & Nelissen, D. (2009). *Technical options to reduce GHG for non-road transport modes*. Paper 3 produced as part of contract ENV.C.3/SER/2008/0053 between European Commission Directorate-General Environment and AEA Technology plc. EU Transport GHG: Routes to 2050? Brussels, Belgium: European Commission.

Hill, N., Branningan, C., Wynn, D., Milnes, R., Van Essen, H., Den Boer, E., et al. (2012). *The role of GHG emissions from infrastructure construction, vehicle manufacturing, and ELVs in overall transport sector emissions*. Brussels, Belgium: European Commission.

IFEU. (2011). *Ecological transport information tool for worldwide transports*. IFEU Heidelberg, Öko-Institut, IVE/RMCON.

IFEU, Öko-Institut, & IVE/RMCON. (2011). *Ecological transport information tool for worldwide transports, methodology and data update*. Retrieved October 13, 2012, from http://www.ecotransit.org/download/ecotransit_background_report.pdf

Knörr, W., Heidt, C., Schmiedt, M., & Notter, B. (2013). *Aktualisierung der Emissionsberechnung für die Binnenschifffahrt und Übertragung der Daten in TREMOD*. Heidelberg, ifeu—Institut für Energie- und Umweltforschung Heidelberg GmbH, INFRAS—Forschung und Beratung

Kruse, C., Protopapas, A., Olson, L. E., & Bierling, D. H. (2009). *Modal comparison of domestic freight transportation—Effects on the general public*. College Station, TX: U.S. Maritime

Administration and the National Waterways Foundation; Texas Transportation Institute, Center for Ports & Waterways, The Texas A&M University System.

Mottschall, M., & Bergmann, T. (2013). Treibhausgas-Emissionen durch Infrastruktur und Fahrzeuge des Straßen-, Schienen- und Luftverkehrs sowie der Binnenschifffahrt in Deutschland—Arbeitspaket 4 des Projektes. *Weiterentwicklung des Analyseinstrumentes Renewbility.*

Myhre, G., Schindell, D., Brèon, F.-M., Collis, W., Fuglestvedt, J., Huang, J., et al. (2013). Anthropogenic and natural radiative forcing. In T. F.Stocker, D. Qin, G.-K. Plattner, M. Tignor, S. K. Allen, J. Baschung, A. Nauels, Y. Xia, V. Bex, & P. M. Midgley (Eds.), Climate change 2013: The physical science basis. Contribution of Working Group I to the Fifth Assessment Report of the Intergovernmental Panel on Climate Change. Cambridge, England: Cambridge University Press.

NEXTLOGIC. (2014). *Towards reliable and competitive inland container shipping.* H. Rotterdam.

OECD/ITF. (2013). *Global transport trends in perspective. Statistics brief—Trends in the Transport Sector.* Paris: International Transport Forum, OECD.

Panteia. (2013). *Contribution to impact assessment of measures for reducing emissions of inland navigation.* Retrieved from http://ec.europa.eu/transport/modes/inland/studies/doc/2013-06-03-contribution-to-impact-assessment-of-measures-for-reducing-emissions-of-inland-navigation.pdf

Pauli, G. (2010). Sustainable transport: A case study of Rhine navigation. *Natural Resources Forum, 34*, 18.

Pauli, G., & Schweighofer, J. (2008). Die Entwicklung der Abgasemissionen in der Binnenschifffahrt. *Binnenschifffahrt—ZfB* (9), 7.

PLANCO. (2007). *Verkehrswirtschaftlicher und ökologischer Vergleich der Verkehrsträger Straße, Bahn und Wasserstraße; Schlussbericht.* Magdeburg, PLANCO Consulting GmbH, Essen, Bundesanstalt für Gewässerkunde, Koblenz.

Renner, V., & Bialonski W. (2004). Technische und wirstschaftliche konzepte für flußangepaßte Binnenschiffe. Duisburg, Versuchsanstalt für Binnenschiffbau e.V.

Riebe, E. (2014). *Third progress report Rhine-Alpine core network corridor.*

Schweighofer, J. (2011). Parallel Workshop 2—Hydrodynamic measures to reduce the CO_2 emissions from inland navigation. In *Inland Navigation CO_2 emissions - How to measure them? How to reduce them?* Strasbourg, France.

TNO. (2010). *Methodologies for estimating shipping emissions in the Netherlands. A documentation of currently used emission factors and related activity data* (BOP Report). Netherlands Research Program on Particulate Matter.

Van Essen, H., & den Boer, E. (2012). *Assessment of external costs of inland waterway transport in the Marco Polo Calculator.* Delft.

WorldBank. (2009). *Sustainable development of inland waterway transport in China* (p. 105). Washington, DC: World Bank.

Wurster, R., & Heidt, C. (2014). *LNG als Alternativkraftstoff für den Antrieb von Schiffen und schweren Nutzfahrzeugen.*

Zöllner, J. (2009). *Strömungstechnische Möglichkeiten zur Reduzierung des Kraftstoffverbrauchs und der CO_2-emissionen von Binnenschiffen.* Bonn, Germany: ZKR Kongress, Rheinschifffahrt und Klimawandel.

Chapter 15
Directions for Further Research

Atle Minsaas and Harilaos N. Psaraftis

Abstract Green transportation logistics is an area that combines the following: (a) it is relatively new in terms of research carried out thus far, (b) it has become increasingly important for both industry and society, and (c) it is rich in topics for further research, both basic and applied. In this final chapter of this book we discuss directions for further research in this area. We do so by taking stock of (1) related recommendations of project SuperGreen, and (2) related activities mainly in European research. Links between research and policy-making as two activities that should go hand in hand are also discussed.

Abbreviations

ALICE	European Technology Platform on Logistics
CGSST	Green Corridor, Sustainable Surface Transport
CNG	Compressed natural gas
CO_2	Carbon dioxide
EC	European Commission
ERRAC	European Rail Research Advisory Council
ERTMS	European Rail Traffic Management System
ERTRAC	European Road Transport Research Advisory Council
ETP	European Technology Platform
EU	European Union
GHG	Green house gas
GNSS	Global Navigation Satellite System
GUI	Graphical user interface

A. Minsaas (✉)
Norwegian Marine Technology Research Institute (Marintek),
POB 4125 Valentinlyst, NO-7450 Trondheim, Norway
e-mail: Atle.Minsaas@marintek.sintef.no

H.N. Psaraftis
Department of Transport, Technical University of Denmark, Bygningstorvet 1,
Building 115, Room 018 2800 Kgs. Lyngby, Denmark
e-mail: hnpsar@transport.dtu.Dk

H.N. Psaraftis (ed.), *Green Transportation Logistics*, International
Series in Operations Research & Management Science 226,
DOI 10.1007/978-3-319-17175-3_15

ICT	Information and communication technologies
ICZM	Integrated coastal zone management
ILO	International Labor Organization
IMO	International Maritime Organization
ISO	International Organization for Standardization
KPI	Key performance indicator
LEIT	Leadership in enabling and industrial technologies
LNG	Liquefied natural gas
MSP	Maritime spatial planning
NMP	Nanotechnologies, advanced materials and advanced manufacturing and processing
OECD	Organization for Economic Cooperation and Development
RIS	River Information Services
R&D	Research and Development
SESAR	Single European Sky Air Traffic Management Research
SME	Small to Medium Enterprise
SRA	Strategic research agenda
SSS	Short sea shipping
TEN-T	Trans European Transport Network
UK	United Kingdom
UN	United Nations
UNCTAD	United Nations Conference on Trade and Development

15.1 Introduction

The previous chapters of this book have hopefully provided the reader with a flavor of the main issues in green transportation logistics. We believe that material in these chapters supports the general conclusion that this is an area that is both relatively new in terms of research carried out thus far, and increasingly important for both industry and society today and in the future. In this final chapter of this book we discuss directions for further research in this area. We shall do so by taking stock of (1) related recommendations of project SuperGreen, and (2) related activities mainly in European R&D which we think are interesting and important. The reference to Europe is mainly dictated by the European orientation of much of the research described in this book (project SuperGreen for instance), however we believe that this causes no loss of generality, as many of the relevant issues are also present in other parts of the world. The link between research and policy-making as two activities that should go hand by hand is also discussed.

Major policy documents such as the EU 2011 White Paper on Transport elaborate on a number of challenges faced by both industry and society in the years ahead, while also laying out corresponding strategies and remedial goals. Known and interrelated challenges such as increasing congestion, growth in trade volumes, energy consumption, and emissions are to be met by improved traffic management systems, infrastructure development, and developments within ICT and other technologies. These are key areas of focus if the target of reducing emissions from transportation by 60 % within 2050 (as compared to 1990 levels), and shifting 30 % of cargo transported beyond 300 km from road to rail and waterborne by 2030 (and more than 50 % by the year 2050), are to be achieved.

In project SuperGreen, a dedicated work package provided recommendations for further R&D on green corridors (see Minsaas et al. (2012) for more details). When summarizing the results from the analysis, the following generic points were noted:

- There must be an increased focus on rectifying the evident lack of data and reliable tools enabling proper benchmarking exercises within the transportation domain.
- A further strengthening of efforts securing integration and implementation of harmonized ICT solutions, also developing new ones (e.g. Single Window concepts).
- Increased focus on corridor and case-by-case specific analysis, both in terms of requirements and tailored solutions.
- Performance of impact studies for assessing potential environmental and cost savings when introducing new ICT and technology solutions.
- Further development of freight flow optimization and traffic management tools.
- Efforts to enhance cargo interchange between transportation modes, including expansion of the technology uptake by industry.
- Development of harmonized transportation documents.

An important part of the SuperGreen R&D analysis was to take on board the so-called 'Strategic Research Agendas' (SRAs) of relevant European Technology Platforms (ETPs). A technology platform tries to identify important issues and develop recommendations for a specific mode (e.g., road, rail, waterborne, etc.) as regards R&D that should be conducted to tackle these issues.

The rest of this chapter augments and updates the R&D recommendations of SuperGreen and is structured as follows. Section 15.2 talks about the SRAs of the main ETPs. Section 15.3 deals with further R&D on green corridors. Section 15.4 presents some mode-specific R&D recommendations. Finally Sect. 15.5 discusses green transportation logistics vis-à-vis Horizon 2020, the new EU program for R&D.

15.2 Strategic Research Agendas (SRAs): The European Scene

Although the significance of previous and on-going R&D efforts has resulted in considerable advances towards increased efficiency and reduced environmental footprint within the European transportation industry, there is still a need for further efforts to support the development of efficient transports and the free movement of goods within the EU.

A closer look at European trade and the region's modal split reveals that road transportation in 2010 accounted for as much as 1,756 Billion tonne-km (Btkm) of all goods transported within the EU. This is 45.8 % of the total Btkm, with rail, inland waterways and short sea shipping accounting for 10.2 %, 3.8 % and 36.9 % respectively (ERF, 2012). This underscores clearly that road is by far the preferred mode of transportation within Europe today. Although the growth in trade has resulted in the development of a more cost effective short-sea shipping (SSS) network, linking Europe's major hub ports to smaller European ports, particularly in the Baltic Sea but also to UK, the Iberian Peninsula, and the Mediterranean Sea and the Black Sea, the current situation in ports and on the European road network gives rise to serious concerns. Due to a lack of transportation capacity from infrastructure struggling to cope with the increase in traffic volume, the European road network is suffering from congestion. The European freight volume is predicted to increase with as much as 50 % by the year 2020. However, the 2011 Transport White Paper (EC, 2011) targets achieving a 60 % reduction in emissions from transportation operations by 2050 (vis-à-vis 1990 levels), and a 50 % shift of road transportation to rail, sea and inland waterways by 2050, clearly signaling that considerable actions are necessary.

For Europe as a region, but also for other parts of the industrialized world, it is vital to continue the long-term efforts towards securing economic growth while at the same time minimizing the impact on the environment and the society at large. Important elements for reaching these targets are a prolonged effort towards establishing innovative infrastructures (e.g. energy-neutral or energy-generating highways), on new organizational concepts (e.g. payload sharing, advanced logistics, supply chain management and e-freight), and methods of working related to their introduction and on innovative vehicle technologies (e.g. modular vans and trucks, electric and diesel-electric vehicles, etc., see for instance the 2011 Transport White Paper (EC, 2011) and the Green Cars Initiative[1]).

This work includes important aspects such as developing measurements of transportation impact on society, establishing a consensus on the measurement framework for transportation and logistics environmental footprint, and on the measurements of transportation and logistics performance.

Another aspect is to secure technology uptake by the industry. An example of the importance of technology uptake by the industry is provided by the Third IMO Green House Gas (GHG) Study (Smith et al., 2014), identifying that 796 million tonnes (2.2%) of global CO_2 emissions on 2012 came from international shipping. A significant potential for a reduction of GHGs through technical and operational measures were acknowledged, and if implemented, these measures could in total increase efficiency and reduce the emissions rate well below the 2012 level.

There are several European instruments promoting R&D recommendations and development strategies for all surface transportation modes, targeted to answer future cargo transportation demands and challenges. To start with, the SRAs of

[1] http://ec.europa.eu/research/transport/road/green_cars/index_en.htm

the ETPs play an important role giving concrete recommendations concerning R&D topics that promote the development of a sustainable and effective European transportation network. These SRAs are developed by a wide range of logistics and transportation stakeholders (such as consumers, manufacturers, suppliers, infrastructure operators and developers, service providers, energy suppliers, research organizations, cities and regions, as well as public authorities at both the EU and national levels).

To that effect, SRAs are developed by ETPs that include:

- the European Road Transport Research Advisory Council—ERTRAC[2]
- the European Rail Research Advisory Council—ERRAC,[3]
- the Waterborne ETP[4] and
- the European Technology Platform on Logistics—ALICE.[5]

The mode-specific focus is given by the nature of the different SRAs, although all documents contain elements for improving transportation interfaces and co-modality. Concerning inland navigation, an SRA has been published as part of the PLATINA project,[6] implementing the strategy of NAIADES[7] action programme.

There are also several other sources relevant for the coordination of the recommended R&D priorities. These include:

- The Transport Research & Innovation Portal[8]
- The EC Research and Innovation website/platform[9]
- The Strategic Transport Technology Plan[10]
- The Joint Research Centre[11]
- The ERA-WATCH platform[12]
- The European Green Car Initiative[13]

SuperGreen aimed at promoting R&D recommendations advancing and reinforcing the ones already identified. As basis for its recommendations, an investigation was carried out to identify possible gaps of R&D and development needs between the 2011 Transport White Paper (EC, 2011), the various policy documents

---

[2] http://cordis.europa.eu/technology-platforms/ertrac_en.html

[3] http://cordis.europa.eu/technology-platforms/errac_en.html

[4] http://cordis.europa.eu/technology-platforms/waterborne_en.html

[5] http://www.etp-logistics.eu/alice/en/home/

[6] http://platina1.naiades.info/platina/downloads

[7] http://ec.europa.eu/transport/modes/inland/promotion/naiades2_en.htm

[8] http://www.transport-research.info/web/

[9] http://ec.europa.eu/research/index.cfm?lg=en

[10] http://ec.europa.eu/transport/research/sttp/sttp_en.htm

[11] https://ec.europa.eu/jrc/

[12] http://erawatch.jrc.ec.europa.eu/erawatch/opencms/about/

[13] http://www.ertrac.org/en/content/european-green-cars-initiative_52/

and SRAs of relevance. A major finding was that the gaps in the various documents are seemingly few and far between. Most likely this is a result of all documents being based on input from industry experts, being subject to recent revisions, but also due to their incorporated support for having a strategic approach to long-term development within each transportation sector. Another main reason is that the different SRAs are mainly mode-specific, and with the provision of more concrete recommendations along with expected research outcome, the recommendations are presented on a more detailed level as compared to the 2011 Transport White Paper (see Minsaas et al. (2012) for more details).

However, some issues of generic nature and common within these SRAs have been identified. These include:

- Efforts towards reducing energy consumption from port and terminal operations should receive continued focus. This also due to the need for reducing externalities from such operations (for instance noise or emissions to air). Supporting the development of technology and solutions for efficient cargo handling technology therefore becomes of key importance.
- In order to support secure transportation solutions, such technology needs to be further developed (also supporting co-modal operations). Among other things this relates to technology fitted for cargo surveillance.
- Key performance indicators (KPIs) and ex-ante project evaluation procedures must receive more focus. The main objective of such work must be to enable proper benchmarking of transportation mode performance, but also to better understand the impact of targeted infrastructure projects (e.g. TEN-T, Inter-Reg, etc.).
- More efforts should be put into the development of a co-modal carbon footprint calculator supporting optimal utilization of available transportation resources. The calculator must avoid polarization of modes, meaning that favoring of modes must be avoided. Further, the tool must be supported by a standardized measurement methodology.
- Investigations on technology uptake within the different transportation modes should be carried out in order to achieve a better overview of current status. Such an overview will also uncover to what extent the different R&D topics have been covered.

15.3 Further R&D on Green Corridors

Based on its findings, the SuperGreen project developed a set of possible 'call-texts' supporting the future development of green corridors. The call texts represent specific examples and suggestions on how the more generic R&D recommendations can be transformed into more specific ones. Each call text is structured as follows:

- Content and scope
- Expected impact

The format of these call texts follows the one used by the European Commission in the Seventh Framework Programme and in the new Framework Programme for Research and Innovation—Horizon 2020 (of which more in Sect. 15.5 of this chapter).

The call texts that are most relevant for this book are listed below. GC.SST stands for Green Corridor Sustainable Surface Transport.

15.3.1 GC.SST.1 Improvement of Green Supply Chain Design and Management

15.3.1.1 Content and Scope

Congestion along main transportation routes on the European continent is a well known and outspoken problem, causing undesirable and negative externalities on local communities as well as on a more global scale. Taking into account the expected growth in transportation volume and cost towards 2050, there is a definite need to establish more sustainable transportation solutions, and hence improve the utilization of modes individually and in combination.

The R&D objectives supporting the specific green corridor development needs and management are closely related to the following activities:

- As a mean to mitigate congestion and emissions to air, more detailed analyses regarding corridor-by-corridor specific needs and requirements are necessary. As part of mapping infrastructure and ICT needs, a study of main transportation systems and modes within the respective corridors is needed.
- Identify tools and mechanisms for accelerating the industry up-take of greening technologies, including identification and dissemination of best practices within the transportation sector.
- Development and implementation of available supply chain management systems and solutions throughout the chain needs to be emphasized. This includes development of business models and/or methodologies supporting a holistic and optimal utilization of transportation assets in co-modal solutions. A prototype of a certified environmental footprint calculator is considered an important deliverable.
- The concept of cluster governance applicable for supply chains needs to be further elaborated, and in particular its value adding potential where a closer integration of the involved actors could lead to improvements in overall supply chain performance.
- Development of reliable tools enabling both data accumulation and proper benchmarking exercises, for use in development of supply chains and for comparative purposes.
- Particularly related to the above, development of systems and solutions for solving the current lack of transportation information on EU-level is expected to be investigated (e.g. volumes, types of goods, flow directions, etc.).

15.3.1.2 Expected Impact

The project is supposed to give important contributions to meet the challenges related to congestions and emissions as defined by the 2011 White Paper on Transport. In particular, new insight to make advances in the development and industry up-take of greening technologies, along with innovative models, methods, and tools for the development of resource efficient transportation operations from a holistic point of view, including co-modality and utilization of hubs, are considered important outcomes of the work.

Identification of development needs to make different corridors more sustainable is also supposed to be among the important deliverables from this project. Thus, identifying solutions for solving current lack of transportation related data is important.

15.3.2 GC.SST.2. ICT for Green Transportation Logistics

15.3.2.1 Content and Scope

Since ICT provides critical information to transportation stakeholders affecting the ability to make decisions, it is clear that the role of ICT in green transportation logistics is paramount. This role has been identified in the 2011 Transport White Paper; however R&D should be conducted to identify the best ways to reach the stated policy goals. Related activities include:

- Formulation of a representative spectrum of transportation logistics problems as optimization problems with the environmental dimension on board. These problems should be at the strategic, tactical and operational levels and should encompass both static and dynamic elements. A taxonomy of relevant KPIs should be identified.
- Identification of role of ICT in terms of data collection, transmission, processing (pre and post), storage, hardware and software, to better handle the problems identified earlier. Public data versus private data. Identification of which data should be collected and best methods of doing so.
- Development of operations research and other methods for the solution of problems identified earlier. Methods include exact approaches and heuristics. Quantification of benefits. Optimization of computational performance.
- Development of ICT tools, both centralized and de-centralized, to help operators and other transportation stakeholders, improve upon solutions and achieve better system performance in terms of the selected KPIs.
- Development of ICT tools for emissions data collection, measurement and reporting, e.g. a certified co-modal carbon footprint calculator.

- Simulation of system performance and real-world demonstration of ICT systems and tools that can achieve documented improvements. Analysis of policy options and policy recommendations.

15.3.2.2 Expected Impact

This work will extend and integrate on-going EU and national research efforts in several disciplines, including ICT technologies, systems and tools, operations research methods for improved transportation logistics, and emissions modelling, inter alia. It will provide important support to reach the sustainability objectives of the 2011 Transport White Paper. It will help identifying the most cost effective ways of deploying ICT systems across Europe, as well as guidance for related strategic decisions. The impact will span all hierarchical levels, including strategic long-run considerations, tactical level planning and operational 'dynamic' environments.

15.3.3 GC.SST.3. Harmonization and Development of Policies and Regulations

15.3.3.1 Content and Scope

Green corridors will not only be affected by transportation policy, but also by other related policies such as environmental, maritime, energy, sustainability, innovation, security and cohesion policies. In the recent years, transportation services have been constrained by contradictory policies, which impose conditions impossible to fulfil all of them at the same time.[14]

Policies are usually implemented through different regulatory initiatives and frameworks, some of them are voluntary: Recommendations and Positions, while others are mandatory: Regulations and Directives. The overlap of international, EU, national and sometime regional regulations, creates additional constraints and bottlenecks in the development of the several and different regulatory frameworks. The policy and regulation implementation supporting this specific green corridor development need are closely related to the following activities:

---

[14] An example, which is not the only one, is the implementation of ICZM—Integrated Coastal Zone Management, a 'land focused' policy Recommendation of 2000 from DG-Environment, vis-a-vis the implementation of the 'sea focused' MSP—Maritime Spatial Planning of 2007 from DG-MARE. These are contradictory due to their difference in perspective (land vs. sea), and number of years in difference.

- Identify technical regulations from ISO, EU and Member States standardization bodies affecting different aspects of green corridors, especially in the technology side, for example, CO_2 calculators.
- Identify legal regulations from UN conventions, EU and Member States legislation bodies, affecting different aspects of green corridors, especially in the implementation side.
- Cluster governance is another related concept, especially for supply chains belonging to cross border co-modal corridors, connecting markets intra-EU and extra-EU from Africa, Asia or non-EU European countries.
- A special focus should be put on passenger transportation services within co-modal Green corridors, which are designed for both cargo and passengers. In the recent years, significant changes in Member States concerning legal and organizational frameworks of public transportation have been observed. These changes aim at improving transparency, economic efficiency and the quality of the service. The European Commission promotes this development through the provision of an appropriate legal framework at European level, as originally suggested in the Citizens' Network Green Paper and later reinforced and clearly indicated in the Communication "Developing the Citizens Network".[15]

15.3.3.2 Expected Impact

Impacts on EU Policies should be expected in the following areas:

- Improved coordination between EU and International regulatory initiatives from organizations like UNCTAD, OECD, ISO, ILO and IMO.
- Improved coordination between EU and Member States regulatory initiatives to avoid contradictory actions to the end-users and customers.
- Simplification of regulatory instruments to avoid negative perception from end-users and customers side.

15.3.4 GC.SST.4. Development and Harmonization of Transportation Infrastructure

15.3.4.1 Content and Scope

Infrastructure and terminals are vital for the overall efficiency of logistics networks, so as for the environmental profile of the co-modal supply chains. Terminals and infrastructure need to be developed in an integrated manner to secure interoperability between modes, and to streamline and harmonize the overall efficiency. This also includes development of technologies and concepts for seamless cargo handling. The harmonized development of the infrastructure should be based on

[15] http://europa.eu/legislation_summaries/transport/mobility_and_passenger_rights/l24215_en.htm

on-going processes defining the best standards to achieve an integrated and streamlined European transportation infrastructure that fulfils environmental and sustainable requirements.

The R&D objectives supporting this specific Green Corridor development need are closely related to the following activities:

- Improvement of intermodal hub equipment and easy cross docking technology to increase productivity and standardized modal shift capability.
- Innovative solutions for more efficient border crossings. Efficient and economically attractive solutions for the upgrading of existing infrastructure as multimodal terminals, sea and river ports, and city logistics centers to make operations more sustainable.
- Innovative solutions to combine freight and passenger transportation infrastructures with green technologies in a realistic and efficient way.
- Advances in the implementation of standard interoperable technologies in railway infrastructure: Signaling system, catenary, axle load, maximum length of passenger and freight trains, gauge, etc.

15.3.4.2 Expected Impact

The work is expected to give important contributions for reaching the objectives related to more harmonized, efficient and sustainable co-modal transportation and is supposed to give guidance for the implementation of novel green technologies in transportation infrastructure (i.e. clean & low-emission energies).

The work will significantly contribute to the deployment of TEN-T infrastructure by gradually integrating modal systems, also considering the new Member States and their specific identified infrastructure gaps, as stated in the 2011 Transport White Paper.

15.3.5 GC.SST.5. Development and Harmonization of Transportation Technology

15.3.5.1 Content and Scope

Green technology is at the heart of future sustainable logistics and the development moves towards '*energy*'-based measures as much as '*activity*'-based measures. For this reason a corridor-based approach would be suitable for this call; looking to address long distance green alternative technology whilst also looking at marketing existing technologies to improve the corridor performance across different transportation modes.

Moreover, development and harmonization of infrastructure, terminals and transportation technology is vital for the overall efficiency of logistics networks and co-modal supply chains. Transportation vehicle technology should be developed alongside the development of terminal and infrastructure so they can easily be integrated ensuring interoperability between modes, and streamlining the overall efficiency. The harmonized development of the transportation technologies for vehicles and infrastructures should refer to the process of defining the best standards and interconnectivity of the European transportation system in the most environmental and sustainable way.

The R&D objectives supporting the development of Green Corridors are closely related to the following activities:

- Gain a better understanding of fuel efficient technology across different modes within a designated corridor.
- Identify an alternative technology within a specific corridor supported by a cost benefit analysis and a number of key stakeholders inputs.
- Enlarge, maintain, and improve a technology knowledge based platform, meeting the demands of long distance freight stakeholder's quality partnership. Transferring successes of city logistics freight quality partnership ventures into longer distance freight operation.
- Improve intermodal hub equipment and cross-docking technology towards increasing productivity and modal shift capability (i.e. supporting seamless co-modal operations).
- Identify innovative solutions for cross-border sections and bottleneck gaps. Providing an efficient and economically attractive solution for the upgrading/up-taking of existing green infrastructure; such as multimodal terminals, maritime and river ports, and city logistic centers.

15.3.5.2 Expected Impact

Through providing a better understanding of each technology's greening potential and impact on fuel efficiency, it is expected that such technologies will be adopted to meet the requirements of the sustainable green economy, thereby contributing to the 2011 Transport White Paper low carbon future goals. While currently country based evidences exist, the corridor level evidence is relatively limited and has potential for further investigation.

Provide opportunities for new insight into how to accelerate development and industry up-take of greening technologies by means of among others innovative models, methods, and tools for the development of resource efficient transportation operations. The efforts should be carried out from a holistic point of view, including co-modality and utilization of hubs. To the extent possible considerations should also try to take into account scenarios beyond the direct outcomes of the work.

15.3.6 *GC.SST.6. Transparency of Information and Increased Cooperation in Co-modal Supply Chains*

15.3.6.1 Content and Scope

There is a considerable cost saving potential inherent in more open information sharing and in the integration of systems along the entire supply chain. This will push the development of co-modal supply chains and further the development of new and integrated ICT systems, along with new methods and tools for information handling.

In the supply chains there are numerous spots where exchange of information takes place, and mutual trust is a prerequisite as well as a facilitator. However, lack of trust causes deviations in transportation chains, unclear responsibilities between different stakeholders, the use of so called 'random subcontractors', and problems related to co-operation and data interchange between transportation clients and operators.

Transparency in information to all stakeholders involved from consignor to consignee is therefore a prerequisite for successful development and implementation of ICT systems. Current systems support and promote co-modal and sustainable logistics, however there is still a need to further develop and implement ICT applications that meet the current demands and those of the future.

Concerning availability and implementation of ICT systems, the main challenge is caused by insufficient harmonization of national and EU policies to implement suitable ICT systems rather than lack of technological solutions and systems. Regulatory measures to implement dedicated ICT systems should be examined and appropriate measures to solve obstacles and enhance willingness to adopt proposed ICT technologies should be implemented.

The R&D objectives supporting these specific Green Corridor development needs are:

- In order to obtain a comprehensive overview of the current status, representative supply chains and their stakeholders should be examined. Transparency in the information sharing throughout the entire supply chains should be aimed at as basis for cross fertilizing between actors and modes. Networking among the relevant parties is essential.
- Closely related to the above, main emphasis should be put on the co-operation and information exchange between different stakeholders within the supply chains to identify problems and challenges as basis for gap analyses. The analytical work should also indicate how networks supporting sustainable transportation operations ought to be structured.
- Standardized activities should be generated based on the data gathered from the case studies. The activities should aim to overcome the issues related to lack of transparency and to improve co-operation within the supply chains. Business models to improve co-operation and plans for data platforms supporting these

models should be generated. The models should cover overall supply chains serving different lines, businesses and industries.
- The standardized activities should be evaluated in order to determine the best solutions with regards to functionality and greening potential. The evaluations should include KPIs related to efficiency, reliability, service quality, environmental sustainability, infrastructural sufficiency, safety, and social issues (e.g. employment, working conditions, traffic safety etc.). Based on the evaluations, the selection of the best business models and data management models as well as standardized activities should be defined.

15.3.6.2 Expected Impact

In particular, the project should contribute to improve the greening of supply chains and hence the greening of the transportation corridors. Parts of the results should be based on access to more open logistics information, thus making it possible to develop and obtain more efficient logistics chains. Business models to improve co-operation and data platforms for different branches of industries as well as more open virtual logistics centers serving several supply chains and industries should be utilized to improve the activities.

The work should also provide new knowledge on how to eliminate overlaps in transportation and logistics operations. Optimal supply chain management systems and platforms should contribute to a reduction in unnecessary transportation operations, increasing capacity utilization of all transportation modes and thus reducing transportation costs. The platforms and business models for operation may contribute to increase business volume for service providers, and promotion of value added services to end customers.

15.4 Mode-Specific R&D Recommendations

Mainly as a reflection of the relevant SRAs (as per Sect. 15.2), the focus of this section is directed towards identifying and elaborating on mode-specific recommendations (i.e. waterborne, intermodal, road and rail). Waterborne transportation and intermodality are grouped together due to the nature of the industry, that is, being fully dependent on operations in port of loading and port of discharge.

Recommendations are broken down into:

- Operational recommendations
- ICT and transportation technology recommendations
- Infrastructural recommendations
- Recommendations related to policies, legislation and regulations.

15.4.1 Waterborne and Intermodal-Specific R&D Recommendations

15.4.1.1 Operational Recommendations

Regarding pure operational bottlenecks related to waterborne operations (i.e. seagoing and inland waterways), and intermodality there are a number of on-going alleviating activities, which also are addressed in the recent updates of the SRAs. In addition to a number of projects that have been launched for improving intermodality operations, research has also focused on developing new vessel concepts that enable a better integration of waterborne transportation modes into intermodal operations. As of this, R&D recommendations include:

- Investigation and impact studies of conditions under which slow steaming in short sea routes seems favorable should be accomplished. This will inevitably necessitate the creation of a ship speed surveillance system which will make sure the rules will be obeyed even in times of high demand for shipping services. Caution should be exercised to avoid reverse modal shifts.
- Related to the above, efforts should be directed towards increasing the understanding of the impact "slow steaming" has on the environment, and on the transportation industry at large.
- Continued focus on further development of cargo handling systems, integrated transportation systems, promotion of best practices, efficient terminal lay-out, and port hinterland connections.
- Development of technology minimizing emissions to air and water (i.e. propulsion technologies, exhaust cleaning systems, and grey/black water treatment technology). This also includes investigation of alternative solutions for propulsion technology (i.e. non-fossil).
- Development of vessel concepts creating a stronger link between shortsea and inland waterways.
- Increase the attractiveness of waterborne professions through developing new training programmes, career opportunities, and technology for improving life at sea and on inland waterways.

15.4.1.2 ICT and Transportation Technology Recommendations

There is a growing concern for developing actions to get better and transparent information of freight flows among the different transportation stakeholders. Hence, policy decisions at European level should reflect the needs of the co modal transportation industry to contribute to promote sustainable and effective logistics. Recommendations for R&D are therefore covering the following:

- Continue work for the development of harmonized ICT and decision support systems.
- Development of ICT systems that promote and support development of improved business models (e.g. new intermodal transportation solutions based

on logistics and co-modality require the deployment of a new range of suitable ICT systems and technologies to become a real option for shippers).

- Efforts for gaining a deeper understanding of future demands within the ICT domain and transportation technology, in addition to definition of ICT system requirements.
- Development of ICT systems that link ship speed optimization with port and terminal capacity availability so as to minimize emissions. Development of ICT for virtual arrival systems (as per considerations of Chap. 9).
- For ice-infested waters there is also a requirement for implementing available decisions support tools for how to assign icebreakers to other vessels, with the use of already existing (communication)-technologies.
- For inland navigation, further actions are needed to close the gaps in national River Information Services (RIS) infrastructure and in the provision of services to logistical RIS users. The European and national legal framework on RIS needs updating and the implementation of the "European RIS" (e.g. European Vessel Certification Database, RIS Data Management System) is recommended.

15.4.1.3 Infrastructural Recommendations

An efficient and seamless transportation system depends on efficient hubs or nodes that enable multimodal interconnections. Recommendations for R&D cover:

- Further development of a network of intermodal terminals across Europe, being part of multimodal supply chains.
- Development of transportation technology and operational solutions targeting alleviation of the growing congestion on European roads and port access.
- Development of approaches and technical solutions for removal of bottlenecks and better utilization of the existing infrastructure.
- Development of new financing and business models for increasing the fleet renewal rate.

15.4.1.4 Recommendations Related to Policies, Legislation and Regulations

There are some interesting measures related to solving the problems concerned with legislation, regulations and procedures in order to make maritime transportation safer, more secure, greener and more competitive. The following recommendations were identified:

- The lack of harmonization of national regulations within various fields between nations.
- The barrier for carriage of dangerous goods due to differing regulations between transportation modes, and thus also being a barrier to free market competition. Currently, the rules and procedures are highly more complex for sea transportation than for land transportation.

15.4.2 Rail-Specific R&D Recommendations

Many of the challenges faced by rail transportation are identified and their anticipated solutions are well known by the industry. However, at least in Europe, there is a glacial response from the railway sector to the recognition and response to these and a reluctance to undertake the sort of macro and micro reforms needed to make rail a more attractive option to shippers.

15.4.2.1 Operational Recommendations

For long distance operation, considerations should be made regarding what the ideal number of large rail freight operators for optimized operation of a rail corridor would be. The rationale is that the appearance of a few more large players in the rail freight market would be helpful to bring the long distance cross border rail freight to be more competitive against road freight. In turn this will invite alliance of smaller players to form as bigger players that consequently creates a competitive market.

15.4.2.2 ICT and Transportation Technology Recommendations

The rail industry needs to focus on developing solutions that produce quick but also sustainable results in terms of additional traffic and revenue, and there needs to be a much more profound adoption of commercially based initiatives. This needs to be matched with measures to constrain and drive down costs commensurate with a drive to raise asset and resource productivity by factor amounts. Moreover, R&D recommendations also include the following:

- In terms of "green technologies" identified by the project (e.g. braking energy recovery for rail), efforts should be targeted towards identifying application areas. This includes carrying out cost/benefit analysis.
- Efforts for shifting focus from over-reliance on technical measures towards more commercially based initiatives should be investigated.
- Identifying where and on which corridors such "green technology" should be promoted and implemented.

Also, the benefit of European Rail Traffic Management System (ERTMS) for advancing interoperability has been around for some time. Within the rail industry there is evidence that countries like UK, Sweden and Germany are setting targets to implement the technology extensively within the next few years. Some rail actors/operators are still in doubt of the benefits of the high investments needed. This is particularly true as high investments will create more expensive services and in turn reduce the competitiveness of the rail services compared to other modes. Further research and development may contribute to make the ERTMS technology and implementation more cost effective and thus reduce the investments needed. However, it has been highlighted that ERTMS would only make economic sense if all the corridors (long distance ones), are equipped with such technology, which then reinforces questions about its deployment and the benefits that flow from its application.

15.4.2.3 Infrastructural Recommendations

The rail sector is too much focused on prioritizing its own supply side measures as a priority. Instead attention should be directed towards meeting customer requirements and expectations making it competitive with road transportation. At the same time there are challenges related to utilization of wagon assets, but also for developing new economic concepts adapted to more sustainable operations. As such, recommendations for R&D activities encompass:

- R&D on developing strategic support and promotional campaigns specifically targeted for the rail industry.
- Development of economic concepts to be applied on different corridors in order to determine necessary investments in infrastructure, bottlenecks, bypasses, technology, rolling stock, longer heavier trains, etc. in order to increase productivity, and generate additional capacity dedicated to freight transportation.
- R&D on how asset management can lead to reduced dwell times and improved commercial activity (i.e. clarification of responsibility for individual assets).

15.4.2.4 Recommendations Related to Policies, Legislation and Regulations

The full use of rail's green endowment should be made ensuring this is contributing to sustainability, both in economic and environmental terms. Further, SuperGreen also supports the recommendations already identified by the NEWOPERA[16] and RETRACK[17] projects, namely:

- Future project supported by EU to sponsor new rail freight services on other corridors to start up in the form of repayable working capital through new mechanisms or via existing instruments such as Marco Polo;
- At the national level, the EU Member States need to ensure that incumbents do not retaliate on pricing to drive away the new entrants;
- Need for a complete through transit ability to track the movement of the train, wagon and cargo module independently of the railway administration to confirm ETA or any revisions.

[16] http://www.newopera.org/

[17] http://www.retrack.eu/

15.4.3 *Road-Specific R&D Recommendations*

15.4.3.1 Operational Recommendations

One of ICT's main scopes is to overcome complexity of modern transportation systems and procedural obstacles. Research should be at a high level aiming to increase visibility in the supply chain management, effortless information transferring, friendlier Graphical User Interfaces (GUIs), simplify operations, unify various stakeholders and different countries technologies. More specifically these R&D recommendations focus on:

- Dynamic vehicle routing using real-time traffic information. Extend the considerations of Chap. 7 (green vehicle routing) in a dynamic setting.
- Dynamic congestion toll charging to control traffic.
- Single window apps along the EU road network.
- Operational aspects (such as toll charging, crossing borders, customs duties, dangerous goods).

15.4.3.2 ICT and Transportation Technology Recommendations

As stated before, ICT's implementation, adaptation and development are very crucial to the overall efficiency of the road transportation. The R&D recommendations are focused in development, integration and implementation of road ICT and technologies:

- Development of new, easily adopted and efficient ICTs to meet the future trends of road transportation.
- Integration of the existing road ICT technologies in order to increase performance and visibility.
- Information flow along the transportation chain.
- Improving the air drag on trucks.
- Improving of combustion technology and exhaust gas cleaning technology.
- Hybrid engine solutions.

15.4.3.3 Infrastructural Recommendations

Infrastructural bottlenecks in road transportation are present and often magnified due to lack of stakeholders willingness for adaptation. To be more precise, the infrastructure sufficiency is dependent on the level of transportation load and characteristics. The increase of transportation load is developing bottlenecks on road transportation and these bottlenecks can be overcome by enhancing road and related infrastructures, by increasing the efficiency and the capacity of existing infrastructure or by encouraging shifts of road cargo to other modes of transportation. Therefore the R&D recommendations for road transportation include:

- Developing forecasting and simulation techniques for future road cargo flows.
- Increasing infrastructure efficiency and performance.
- Increasing road modes efficiency, performance and payload.
- Enhancing modal shift from road transportation to rail, sea and inland waterways.
- Specific areas of road transportation such as expert pollution charging systems in trans European road networks or congestion avoidance systems.

15.4.3.4 Recommendations Related to Policies, Legislation and Regulations

It is a commonly accepted fact that ICTs in general are value added investments in the transportation industry. Additionally those systems exhibit an eco-friendly impact on the environment either by directly decreasing emissions and other polluters or by increasing levels of safety and minimizing accidental pollution. To that extent, proposed R&D recommendations include:

- Seeking for and designing appropriate funding methods to develop, install and operate ICTs especially for road.
- Aiming to enhance homogenizing standards and procedures for road transportation among EU countries.
- Adopt a single window intra-European application dedicated to road transportation.
- Regulatory framework of selection of critical ICTs and transforming them voluntary to mandatory use.
- Enhance the use of common EU policies and regulations over the nationals so as if to avoid multi-legislation issues.

15.5 Green Transportation Logistics in Horizon 2020

15.5.1 Introduction

Horizon 2020[18] is the new EU research framework programme planned for the period 2014–2020, in continuation of the Seventh Framework Programme. Three main research areas are described: Societal challenges, industrial development and strengthened R&D:

[18] Horizon 2020—The Framework Programme for Research and Innovation, COM (2011) 808 final, November 2011. http://ec.europa.eu/programmes/horizon2020/

- Societal challenges include research on health, demographic changes and wellness, food security and bio-based economy, safe, clean and effective energy, smart, green and integrated transport, access to raw materials, resource efficiency and climate handling, including innovative and safe society.
- Industrial development includes leadership in generic and industrial technologies, access to risk capital and innovation in small enterprises.
- Strengthened R&D includes strengthened basic research, future and emerging technologies, qualification and carrier development and research infrastructure.

Particularly relevant for green transportation logistics is how the new research program aims to secure development of "smart, green and integrated transportation solutions" for contributing towards the development of a resource-efficient, environmentally-friendly, safe and seamless transportation operations. The Specific Programme is structured in four broad lines of activities aiming at:

(a) **Resource efficient transport that respects the environment**. The aim is to minimize transport systems' impact on climate and the environment (including noise and air pollution) by improving its efficiency in the use of natural resources, and by reducing its dependence on fossil fuels.
(b) **Better mobility, less congestion, more safety and security**. The aim is to reconcile the growing mobility needs with improved transport fluidity, through innovative solutions for seamless, inclusive, affordable, safe, secure and robust transport systems.
(c) **Global leadership for the European transport industry**. The aim is to reinforce the competitiveness and performance of European transport manufacturing industries and related services including logistic processes and retain areas of European leadership (e.g. such as aeronautics).
(d) **Socio-economic and behavioural research and forward looking activities for policy making**. The aim is to support improved policy making which is necessary to promote innovation and meet the challenges raised by transport and the societal needs related to it.

These activities are addressed in three areas for calls for proposals:

1. Mobility for Growth
2. Green Vehicles
3. Small Business Innovation for Transport

The total R&D budget (EC contribution) allocated to these activities is 578.91 million euros for 2014 and 302.57 million euros for 2015.

Some details of calls of the first two areas 'Mobility for Growth' and 'Green Vehicles' that are more relevant for this book and for the period 2014–2015 are presented below.

15.5.2 Mobility for Growth

According to this call, transportation is on the brink of a new era of "smart mobility" where infrastructure, transportation means, travelers and goods will be increasingly interconnected to achieve optimized door-to-door mobility, higher safety, less environmental impact and lower operations costs. In order to achieve efficiency at system-level, targeted efforts are needed to develop and validate new solutions that can be rapidly deployed, notably on corridors and in urban areas. They will address transportation means and infrastructure and integrate them into a user friendly European transportation system of smart connected mobility and logistics. Research and innovation on equipment and systems for vehicles, aircraft and vessels will make them smarter, more automated, cleaner and quieter, while reducing the use of fossil fuels. Research and innovation on smart infrastructure solutions is necessary to deploy innovative traffic management and information systems, advanced traveler services, efficient logistics, construction and maintenance technologies.

A thorough and mature research and innovation agenda for this call has been defined taking into account the other calls and initiatives where the Transport Challenge is concerned, i.e. the calls on 'Green Vehicles', 'Small Business and Fast Track Innovation for Transport', 'Blue Growth', and 'Smart Cities and Communities', and the 'Clean Sky 2', the 'Single European Sky Air Traffic Management Research (SESAR)', the 'Shift2Rail' and 'Fuel Cells and Hydrogen 2' joint undertakings (in different degrees of preparation). In addition, the European Global Navigation Satellite System (GNSS) will provide new opportunities for the localization and the guidance of vehicles. It is intended to create synergies with all these initiatives as well as with other parts of Horizon 2020, namely ICT, Energy and Space. Special attention is dedicated to innovation aspects not covered in the other parts of the Transport Challenge, as well as to SMEs.

As indicated in the Specific Programme, the "activities" will be organized in such a way as to allow for an integrated and mode-specific approach as appropriate. Therefore, the contents of the "Mobility for Growth" call have been organized as follows:

15.5.2.1 Aviation

MG.1.1-2014. Competitiveness of European aviation through cost efficiency and innovation

MG.1.2-2015 Enhancing resource efficiency of aviation

MG.1.3-2014 Seamless and customer oriented air mobility

MG.1.4-2014. Coordinated research and innovation actions targeting the highest levels of safety for European aviation.

MG.1.5-2014 Breakthrough innovation for European aviation

MG.1.6-2014. Improving skills and knowledge base in European aviation

MG.1.7-2014. Support to European aviation research and innovation policy

MG.1.8-2014-2015. International cooperation in aeronautics with Japan
MG.1.9-2015. International cooperation in aeronautics with Canada
MG.1.10-2015. International cooperation in aeronautics with China

15.5.2.2 Rail

MG.2.1-2014. I^2I—Intelligent Infrastructure
MG.2.2-2014. Smart rail services
MG.2.3-2014. New generation of rail vehicles

15.5.2.3 Road

MG.3.1-2014. Technologies for low emission powertrains
MG.3.2-2014. Advanced bus concepts for increased efficiency
MG.3.3-2014. Global competitiveness of automotive supply chain management
MG.3.4-2014. Traffic safety analysis and integrated approach towards the safety of Vulnerable Road Users
MG.3.5-2014. Cooperative ITS for safe, congestion-free and sustainable mobility
MG.3.6-2015. Safe and connected automation in road transport

15.5.2.4 Waterborne

MG.4.1-2014. Towards the energy efficient and very-low emission vessel
MG.4.2-2014. Safer and more efficient waterborne operations through new technologies and smarter traffic management
MG.4.3-2015. System modelling and life-cycle cost and performance optimization for waterborne assets
MG.4.4-2014. Advancing innovation in the Inland Waterways Transport (IWT) sector

15.5.2.5 Urban Mobility

MG.5.1-2014. Transforming the use of conventionally fuelled vehicles in urban areas
MG.5.2-2014. Reducing impacts and costs of freight and service trips in urban areas
MG.5.3-2014. Tackling urban road congestion
MG.5.4-2015. Strengthening the knowledge and capacities of local authorities
MG.5.5-2015. Demonstrating and testing innovative solutions for cleaner and better urban transport and mobility

15.5.2.6 Logistics

MG.6.1-2014. Fostering synergies alongside the supply chain (including e-commerce)
MG.6.2-2014. De-stressing the supply chain
MG.6.3-2015. Common communication and navigation platforms for pan-European logistics applications

15.5.2.7 Intelligent Transport Systems

MG.7.1-2014. Connectivity and information sharing for intelligent mobility
MG.7.2-2014. Towards seamless mobility addressing fragmentation in ITS deployment in Europe

15.5.2.8 Infrastructure

MG.8.1-2014. Smarter design, construction and maintenance
MG.8.2-2014. Next generation transport infrastructure: resource efficient, smarter and safer
MG.8.3-2015. Facilitating market take up of innovative transport infrastructure solutions
MG.8.4-2015. Smart governance, network resilience and streamlined delivery of infrastructure innovation

15.5.2.9 Socioeconomic and Behavioural Research and Forward Looking Activities for Policy Making

MG.9.1-2015. Transport societal drivers
MG.9.2-2014. User behaviour and mobility patterns in the context of major societal trends
MG.9.3-2014. Analysis of funding schemes for transport infrastructure
MG.9.4-2014. Research, technology development and market prospects for the European transport industries
MG.9.5-2015. Fostering transnational cooperation in European transport research and innovation—National Contact Point (NCP) network
MG.9.6-2014. Strengthening the research and innovation strategies of the transport industries in Europe
MG.9.7-2014. Innovation awards for students and researchers in the context of the Transport Research Arena conference—TRA 2016

15.5.3 Green Vehicles

This call of the Transport Challenge represents an essential component of road transportation research and innovation. It includes research, technological developments, innovation and demonstration in support of improvements in energy efficiency of road transportation vehicles and the use of new types of non-conventional energies into road transportation such as electricity, CNG and LNG, renewable and tailored fuels.

The scope of the activities includes both advanced power-train technologies and new vehicle architectures, weight reduction, improved aerodynamics and rolling resistance and component development for alternative fuel vehicles. Concerning new forms of energy, the interfaces between the vehicles and the recharging infrastructure will also need to be taken into account with particular attention to standardization issues. Demonstration activities will play an essential role in ensuring a proper and timely deployment of the new technologies. In this respect, innovation activities linked with other EU funding mechanisms such as cohesion and regional funds should be considered.

This call has been defined taking into account the other calls and initiatives where the transport challenge is concerned, particularly the calls on 'Mobility for Growth' and 'Smart Cities and Communities', and the 'Fuel Cells and Hydrogen 2' joint undertakings. Multi-sectorial research involving other research and innovation areas such as energy and environment coupled with research on new materials, advanced production and information and communication technologies will be encouraged, particularly in fields such as advanced energy storage systems and interfaces between vehicles and energy recharging infrastructures.

In addition to the topics of this call, a topic on post lithium ion batteries for electric automotive applications" (NMP 17—2014) is included in "Nanotechnologies, Advanced Materials and Advanced Manufacturing and Processing (NMP)" under "Leadership in Enabling and Industrial Technologies" (LEIT).

The contents of the "Green Vehicles" call have been organized as follows for the period 2014–2016:

GV.1-2014. Next generation of competitive lithium ion batteries to meet customer expectations
GV.2-2014. Optimized and systematic energy management in electric vehicles
GV.3-2014. Future natural gas powertrains and components for cars and vans
GV.4-2014. Hybrid light and heavy duty vehicles
GV.5-2014. Electric two-wheelers and new light vehicle concepts
GV.6-2015. Powertrain control for heavy-duty vehicles with optimised emissions
GV.7-2014. Future natural gas powertrains and components for heavy duty vehicles
GV.8-2015. Electric vehicles' enhanced performance and integration into the transport system and the grid

References

Smith, T. W. P., Jalkanen, J. P., Anderson, B. A., Corbett, J. J., Faber, J., Hanayama, S., et al. (2014). Third IMO GHG Study 2014. London, England: International Maritime Organization (IMO).

EC. (2011, March 28). *White Paper. Roadmap to a single European transport area—Towards a competitive and resource efficient transport system.* COM (2011) 144 final, Brussels, Belgium.

ERF. (2012). *European road statistics 2012*. Brussels, Belgium: European Union Road Federation. Retrieved from http://www.irfnet.eu/images/Statistics/ER_Statistics_Final_2012.pdf

Minsaas, A., Holte, E. A., Salanne, I., Rönkkö, S., Aditjandra, P. T., Berd, R. G., et al. (2012, January 18). *Define and submit R&D recommendations*. SuperGreen project deliverable 5.2.

Annex I: The SuperGreen Project

Project Identity

• Project full title	Supporting EU's Freight Transport Logistics Action Plan on Green Corridors Issues
• Type of project	Coordination and Support Action
• Financed through	7th Framework Programme
• Duration	15 Jan 2010–14 Jan 2013
• Consortium	22 partners from 13 countries
• Leader	National Technical University of Athens
• Total budget	3,453,747 EUR
• EC contribution	2,634,698 EUR
• Web site	www.supergreenproject.eu

Project Partners

Partner number	Partner name	Partner short name	Country
1 (Coordinator)	National Technical University of Athens	NTUA	Greece
2	Norsk Marinteknisk Forskningsinstitutt AS, MARINTEK	MAR	Norway
3	Sito Ltd (Finnish Consulting Engineers Ltd)	SITO	Finland
4	D'Appolonia S.p.A.	DAPP	Italy
5	Autoridad Portuaria de Gijon Gijon Port Authority	PAG	Spain

(continued)

H.N. Psaraftis (ed.), *Green Transportation Logistics*, International Series in Operations Research & Management Science 226,
DOI 10.1007/978-3-319-17175-3

Partner number	Partner name	Partner short name	Country
6	DNV—Det norske Veritas	DNV	Norway
7	via donau Osterreichische WasserstraBen-Gesellschaft mbH	VIA	Austria
8	NewRail—Newcastle University	UNEW	UK
9	CONSULTRANS	CONS	Spain
10	PSA Sines	PSAS	Portugal
11	Finnish Transport Agency	FMA	Finland
12	Straightway Finland Ry	SWAY	Finland
13	SNCF Fret Italia	SFI	Italy
14	Procter and Gamble Eurocor	PG	Belgium
15	VR Group	VRG	Finland
16	Lloyd's Register—Fairplay Research	LRFR	Sweden
17	Hellenic Shortsea Shipowners Association	HSSA	Greece
18	Dortmund University of Technology	DUT	Germany
19	TES Consult Ltd	TES	Ukraine
20	Turkish State Railways	TCDD	Turkey
21	DB Schenker AG	SCH	Germany
22	Norwegian Public Road Administration	NPRA	Norway

Project Objectives

- Give overall support and recommendations on green corridors to EU's Freight Transport Logistics Action Plan.
- Encourage co-modality for sustainable solutions.
- Provide a schematic for overall benchmarking of green corridors based on selected KPIs covering all aspects of transport operations and infrastructure (emissions, internal and external costs).
- Conduct a programme of networking activities between stakeholders (public and private) and ongoing EU and other research and development projects to facilitate information exchange, research results dissemination, communication of best practices and technologies at a European, national, and regional scale, thus adding value to ongoing programmes.
- Deliver policy recommendations at a European level for the further development of green corridors.
- Provide recommendations concerning new calls for R&D proposals to support development of green corridors.

Modes Covered

- Road
- Rail
- Short sea shipping
- Deep sea shipping
- Inland waterway transport
- Intermodal transport

Project Structure

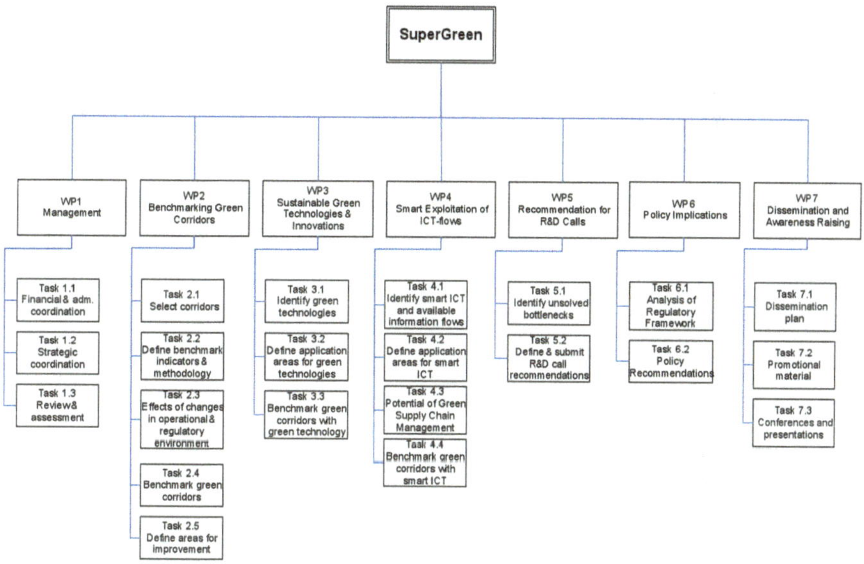

SuperGreen Corridors

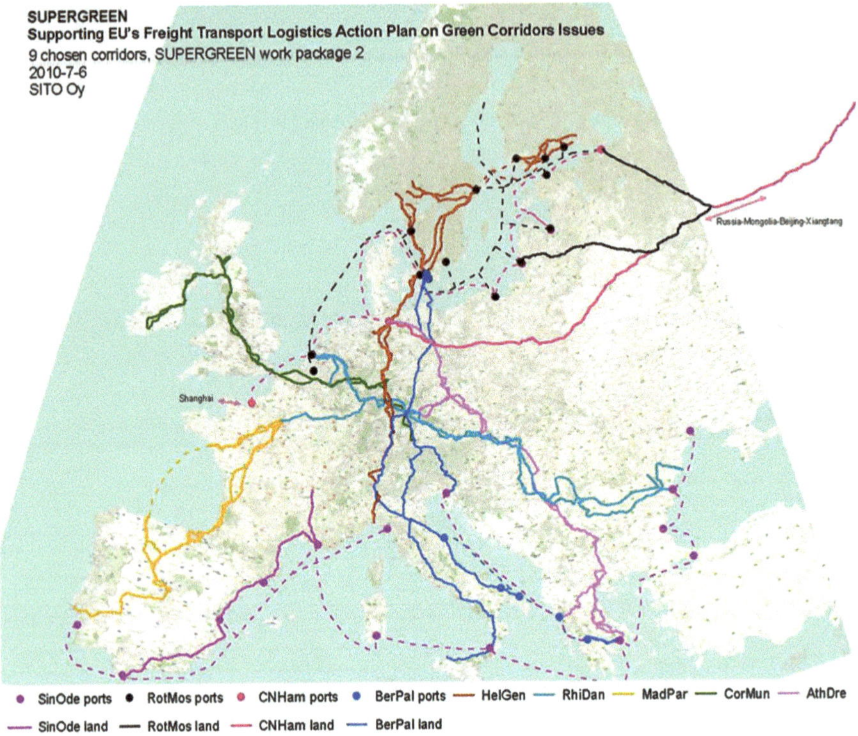

Nicknames	Acronym	Corridor Description
Brenner	BerPal	Malmö-Trelleborg-Rostock/Sassnitz-Berlin-Munich-Salzburg-Verona-Bologna-Naples-Messina-Palermo Branch A: Salzburg-Villach-Trieste (Tauern axis) Branch B: Bologna-Ancona/Bari/Brindisi-Igoumenitsa/Patras-Athens
Finis Terrae	MadPar	Madrid-Gijon-Saint Nazaire-Paris Branch A: Madrid-Lisboa
Cloverleaf	CorMun	Cork-Dublin-Belfast-Stranraer Branch A: Munchen-Friedewald-Nuneaton Branch B: West Coast Main line
Edelweiss	HelGen	Helsinki-Turku-Stockholm-Oslo-Göteborg-Malmö-Copenhagen (Nordic triangle including the Oresund fixed link)- Fehmambelt - Milan - Genoa
Nureyev	RotMos	Motorway of Baltic sea Branch: St. Petersburg-Moscow-Minsk-Klapeida
Strauss	RhiDan	Rhine/Meuse-Main-Danube inland waterway axis Branch A: Betuwe line Branch B: Frankfurt-Paris
Two Seas	AthDre	Igoumenitsa/Patras-Athens-Sofia-Budapest-Vienna-Prague-Nurnberg/Dresden-Hamburg
Mare Nostrum	SinOde	Odessa-Constanta-Bourgas-Istanbul-Piraeus-Gioia Tauro-Cagliari-La Spezia-Marseille-(Barcelona/Valencia)-Sines Branch A: Algeciras-Valencia-Barcelona-Marseille-Lyon
Silk Way	CNHam	Shanghai-Le Havre/Rotterdam-Hamburg/Gothenburg-Gdansk-Baltic ports-Russia Branch:Xiangtang-Beijing-Mongolia-Russia-Belarus-Poland-Hamburg

Index

H.N. Psaraftis (ed.), *Green Transportation Logistics*, International Series in Operations Research & Management Science 226, DOI 10.1007/978-3-319-17175-3

D

J

K

L

N

O

P

V

W

FSC
www.fsc.org
MIX
Papier | Fördert gute Waldnutzung
FSC® C083411

Kolibri 360 GmbH
Ferdinand-Jühlke-Straße 7
99095 Erfurt, Deutschland
produktsicherheit@kolibri360.de